CONTROL SYSTEMS

CONTROL SYSTEMS

(FOURTH EDITION)

Naresh K. Sinha
Former Professor Emeritus
Department of Electrical and Computer Engineering
McMaster University, Hamilton
Canada

Publishing Globally
NEW AGE INTERNATIONAL (P) LIMITED, PUBLISHERS
New Delhi • Bangalore • Chennai • Cochin • Guwahati
Hyderabad • Kolkata • Lucknow • Mumbai
Visit us at **www.newagepublishers.com**

Published by New Age International (P) Ltd., Publishers
First Edition: 1994
Fourth Edition: 2013
Reprint: 2015

BRANCHES

- **Bangalore** 37/10, 8th Cross (Near Hanuman Temple), Azad Nagar, Chamarajpet, Bangalore-560 018 Tel.: (080) 26756823, Telefax: 26756820, **E-mail: bangalore@newagepublishers.com**
- **Chennai** 26, Damodaran Street, T. Nagar, Chennai-600 017, Tel.: (044) 24353401 Telefax: 24351463, **E-mail: chennai@newagepublishers.com**
- **Cochin** CC-39/1016, Carrier Station Road, Ernakulam South, Cochin-682 016 Tel.: (0484) 2377303, Telefax: 4051303, **E-mail: cochin@newagepublishers.com**
- **Guwahati** Hemsen Complex, Mohd. Shah Road, Paltan Bazar, Near Starline Hotel, Guwahati-781 008 Tel.: (0361) 2513881, Telefax: 2543669, **E-mail: guwahati@newagepublishers.com**
- **Hyderabad** 105, 1st Floor, Madhiray Kaveri Tower, 3-2-19, Azam Jahi Road, Near Kumar Theater Nimboliadda, Kachiguda, Hyderabad-500 027, Tel.: (040) 24652456, Telefax: 24652457 **E-mail: hyderabad@newagepublishers.com**
- **Kolkata** RDB Chambers (Formerly Lotus Cinema) 106A, 1st Floor, S.N. Banerjee Road, Kolkata-700 014 Tel.: (033) 22273773, Telefax: 22275247, **E-mail: kolkata@newagepublishers.com**
- **Lucknow** 16-A, Jopling Road, Lucknow-226 001, Tel.: (0522) 2209578, 4045297, Telefax: 2204098 **E-mail:lucknow@newagepublishers.com**
- **Mumbai** 142C, Victor House, Ground Floor, N.M. Joshi Marg, Lower Parel, Mumbai-400 013 Tel.: (022) 24927869, Telefax: 24915415, **E-mail: mumbai@newagepublishers.com**
- **New Delhi** 22, Golden House, Daryaganj, New Delhi-110 002, Tel.: (011) 23262368, 23262370 Telefax: 43551305, **E-mail: sales@newagepublishers.com**

ISBN: 978-81-224-3353-1

C-15-08-8702

Printed in India at Kanishik Printing Press, Delhi.
Typeset at ABRO Enterprises, Delhi.

PUBLISHING GLOBALLY
NEW AGE INTERNATIONAL (P) LIMITED, PUBLISHERS
7/30 A, Daryaganj, New Delhi-110002
Visit us at **www.newagepublishers.com**

Preface to the Fourth Edition

After the good response of the third edition it is a great honour for us to present the fourth edition of the book. The third edition was well accepted by the readers as well as the teachers.

In this edition we have given a new look by changing the style pattern of the book. All the illustrations have been redrawn with clarity and accuracy. In addition, we have given our best effort to correct the errors and the misprints of the third edition. We are also thankful to those who have given us suggestions and criticism for the improvement of the book. Lastly, we hope that this edition will also be more useful than the earlier edition.

—PUBLISHERS

Preface to the First Edition

Teaching control theory has always been a challenging task. Although real control systems are often quite complicated, they must be presented in a simplified form so that they are mathematically tractable. In the past, there has been a tendency to oversimplify, mainly because real-life problems require a great deal of computation. This is no longer a problem now since engineers and engineering students have easy access to personal computers, in fact most of them own such a computer. The main difficulty in taking full advantage of the situation has been lack of availability of computer programs suitable for use by the students of control theory. Some commercially available programs are quite good, but they are usually expensive and most students cannot afford them. This book is an attempt to remedy this situation by providing such programs, designed especially so that they can aid the student by removing the drudgery out of the numerical computations while still requiring proper understanding of the subject in order to solve many challenging and realistic problems given at the end of each chapter. In addition, there are numerous worked examples and drill problems with answers to help the student in mastering the subject.

This book is intended as an introduction to control systems. The object is to provide the reader with the basic concepts of control theory as developed over the years in both the frequency domain and the time domain. An attempt has been made to retain the classical concepts while introducing modern ideas. The effect of the availability of inexpensive computers has been kept in mind.

Chapter 1 gives a brief history of control theory and introduces the students to the subject through some common examples. Chapters 2 and 3 present a unified treatment of modeling of dynamic systems, including transfer function as well as state-space approach. Chapters 4 and 5 present the classical material on the performance of feedback systems based on the transfer function approach. Chapter 6 is concerned with the stability of linear systems and presents the Routh-Hurwitz criterion in detail. Chapter 7 deals with the root locus plots. Although these concepts were developed in the 1950s, they have become even more useful when the student can use a computer to obtain a quick and accurate plot. With the help of these plots the student can easily visualize the effect of the variation of one parameter of the transfer function of the system. Chapter 8 introduces the student to various types of frequency response plots. These provide a great deal of information about the system. Again, the value of these plots towards understanding the system under study is greatly increased if a computer can be used for quick and accurate plots. Chapter 9 shows how the frequency response of the open-loop transfer function can be used to predict the stability of the closed-loop system though the classical concepts of Nyquist stability theory. Chapter 10, on the compensation of control systems, has been presented in a logical and unified manner. In particular, the trail-and-error approach to the design of lead compensators, as found in most text-books, has been replaced by a direct method developed in the late 1970s. Moreover, the design of pole-placement compensators using transfer functions, which is the counterpart of the combined observer and state-feedback controller, has been included for the first time in a book appropriate for undergraduate and practising engineers. Chapter 11, digital control, is an up-to-date treatment of a rapidly developing and popular area, presented in a manner suitable for a first study of the subject, with many practical examples to provide motivation. Chapter 12 provides a good treatment of the state-space approach to the design of controllers, presented in a manner that will make it attractive to both undergraduates and professionals. The concepts of controllability and observability are introduced using the theory of discrete-time systems, already developed in chapter 11. This is followed by the theory of state variable feedback. Both state-space and transfer function approaches for determining the state feedback vector are presented. The asymptotic state observer is then introduced and its design described. The combined state feedback and observer design is also discussed and related to the transfer function approach presented in chapter 10.

It is appreciated that in most engineering disciplines one has to solve many numerical examples in order to learn various details and subtleties and to be able to handle engineering design problems. To provide motivation, numerous realistic problems have been included at the end of each chapter. Executable versions of computer programs, ready to run on any IBM compatible personal computer, can be obtained on floppy disks from the publishers at a nominal cost. These programs are described briefly in Appendix E, and have been designed to remove the computational burden from the student, but do not replace the thinking or learning process. In order to use these programs effectively, the student must have basic understanding of the subject. It may be said that these programs aid the learning process and do nor replace it. A Solutions Manual is also available for instructors of the course. The first edition of the book, published in 1986 by Holt, Rinehart and Winston (third printing in 1988) has been used in departments of electrical as well as mechanical engineering in over 50 engineering schools in North America, the United Kingdom, India, Pakistan, Hong Kong, Brazil, Australia and South Africa. This second edition has been modified on the basis of the valuable feedback from several professors who have used this book in the past.

The first edition of the book was designed for a one-semester senior level course. Since then, in many universities, the subject matter in the first nine chapters of the book is taught at the junior level. These schools also have a senior level course in control systems. Chapters 10 to 13 have been enlarged so that these contain sufficient material for a one-semester course at the senior level. This is the way, the second edition is currently being used in the Department of Electrical and Computer Engineering of McMaster University, where each semester consists of 13 weeks with three one-hour sessions every week.

—Author

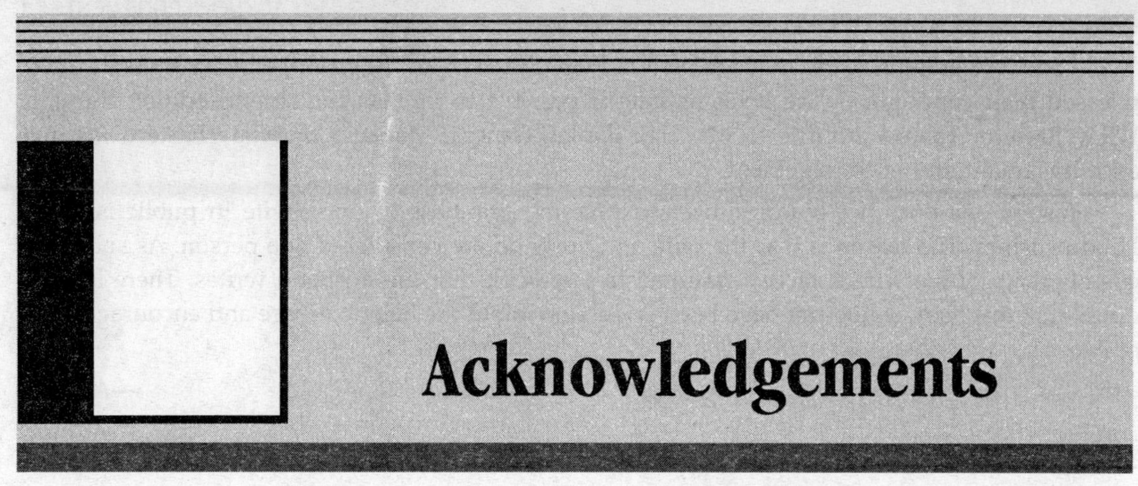

Acknowledgements

Several reviewers had helped polish the first edition of the book in many ways. Among these were Professors

- M.E. Van Valkenberg of the University of Illinois
- Susan Reidel of Marquette University
- Edgar Tacker of the University of Telsa
- Don Kirk of the U.S. Naval Postgraduate School
- Lincoln Jones of San Jose University
- M. Sunderashan of the University of Arizona
- Violet Haas of Purdue University

Many other professors who have used the first edition have made valuable suggestions for improving the text. Most of these suggestions have been incorporated in the second edition. Among these were

- Professor G.J. Lastman, University of Waterloo, Canada.
- Professor Ian Shaw, University of Witwaterstrand, South Africa.
- Professor Kishore Dabke, Monash University, Australia.
- Professor Antonio Pertence of César Rodrigues Senai, Brazil.
- Dr. Y. Dableh, McMaster University.

Several other friends and colleagues have helped me in preparing the manuscript in many ways. In particular, it is a pleasure to acknowledge that my son Anand help me in proof-reading the first edition of this book, and my ex-student Dr. Sarat Puthenpura (currently a distinguished member of the technical staff of the AT&T Bell Laboratories) helped me in preparing the Solutions Manual for the first edition. I am especially grateful to Professor G.J. Lastman who helped polish most of the computer programs which are made available with the book.

I am very grateful to the management of Holt, Rinehart and Winston, who have very kindly released their copyright of the book making it possible to publish the second edition through Wiley Eastern. Thanks are due to Mr. H.S. Poplai, General Manager of Wiley Eastern for this encouragement and encouragement.

Just as one does not thank himself, expressing gratitude to one's wife in public is not a Hindu custom. The reason is that the wife and husband are considered one person. As such, the coauthorship of the wife is tacitly assumed in any book that the husband writes. There is little doubt that this book would not have been possible without the help, patience and encouragement of Meena.

—Author

Contents

CHAPTER

1 Introduction

The subject of control systems is very important to all engineers. The objective is to free human beings from boring repititive tasks that can be done easily and more economically by automatic control devices. Recent developments in large-scale integration of semiconductor devices and the resulting availability of inexpensive microprocessors have made it practical to use computers as integral parts of control systems, making them cheaper as well as more sophisticated.

Historically, the first automatic control device used in industry was the Watt fly-ball governor, invented in 1767 by James Watt, who was also the inventor of the steam engine. The object of this device was to keep the speed of the engine constant by regulating the supply of steam to the engine. A schematic diagram is shown in Fig. 1.1. The two fly balls in the governor rotate about a vertical axis at a speed proportional to the speed of the engine. Due to the centrifugal force acting on them, they tend to move out. This movement controls the supply of steam to the engine through a mechanical linkage to the steam flow valve. This is done so that the steam supply is reduced when the speed is high and increased when the speed is low.

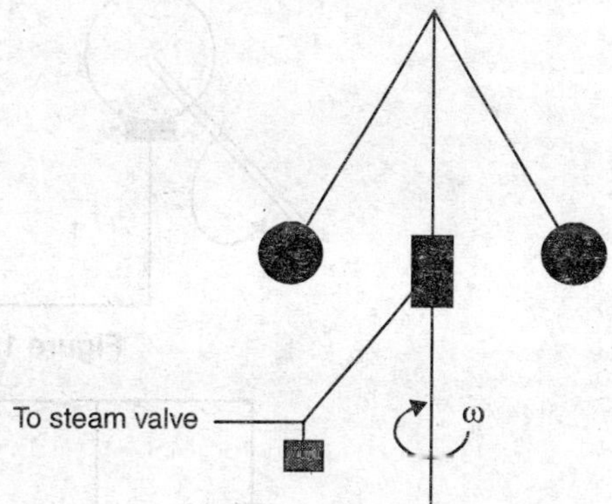

Figure 1.1. *The watt flyball governor*

It was found that by a proper design of the governor, the speed could be kept within narrow limits of a specified value. It was also observed that if one tried to increase the sensitivity of the governor by increasing the gear ratio between the engine shaft and the governor, it tended to 'hunt' or oscillate about the desired setting. About 100 years later, James Clerk Maxwell (more well known for his contributions to electromagnetic field theory) made a complete mathematical analysis of this problem.

Much later, it was realized that all automatic control systems worked on the principle of feedback. By coincidence, electrical engineers, who had been concerned with transmitting telephone signals over long distances, had developed the theory of feedback amplifiers at about the same time. In particular, one may mention the Nyquist criterion of stability in developed about 1930. A great impetus to theory of automatic control came during World War II when servo-mechanisms were used for the control of antiaircraft guns. After World War II many peacetime applications followed. Some of these are the 'autopilot' for aircraft, automatic control of machine tools, automatic control of chemical processes, and automatic regulation of voltage at electrical power plants. Although originally the theory was based on frequency response and Laplace transform methods, in the 1960s the impact of the digital computer led to the development of time-domain theory using state variables. This was especially useful as more sophisticated multivariable control systems were developed for more complex processes. As computers have become cheaper and more compact, they have been used as components of more advanced control systems.

Let us consider some simple examples of control systems. Figure 1.2 shows the scheme for controlling the voltage at an electric power station in the 1940s. A human operator was required to watch a voltmeter connected to the busbars and adjust the field rheostat to keep the voltage close to the specified value. A scheme for automatic voltage regulation is depicted in Fig. 1.3 and shows that it works by comparing the actual value of the voltage with the desired value. The difference or 'error' is applied to a servomotor, after suitable amplification. This servomotor drives a shaft coupled to the field rheostat to alter the resistance in the field winding in such a way that the error is reduced. Thus, it may be said that 'feedback' is utilized to obtain automatic

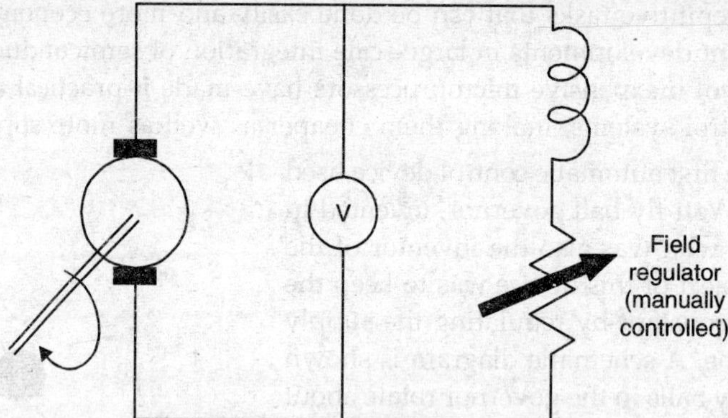

Figure 1.2. *A voltage control system*

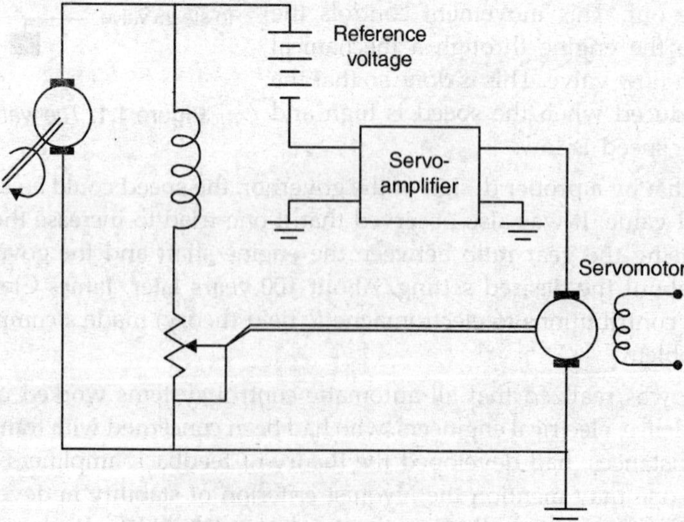

Figure 1.3. *An automatic voltage regulator*

control. In fact, all automatic control systems use feedback and can be represented by the block diagram shown in Fig. 1.4. It can be seen that the controlled output is fed back and compared with the reference input. The difference, called the 'error,' is then utilized to drive the system in such a manner that the output approaches the desired value (*i.e.*, the reference input).

Another example is the home heating system. A thermostat senses the temperature and if it is lower than a set value, the furnace is turned on. The furnace is turned off when the temperature exceeds another set value. The block diagram is shown in Fig. 1.5. Although it is similar to Figs. 1.3 and 1.4, it may be noted that this is an on/off type control systems, whereas the voltage regulator is a continuous-type system.

It was noticed at the very outset that if one tried to improve the accuracy of a control system by increasing its loop gain, it led to instability, or hunting. It is caused by the fact that although the system is designated with negative feedback, due to inherent time lags, it may change into positive feedback at some frequency. Therefore, oscillations may be produced at this frequency if the gain is increased sufficiently. The Nyquist criterion of stability, developed for feedback amplifiers, provides a good understanding of this topic. We shall later see how one can increase the sensitivity (or accuracy) without causing instability.

The components used for control systems are usually of a wide variety. For example, these may be electromechanical, electronic, thermal hydraulic, or pneumatic. To analyze the response of the various components, we replace them by their mathematical models. Although the input and the output of these devices are generally related through nonlinear differential equations, it is customary to obtain simplified linear models about the operating points because such models are easier to analyze. Transfer function and state variable models are most commonly employed.

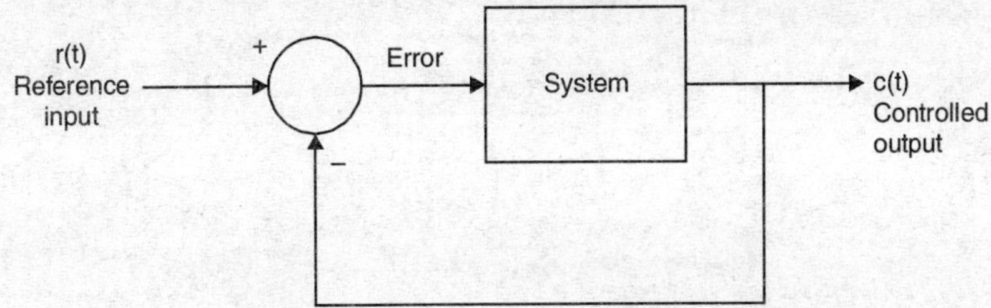

Figure 1.4. *A typical feedback control system*

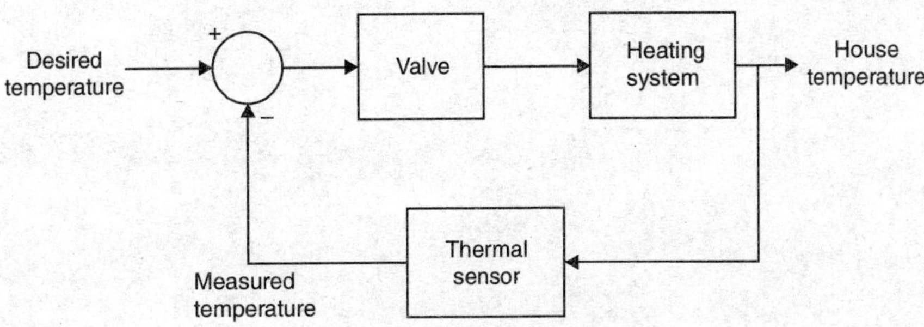

Figure 1.5. *A home-heating system*

In our development of control theory, we shall generally be carrying out the analysis and design in terms of the mathematical models. Although this approach may sometimes appear abstract, one must appreciate that these models represent real systems. To a certain extent, this abstraction is sometimes necessary for developing a unified theory of automatic control systems despite the great variety of components. One important aspect is the problem of obtaining mathematical models for different types of physical systems. This will be discussed in Chapter 2. It will be assumed that the reader is familiar with Laplace transforms. For the sake of completeness, a review of Laplace transforms is given in Appendix A.

We shall close this chapter by mentioning some areas in which the theory of control systems has been applied. These include the use of robots in automated manufacturing, automatic control of large-scale power systems, numerical control of machine tools, autopilots for aircraft, prosthetic devices for handicapped persons, and the steering control of ocean liners. An important consequence of the development of control theory has been the increased use of automation in industry with a view to increasing productivity. The concept of modelling developed by control engineers has been applied to many diverse areas, including biomedical systems, socio-economical systems, and ecological systems. With the rapid advances in microelectronics and the exciting possibility of using inexpensive computers as parts of control systems, it can truly be said that the applications of control theory are limited only by human imagination. For example, one of the great challenges for the next decade will be the use of computers to develop truly 'intelligent' control systems with the capability of learning to control complex systems. Another is the development of 'intelligent' robots which can carry out routine household and industrial tasks without supervision from humans.

CHAPTER 2

Transfer Function Models of Physical Systems

2.1 INTRODUCTION

A crucial problem in engineering design and analysis is the determination of a mathematical model of a given physical system. This model must relate in a quantitative manner the input and the output of the system which may be regarded as the 'cause' and the 'effect'. A model may be defined as "a representation of the essential aspects of the system which presents knowledge of the system in a usable manner." To be useful, the model must not be so complicated that it cannot be understood and thereby be unsuitable for analysis; at the same time, it must not be oversimplified and trivial to the extent that predictions of the behaviour of the system based on this model are grossly inaccurate.

While dealing with control systems, we shall be concerned mostly with *dynamic* systems. The behaviour of such systems is described in the form of differential equations. Although these will normally be nonlinear, it is customary to linearize them about an operating point to obtain linear diffcrntial equations. This is done in order that the analysis can be carried out conveniently. It should, however, be borne in mind that such linear models, although useful for preliminary analysis and design, are valid only over a limited operating range. Nevertheless, they are employed extensively in engineering, utilizing Laplace transform, frequently response, and state-space techniques. It is expected that the reader is familiar with basic concepts of Laplace transforms. These are reviewed briefly in Appendix A.

2.2 DIFFERENTIAL EQUATIONS AND TRANSFER FUNCTIONS FOR PHYSICAL SYSTEMS

The components of a control system are diverse in nature and may include electrical, mechanical, thermal and fluidic devices. The differential equations relating the input and output quantities for these devices are obtained using the basic laws of physics. These include balancing forces, energy and mass. In practice, some simplifying assumptions are often made to obtain linear differential equations with constant coefficients, although in most cases exact analysis would lead to nonlinear partial differential equations. For most physical devices one may classify the variables as either 'through' or 'across' variables, in the sense that the former refer to a point while the latter are measured between two points. A list of analogous variables for different systems is given in Table 2.1.

TABLE 2.1: Analogous Variables for Physical Systems

System	*Through-variable*	*Across-variable*
Electrical	Current, i	Potential difference or voltage v
Mechanical (translation)	Force, f	Relative velocity, v
Mechanical (rotational)	Torque, T	Relative angular velocity, ω
Thermal	Rate of flow of heat energy, q	Difference in temperature, T
Fluidic	Volumetric rate of fluid flow, Q	Difference in pressure, P

It may be noted that the equations of equilibrium in different systems are based on similar principles. For example, Kirchhoff's current law for an electrical circuit, equating the algebraic sum of currents meeting at a point to zero, is analogous to D' Alembert's principle in mechanics, which equates the algebraic sum of forces at a point to zero. In fact, it is often convenient to determine the differential equation for a mechanical or thermal system by first obtaining an equivalent electrical circuit. This is helpful because most people are familiar with the procedures for writing loop or node equations for electrical networks. By taking the Laplace transforms of these differential equations, we obtain an algebraic equation in terms of the complex frequency variables, relating the through-and across-variables. These algebraic equations can then be easily manipulated to obtain the *transfer function* of the system, defined as the "ratio of the Laplace transforms of the output and the input under zero initial conditions."

This transfer function represents a linear model of the system, and is usually shown in the form of a block diagram as indicated in Fig. 2.1. It may be noted that the block is 'unidirectional'.

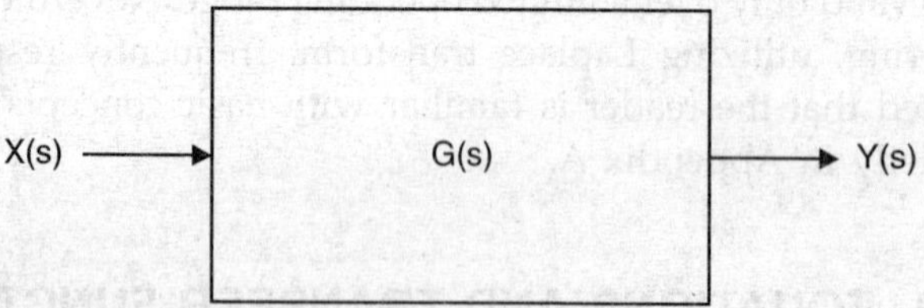

Figure 2.1. *Block Diagram of a linear system*

The input may be regarded as the 'cause' and the output as the 'effect'. The block diagram is unidirectional since the 'effect' cannot produce the 'cause'. The transfer function $G(s)$ relates the Laplace transforms $Y(s)$ of the output $y(t)$ to the Laplace transform $X(s)$ of the input $x(t)$ through the relationship.

$$Y(s) = G(s)X(s) \quad ...(2.1)$$

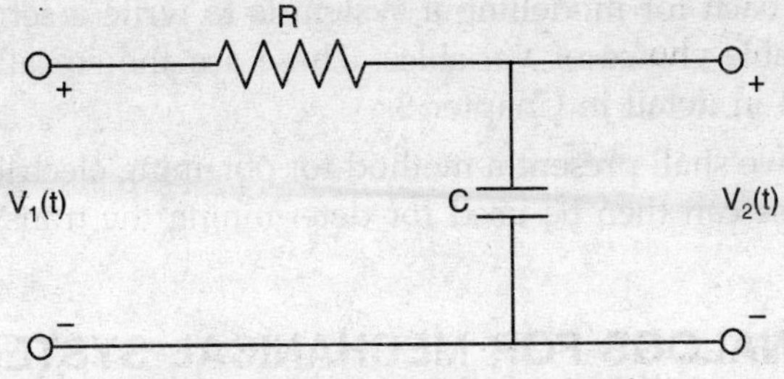

Figure 2.2. *A simple lag network*

For example, consider the simple electrical circuit shown in Fig. 2.2, in which we apply an input voltage $v_1(t)$ to an *RC* network. The output voltage $v_2(t)$ is related to the input through the differential equation

$$v_1(t) = RC\frac{dv_2}{dt} + v_2(t) \qquad \text{...(2.2)}$$

which has been obtained by applying Kirchhoff's voltage law. Taking the Laplace transform of both sides of Equation (2.2) and assuming zero initial conditions, we have

$$V_1(s) = RCsV_2(s) + V_2(s) \qquad \text{...(2.3)}$$

We can solve Equation (2.3) to obtain the transfer function

$$G(s) = \frac{V_2(s)}{V_1(s)} = \frac{1}{1+sCR} \qquad \text{...(2.4)}$$

DRILL PROBLEM 2.1

Determine the transfer functions of the RC networks shown in Figure 2.3.

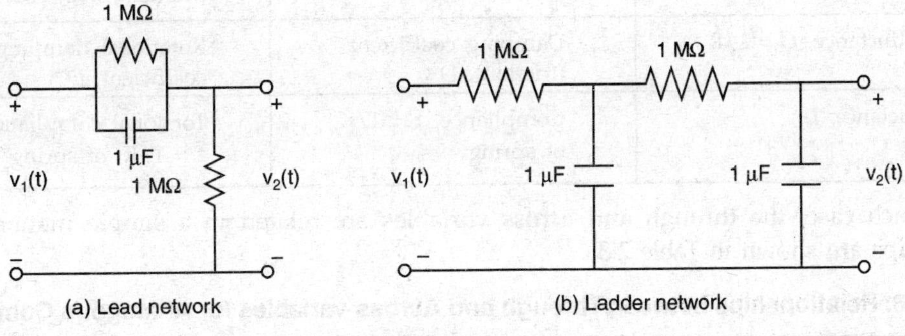

(a) Lead network

(b) Ladder network

Figure 2.3. *Networks*

Ans. (*a*) $\dfrac{s+1}{s+2}$ (*b*) $\dfrac{1}{s^2+3s+1}$

Often we have more complicated systems with several independent inputs and outputs. In such cases, we can represent the input-output relationship through a transfer function matrix. Each element of this matrix is a transfer function relating a particular output to a specific input, assuming that all other inputs are zero, and obtained in the same manner as above. This is justified through the principle of superposition for linear systems.

An alternative approach for modelling a system is to write a set of first-order differential equation through a suitable choice of variables. These are the so-called *state variables* for the system. This is discussed in detail in Chapter 3.

In the next section, we shall present a method for obtaining electrical equivalent circuits for mechanical systems. These can then be used for determining the transfer function.

2.3 ELECTRICAL ANALOGS FOR MECHANICAL SYSTEMS

It is often very convenient to obtain an electrical equivalent circuit analogous to a mechanical system. We can then apply Kirchhoff's laws to write the circuit equations and determine the transfer function. It is also possible to write these equations directly in terms of the Laplace transforms of the currents and voltages. Furthermore, one can use network theorems to simplify the circuits. Electrical analogs for mechanical systems have also been used for simulation.

In particular, if one uses the *force-current analogy* (or force-torque for a rotational system), the topology of the electrical analog is very similar to that of the mechanical system. Table 2.2 gives the list of analogous quantities for this case.

TABLE 2.2: Analogous Quantities in Electrical and Mechanical Systems

Electrical	*Mechanical (translation)*	*Mechanical (rotation)*
Current, i	Force, f	Torque, T
Voltage, v	Velocity, v	Angular velocity, ω
Flux linkages, $N\phi$	Displacement, x	Angular displacement, θ
Capacitance, C	Mass, M	Moment of inertia, J
Conductance, $G = 1/R$	Damping coefficient (friction), D	Rotational damping, coefficient, D
Inductance, L	Compliance, $\tau = 1/K$, of spring	Torsional compliance, $\tau = 1/K$, of spring

In each case, the through and across variables are related in a simple manner. These relationships are shown in Table 2.3.

TABLE 2.3: Relationships between Through and Across-variables for Analogous Components

Electrical	*Mechanical (translation)*	*Mechanical (rotation)*
$i = C\dfrac{dv}{dt}$	$f = M\dfrac{dv}{dt}$	$T = J\dfrac{d\omega}{dt}$
$i = Gv$	$f = Dv$	$T = Dv$
$i = \dfrac{1}{L}N\phi = \dfrac{1}{L}\int v\,dt$	$f = Kx = K\int v\,dt$	$T = K\theta = K\int \omega\,dt$

Following Cheng [1][1], the rule for drawing the equivalent circuit can be stated as follows:

> Each junction in the mechanical system corresponds to a node in the equivalent electrical circuit, joining excitation sources and passive elements. All points on a rigid mass are considered as the same junction, and one terminal of the capacitor analogous to the mass is always connected to the ground in the electrical circuit.

The reason for connecting one terminal of the capacitor to the ground is that the velocity (or displacement) of a mass is always referred to the earth.

The following examples will illustrate the procedure:

EXAMPLE 2.1

The electrical analog of a mass on wheels, coupled to a rigid body is shown in Fig. 2.4.

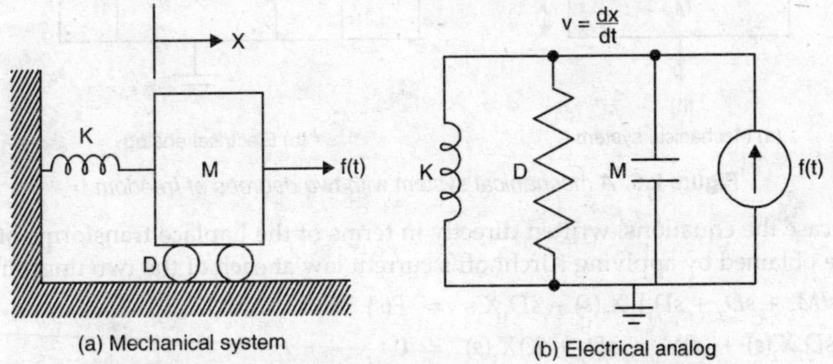

Figure 2.4. *A mechanical system with one degree of freedom*

It will be seen that both the mechanical system and its electrical equivalent circuit have two nodes, the ground and the node corresponding to the displacement x, or the voltage v. The differential equation for both systems is given as Equation (2.5). In the case of the electrical network, the equation was obtained by applying Kirchhoff's current law at the node v, and is identical to the equation that would have been obtained by applying D' Alembert's principle to the mechanical system.

$$M\frac{d^2x}{dt^2} + D\frac{dx}{dt} + Kx = f(t) \quad ...(2.5)$$

Taking Laplace transforms of both sides of Equation (2.5), assuming zero initial conditions, we get the transfer function

$$G(s) = \frac{X(s)}{F(s)} = \frac{1}{Ms^2 + Ds + K} \quad ...(2.6)$$

EXAMPLE 2.2

A mechanical system with two degrees of freedom and its electrical equivalent circuit are shown in Figure 2.5, where D_1 and D_2 represent dashpots.

[1] Numbers in brackets correspond to numbers in the Reference section at the end of the chapter.

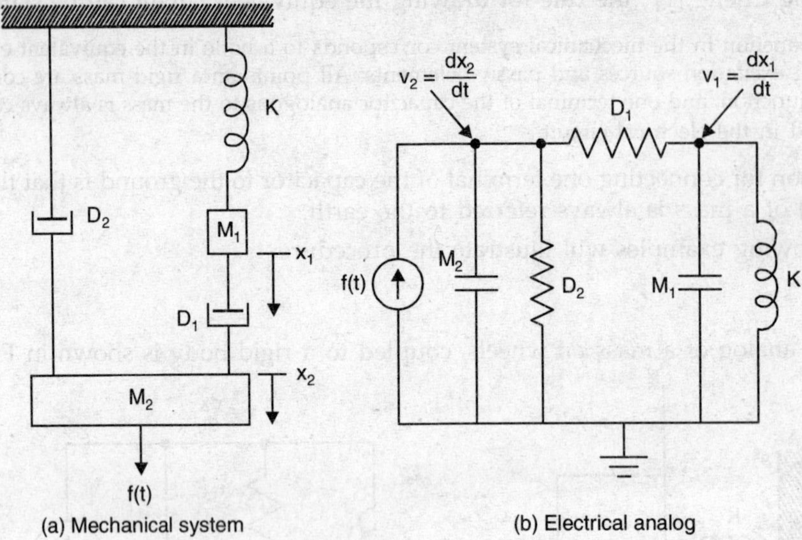

(a) Mechanical system (b) Electrical analog

Figure 2.5. *A mechanical system with two degrees of freedom*

In this case the equations, written directly in terms of the Laplace transforms of the various variables, are obtained by applying Kirchhoff's current law at each of the two ungrounded nodes.

$$(s^2M_2 + sD_2 + sD_1)\ X_2(s) - sD_1X_1s \ = \ F(s) \qquad ...(2.7)$$
$$-sD_1X_2(s) + (s^2M_1 + sD_1 + K)X_1(s) \ = \ 0$$

The convenience of writing Equation (2.7) in terms of node voltages is evident. If we want to obtain the transfer function relating $X_2(s)$ to F(s), we simply solve Equation (2.7) for $X_2(s)$. This gives us

$$\frac{X_2(s)}{F(s)} = \frac{s^2M_1 + sD_1 + K}{\left(s^2M_1 + sD_1 + K\right)\left(s^2M_2 + sD_2 + sD_1\right) - s^2D_1^2} \qquad ...(2.8)$$

DRILL PROBLEM 2.2

Determine the transfer function relating $X_1(s)$ to F(s) for the mechanical system shown in Figure 2.6.

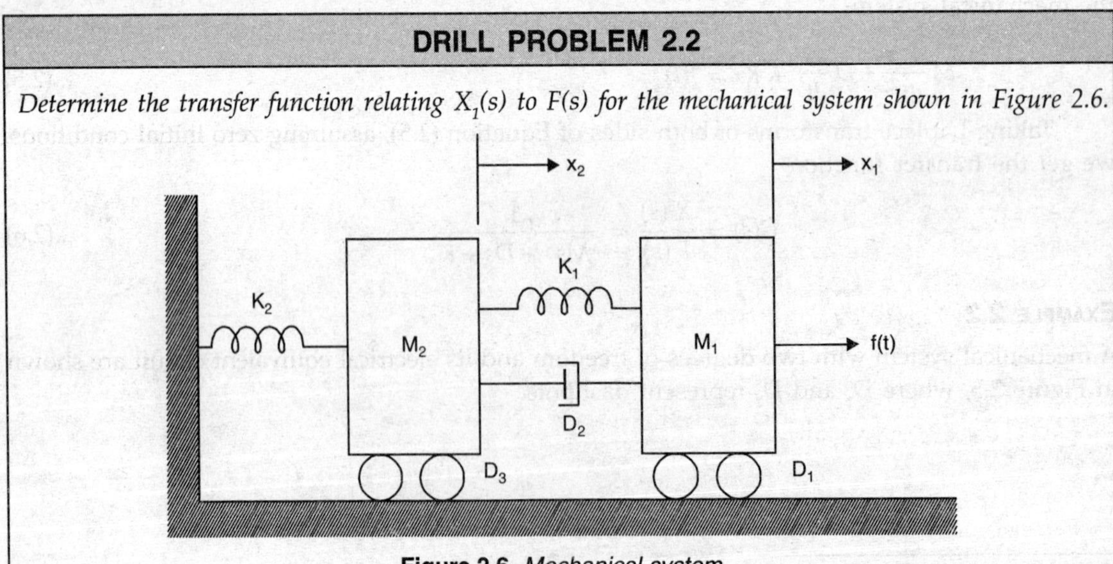

Figure 2.6. *Mechanical system*

Ans. $$\frac{X_1(s)}{F(s)} = \frac{M_2s^2 + D_2s + D_2s + K_1 + K_2}{\left(M_1s^2 + D_1s + D_2s + K_1\right)\left(M_2s^2 + D_2s + D_3s + K_1 + K_2\right) - \left(D_2s + K_1\right)^2}$$

EXAMPLE 2.3

A mechanical (rotational) system is shown in Figure 2.7. The rotor of a motor (with moment of inertia, J_1) is coupled to a load with moment of inertia J_2, through gears. A driving torque $T(t)$ is developed by the motor. The equivalent electrical circuit is also shown. It may be noted that gears in a mechanical system are equivalent to transformers in electrical networks. Here the gears will be assumed to be frictionless and thus equivalent to ideal transformers.

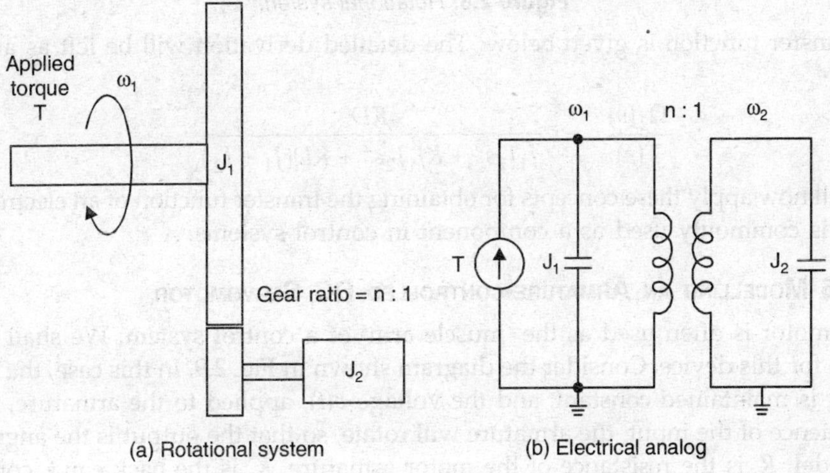

Figure 2.7. *Rotational system with gears*

The transfer function relating ω_2 to T is obtained easily from the electrical analog.

$$\frac{\Omega_2(s)}{T(s)} = \frac{1}{\left(J_1 + n^2 J_2\right)s} \qquad ...(2.9)$$

EXAMPLE 2.4

A mechanical (rotational) system is shown in Fig. 2.8. The rotor of a motor (with moment of inertia J_1) is coupled to a load with moment of inertia J_2, through a fluid coupling with viscous friction coefficient D_1 and a shaft with spring constant K. A driving torque $T(t)$ is developed by the motor. The friction coefficient of the load may be taken as D_2. The equivalent electrical circuit is also shown. There are three node voltages with reference to ground, corresponding to the angular velocities of the rotor, the coupling and the load shaft. The transfer function relating the angular velocity of the output shaft to the applied torque is obtained conveniently from the electrical analog by using the concept of the current divider.

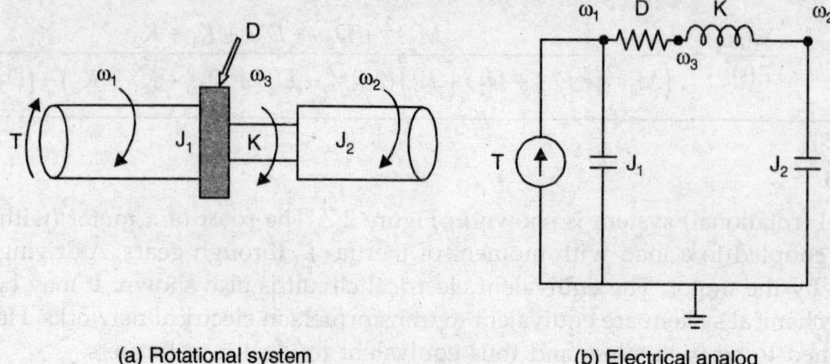

(a) Rotational system (b) Electrical analog

Figure 2.8. *Rotational system*

The transfer function is given below. The detailed derivation will be left as an exercise to the student.

$$\frac{\Omega_2(s)}{T(s)} = \frac{KD}{J_1J_2s^3 + KJ_1J_2s^2 + KD(J_1 + J_2)} \qquad \text{...(2.10)}$$

We shall now apply these concepts for obtaining the transfer function of an electro-mechanical device; that is commonly used as a component in control systems.

Example 2.5 Modelling an Armature-controlled DC Servomotor

A d.c. servomotor is often used as the 'muscle arm' of a control system. We shall determine a linear model for this device. Consider the diagram shown in Fig. 2.9. In this case, the field current of the motor is maintained constant, and the voltage $v(t)$, applied to the armature, is the input. As a consequence of the input, the armature will rotate, so that the output is the angular position θ. It this model, R_a is the resistance of the motor armature, K_b is the back e.m.f. constant of the motor, J is the moment of inertia of the motor and load, F is frictional torque coefficient opposing the motion, and K_t is the torque constant of the motor. It should be noted that the following assumptions have been made to simplify the model:

1. The inductance of the armature winding has been neglected.
2. It is assumed that the magnetic flux is unaffected by armature reaction.
3. Frictional torque is assumed to be linearly related to velocity.

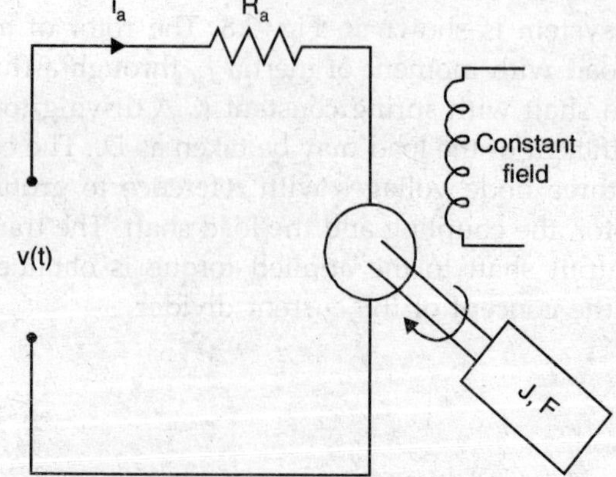

Figure 2.9. *An armature-controlled dc servomotor*

The following equations are easily obtained

$$v - R_a i_a = v_b = K_b \frac{d\theta}{dt} \qquad ...(2.11)$$

$$J\frac{d^2\theta}{dt^2} + F\frac{d\theta}{dt} = K_t i_a$$

The second Equation (2.11) equates the electrical torque developed by the motor with the mechanical torque required to turn the armature and the load. Solving the first equation in (2.11) for i_a and substituting in the second, we obtain

$$J\frac{d^2\theta}{d\theta^2} + F\frac{d\theta}{dt} = k_t\left[\frac{v}{R_a} - \frac{K_b}{R_a}\frac{d\theta}{dt}\right] \qquad ...(2.12)$$

which can be rearranged as below.

$$J\frac{d^2\theta}{dt^2} + \left(F + \frac{K_t K_b}{R_a}\right)\frac{d\theta}{dt} = \frac{K_t}{R_a}v \qquad ...(2.13)$$

Finally, taking Laplace transforms, we obtain the transfer function of the motor as

$$\frac{\theta(s)}{V(s)} = \frac{A}{s(s+\alpha)} \qquad ...(2.14)$$

where the constants A and α are given by

$$A = \frac{K_t}{JR_a} \qquad ...(2.15)$$

$$\alpha = \frac{F}{J} + \frac{K_t K_b}{JR_a}$$

Note that if we consider the angular velocity, ω, as the output, we obtain the transfer function.

$$\frac{\Omega(s)}{V(s)} = \frac{A}{s+\alpha} \qquad ...(2.16)$$

The derivation of the transfer function can be visualized better through the block diagram shown in Fig. 2.10, obtained by taking the Laplace transforms of Equation (2.11). It may be noted that if the armature inductance cannot be neglected, we can take this into account by simply replacing $1/R_a$ in the block diagram with $1/(R_a + sL_a)$.

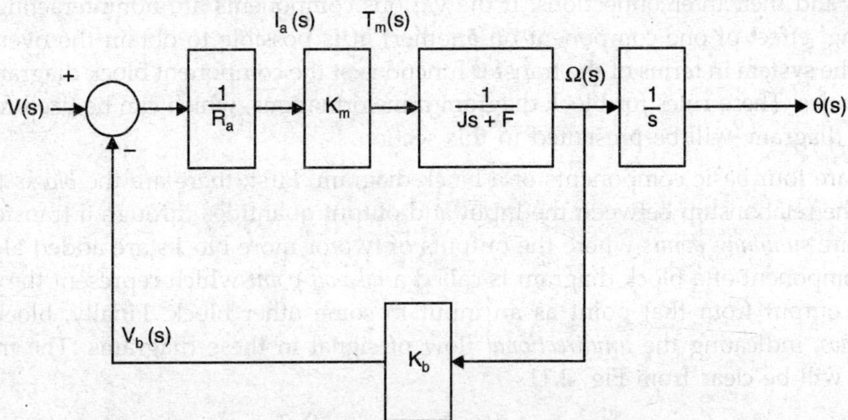

Figure 2.10. *Block diagram for the armature controlled dc servomotor*

This simple example illustrates the power of the block diagram representation. It can be utilized fully through certain rules of block diagram algebra. These will be discussed in the following section.

It must be emphasized that the transfer function that we have obtained for the dc servomotor is based on several simplifying assumptions. Not only have we neglected the inductance of the armature, but we have also neglected the armature reaction in the motor and the voltage drops in the brushes. In addition, we have assumed that the frictional torque is linearly related to the angular velocity. This is seldom true in practice. A more detailed and accurate model for the dc servomotor is nonlinear and complicated [2]. Nevertheless, the linear model obtained above is quite useful for a preliminary design of control systems utilizing dc servomotors.

DRILL PROBLEM 2.3

For a particular motor, the parameters of the transfer function are $A = 10$ and $\alpha = 1$. Calculate the angular velocity of the motor as a function of time if the input voltage is a step function of magnitude 5 volts. Assume that the motor is initially at rest.

Ans. $\omega(t) = 50\,(1 - e^{-1})$ radians per second

DRILL PROBLEM 2.4

Determine the transfer function of the d.c. servomotor if the input voltage is applied to the field winding, with inductance L_f and resistance R_f, while the armature current is maintained at a constant value. This is called the field control mode.

Ans. $$\frac{\theta(s)}{V(s)} = \frac{K_m}{s\,(Js + F)\left(L_f s + R_f\right)}$$

2.4 SIMPLIFICATION OF BLOCK DIAGRAMS

In the analysis of control systems, it is very convenient to obtain the block diagrams of different components and their interconnections. If the various components are noninteracting (*i.e.*, there is no 'leading' effect of one component on another), it is possible to obtain the overall transfer function of the system in terms of the transfer functions of the component block diagrams utilizing some basic rules. These rules for block diagram transformations, which can be used for reducing the original diagram, will be presented in this section.

There are four basic components of a block diagram. First, there are the *blocks* themselves, describing the relationship between the input and output quantities through a transfer function. Then there are *summing points* where the outputs of two or more blocks are added algebraically. The third component of a block diagram is called a *take-off point* which represent the application of the total output from that point as an input to some other block. Finally, block diagrams contain *arrows*, indicating the *unidirectional* flow of signal in these diagrams. The meanings of these terms will be clear from Fig. 2.11.

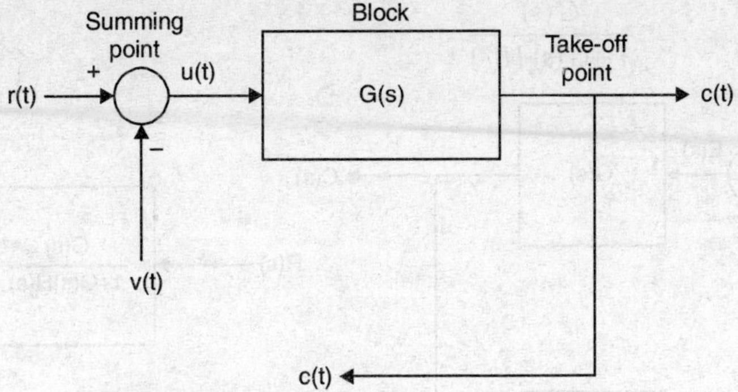

Figure 2.11. *Components of a block diagram*

The following relationships are immediately obvious:

$$u(t) = r(t) - v(t) \quad \text{...(2.17)}$$

$$C(s) = G(s)\, U(s)$$

We shall now discuss the rules of block diagram algebra.

1. Combining Cascade Blocks

Blocks connected in cascade can be replaced by a single block with transfer function equal to the product of the respective transfer functions, as shown in Fig. 2.12.

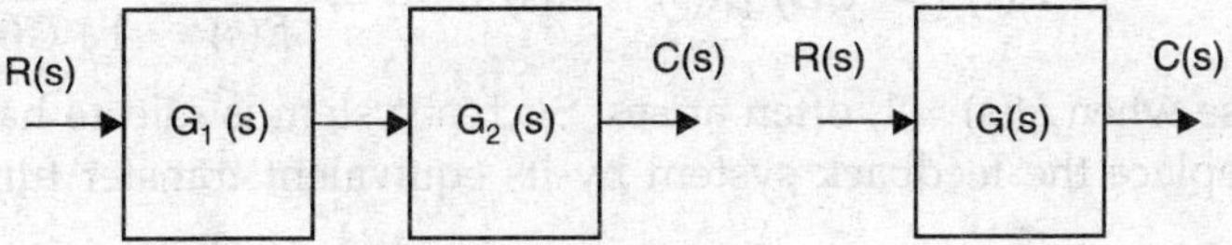

Figure 2.12. *Cascade blocks*

that is, $$G(s) = G_1(s)\, G_2(s) \quad \text{...(2.18)}$$

Comment: Note that this is valid only if there is no loading effect on first block due to the second block.

As an example, consider the R-C ladder networks shown in Fig. 2.13. It is easily demonstrated that its transfer function is

$$\frac{V_2(s)}{V_1(s)} = \frac{1}{1+3sCR+s^2C^2R^2}$$

whereas the transfer function of one section of the R-C network is

$$G_1(s) = G_2(s) = \frac{1}{1+sCR}$$

if the loading effect is not taken into account.

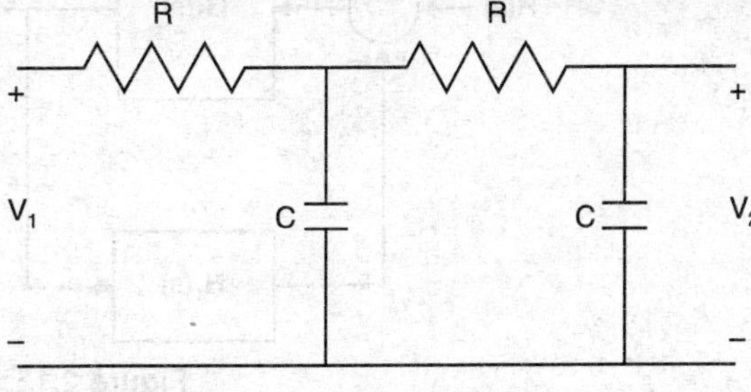

Figure 2.13. *R-C ladder network*

2. Eliminating a Feedback Loop

The transfer function of a simple feedback loop (see Figure 2.15), where $G(s)$ is the transfer function of the forward path and $H(s)$ is the transfer function of the (negative) feedback path is given by

$$\frac{G(s)}{1+G(s)\,H(s)} \quad ...(2.19)$$

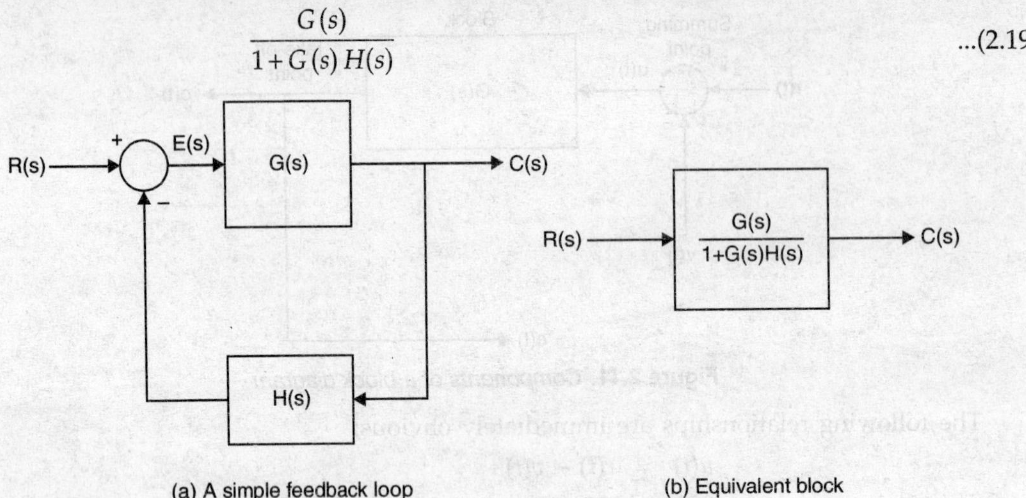

(a) A simple feedback loop (b) Equivalent block

Figure 2.14. *A simple feedback system*

This is proved through the following equations:

$$C(s) = G(s)\ E(s)$$

$$E(s) = R(s) - H(s)\ C(s)$$

$$C(s) = G(s)\ [R(s) - H(s)\ C(s)] \Rightarrow \frac{C(s)}{R(s)} = \frac{G(s)}{1+G(s)\,H(s)}$$

The special case when $H(s) = 1$, often arises. Such a system is said to have *unity feedback.* In all cases, one can replace the feedback system by its equivalent transfer function.

DRILL PROBLEM 2.5

Determine the overall transfer function of the system shown in Figure 2.15.

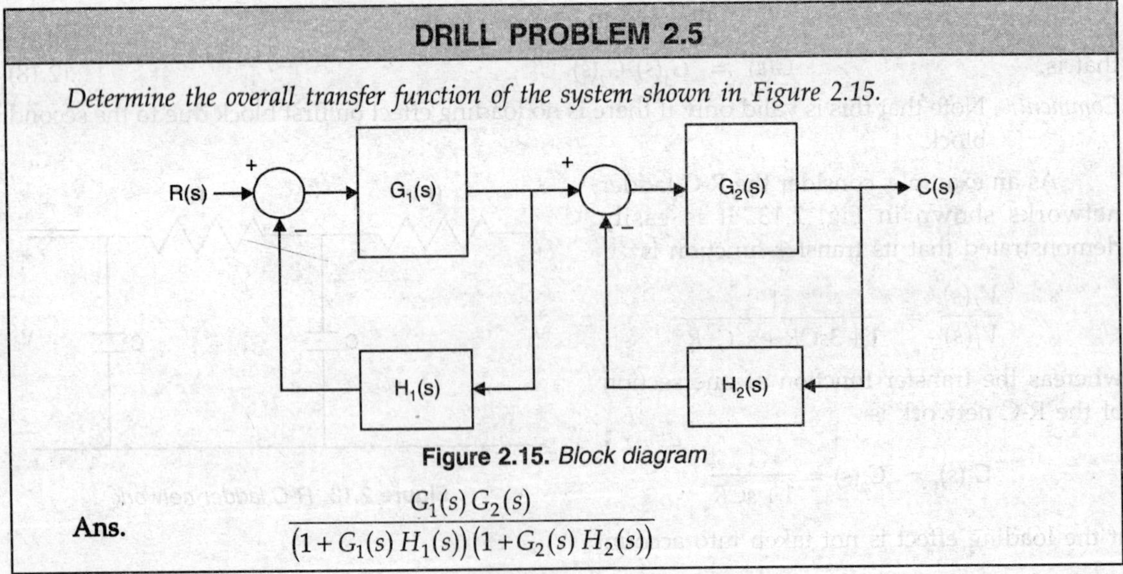

Figure 2.15. *Block diagram*

Ans. $\dfrac{G_1(s)\,G_2(s)}{(1+G_1(s)\,H_1(s))\,(1+G_2(s)\,H_2(s))}$

3. Parallel Blocks

Blocks are said to be in parallel if they have a common input and the overall output is the sum of the individual outputs, as shown in Fig. 2.16. The transfer function of the parallel combination is simply the sum of the transfer functions of the parallel blocks. The proof is straightforward and will be left as an exercise for the student.

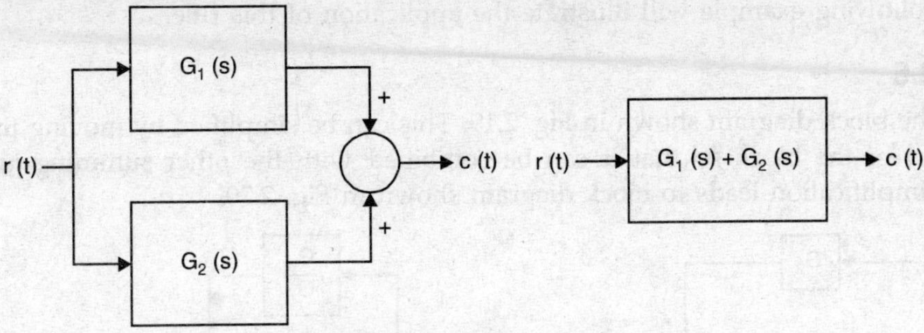

Figure 2.16. *Parallel blocks*

DRILL PROBLEM 2.6

Determine the overall transfer function for the following block diagram:

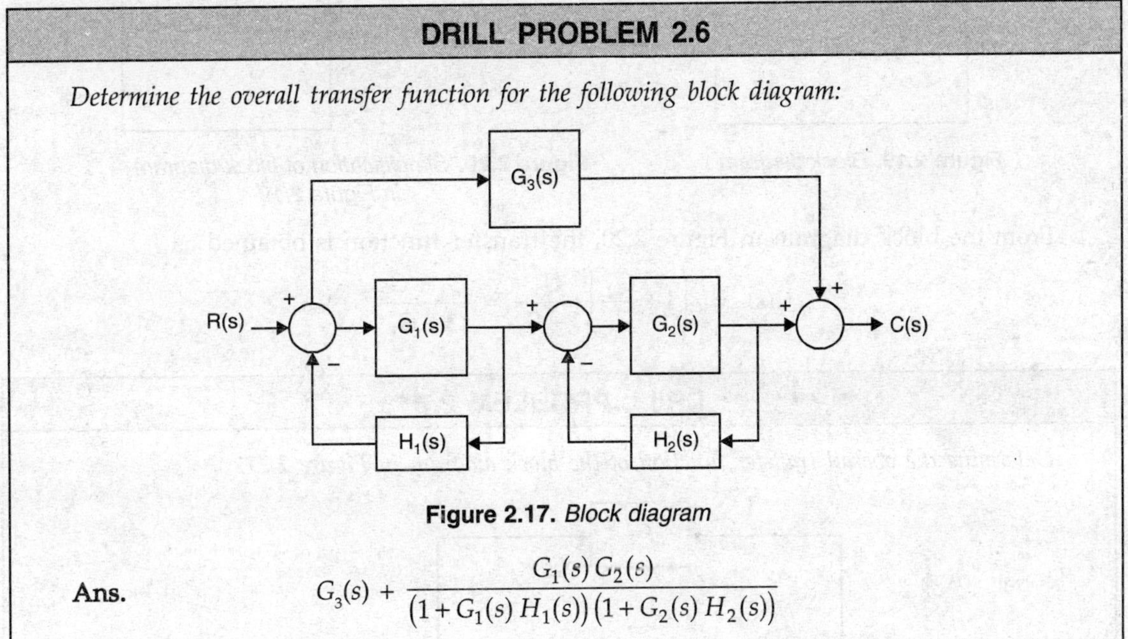

Figure 2.17. *Block diagram*

Ans.
$$G_3(s) + \frac{G_1(s)\,G_2(s)}{\left(1+G_1(s)\,H_1(s)\right)\left(1+G_2(s)\,H_2(s)\right)}$$

4. Moving a Summing Point Ahead of a Block

It is sometimes necessary to move a summing point either ahead of a block to simplify the block diagram. This can be done provided the transfer function of the blocks are modified accordingly. Figure 2.18 illustrates the necessary modification.

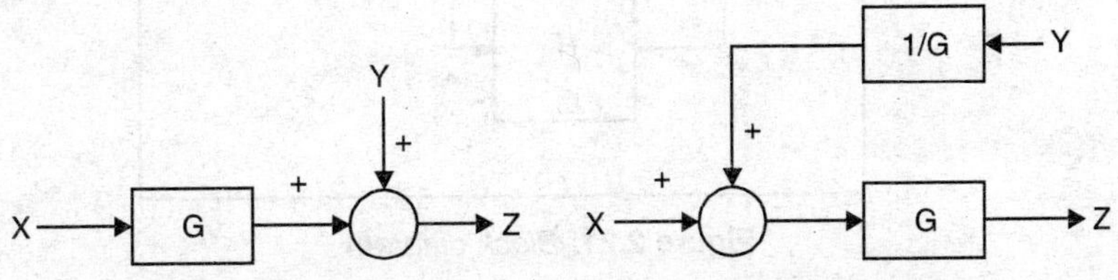

Figure 2.18. *Moving a summing point ahead of a block*

It can be noted that for both the block diagrams shown above, the output is given by

$$Z = GX + Y$$

The following example will illustrate the application of this rule.

EXAMPLE 2.6

Consider the block diagram shown in Fig. 2.19. This can be simplified by moving the summing point ahead of the block so that it can be combined with the other summing junction. The resulting simplification leads to block diagram shown in Fig. 2.20.

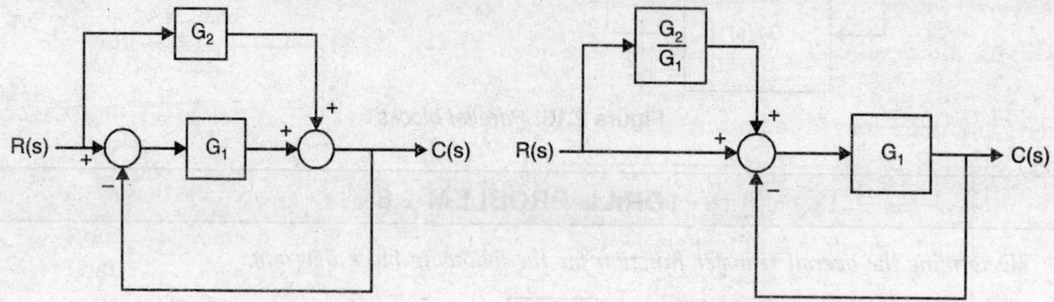

Figure 2.19. *Block diagram*

Figure 2.20. *Simplification of block diagram in Figure 2.19*

From the block diagram in Figure 2.20, the transfer function is obtained as

$$G(s) = \left(1+\frac{G_2}{G_1}\right)\frac{G_1}{1+G_1} = \frac{G_1+G_2}{1+G_1}$$

DRILL PROBLEM 2.7

Determine the overall transfer function of the block diagram in Figure 2.21.

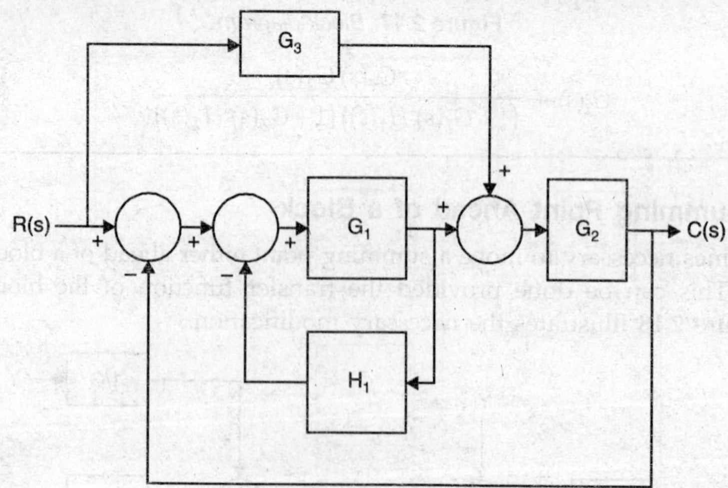

Figure 2.21. *Block diagram*

Ans. $\dfrac{G_1G_2 + G_2G_3 + G_1G_2\ G_3\ H_1}{1+G_1H_1+G_1G_2}$

5. Moving a Summing Point Behind a Block

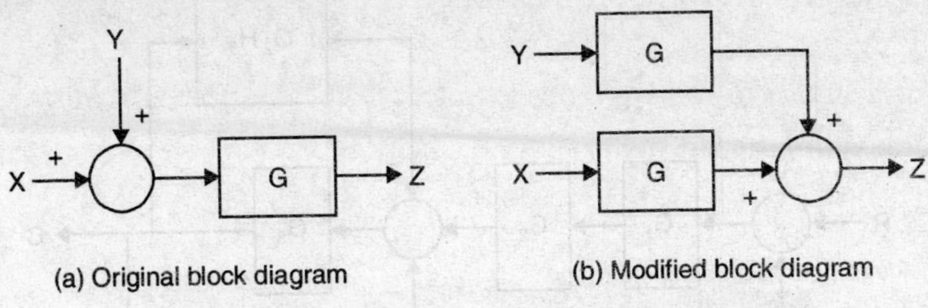

Figure 2.22. *Moving a summing point behind a block*

This is the inverse of rule 4, as indicated in Figure 2.22. It will be seen that in both cases, the output is unchange, i.e.,

$$Z = G(X + Y)$$

EXAMPLE 2.7

Consider the block diagram shown in Fig. 2.23.

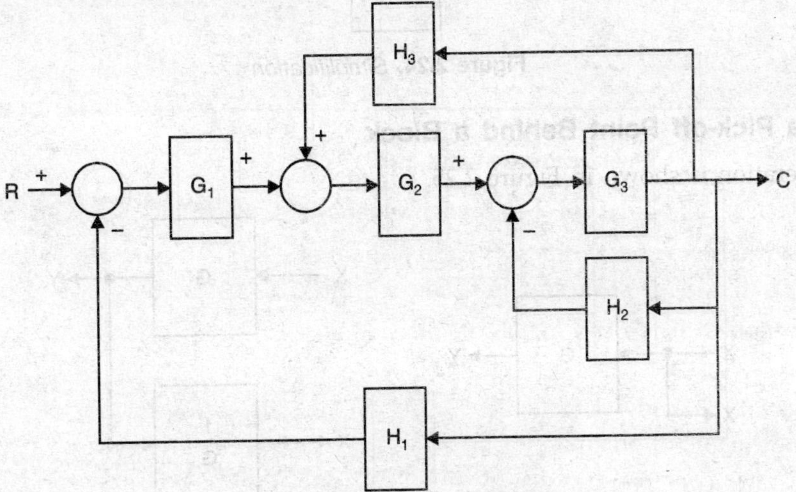

Figure 2.23. *Block diagram*

Moving the second summing junction behind the block G_2 results in the block diagram shown in Fig. 2.24. This makes it possible to combine the blocks G_1 and G_2, which are now in cascade, in addition to combining two feedback paths. Consequently, the following overall transfer function is obtained.

$$G = \frac{G_1G_2G_3}{1 + G_1G_2G_3H_1 + G_3H_2 - G_2G_3H_3}$$

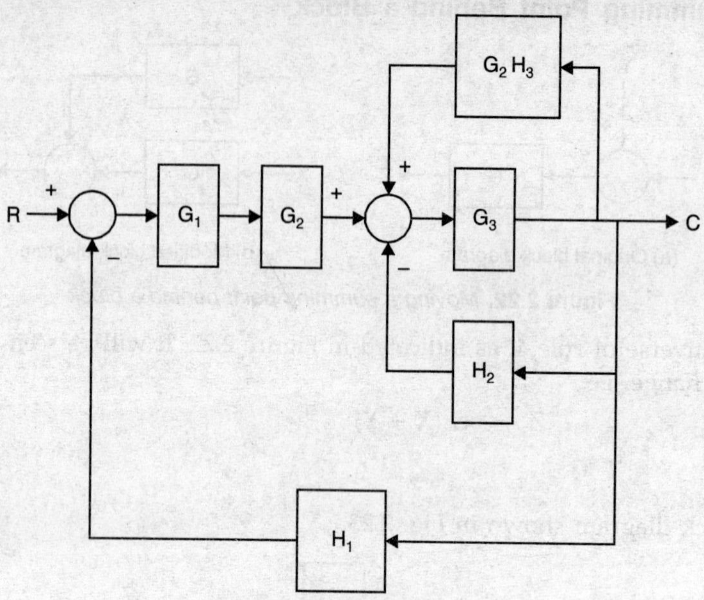

Figure 2.24. *Simplification*

6. Moving a Pick-off Point Behind a Block

This operation is shown in Figure 2.25.

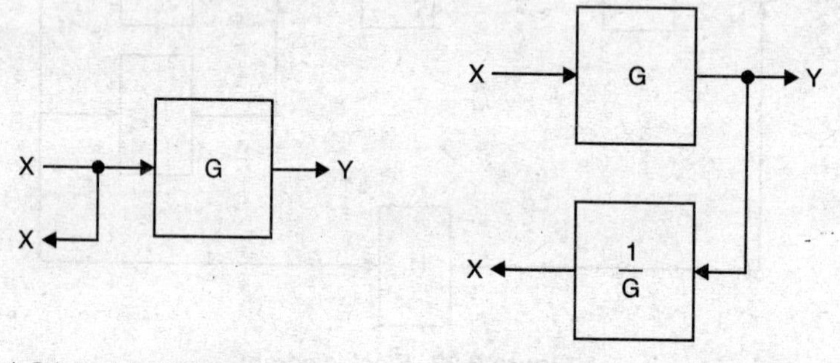

Figure 2.25. *Moving a pick-off point behind a block*

EXAMPLE 2.8

As an example, consider the block diagram shown in Fig. 2.26.

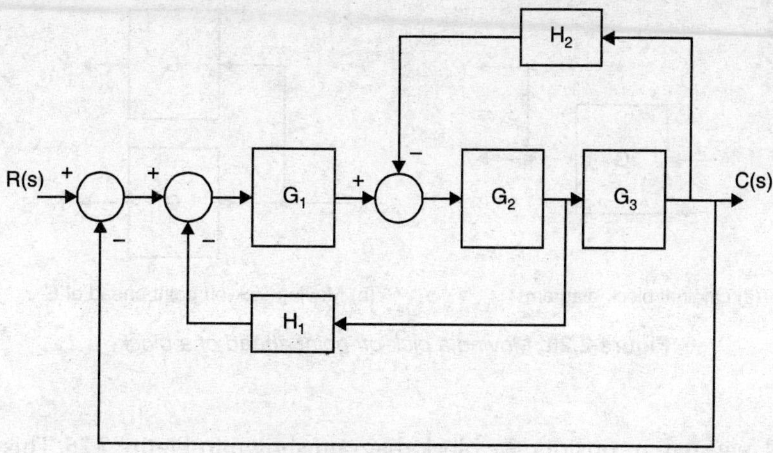

Figure 2.26. *Block diagram*

This can be simplified by moving the pick-off point currently ahead of the block containing G_3 to behind it. This puts the blocks G_2 and G_3 in cascade and also the two outer feedback paths in parallel. The resulting block diagram is shown in Fig. 2.27.

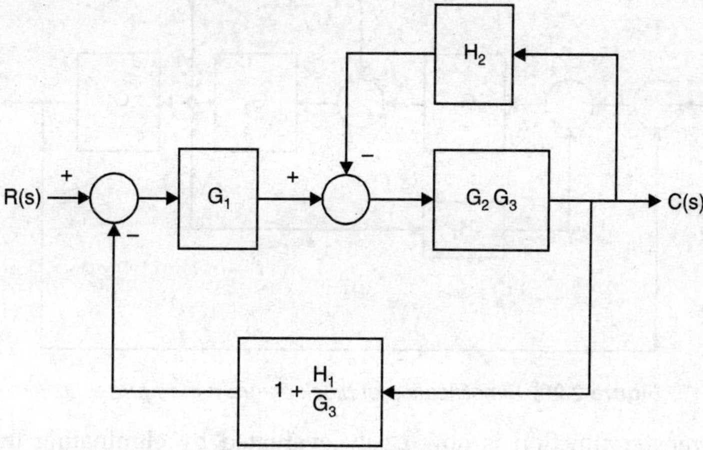

Figure 2.27. *Simplification of block diagram in Figure 2.26*

The overall transfer function is now obtained as

$$G(s) = \frac{G_1G_2G_3}{1 + G_1G_2G_3 + G_2G_3H_2 + G_1G_2H_1}$$

7. Moving a Pick-off Point Ahead of a Block

This is shown in Fig. 2.28.

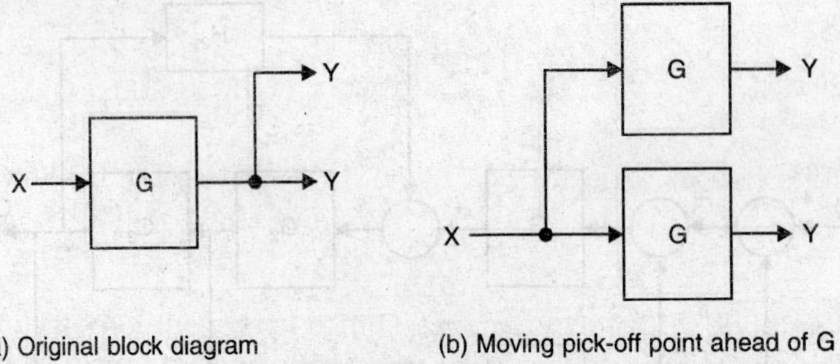

(a) Original block diagram (b) Moving pick-off point ahead of G

Figure 2.28. *Moving a pick-off point ahead of a block*

EXAMPLE 2.9

To illustrate this, we shall reconsider the block diagram shown in Figure 2.26. This time, we shall simplify it moving the pick-off point after the block G_3 to ahead of this block, resulting in the block diagram shown in Figure 2.29.

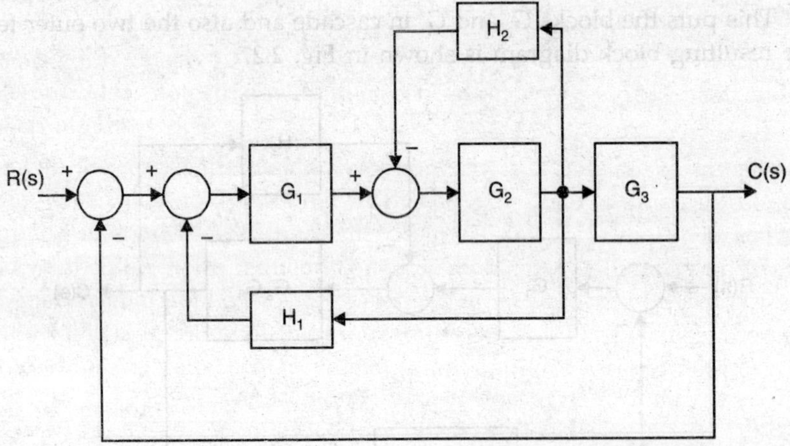

Figure 2.29. *Simplification of block diagram in Figure 2.26*

The overall transfer function is now easily evaluated by eliminating the inner feedback loops.

2.5 MASON'S RULE

It is also possible to determine the transfer function of a complicated block diagram directly using Mason's gain formula [3]. It was originally developed using topological properties of signal flow graphs but is also applicable to block diagrams. Here we shall state the rule in terms of block diagrams. The first step is to determine all possible *loops* in the block diagram. A loop may be defined as "a closed path in the direction of the arrows that does not go through any point more than once". Two loops are said to be 'non-touching' if they do not have any common point. A forward path is defined as a direct path from the input to the output which does not retrace its step at any point.

The following equation can be used for determining the overall transmittance.

$$G(s) = \frac{C(s)}{R(s)} = \sum_k \frac{T_k(s)\Delta_k(s)}{\Delta(s)} \qquad ...(2.20)$$

where $T_k(s)$ ≜ transfer function of the kth forward path from the input to the output; a forward path cannot contain any feedback loop.

$\Delta(s)$ ≜ determinant of the block diagram:

= 1 – sum of all individual loop transfer functions

+ sum of the products of the transfer functions of all possible sets of two non-touching loops

– sum of the products of the transfer functions of all possible sets of three non-touching loops

+ ...

$\Delta_k(s)$ = all terms in $\Delta(s)$ that do not have elements or paths common with an element or path in $T_k(s)$.

EXAMPLE 2.10

Consider the block diagram shown in Fig. 2.19. Here, we have two forward paths, with transmittances G_1 and G_2, and only one loop with transmittance $-G_1$, which is touched by both forward paths. Consequently, the application of Mason's rule gives us the following overall transfer function

$$\frac{C(s)}{R(s)} = \frac{G_1 + G_2}{1 + G_1}$$

which is seen to be identical to that obtained in Example 2.6, by moving a summing point ahead of a block.

EXAMPLE 2.11

Consider the block diagram shown in Figure 2.21 (Drill Problem 2.7). Here, we have two forward paths, with transfer functions G_1G_2 and G_3G_2, respectively. Also, there are two loops, with transmittances $-G_1H_1$ and $-G_1G_2$. We also see that the forward path G_3G_2 is not touching the first loop. Furthermore, the two loops are touching each other. Hence, application of Mason's rule gives us the following overall transfer function.

$$G(s) = \frac{G_1G_2 + G_3G_2(1 + G_1H_1)}{1 + G_1H_1 + G_1G_2} = \frac{G_1G_2 + G_2G_3 + G_1G_2G_3H_1}{1 + G_1H_1 + G_1G_2}.$$

EXAMPLE 2.12

Consider the block diagram shown in Fig. 2.30.

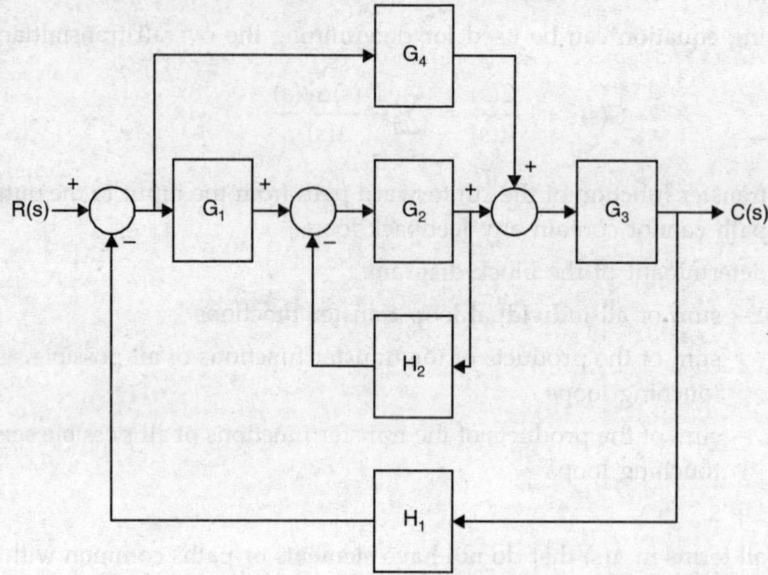

Figure 2.30. *A block diagram with feedforward and feedback loops*

The overall transfer function can be determined by block diagram simplification. This is done most conveniently by replacing the inner feedback loop with its equivalent and then adding the two parallel branches to obtain a single-loop diagram. We shall apply Mason's rule, and the result can be verified by students by using the procedure suggested above.

Here, we have two forward paths from $R(s)$ to $C(s)$, and

$$T_1(s) = G_1G_2G_3$$

$$T_2(s) = G_4G_3$$

We also have three feedback loops with transmittances $-G_2H_2$, $-G_1G_2G_3H_1$ and $-G_4G_3H_1$, where the first and the third loops are non-touching. Furthermore, T_1 is touching all loops, but T_2 is not touching the second loop. Therefore, we have

$$\Delta(s) = 1 + (G_2H_2 + G_1G_2G_3H_1 + G_4G_3H_1) + G_2H_2G_4G_3H_1$$

$$\Delta_1(s) = 1$$

$$\Delta_2(s) = 1 + G_2H_2$$

Hence, the overall transfer function is given by

$$\frac{C(s)}{R(s)} = \frac{G_1G_2G_3 + G_4G_3(1+G_2H_2)}{1+G_2H_2 + G_1G_2G_3H_1 + G_4G_3H_1 + G_2H_2G_4G_3H_1}.$$

EXAMPLE 2.13

Consider the block diagram shown in Fig. 2.31.

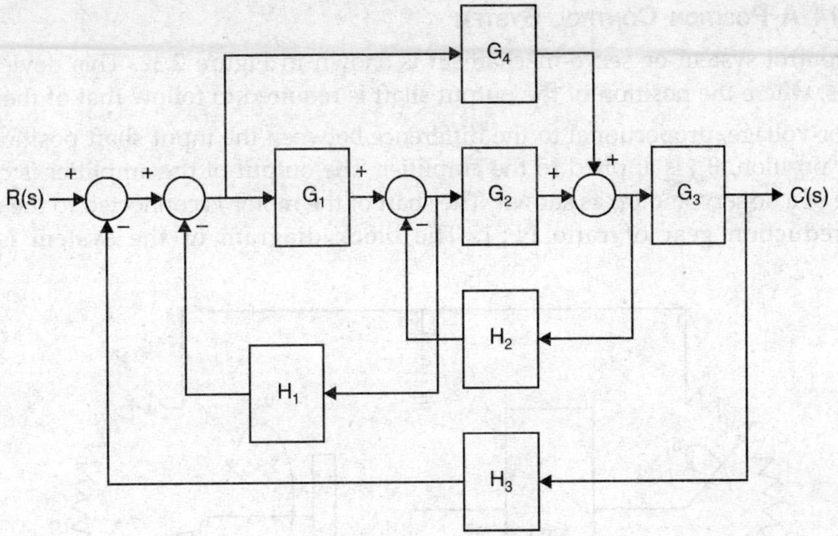

Figure 2.31. *Block diagram*

Here, we have two forward paths with transmittance given by

$$T_1 = G_1G_2G_3 \text{ and } T_2 = G_4G_3$$

Also, there are five loops, with transmittances $-G_1H_1$, $-G_2H_2$, $-G_1G_2G_3H_3$, $-G_4G_3H_3$ and $G_4H_2H_1$, and they are all touching each other. Accordingly, we have

$$\frac{C(s)}{R(s)} = \frac{G_1G_2G_3 + G_4G_2}{1 + G_1H_1 + G_2H_2 + G_1G_2G_3H_3 + G_4G_3H_3 - G_4H_2H_1}.$$

DRILL PROBLEM 2.8

Determine the overall transfer function for the block diagram in Fig. 2.23 using Mason's rule. Compare your answer with that given in Example 2.7.

DRILL PROBLEM 2.9

Determine the overall transfer function for the block diagram in Fig. 2.26 using Mason's rule. Compare your answer with that given in Example 2.8.

2.6 MODELS FOR SOME CONTROL SYSTEMS

In this section, we shall derive models for some typical control systems. In each case, we shall first develop a block diagram depicting the transfer functions of the various components and their interconnections. After that, Mason's rule will be utilized for obtaining the overall transfer function.

The various examples used here are intended to provide an insight to students into the methodology commonly used.

EXAMPLE 2.14 A POSITION CONTROL SYSTEM

A position control system or servo-mechanism is shown in Figure 2.32. This device has many practical uses, where the position of the output shaft is required to follow that of the input shaft.

An error voltage, proportional to the difference between the input shaft position θ_i and the output shaft position, θ_o, is applied to the amplifier. The output of the amplifier is connected to the armature of a dc servomotor, as shown. The shaft of the motor is connected to the output shaft through a reduction gear of ratio N : 1. The block diagram of the system is shown in Figure 2.33.

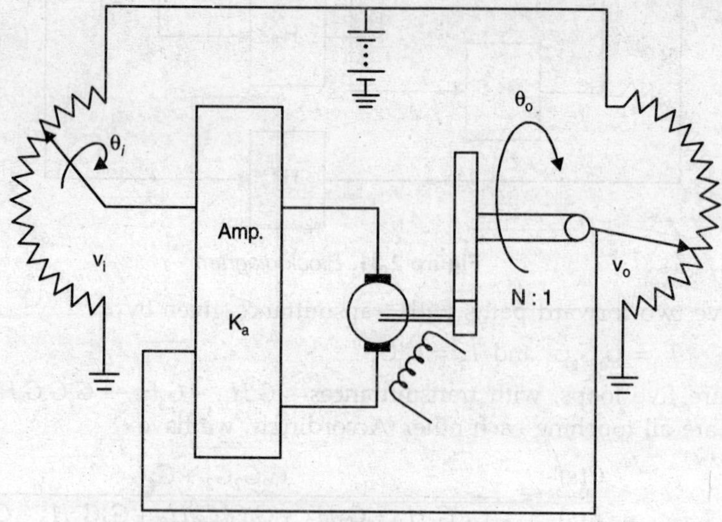

Figure 2.32. *A dc position control system*

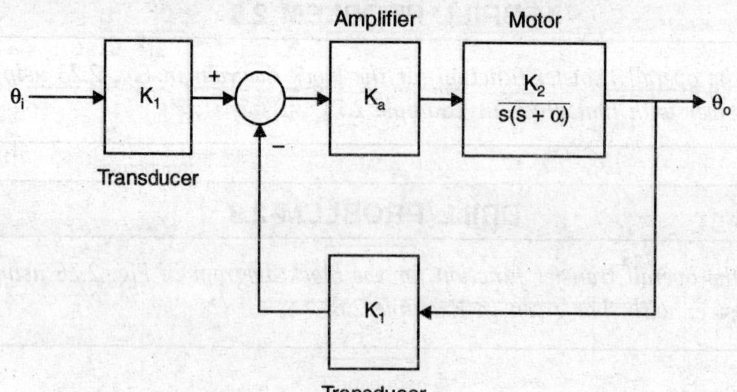

Figure 2.33. *Block diagram of the dc position control system*

The following relations are easily derived.

$$v_i = K_1\theta_i \qquad ...(2.21)$$

$$v_o = K_1\theta_i \qquad ...(2.22)$$

where K_1 is the constant of the position transducer (the potentiometer in this case).

Similarly, the load torque on the motor is given by

$$T = J\frac{d^2\theta_o}{dt^2} + D\frac{d\theta_o}{dt} \qquad ...(2.23)$$

where J is the effective moment of inertia and D the effective damping. This torque must equal the electrical torque developed by the motor.

It will be seen that in developing the block diagram, we have taken advantage of our knowledge of the transfer function of a dc servomotor, derived in Example 2.5. This is the main point in using the block diagram, since it shows the interconnection of various components. Also, we can determine the transfer function of each component separately. It may be noted that we have assumed that there is no loading effect of one block on the other. If this is not the case, then we must make necessary modifications. For instance, if the loading effect of the motor on the amplifier cannot be neglected, we would have to determine the transfer function of the amplifier-motor combination from the first principles, rather than simply multiplying the two transfer functions.

From the block diagram, using Mason's rule, we get,

$$\frac{\theta_o(s)}{\theta_i(s)} = \frac{K_1K_a\dfrac{K_2}{s(s+\alpha)}}{1+\dfrac{K_1K_aK_2}{s(s+\alpha)}} = \frac{K_1K_aK_2}{s^2+\alpha s+K_1K_aK_2} \qquad ...(2.24)$$

DRILL PROBLEM 2.10

For a particular position of control system, the product of the constants K_1, K_a and K_2 was found to be 8, while the constant α was found to be 4. Assuming that the system is initially at rest, determine the expression for the position of the output shaft as a function of time, if the input is a unit step.

Ans. $\theta_0(t) = 1 - 4\,e^{-2t}\cos 2t$

EXAMPLE 2.15 A WARD-LEONARD SPEED CONTROL SYSTEM

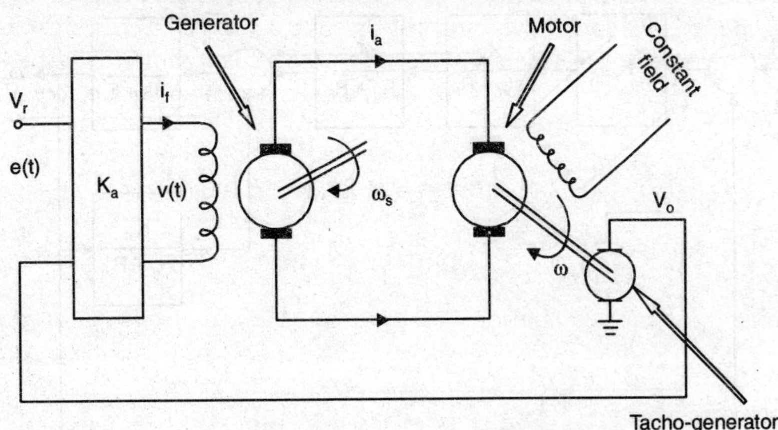

Figure 2.34. *A speed control system*

Figure 2.34 shows a typical control system used for controlling the speed of a large motor. It is used extensively in steel rolling mills. As will be seen, a voltage proportional to the desired speed is compared with the output of a tacho-generator coupled to the motor the speed of which is to be controlled. The 'error' voltage, which is a measure of the difference between the actual and desired speeds, is applied, after amplification to the field winding of a dc generator that is driven at a constant speed by a synchronous motor. The output voltage from this generator is then applied to the armature of the dc motor which drives the main load. The field current in this motor is maintained constant. Thus, the transfer function relating the speed of the motor to the voltage applied to it will be of the same from as that of the dc servomotor discussed earlier. The e.m.f. induced in the generator depends on the current in its field winding, which is related to the error voltage.

With some simplifying assumptions, we obtain the following equations for the system, where

K_s = Proportionality constant of the velocity sensor,
K_a = Gain constant of the amplifier,
K_g = Constant relating the field current to the e.m.f. of the generator,
K_m = Constant relating the field current to the motor back e.m.f.,
K_t = Constant relating the motor torque to the armature current,
R_f and L_f are the resistance and the inductane of the field winding,

and R_g, R_m are the resistances of the armature of the generator and motor, respectively.

$$e(t) = K_s(\omega_r - \omega_o)$$

$$v(t) = K_o e(t)$$

$$L_f \frac{di_f(t)}{dt} + R_f i_f = v(t)$$

$$i_a = \frac{K_g i_f - K_m \omega_o}{R_g + R_m}$$

$$J\frac{d\omega_o}{dt} + F\omega_o = K_t i_a$$

These equations lead to the block diagram shown in Fig. 2.35.

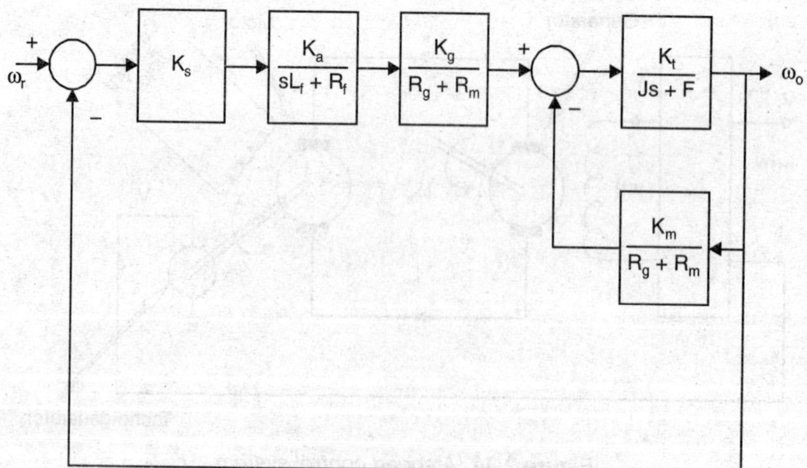

Figure 2.35. *Block diagram for speed control system*

One may now use Mason's rule for determining the overall transfer function of the system. The final result is given below.

$$\frac{\Omega_o(s)}{\Omega_r(s)} = \frac{K_s K_a K_g K_t}{(sL_f + R_f)(Js + F)(R_g + R_m) + K_t K_m (sL_f + R_f) + K_s K_a K_g K_t} \quad \text{...(2.25)}$$

SUMMARY

- In this chapter, we have discussed the general procedure for obtaining mathematical models for physical systems. This is often the most important and crucial step in being able to control a system. We may paraphrase Aristotle to say that we can control any system provided that we have a suitable mathematical model for it.
- The control engineer often has to obtain the mathematical model of a variety of components of diverse nature, such as electrical, mechanical, fluidic, and thermal, as well as various combinations of these. In all cases, one must first identify the system variables, which may be either through-variables or cross-variables. Often it is helpful to draw an analogous electrical circuit since in this case, the differential equations are obtained easily by applying Kirchhoff's laws.
- We start by writing the differential equations which relate the variables in each component of the system, as well as the interconnection equations. Although in most cases these differential equations are nonlinear, they are linearized about the operating points. It is most convenient to identify one variable as the input, or cause, and another variable as the output, or effect, for each component. This leads to the block diagram for each component, which relates the Laplace transforms of the input and the output through the transfer function. The overall transfer function of the system is then obtained either by simplifying the block diagram or by using Mason's rule.
- In many practical situations, the model obtained by using the procedure described above may be rather complicated. Furthermore, it may be difficult to measure accurately some of the parameters of the model. In such cases, an alternative approach commonly described as 'system identification' [4], is often used. The measurements of the frequency response or the response to some standard test functions like a unit step or a unit rectangular pulse are utilized for estimating the transfer function. It can be regarded as a type of curve-fitting since the transfer function is selected to fit the available data in an optimum manner. The details are beyond the scope of this book. Problem 4, at the end of this chapter, involves estimation of the transfer function of a dc servomotor from its step response, and gives the general idea.

Problems

1. Determine the transfer function for the electrical networks shown in Fig. P2.1 (*a*) and (*b*), where the input $v_1(t)$ and the output is $v_2(t)$.

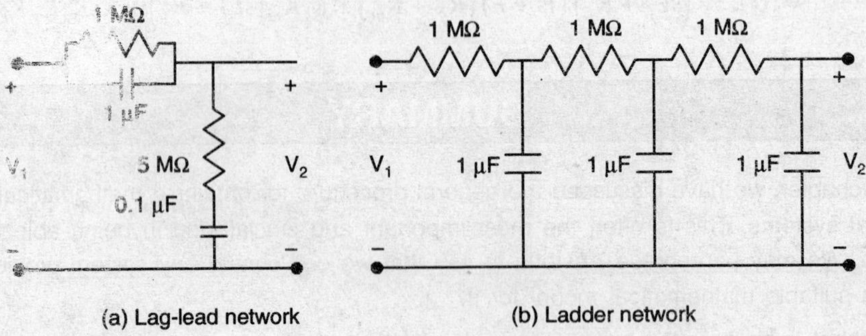

Figure P2.1. *RC networks*

2. Compare the transfer function of the network in Fig. P2.1 (*b*) with that obtained by taking the cube of the network shown in Fig. P2.2. Why are they different?

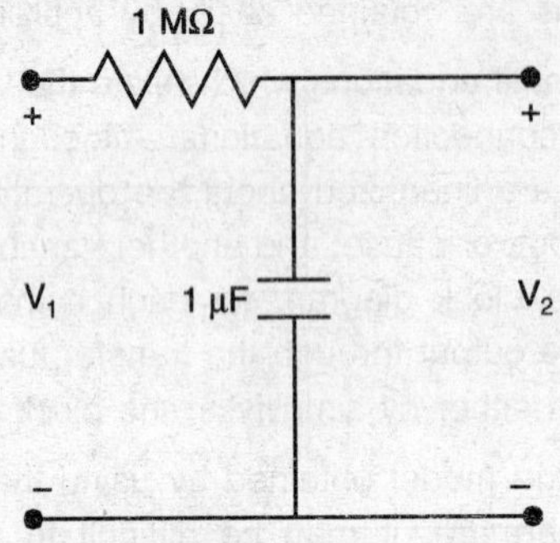

Figure P2.2. *RC network*

3. For each of the networks in problem 1, determine the steady-state component of $v_2(t)$ if the input is given by $v_2(t) = 2 + 3\cos(2t + \pi/6)$. (*Hint:* Use superposition and the concept of frequency response from transfer function. See Section A.6 in Appendix A.)
4. The transfer function relating the angular velocity to the input voltage for an armature-controlled dc servomotor was derived in Example 2.4 and is given in Equation (2.16). A practical method for determining the values of the constants K and α is to apply a constant voltage to the armature, and measure the steady-state speed of the motor, with the field current maintained constant. In addition, one must also measure the time required for the motor, starting from rest, to reach a certain fraction of the steady-state speed, say 50 per cent or 63.2 per cent. This is based on the fact that the response to a unit step input is given by

$$\omega(t) = \frac{K}{\alpha}\left(1 - e^{-\alpha t}\right)$$

In such an experiment, it was found that with a step input of 10 volts applied to the motor, its steady-state speed was 1200 revolutions per minute, and it required 12 seconds to reach 50 per cent of this speed. Determine the transfer function of the motor. (*Hint:* Note that in the expression for the transfer function, the angular velocity is in radiations per second.)

5. One form of a self-balancing potentiometer is shown in Figure P2.5. The input is $e_i(t)$ in volts and the output is $x(t)$ in centimeters. The amplifier gain $K = 10$. The lead screw advances 1 mm per revolution. The output of the movable arm of the potentiometer increases by 0.4 volt for each centimeter movement of x. With a step input of 10 volts applied to the motor (with the load), its steady-state speed is 1000 revolutions per minute, and it takes 0.5 second to reach 63.2 per cent of this value. Draw a block diagram for the system and determine the transfer function relating $X(s)$ to $E_i(s)$. Also, determine $x(t)$ if $e_i(t)$ is a unit step of 2 volts.

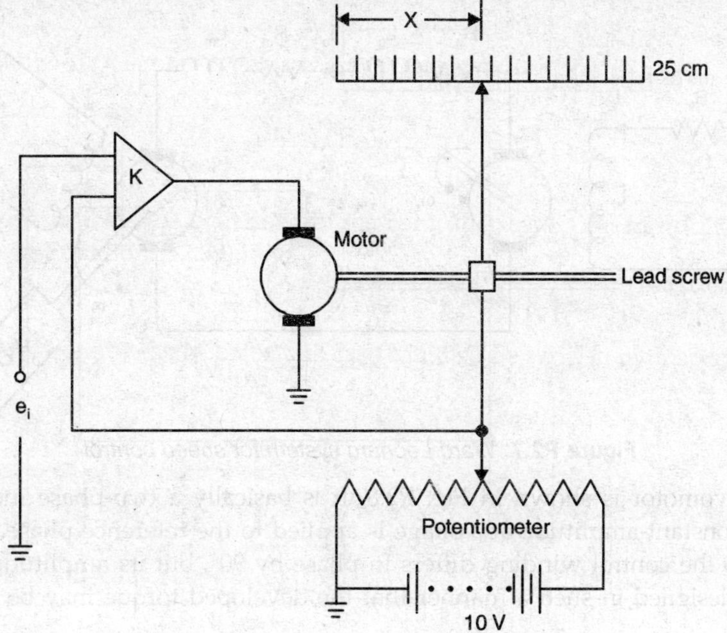

Figure P2.5. *A self-balancing potentiometer*

6. A dc servomotor is sometimes used in the field-control mode. In this case, the armature current is maintained constant and the input voltage is applied to the field winding. A schematic diagram is shown in Fig. P2.6. Determine the transfer function relating the output position to the input voltage. (**Hint:** Note that the inductance of the field winding cannot be neglected. Draw a block diagram similar to Fig. P2.6.)

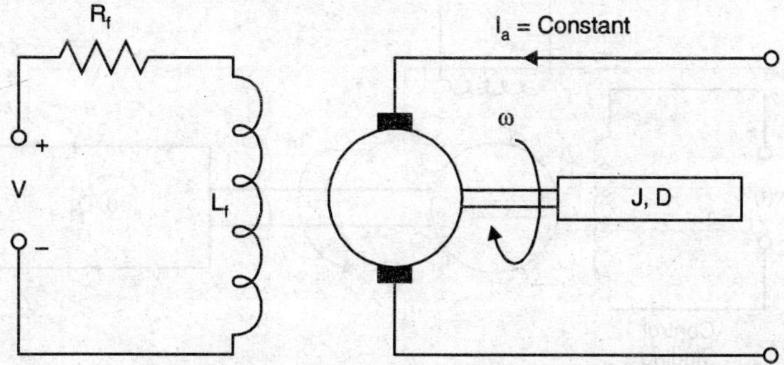

Figure P2.6. *A field-controlled dc servomotor*

7. The Ward-Leonard scheme for controlling the speed of heavy loads (for example, in a steel rolling mill) is shown in Fig. P2.7. A dc generator, driven at a constant speed by a three-phase synchronous motor, drives an armature-controlled dc motor. Draw a block diagram for this system and determine the transfer function relating ω_o to v if the constants of the machine are as given below.

Generator : $R_f = 1\ \Omega$, $L_f = 0.2$ H, $R_g = 0.3\ \Omega$, $L_g = 0.01$ H, $K_g = 2$ V/A

Motor : $R_m = 0.5\ \Omega$, $L_m = 0.02$ H, $K_m = 0.1$ V/rad/s

Load : $J = 4$ kg–m^2, $D = 0.1$ N–m/rad/s

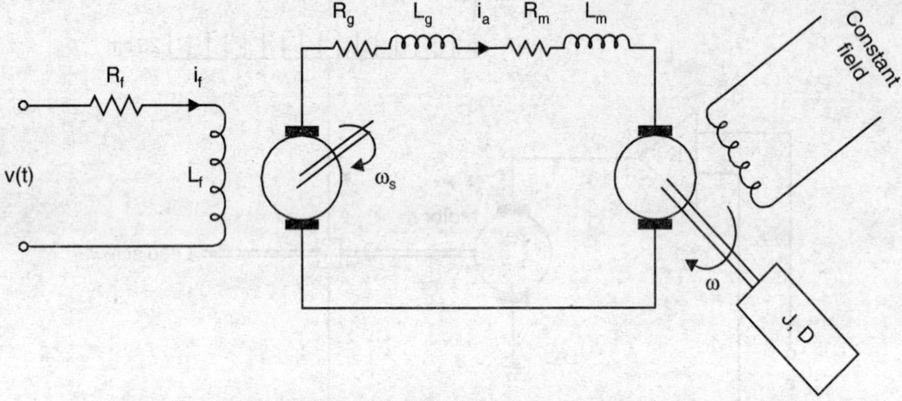

Figure P2.7. *Ward-Leonard system for speed control*

8. An ac servomotor is shown in Fig. P2.8. It is basically a two-phase induction motor in which a constant-amplitude ac voltage is applied to the reference phase. The voltage $v(t)$, applied to the control winding differs in phase by 90°, but its amplitude is variable. The motor is designed in such a manner that the developed torque may be approximated as

$$T = K_1 v - K_2 \omega$$

where ω is the angular velocity of the motor shaft. Draw a block diagram and derive the transfer function relating the angular position θ to the applied voltage v.

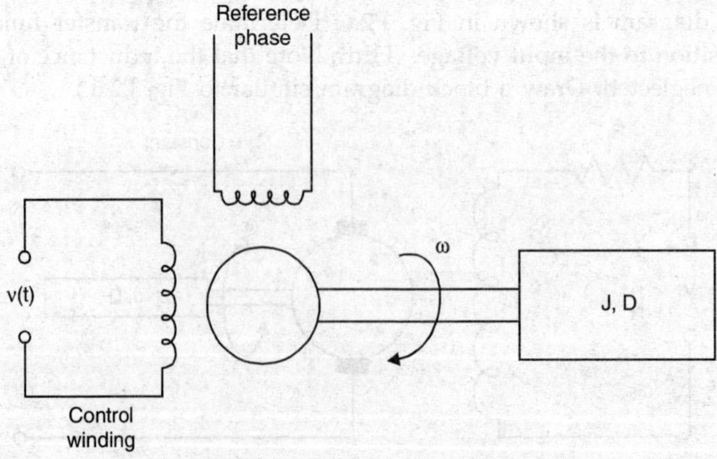

Figure P2.8. *An ac servomotor*

9. Due to the lack of proper dynamic balance, many machines (including automobiles) exhibit vibrations at certain frequencies. A scheme for eliminating this vibration at a certain frequency is shown in Fig. P2.9. The object is to prevent the mass M_1 from vibrating at a frequency ω_0. This is done by including the spring K_2 and mass M_2 as shown. By obtaining the equivalent electrical circuit, determine the relationship between K_2 and M_2 that will meet this objective. Also, determine the transfer function relating $x_1(t)$ to $f(t)$. (*Hint:* Examine the effect of the analogous of M_2 and K_2 on the voltage analogous to dx_1/dt.)

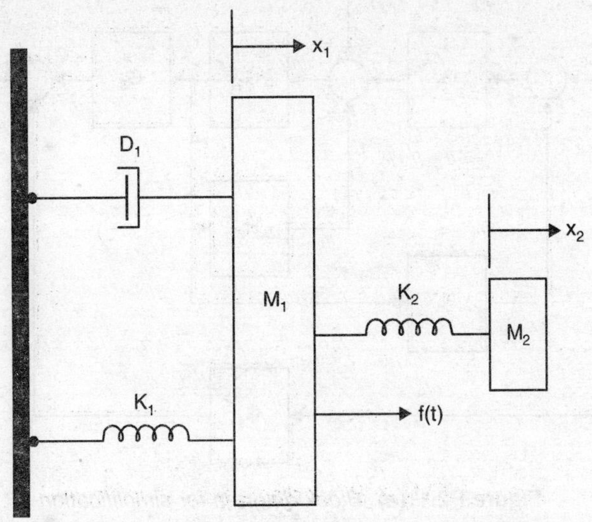

Figure P2.9. *A dynamic vibration absorber*

10. An analog computer is often used for simulating dynamic systems and studying the effect of adjusting the values of certain parameters. Its basic components are operational amplifiers and potentiometers. The former can be used as adders, integrators and inverters, while the latter are used by multiplying by constants less than one. The simulation diagram for a linear system is shown in Fig. P2.10, where three types of blocks are shown. These are adders, integrators and multipliers (by a constant). The block diagram can be drawn simply be replacing each integrator by its transfer function,

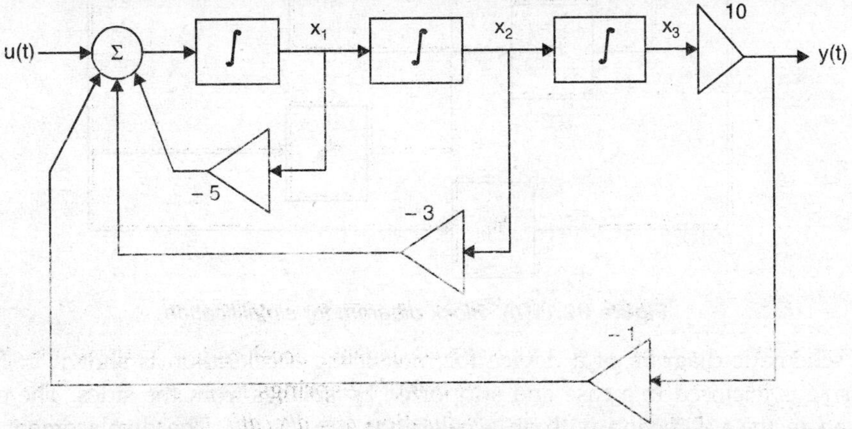

Figure P2.10. *Analog computer simulation diagram*

11. For the block diagrams shown in Fig. P2.11(a) and P2.11(b), determine the overall transfer function, $C(s)/R(s)$, by block diagram simplification and verify using Mason's rule.

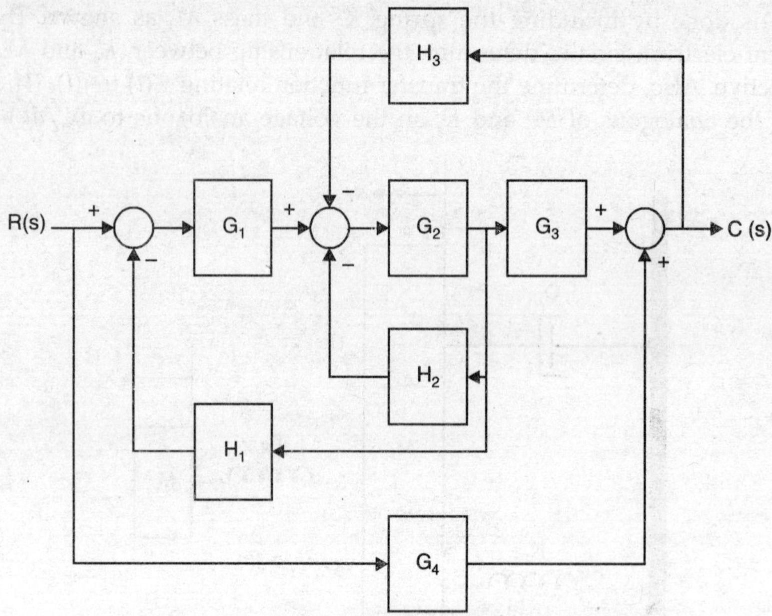

Figure P2.11(*a*). *Block diagram for simplification*

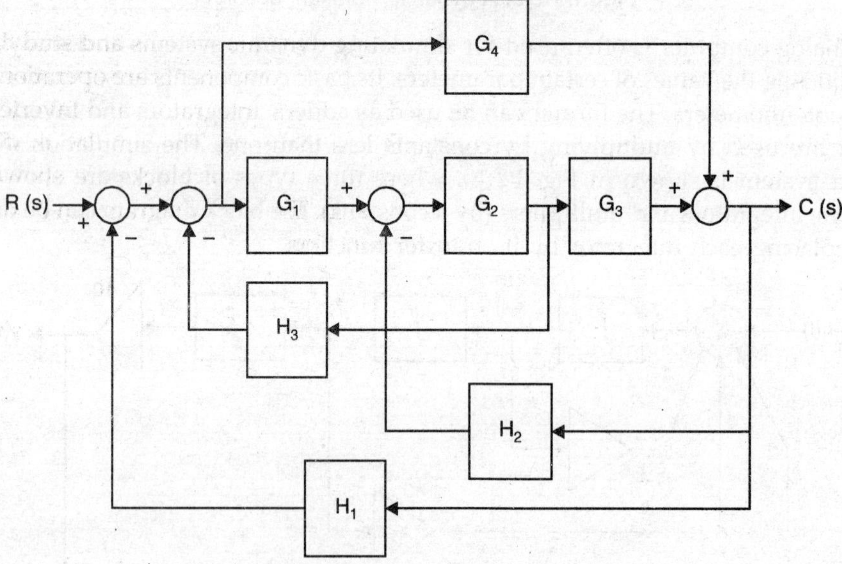

Figure P2.11(*b*). *Block diagram for simplification*

12. The schematic diagram of a device for measuring acceleration is shown in Fig. P2.12. A mass M is enclosed in a case and supported by springs from the sides. The entire case is moved in the x direction with an acceleration $a = d^2x/dt^2$. The displacement of the mass M, relative to the case, is proportional to the acceleration. Draw the equivalent electrical circuit and determine the transfer function relating the output $y(t)$ to the acceleration a.

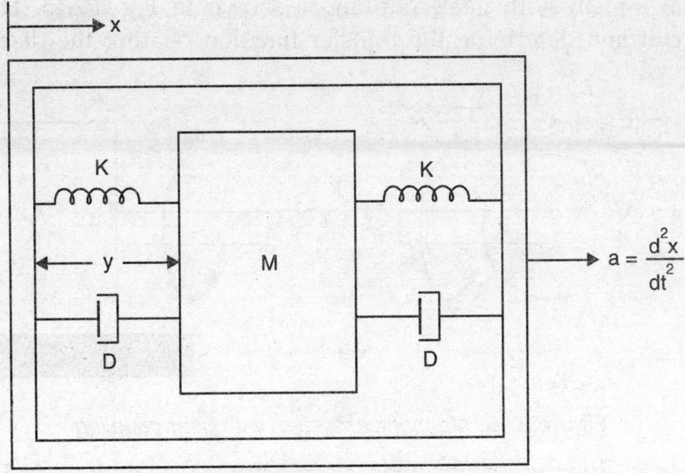

Figure P2.12. *Accelerometer*

13. Consider the block diagrams shown in Fig. P2.13. Determine the values of G_e and H_e so that the two may be equivalent.

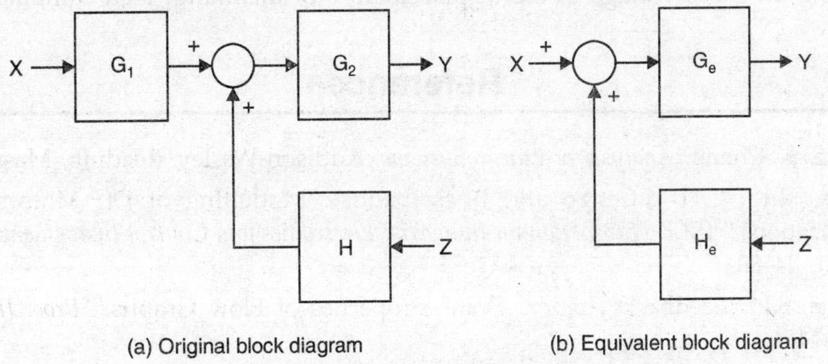

Figure P2.13. *Equivalence of block diagrams*

14. Consider the block diagrams shown in Fig. P2.14. Determine the values of G_e and H_e so that the two may be equivalent.

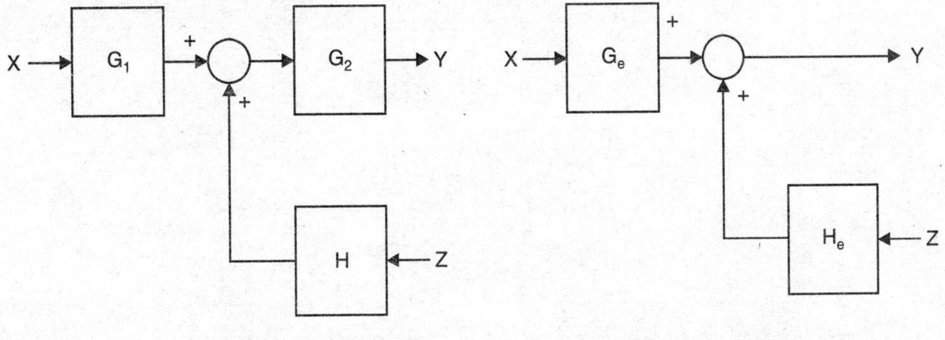

Figure P2.14. *Block diagram equivalence*

15. A mechanical system with gear-coupling is shown in Fig. P2.15. Draw an equivalent electrical circuit and determine the transfer function relating the displacement θ_3 to the input torque τ.

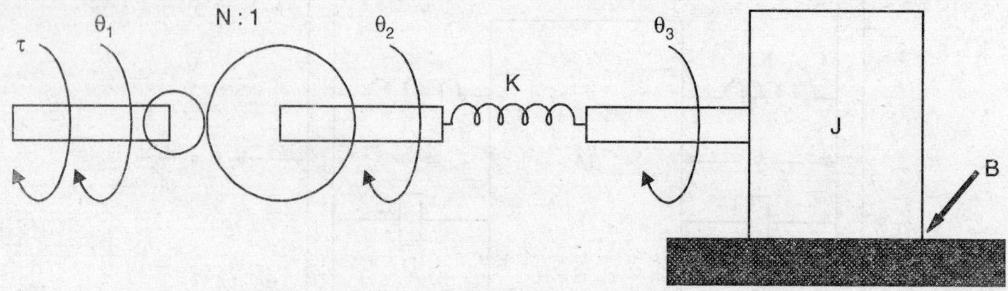

Figure 2.15. *Mechanical system with gear coupling*

16. The model for an armature-controlled dc servomotor was derived in Example 2.5, where it was considered that the load was coupled rigidly and directly to the shaft of the motor. Now, consider the case when the load is coupled to the motor through gears and can be modelled, as in the previous problem (Fig. P2.15). Determine the overall transfer function relating the input voltage to the displacement θ of this motor-load combination.

References

1. David, K. Cheng, *Analysis of Linear Systems*, Addison-Wesley, Reading, Mass., 1959.
2. Sinha, N.K., C.D. diCenzo and B. Szabados, "Modelling of DC Motors for Control Applications," *IEEE Transaction on Industrial Electronics and Control Instrumentation*, **IECI-21** (1974), 84–88.
3. Mason, S.J., "Feedback Theory—Some Properties of Flow Graphs," *Proc. IRE*, *41* (1953), 1144–1156.
4. Sinha, N.K., and B. Kuszta, *Modelling and Identification of Dynamic Systems*, Van Nostrand-Reinhold, New York, 1983.

CHAPTER 3

State-Space Models

3.1 INTRODUCTION

In the previous chapter, we studied the use of transfer function models for linear time-invariant continuous-time systems. These have also been called *frequency response* models due mainly to the interpretation of the Laplace transform variable s as complex frequency, in contrast with differential Equation models, which are *time-domain models.* Although the transfer function model provides an important tool for the analysis and design of control systems, it has certain basic limitations. For example, we cannot apply such models to nonlinear or linear time-varying systems. Furthermore, these models cannot be used efficiently for systems of high order or for multivariable systems— systems with many inputs and outputs.

In this chapter, we shall return to a particular class of time-domain models, known as *state-space models,* which are especially suitable for use with computers. An important advantage of the state-space representation is that it allows us to study multivariable systems without the need for changing the notational framework used with single-input single-output systems. Furthermore, it is much easier to study discrete-time systems with the help of digital computers when the state-space representation is applied. These models can also be used for the study of nonlinear or time-varying systems.

An important feature of the state-space representation is that it gives us information about the internal behaviour of the system, as well as the input-output behaviour. In contrast, the transfer function model tells us only about the input-output behaviour of the system. The concept of state will be presented in the next section.

3.2 THE CONCEPT OF STATE

The state of a system is defined as *the smallest set of variables that must be known at any given instant in order that the future response of the system to any specified input may be calculated from the given dynamic equation.* Thus, the state can be regarded as a compact representation of the past history of the system, which can be utilized for predicting its future behaviour in response to an external stimulus. Since the complete solution of a differential Equation of order n requires precisely n initial conditions, it follows that the state of such a system will be specified by the values of n quantities, called the *state variables.*

The number of state variables for a system will be precisely equal to the order of the differential Equation required to model it. As an example, in an electrical network, we must know the value of the voltage across each capacitor and the current through each inductor at a given instant of time in order to predict the future values of currents and voltages in response to a given set of inputs. As expected, the number of independent state variables in an electrical network will *normally* be equal to the number of energy storage elements contained in it. However, there will be fewer state variables if there are loops containing only capacitors and voltage sources and cut-sets containing only inductors and current sources. It should, however, be emphasized that, in general, the state-space representation is not unique; there are many different sets of state variables that can be utilized for a given system. This will be illustrated through some examples later in this section. In practice, it is desirable to select those physical variables as the states, which may be measured easily for use in feedback loops.

State Equations are arranged as a set of first-order differential equations, and have the following form for a linear time-invariant system

$$\dot{x} = Ax + Bu \qquad \text{...(3.1)}$$

where $x(t)$ is the n-dimensional state vector, the dot over x represents the derivative of x with respect to time, and $u\,(t)$ is the m-dimensional input vector. A and B are matrices of dimensions $n \times n$ and $n \times m$, respectively with constant elements. The output of the system is given by

$$y(t) = Cx\,(t) + Du\,(t) \qquad \text{...(3.2)}$$

where $y(t)$ is a vector of dimension p, the number of outputs, and C and D are constant matrices of dimensions $p \times n$ and $p \times m$ respectively.

Although Equations (3.1) and (3.2) are meant for the general case of multivariable systems, in this book we shall only study the special and simple case of single-input single-output systems, for which $m = 1$ and $p = 1$. Consequently, B will be a matrix of dimension $n \times 1$, that is a column vector of dimension n. Similarly, C will be a row vector of dimension n, and D will be a 1×1 matrix, or a scalar.

We shall now consider some examples showing how one can write state equations for physical networks. We shall start with the simple series RLC network, shown in Fig. 3.1.

EXAMPLE 3.1: A SERIES RLC NETWORK

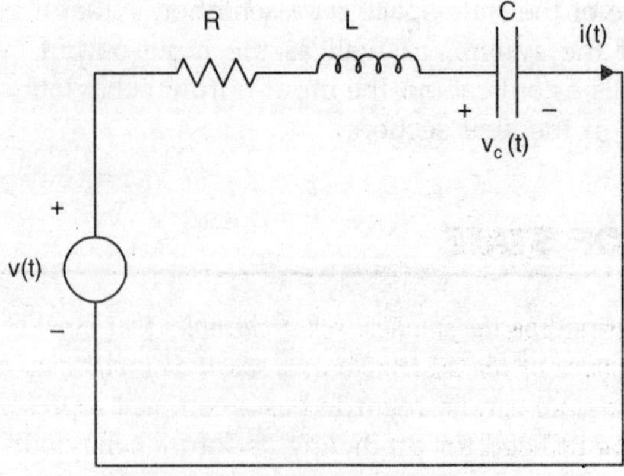

Figure 3.1. *A simple RLC network*

Application of Kirchhoff's voltage law to the circuit shown in Fig. 3.1 leads to the following integro-differential equation:

$$L\frac{di}{dt} + Ri + \int_{-\infty}^{t} \frac{1}{C} i\, dt = v(t) \qquad ...(3.3)$$

Differentiating both sides of the above equation with respect to time, we get

$$L\frac{d^2 i}{dt^2} + R\frac{di}{dt} + \frac{1}{C} i = \frac{dv}{dt} \qquad ...(3.4)$$

One approach is to select i and $\frac{di}{dt}$ as the state variables. Let

$$x_1 = i$$

$$x_2 = \frac{di}{dt} \qquad ...(3.5)$$

We can now write Equation (3.4) as

$$L\dot{x}_2 + Rx_2 + \frac{1}{C} x_1 = \dot{v} \qquad ...(3.6)$$

Combining Equation (3.6) with (3.5) and rearranging, we obtain the following set of two first-order differential equations

$$\dot{x}_1 = x_2$$

$$\dot{x}_2 = -\frac{R}{L} x_2 - \frac{1}{LC} x_1 + \frac{1}{L} \dot{v} \qquad ...(3.7)$$

Equation (3.7) can be arranged in the matrix form that was shown in Equation (3.1), and constitute one set of state equations for this network, with the state variables as defined in Equation (3.5). The presence of the derivative of the input in these equations makes them undesirable. Another set of state equations, not containing this derivative, may be obtained by selecting i and v_c as the state variables. Letting

$$x_1 = i \qquad ...(3.8)$$

$$x_2 = v_c$$

we obtain the following from Equation (3.3):

$$\dot{x}_1 = -\frac{R}{L} x_1 - \frac{1}{L} x_2 + \frac{1}{L} v(t) \qquad ...(3.9)$$

$$\dot{x}_2 = \frac{1}{C} x_1$$

Thus, we have obtained two sets of state equations for this network. Although both of them are equally valid, Equation (3.9) utilize the physical variables i and v_c and do not require differentiation of the input. These equations can be solved more directly if the initial conditions are given as $i(0)$ and $v_c(0)$. Furthermore, it is possible to write state equations of electrical networks directly using some basic concepts of network topology if we select the voltages across capacitors and currents to through inductors as the state variables. This is described briefly in Appendix *D*.

Example 3.2: An Armature-Controlled DC Servomotor

The following differential equation was derived in Example 2.5:

$$\frac{d^2\theta}{dt^2} + \alpha\frac{d\theta}{dt} = Kv \qquad ...(3.10)$$

This second-order differential equation is easily changed into two first-order differential equations by defining the state variables as

$$x_1 = \theta$$

$$x_2 = \dot{x}_1 = \frac{d\theta}{dt} \qquad \text{...(3.11)}$$

so that Equation (3.10) can be written as

$$\dot{x}_2 = Kv - \alpha x_2 \qquad \text{...(3.12)}$$

Equations (3.11) and (3.12) can now be combined to form the following set of state equations, which have been expressed in the form of a matrix.

$$\begin{bmatrix} \dot{x}_1 \\ \dot{x}_2 \end{bmatrix} = \begin{bmatrix} 0 & 1 \\ 0 & -\alpha \end{bmatrix} \begin{bmatrix} x_1 \\ x_2 \end{bmatrix} + \begin{bmatrix} 0 \\ K \end{bmatrix} v \qquad \text{...(3.13)}$$

$$y = \begin{bmatrix} 0 & 1 \end{bmatrix} \begin{bmatrix} x_1 \\ x_2 \end{bmatrix}$$

where, the output y is angular position θ.

Example 3.3: A Field-Controlled DC Servomotor

Consider the field-controlled *dc* servomotor shown in Fig. 3.2.

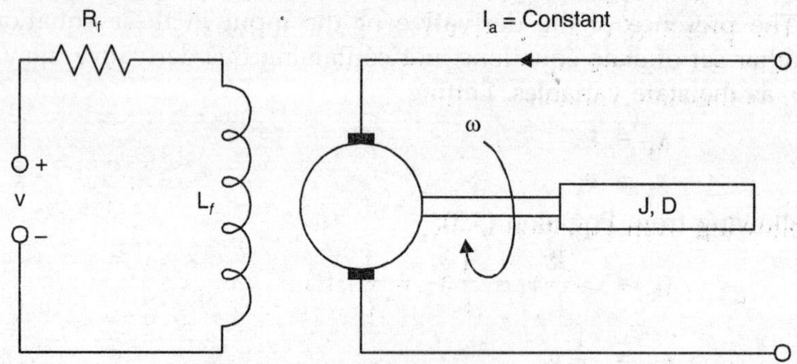

Figure 3.2. *A field-controlled dc servomotor*

The following differential equations are obtained

$$v = L\frac{di}{dt} + Ri$$

$$Ki = J\frac{d\omega}{dt} + D\omega \qquad \text{...(3.14)}$$

$$\omega = \frac{d\theta}{dt}$$

These are already a set of three first-order differential equations. Hence, if we let $x_1 = \theta$, $x_2 = \omega$, and $x_3 = i$, we obtain the following state equations:

$$\begin{bmatrix} \dot{x}_1 \\ \dot{x}_2 \\ \dot{x}_3 \end{bmatrix} = \begin{bmatrix} 0 & 1 & 0 \\ 0 & -\dfrac{D}{J} & \dfrac{K}{J} \\ 0 & 0 & -\dfrac{R}{L} \end{bmatrix} \begin{bmatrix} x_1 \\ x_2 \\ x_3 \end{bmatrix} + \begin{bmatrix} 0 \\ 0 \\ \dfrac{1}{L} \end{bmatrix} v \qquad \text{...(3.15)}$$

$$y(t) = [1 \quad 0 \quad 0] \begin{bmatrix} x_1 \\ x_2 \\ x_3 \end{bmatrix}$$

Another set of state equations can be obtained by selecting the angular position θ, and its first and second derivatives as the state variables. If we differentiate the second equation of the set (3.14) and eliminate i and its derivative from these equations, we obtain

$$\begin{bmatrix} \dot{x}_1 \\ \dot{x}_2 \\ \dot{x}_3 \end{bmatrix} = \begin{bmatrix} 0 & 1 & 0 \\ 0 & 0 & 1 \\ 0 & -\dfrac{DR}{JL} & -\left(\dfrac{D}{J}+\dfrac{R}{L}\right) \end{bmatrix} \begin{bmatrix} x_1 \\ x_2 \\ x_3 \end{bmatrix} + \begin{bmatrix} 0 \\ 0 \\ \dfrac{K}{JK} \end{bmatrix} v \qquad \text{...(3.16)}$$

$$y(t) = [1 \quad 0 \quad 0] \begin{bmatrix} x_1 \\ x_2 \\ x_3 \end{bmatrix}$$

where $x_1 = \theta$, $x_2 = \omega$ and $x_3 = \dfrac{d\omega}{dt}$.

Although both the sets of state equations are equally valid, the variables selected in Equation (3.15) are easier to measure.

Example 3.4: A Mechanical System with Two Degrees of Freedom

Consider the mechanical system with two degrees of freedom, shown in Fig. 3.3. This was also studied in Example 2.2. The following differential equations are easily derived:

$$M_2 \frac{d^2x_2}{dt^2} + (D_1 + D_2)\frac{dx_2}{dt} - D_2\frac{dx_1}{dt} = f(t) \qquad \text{...(3.17)}$$

$$M_1 \frac{d^2x_1}{dt^2} + D_2\frac{dx_1}{dt} + Kx_1 - D_2\frac{dx_2}{dt} = 0$$

These are second-order differential equations, which are coupled to each other, as is typical with mechanical systems. However, we can transform them into a set of four first-order differential equations in a very straight-forward manner. We define two more state variables, which are the derivatives of the two displacements (*i.e.*, the corresponding velocities) as below:

$$x_3 = \frac{dx_1}{dt} \quad ...(3.18)$$

$$x_4 = \frac{dx_2}{dt}$$

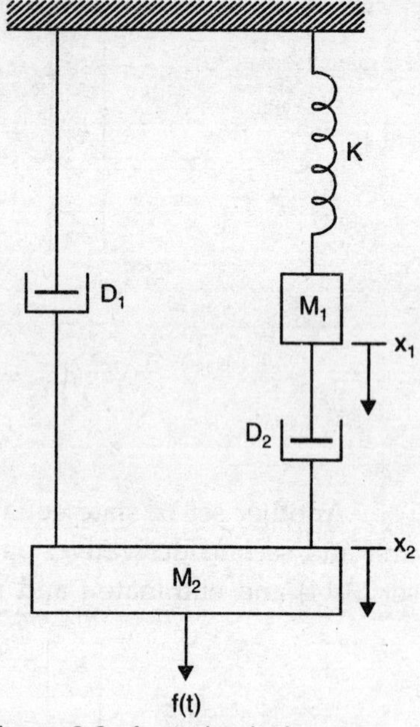

Figure 3.3. *A mechanical system*

Substituting these into Equation (3.17), we obtain two first-order differential equations relating the derivatives of these velocities to the other state variables, as shown below:

$$M_2 \frac{dx_4}{dt} + (D_1 + D_2)x_4 - D_2x_3 = f(t) \quad ...(3.19)$$

$$M_1 \frac{dx_3}{dt} + D_2x_3 + Kx_1 - D_2x_4 = 0$$

We can now solve Equation (3.19) for the derivative terms and combine them with Equation (3.18) to obtain the set of four first-order equations shown in Equation (3.20). It may be noted that the four states are the two displacements and the corresponding velocities.

$$\begin{bmatrix} \dot{x}_1 \\ \dot{x}_2 \\ \dot{x}_3 \\ \dot{x}_4 \end{bmatrix} = \begin{bmatrix} 0 & 0 & 1 & 0 \\ 0 & 0 & 0 & 1 \\ -\frac{K}{M_1} & 0 & -\frac{D_2}{M_1} & \frac{D_2}{M_1} \\ 0 & 0 & \frac{D_2}{M_2} & -\frac{D_1 + D_2}{M_2} \end{bmatrix} \begin{bmatrix} x_1 \\ x_2 \\ x_3 \\ x_4 \end{bmatrix} + \begin{bmatrix} 0 \\ 0 \\ 0 \\ \frac{1}{M_2} \end{bmatrix} \bar{f}(t) \quad ...(3.20)$$

DRILL PROBLEM 3.1

Write a set of state equations for the mechanical system shown in Figure 3.4.

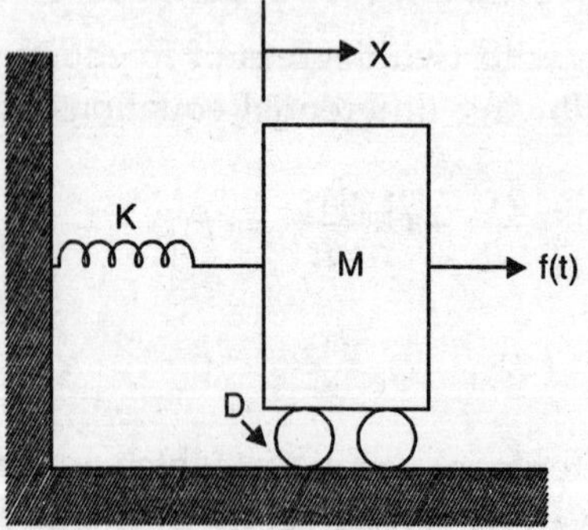

Figure 3.4. *Mechanical system*

Ans.
$$\begin{bmatrix} \dot{x}_1 \\ \dot{x}_2 \end{bmatrix} = \begin{bmatrix} 0 & 1 \\ -\dfrac{K}{M} & -\dfrac{D}{M} \end{bmatrix} \begin{bmatrix} x_1 \\ x_2 \end{bmatrix} + \begin{bmatrix} 0 \\ \dfrac{1}{M} \end{bmatrix} f$$

DRILL PROBLEM 3.2

Write a set of state equations for the mechanical system shown in Fig. 3.5.

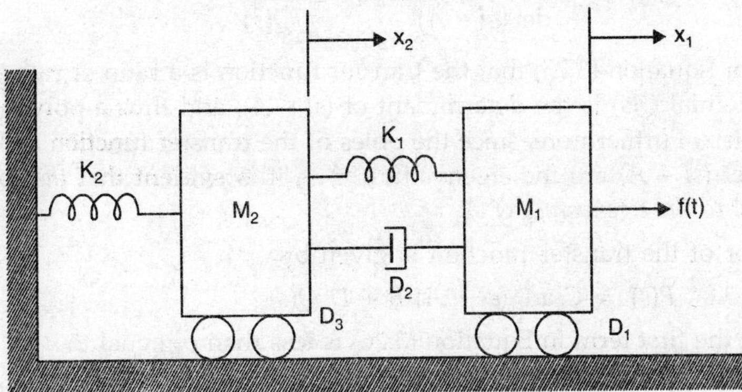

Figure 3.5. *Mechanical system*

Ans.

$$\begin{bmatrix} \dot{x}_1 \\ \dot{x}_2 \\ \dot{x}_3 \\ \dot{x}_4 \end{bmatrix} = \begin{bmatrix} 0 & 0 & 1 & 0 \\ 0 & 0 & 0 & 1 \\ -\dfrac{K_1}{M_1} & \dfrac{K_1}{M_1} & -\dfrac{D_1+D_2}{M_1} & \dfrac{D_2}{M_1} \\ \dfrac{K_1}{M_2} & -\dfrac{K_1+K_2}{M_2} & \dfrac{D_2}{M_2} & -\dfrac{D_2+D_3}{M_2} \end{bmatrix} \begin{bmatrix} x_1 \\ x_2 \\ x_3 \\ x_4 \end{bmatrix} + \begin{bmatrix} 0 \\ 0 \\ \dfrac{1}{M_1} \\ 0 \end{bmatrix} f(t)$$

3.3 TRANSFER FUNCTION FROM STATE EQUATIONS

As mentioned earlier, state equations represent the complete internal description of a system, whereas the transfer function is only the input-output representation. Consequently, the transfer function can be obtained uniquely from the state equations. Recall that the transfer function of a single-input single-output system was defined as the ratio of the Laplace transforms of the output and the input under zero initial conditions. Taking the Laplace transform of both sides of Equation (3.1) with $x(0)$ set equal to zero, we get

$$sX(s) = AX(s) + BU(s) \quad ...(3.21)$$

which can be solved for $X(s)$ to obtain

$$X(s) = (sI - A)^{-1} BU(s) \quad ...(3.22)$$

Substituting this expression in the Laplace transform of Equation (3.2), we have

$$Y(s) = [C(sI - A)^{-1} B + D]\ U(s) \quad ...(3.23)$$

The transfer function is now obtained as

$$G(s) = \frac{Y(s)}{U(s)} = C(sI - A)^{-1} B + D \quad ...(3.24)$$

Recalling that the inverse of a matrix can be obtained by dividing its adjoint by its determinant, we may rewrite Equation (3.24) as

$$G(s) = \frac{C \operatorname{adj}(sI - A) B}{\det(sI - A)} + D = \frac{P(s)}{Q(s)} \quad ...(3.25)$$

It follows from Equation (3.25) that the transfer function is a ratio of two polynomials. The denominator polynomial $Q(s)$ is the determinant of $(sI - A)$, and thus a polynomial of degree n if A is an $n \times n$ matrix. Furthermore, since the poles of the transfer function are the roots of $Q(s)$ and the roots of det $(sI - A)$ are the eigenvalues of A, it is evident that *the poles of the transfer function are identical to the eigenvalues of A.*

The numerator of the transfer function is given by

$$P(s) = C \operatorname{adj}(sI - A) B + D\ Q(s) \quad ...(3.26)$$

The degree of the first term in Equation (3.26) is less than or equal to n–1, while that of the second term will be n, since D is a scalar. Hence, the transfer function will always be *proper* if D is nonzero and will be *strictly proper* if $D = 0$.

It will be seen from Equation (3.26) that we have to calculate the inverse of the matrix $(sI - A)$ to determine the transfer function from the state equations. This can be done conveniently by using Leverrier's algorithm (also called the Souriau-Frame algorithm).

Let $$(sI - A)^{-1} = \frac{\operatorname{adj}(sI - A)}{\det(sI - A)} \quad ...(3.27)$$

$$= \frac{P_1 s^{n-1} + P_2 S^{n-2} + \ldots + P_{n-1} s + P_n}{s^n + a_1 s^{n-1} + \ldots + a_{n-1} s + a_n}$$

where P_i are $n \times n$ matrices and a_i are scalars. The elements of P_i are real and so are the a_i if the elements of A are real, as is true for all physical systems. Leverrier's algorithm allows us to calculate P_i and a_i in a recursive manner as described below:

$$P_1 = I$$

$$a_1 = -\operatorname{tr}(A)$$

$$P_2 = P_1 A + a_1 I$$

$$a_2 = -\frac{1}{2} \operatorname{tr}(P_2 A) \quad ...(3.28)$$

$$\vdots$$

$$P_n = P_{n-1} A + a_{n-1} I$$

$$a_n = -\frac{1}{n} \operatorname{tr}(P_n A)$$

where $tr(A)$ represents the *trace* of the matrix A, defined as the sum of the elements on the main diagonal of the matrix.

The above equation is suitable for use on a digital computer and can be easily programmed. The programme 'TRANSFER.EXE' on the disk available with this book uses this method to calculate the transfer function of a system from its state equations.

Since digital computers can carry only a finite number of digits, they often introduce errors due to truncation. Hence, it is desirable to verify the results obtained from the algorithm by using the following relationship.

$$P_n A + a_n I = 0 \qquad \text{...(3.29)}$$

EXAMPLE 3.5

We shall use the above algorithm to determine the transfer function for the following case:

$$A = \begin{bmatrix} -2 & 0 & 1 \\ 1 & -2 & 0 \\ 1 & 1 & 1 \end{bmatrix} \quad B = \begin{bmatrix} 1 \\ 0 \\ 1 \end{bmatrix} \quad C = [2 \;\; 1 \;\; -1] \quad D = 0 \qquad \text{...(3.30)}$$

In this case, we have $n = 3$. Consequently, we obtain

$$P_1 = I$$

$$a_1 = -\,tr(A) = 5$$

$$P_2 = P_1 A + a_1 I = \begin{bmatrix} -2 & 0 & 1 \\ 1 & -2 & 0 \\ 1 & 1 & -1 \end{bmatrix} + \begin{bmatrix} 5 & 0 & 0 \\ 0 & 5 & 0 \\ 0 & 0 & 5 \end{bmatrix}$$

$$= \begin{bmatrix} 3 & 0 & 1 \\ 1 & 3 & 0 \\ 1 & 1 & 4 \end{bmatrix}$$

$$a_2 = -\frac{1}{2}\,tr(P_2 A) = -\frac{1}{2}\,tr\left\{\begin{bmatrix} 3 & 0 & 1 \\ 1 & 3 & 0 \\ 1 & 1 & 4 \end{bmatrix}\begin{bmatrix} -2 & 0 & 1 \\ 1 & -2 & 0 \\ 1 & 1 & -1 \end{bmatrix}\right\}$$

$$= -\frac{1}{2}\,tr\begin{bmatrix} -5 & 1 & 2 \\ 1 & -6 & 1 \\ 3 & 2 & -3 \end{bmatrix} = 7$$

$$P_3 = P_2 A + a_2 I = \begin{bmatrix} -5 & 1 & 2 \\ 1 & -6 & 1 \\ 3 & 2 & -3 \end{bmatrix} + \begin{bmatrix} 7 & 0 & 0 \\ 0 & 7 & 0 \\ 0 & 0 & 7 \end{bmatrix} = \begin{bmatrix} 2 & 1 & 2 \\ 1 & 1 & 1 \\ 3 & 2 & 4 \end{bmatrix}$$

$$a_3 = -\frac{1}{3}\, tr\,(P_3\, A) = -\frac{1}{3}\, tr\left\{\begin{bmatrix} 2 & 1 & 2 \\ 1 & 1 & 1 \\ 3 & 2 & 4 \end{bmatrix}\begin{bmatrix} -2 & 0 & 1 \\ 1 & -2 & 0 \\ 1 & 1 & -1 \end{bmatrix}\right\}$$

$$= -\frac{1}{3}\, tr \begin{bmatrix} -1 & 0 & 0 \\ 0 & -1 & 0 \\ 0 & 0 & -1 \end{bmatrix} = 1$$

We can now easily verify that

$$P_3\, A + a_3\, I = 0$$

Thus, we have the denominator polynomial given by

$$Q\,(s) = s^3 + a_1 s^2 + a_2 s + a_3 = s^3 + 5s^2 + 7s + 1,$$

and the adjoint of $(sI - A)$ is obtained as

$$\text{adj}\,(sI - A) = P_1 s^2 + P_2 s + P_3$$

$$= \begin{bmatrix} s^2 + 3s + 2 & 1 & s + 2 \\ s + 1 & s^2 + 3s + 1 & 1 \\ s + 3 & s + 2 & s^2 + 4s + 4 \end{bmatrix}$$

Since $D = 0$, the numerator polynomial is obtained as

$$P\,(s) = C\ \text{adj}\,(sI - A)\, B$$

$$= [2 \quad 1 \quad -1] \begin{bmatrix} s^2 + 3s + 2 & 1 & s + 2 \\ s + 1 & s^2 + 3s + 1 & 1 \\ s + 3 & s + 2 & s^2 + 4s + 4 \end{bmatrix}\begin{bmatrix} 1 \\ 0 \\ 1 \end{bmatrix}$$

$$= s^2 + 4s + 3$$

The transfer function may now be written as

$$G(s) = \frac{P(s)}{Q(s)} = \frac{s^2 + 4s + 3}{s^3 + 5s^2 + 7s + 1} \qquad ...(3.31)$$

DRILL PROBLEM 3.3

Determine the transfer function of the state equations in Example 3.4 if A, B and C are unchanged but D = 1.

Ans. $\dfrac{s^3 + 6s^2 + 11s + 4}{s^3 + 5s^2 + 7s + 1}$

DRILL PROBLEM 3.4

Calculate the transfer function of a linear system described by the following state equation:

$$\dot{x} = \begin{bmatrix} -2 & -5 & -4 \\ 1 & -1 & 0 \\ 0 & 1 & 0 \end{bmatrix} x + \begin{bmatrix} 0 \\ 1 \\ 1 \end{bmatrix} u \qquad \text{...(3.32)}$$

$$y = [3 \quad 1 \quad 0]\, x$$

Ans. $$\frac{s^2 - 25s - 28}{s^3 + 3s^2 + 7s + 4}$$

3.4 STATE EQUATIONS FROM TRANSFER FUNCTIONS

We shall now study the inverse problem of determining the state equations of a system from its transfer function. State equations represent the complete description of the system, whereas the transfer function represents only its input-output behaviour. Consequently, as expected, the state equations for a given transfer function are not unique, and can be obtained in several ways. One way to approach these, is through analog computer simulation diagrams for a transfer function, which are also called 'realizations' of the transfer function. There are some standard forms, known are *'canonical realizations'*, because these minimize the number of potentiometers (multipliers) required.

These will be illustrated through an example of a third-order transfer function. This is easily generalized to transfer functions of higher order.

EXAMPLE 3.6: SOME CANONICAL REALIZATIONS

Consider a system described by the transfer function given below:

$$\frac{Y(s)}{U(s)} = \frac{b_1 s^2 + b_2 s + b_3}{s^3 + a_1 s^2 + a_2 s + a_3} \qquad \text{...(3.33)}$$

The analog computer simulation diagram shown in Fig. 3.6 is one way to realize this system, and is called *direct realization.* This is easily verified by applying Mason's rule to determine the transfer function relating $Y(s)$ to $U(s)$, noting that the transfer function of an integrator is s^{-1}.

If we label the output of each integrator as a state variable, as shown in the Fig., we get the following set of differential equations relating the input to each integrator to the various variables:

$$\dot{x}_1 = -a_1 x_1 - a_2 x_2 - a_3 x_3 + u$$

$$\dot{x}_2 = x_1 \qquad \text{...(3.34)}$$

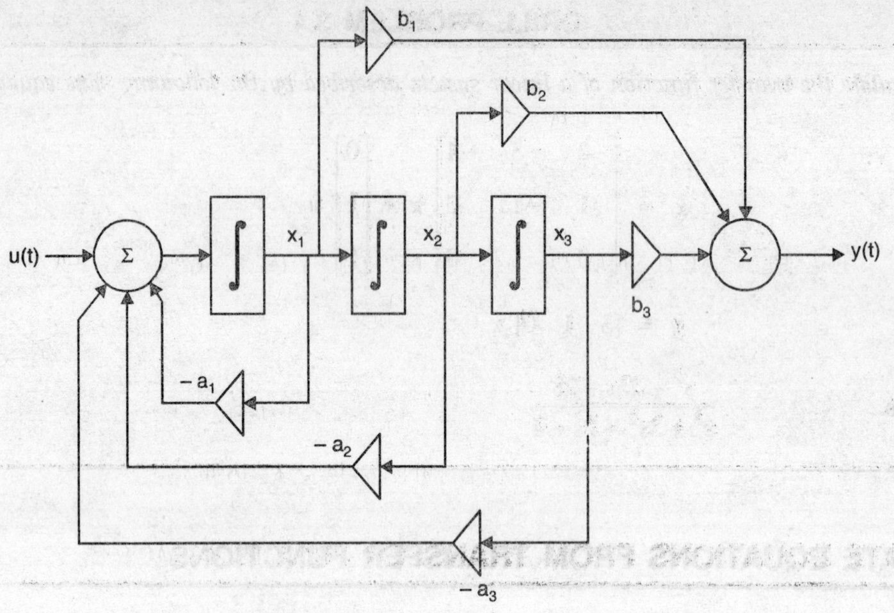

Figure 3.6. *Controller form realization*

$$\dot{x}_3 = x_2$$

$$y = b_1x_1 + b_2x_2 + b_3x_3$$

These can be arranged in the matrix form to obtain Equation (3.35). The state equations in this form are known in the control theory literature as the *controller canonical form,* as these are useful for designing controllers utilizing state variable feedback. This form will be discussed in detail in Chapter 12.

$$\begin{bmatrix} \dot{x}_1 \\ \dot{x}_2 \\ \dot{x}_3 \end{bmatrix} = \begin{bmatrix} -a_1 & -a_2 & -a_3 \\ 1 & 0 & 0 \\ 0 & 1 & 0 \end{bmatrix} \begin{bmatrix} x_1 \\ x_2 \\ x_3 \end{bmatrix} + \begin{bmatrix} 1 \\ 0 \\ 0 \end{bmatrix} u \qquad \text{...(3.35)}$$

$$y = [b_1 \quad b_2 \quad b_3] \begin{bmatrix} x_1 \\ x_2 \\ x_3 \end{bmatrix}$$

It is important to relate the structure of the matrices *A, B* and *C* in Equation (3.35) to the transfer function in Equation (3.33). By examining this carefully, it is possible to write these matrices directly for any given transfer function.

The analog computer simulation diagram shown in Fig. 3.7 is another way to realize the given transfer function, as can be verified again by applying Mason's rule.

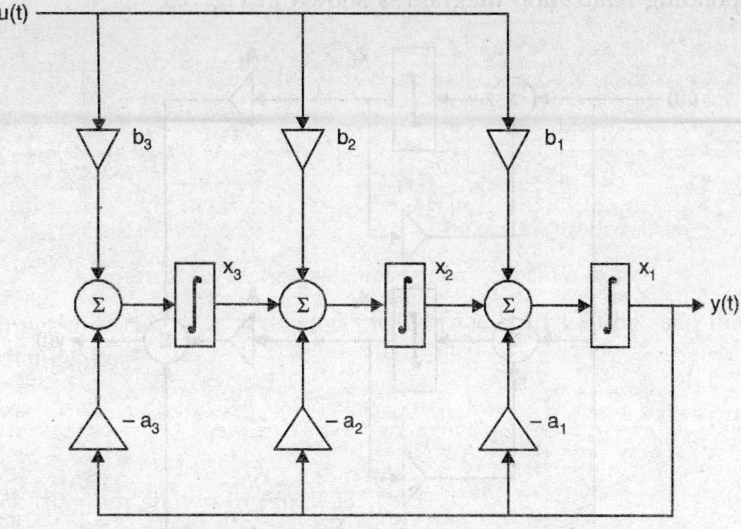

Figure 3.7. *Observer form realization*

Labelling the output of each integrator as a state variable, we again get a set of first-order differential equations relating the input to each integrator. Arranging them in the matrix form leads to the state equations given in Equation (3.36). It is interesting to compare Equation (3.36) with (3.35). It is seen that the A-matrices for the two cases are transposes of each other, and that the B and C matrices have been interchanged. In both cases, it is possible to write the state equations directly from the transfer function.

$$\begin{bmatrix} \dot{x}_1 \\ \dot{x}_2 \\ \dot{x}_3 \end{bmatrix} = \begin{bmatrix} -a_1 & 1 & 0 \\ -a_2 & 0 & 1 \\ -a_3 & 0 & 0 \end{bmatrix} \begin{bmatrix} x_1 \\ x_2 \\ x_3 \end{bmatrix} + \begin{bmatrix} b_1 \\ b_2 \\ b_3 \end{bmatrix} u \qquad \text{...(3.36)}$$

$$y = [1 \quad 0 \quad 0] \begin{bmatrix} x_1 \\ x_2 \\ x_3 \end{bmatrix}$$

The state equations in (3.36) are said to be in *observer canonical form.* This form is useful for designing state observers, and will be discussed again in Chapter 12.

Another realization is based on performing a partial fraction expansion of the transfer function, expressing it as a sum of a set of first- or second-order transfer functions (the latter will be required for the case of complex conjugate poles in order to avoid complex numbers). Let us assume that the transfer function in Equation (3.33) can be written as

$$\frac{Y(s)}{U(s)} = \frac{A_1}{s+\lambda_1} + \frac{A}{s+\lambda_2} + \frac{A_3}{s+\lambda_3} \qquad \text{...(3.37)}$$

The corresponding realization diagram is shown in Fig. 3.8.

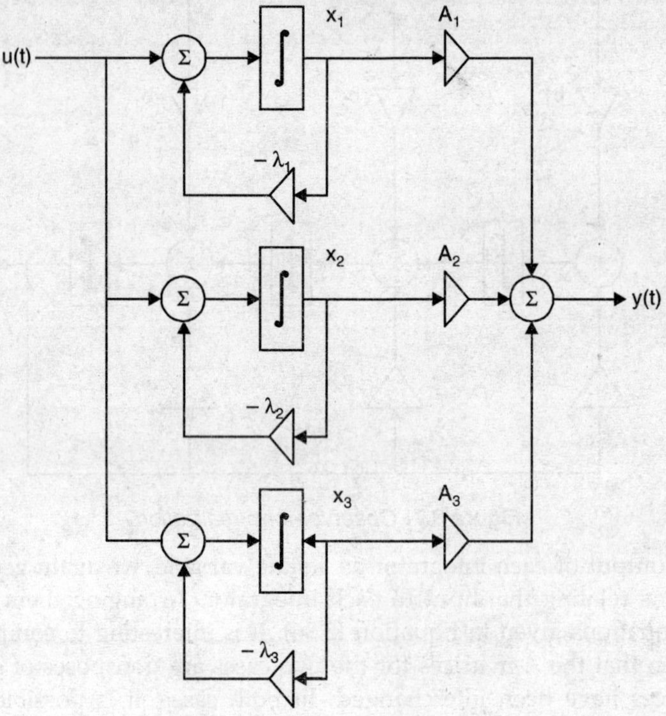

Figure 3.8. *Parallel realization*

The following state equations are obtained from Fig. 3.8.

$$\begin{bmatrix} \dot{x}_1 \\ \dot{x}_2 \\ \dot{x}_3 \end{bmatrix} = \begin{bmatrix} -\lambda_1 & 0 & 0 \\ 0 & -\lambda_2 & 0 \\ 0 & 0 & -\lambda_3 \end{bmatrix} \begin{bmatrix} x_1 \\ x_2 \\ x_3 \end{bmatrix} + \begin{bmatrix} 1 \\ 1 \\ 1 \end{bmatrix} u \quad \text{...(3.38)}$$

$$y = [A_1 \quad A_2 \quad A_3] \begin{bmatrix} x_1 \\ x_2 \\ x_3 \end{bmatrix}$$

It is interesting to note that in this case, the A-matrix is of the *diagonal* form. Consequently, the three differential equations are *decoupled*, and can be solved more easily. We shall return to this topic when we study methods for solving state equations in Section 3.6.

Here, we had assumed that all the poles of the transfer function are real. Let us now consider the case when one pair of poles is complex, so that we have

$$\frac{Y(s)}{U(s)} = \frac{A_1}{s+\lambda_1} + \frac{A_2 s + A_3}{s^2 + \alpha s + \beta} \quad \text{...(3.39)}$$

A parallel realization for this system is shown in Fig. 3.9.

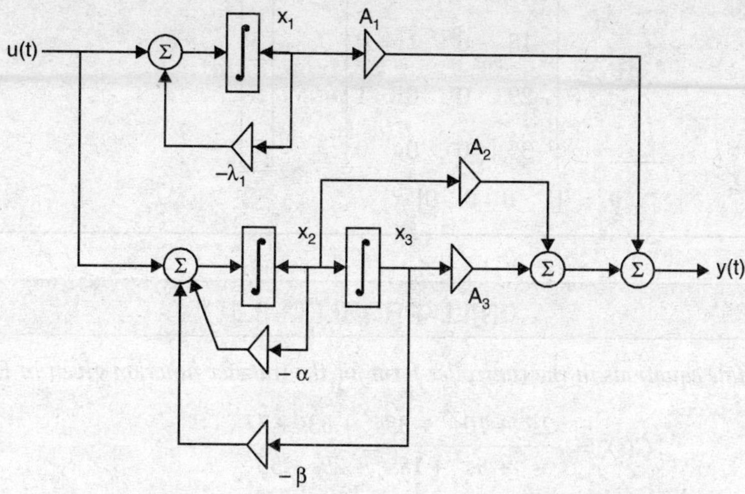

Figure 3.9. *Parallel realization with real and complex poles*

The resulting state equations are shown in (3.40). It will be seen that although A is not a diagonal matrix, it is of the block diagonal form.

$$\begin{bmatrix} \dot{x}_1 \\ \dot{x}_2 \\ \dot{x}_3 \end{bmatrix} = \begin{bmatrix} -\lambda_1 & 0 & 0 \\ 0 & -\alpha & -\beta \\ 0 & 1 & 0 \end{bmatrix} \begin{bmatrix} x_1 \\ x_2 \\ x_3 \end{bmatrix} + \begin{bmatrix} 1 \\ 1 \\ 0 \end{bmatrix} u \quad \text{...(3.40)}$$

$$y = [A_1 \quad A_2 \quad A_3] \begin{bmatrix} x_1 \\ x_2 \\ x_3 \end{bmatrix}$$

DRILL PROBLEM 3.5

Obtain the state equations in (a) the controller form and (b) the observer form for the transfer function given in Equation (3.41).

$$G(s) = \frac{3s^2 + 5s + 7}{s^4 + 5s^3 + 18s^2 + 29s + 35} \quad \text{...(3.41)}$$

Ans.

(*a*)
$$\dot{x} = \begin{bmatrix} -5 & -18 & -29 & -35 \\ 1 & 0 & 0 & 0 \\ 0 & 1 & 0 & 0 \\ 0 & 0 & 1 & 0 \end{bmatrix} x + \begin{bmatrix} 1 \\ 0 \\ 0 \\ 0 \end{bmatrix} u$$

$$y = [0 \quad 3 \quad 5 \quad 7]\, x$$

(b)
$$\dot{x} = \begin{bmatrix} -5 & 1 & 0 & 0 \\ -18 & 0 & 1 & 0 \\ -29 & 0 & 0 & 1 \\ -35 & 0 & 0 & 0 \end{bmatrix} x + \begin{bmatrix} 0 \\ 3 \\ 5 \\ 7 \end{bmatrix} u$$
$$y = [1 \;\; 0 \;\; 0 \;\; 0]\, x$$

DRILL PROBLEM 3.6

Obtain the state equations in the controller form for the transfer function given in Equation (3.42).

$$G(s) = \frac{2s^4 + 10s^3 + 39s^2 + 63s + 77}{s^4 + 5s^3 + 18s^2 + 29s + 35} \qquad \text{...(3.42)}$$

Ans.

$$\dot{x} = \begin{bmatrix} -5 & -18 & -29 & -35 \\ 1 & 0 & 0 & 0 \\ 0 & 1 & 0 & 0 \\ 0 & 0 & 1 & 0 \end{bmatrix} x + \begin{bmatrix} 1 \\ 0 \\ 0 \\ 0 \end{bmatrix} u$$
$$y = [0 \;\; 3 \;\; 5 \;\; 7]\, x + 2u$$

DRILL PROBLEM 3.7

Obtain the state equations for the transfer function in Equation (3.43) using parallel realization.

$$G(s) = \frac{10}{s+1} + \frac{4s+5}{s^2+4s+8} + \frac{1}{s+2} + \frac{3}{(s+2)^2} \qquad \text{...(3.43)}$$

Ans.

$$\dot{x} = \begin{bmatrix} -1 & 0 & 0 & 0 & 0 \\ 0 & -4 & -8 & 0 & 0 \\ 0 & 1 & 0 & 0 & 0 \\ 0 & 0 & 0 & -4 & -4 \\ 0 & 0 & 0 & 1 & 0 \end{bmatrix} x + \begin{bmatrix} 1 \\ 1 \\ 0 \\ 1 \\ 0 \end{bmatrix} u$$
$$y = [10 \;\; 4 \;\; 5 \;\; 1 \;\; 5]\, x$$

In addition to the three canonical forms described above, many other forms of state equations can be obtained for any given transfer function by performing similarity transformation of the state vector. This will be discussed in the next section.

3.5 SIMILARITY TRANSFORMATIONS

One reason for the nonuniqueness of the state equations for a given transfer function is that the input-output relationship of a system is invariant under linear transformation of the state. For instance, consider such a transformation, given by

$$w = P^{-1} x \qquad ...(3.44)$$

where P is an $n \times n$ nonsingular constant matrix. This is known as a *similarity transformation.*

Substituting Equation (3.44) into Equations (3.1) and (3.2), we obtain a new set of equations in terms of the transformed state vector w. These are given below:

$$\dot{w} = \bar{A}w + \bar{B}u \qquad ...(3.45)$$

$$y = \bar{C}w + Du$$

where

$$\bar{A} = P^{-1} AP$$
$$\bar{B} = P^{-1} B \qquad ...(3.46)$$
$$\bar{C} = CP$$

We shall now show that the transfer function in terms of the transformed state is identical to that obtained from the original state.

$$\begin{aligned}\bar{G}(s) &= \bar{C}\left(sI - \bar{A}\right)^{-1}\bar{B} + D = CP(sI - P^{-1} AP)^{-1} P^{-1} B + D \\ &= CP\,(sP^{-1} P - P^{-1} AP)^{-1} P^{-1} B + D \qquad ...(3.47) \\ &= CPP^{-1}(sI - A)^{-1} PP^{-1} B + D \\ &= C\,(sI - A)^{-1} B + D = G(s)\end{aligned}$$

where we have used the matrix identity

$$(PQ)^{-1} = Q^{-1} P^{-1} \qquad ...(3.48)$$

It is evident that different choices of P will give us many sets of state equations. An important application of these transformations is to obtain canonical forms of state equations from non-canonical forms. In particular, one can use similarity transformations to transform the state equations into a diagonal form if the eigenvalues of A are distinct. As we shall see in the next section, the diagonal form is very useful in solving the vector differential Equation (3.1).

3.6 SOLUTION OF STATE EQUATIONS

We shall now study how to solve the state Equation (3.1) for a given input $u(t)$ and a given initial condition $x(t_0)$. Two methods can be used for the solution; either one can do it in the time domain or one can use Laplace transforms. We shall start with the time domain method, which is based on using the matrix exponential, defined as

$$e^{At} \triangleq I + At + \frac{1}{2!}(At)^2 + \frac{1}{3!}(At)^3 + \ldots \qquad ...(3.49)$$

This looks like the scalar exponential e^{at} and also has similar properties. In particular,

$$\frac{d}{dt}\left(e^{At}\right) = Ae^{At} = e^{At} A \qquad \text{...(3.50)}$$

as can be proved by differentiating the infinite series on the right-hand side of the Equation (3.49) term by term. It is also seen that A and e^{At} commute with each other, which is the consequence of the fact that

$$A^m A^n = A^n A^m = A^{m+n} \qquad \text{...(3.51)}$$

To solve the differential Equation (3.1), we shall first rewrite it as

$$\dot{x} - Ax(t) = Bu(t) \qquad \text{...(3.52)}$$

We now premultiply both sides of Equation (3.52) by e^{-At} so that the left-hand side becomes an exact differential, that is,

$$e^{-At}\dot{x} - e^{-At}Ax = \frac{d}{dt}(e^{-At}x) = e^{-At}Bu(t) \qquad \text{...(3.53)}$$

Multiplying both sides by dt and integrating between the limits t_0 and t, we obtain

$$e^{-At}x(t) - e^{-At_0}x(t_0) = \int_{t_0}^{t} e^{-A\tau} Bu(\tau)\, d\tau \qquad \text{...(3.54)}$$

where the dummy variable τ has been used in the integral to avoid confusion with t. After some rearrangement, we get the final solution.

$$x(t) = e^{A(t-t_0)}x(t_0) + \int_{t_0}^{t} e^{A(t-\tau)} Bu(\tau)\, d\tau \qquad \text{...(3.55)}$$

The first term on the right-hand side of Equation (3.55) is seen to be the complementary function or *natural response* of the system, while the second term, which is the convolution integral, is the particular solution or *forced response*.

We shall now solve Equation (3.52) by taking the Laplace transform of both sides. This has given us

$$sX(s) - x(0) - AX(s) = BU(s) \qquad \text{...(3.56)}$$

Solving Equation (3.56) for $X(s)$, we get

$$X(s) = (sI - A)^{-1}x(0) + (sI - A)^{-1}BU(s) \qquad \text{...(3.57)}$$

$x(t)$ can be obtained by taking the inverse Laplace transform of Equation (3.57). Comparing Equation (3.57) with (3.55), we recognize that e^{At} is the inverse Laplace transform of $(sI - A)^{-1}$. Similarly, the inverse Laplace transform of the second term on the right-hand side of Equation (3.57) is recognized as the convolution integral in Equation (3.55).

Thus, we have two alternative approaches to solving the state equation. To use the Laplace transform method, we first determine $X(s)$ as in Equation (3.57) and then find $x(t)$ by taking the inverse Laplace transform of $X(s)$. Not only does this require inversion of the matrix $(sI - A)$, but after that the inverse Laplace transform must be obtained of $(sI - A)^{-1}$, as well as of the product $(sI - A)^{-1}BU(s)$. On the other hand, Equation (3.55) is ideally suited for numerical solution, provided that the matrix e^{At} is known. This matrix is known as the *state transition matrix* of the system. It is also called the *fundamental matrix* of the system and is often denoted by the symbol $\phi(t)$. In the next section, we shall describe two methods for evaluating this matrix.

3.7 EVALUATION OF e^{At}

There are several methods for evaluating the fundamental matrix. Here, we shall present two of the most straightforward methods.

Use of the Laplace Transform

Conceptually, the most direct approach is to use the relationship

$$e^{At} = \mathcal{L}^{-1}[(sI - A)^{-1}] \qquad \text{...(3.58)}$$

This requires first evaluating the inverse of the matrix $(sI - A)$ and then determining the inverse Laplace transform of each term of the matrix. Use of Leverrier's algorithm, described in Section 3.3, makes the matrix inversion quite straightforward and can be done using a computer. The following example will illustrate this:

EXAMPLE 3.7

Consider the A-matrix shown below.

$$A = \begin{bmatrix} -2 & -5 & -5 \\ 1 & -1 & 0 \\ 0 & 1 & 0 \end{bmatrix} \qquad \text{...(3.59)}$$

Using Leverrier's algorithm, we obtain the following expressions for the adjoint and the determinant of $(sI - A)$.

$$\text{adj}(sI - A) = P_1 s^2 + P_2 s + P_3$$

$$= \begin{bmatrix} s^2 + s & -5s - 5 & -5s - 5 \\ s & s^2 + 2s & -5 \\ 1 & s + 2 & s^2 + 3s + 7 \end{bmatrix} \qquad \text{...(3.60)}$$

$$\det(sI - A) = s^3 + a_1 s^2 + a_2 s + a_3$$

$$= s^3 + 3s^2 + 7s + 5 = (s + 1)(s^2 + 2s + 5)$$

The inverse of the matrix is obtained by dividing its adjoint by its determinant. Cancellation of the common factors between the numerator and denominator of the various elements leads to some simplification. The resulting matrix is given in Equation (3.61).

$$(sI - A)^{-1} = \begin{bmatrix} \dfrac{s}{s^2 + 2s + 5} & \dfrac{-5}{s^2 + 2s + 5} & \dfrac{-5}{s^2 + 2s + 5} \\ \dfrac{s}{s^3 + 3s^2 + 7s + 5} & \dfrac{s^2 + 2s}{s^3 + 3s^2 + 7s + 5} & \dfrac{-5}{s^3 + 3s^2 + 7s + 5} \\ \dfrac{1}{s^3 + 3s^2 + 7s + 5} & \dfrac{s + 2}{s^3 + 3s^2 + 7s + 5} & \dfrac{s^2 + 3s + 7}{s^3 + 3s^2 + 7s + 5} \end{bmatrix} \qquad \text{...(3.61)}$$

Finally, taking the inverse Laplace transform of each element of this matrix, we obtain the fundamental matrix given as follows:

$$\begin{bmatrix} e^{-t}(\cos 2t - 0.5\sin 2t) & -2.5\,e^{-t}\sin 2t & -2.5\,e^{-t}\sin 2t \\ 0.25\,e^{-t}(\cos 2t + \sin 2t - 1) & 0.25\,e^{-t}(5\cos 2t - 1) & 1.25\,e^{-t}(\cos 2t - 1) \\ 0.25\,e^{-t}(1-\cos 2t) & 0.25\,e^{-t}(1 - \cos 2t + 2\sin 2t) & 0.25\,e^{-t}(5 - \cos 2t + 2\sin 2t) \end{bmatrix}$$

It is evident that this method is rather tedious. Although it is possible to use a computer programme based on Leverrier's algorithm for evaluating the adjoint as well as the determinant of $(sI - A)$, the evaluation of the inverse Laplace transform can still be quite time consuming. The main problem is that it is hard to automate the complete procedure. Hence, it is not very convenient to use this method if the dimension of A is larger than 3 × 3. The method to be described next is more suitable for use with a digital computer.

DRILL PROBLEM 3.8

Determine the fundamental matrix for the A-matrix shown below.

$$A = \begin{bmatrix} 1 & 2 \\ -3 & 4 \end{bmatrix}$$

Ans. $\begin{bmatrix} 3e^{-t} - 2e^{-2t} & 2e^{-t} - 2e^{-2t} \\ -3e^{-t} + 3e^{-2t} & -2e^{-t} + 3e^{-2t} \end{bmatrix}$

DRILL PROBLEM 3.9

Determine the fundamental matrix for the A-matrix shown below.

$$A = \begin{bmatrix} -1 & 2 & -8 \\ 0 & -2 & 4 \\ 0 & 0 & -4 \end{bmatrix}$$

Ans. $\begin{bmatrix} e^{-t} & 2e^{-t} - 2e^{-2t} & -4e^{-2t} + 4e^{-4t} \\ 0 & e^{-2t} & 2e^{-2t} - 2e^{-4t} \\ 0 & 0 & e^{-4t} \end{bmatrix}$

Use of Diagonal or Jordan Forms

The computation of e^{At} is quite straightforward for the special case when A is a diagonal matrix, as will be evident from examining Equation (3.49). Every term in the series for the exponential will also be a diagonal matrix. Thus, the various terms can be added to form another diagonal matrix the elements of which are exponentials corresponding to the diagonal elements of A. For example, consider

$$A = \begin{bmatrix} \lambda_1 & 0 & 0 \\ 0 & \lambda_2 & 0 \\ 0 & 0 & \lambda_3 \end{bmatrix}$$

Using the series for e^{At} from Equation (3.49), we get

$$e^{At} = \begin{bmatrix} 1 & 0 & 0 \\ 0 & 1 & 0 \\ 0 & 0 & 1 \end{bmatrix} + \begin{bmatrix} \lambda_1 t & 0 & 0 \\ 0 & \lambda_2 t & 0 \\ 0 & 0 & \lambda_3 t \end{bmatrix} + \begin{bmatrix} \frac{(\lambda_1 t)^2}{2!} & 0 & 0 \\ 0 & \frac{(\lambda_2 t)^2}{2!} & 0 \\ 0 & 0 & \frac{(\lambda_3 t)^2}{2!} \end{bmatrix} + \ldots$$

$$= \begin{bmatrix} e^{\lambda_1 t} & 0 & 0 \\ 0 & e^{\lambda_2 t} & 0 \\ 0 & 0 & e^{\lambda_3 t} \end{bmatrix}$$

This is intuitively obvious, since a diagonal A-matrix represents a set of *uncoupled* first-order differential equations, that is, every equation of the set contains only one of the state variables on the right-hand side and its derivative is on the left-hand side. Consequently, each differential equation can be solved separately, without having to use a matrix.

As the A-matrix for a system is not usually diagonal, it will be desirable to find out if it is possible to transform it to the diagonal form by using a similarity transformation of the type described in Section 3.5. We shall now show that this can always be done if all the eigenvalues of A are distinct. Let λ_i be an eigenvalue of A and let v_i be the corresponding eigenvector. Then, for each of the n eigenvalues of A, we obtain the following equations from the definitions of eigenvalues and eigenvectors:

$$\begin{aligned} Av_1 &= \lambda_1 v_1 \\ Av_2 &= \lambda_2 v_2 \\ &\vdots \\ Av_n &= \lambda_n v_n \end{aligned} \qquad \text{...(3.62)}$$

These equations can be combined to obtain

$$AM = M\Lambda \qquad \text{...(3.63)}$$

where Λ is a diagonal matrix with the eigenvalues on its main diagonal and the columns of M are the n eigenvectors of A, that is,

$$\Lambda = diagonal\ \{\lambda_1, \ \lambda_2, \ \ldots, \ \lambda_n\} \qquad \text{...(3.64)}$$

$$M = [v_1 \ \ v_2 \ \cdots \ \ v_n]$$

M is also called the *modal* matrix and it will be non-singular if and only if the n eigenvectors are linearly independent. This will always be the case if the eigenvalues of A are distinct. Hence, we can solve Equation (3.63) to obtain

$$\Lambda = M^{-1} AM \qquad \text{...(3.65)}$$

It can now be easily shown that

$$e^{\Lambda t} = M^{-1}\, e^{At}\, M \qquad \text{...(3.66)}$$

and

$$e^{At} = M e^{\Lambda t} M^{-1} \qquad \text{...(3.67)}$$

EXAMPLE 3.8

Consider the following A-matrix

$$A = \begin{bmatrix} 0 & 1 \\ -3 & -4 \end{bmatrix}$$

The determinant of $(\lambda I - A)$ is obtained as

$$\lambda_2 + 4\lambda + 3$$

From the roots of the determinant, the eigenvalues of A are found to be $\lambda_1 = -1$ and $\lambda_2 = -3$. The corresponding eigenvectors are easily obtained as

$$v_1 = \begin{bmatrix} 1 \\ -1 \end{bmatrix} \quad \text{and} \quad v_2 = \begin{bmatrix} 1 \\ -3 \end{bmatrix}$$

so that the modal matrix and its inverse are as given below:

$$M = \begin{bmatrix} 1 & 1 \\ -1 & -3 \end{bmatrix}, \quad M^{-1} = \begin{bmatrix} 1.5 & 0.5 \\ -0.5 & -0.5 \end{bmatrix}$$

Thus, we get the following equations:

$$\Lambda = \begin{bmatrix} -1 & 0 \\ 0 & -3 \end{bmatrix}$$

$$e^{At} = M e^{\Lambda t} M^{-1}$$

$$= \begin{bmatrix} 1 & 1 \\ -1 & -3 \end{bmatrix} \begin{bmatrix} e^{-t} & 0 \\ 0 & e^{-3t} \end{bmatrix} \begin{bmatrix} 1.5 & 0.5 \\ -0.5 & -0.5 \end{bmatrix}$$

$$= \begin{bmatrix} 1.5e^{-t} - 0.5e^{-3t} & 0.5e^{-t} - 0.5e^{-3t} \\ 1.5e^{-3t} - 1.5e^{-t} & 1.5e^{-3t} - 0.5e^{-t} \end{bmatrix}$$

Equation (3.67) provides us with a procedure that is especially suitable for use with a digital computer if A can be diagonalized. Several standard routines for calculating the eigenvalues and eigenvectors of a matrix with real elements are available and can be used for this purpose. The most well-known is the *Q-R* algorithm. The corresponding computer programme, EISPACK, is available in most computer libraries. It is important to note that it is possible to diagonalize an $n \times n$ matrix A if all the eigenvectors, corresponding to the n eigenvalues of the matrix, are linearly independent, and this is guaranteed only if all the eigenvalues are distinct. In the case of repeated eigenvalues, it is possible to transform A into a *Jordan* matrix, using the concept of generalized eigenvectors. This is described in detail in Appendix *B*. The Jordan form looks very similar to the diagonal form, except that, depending on the number of repeated eigenvalues, some of the entries on the diagonal above the main diagonal are one instead of zero. The exponential of the Jordan form matrix can be obtained almost as easily as that of diagonal matrix. We can then use a transformation, similar to that in Equation (3.67) to calculate the fundamental matrix.

A Jordan form and its fundamental matrix are illustrated in Example 3.9.

EXAMPLE 3.9

Consider the Jordan form matrix shown below, which has one eigenvalue equal to –1 and two igenvalues equal to –3.

$$J = \begin{bmatrix} -1 & 0 & 0 \\ 0 & -3 & 1 \\ 0 & 0 & -3 \end{bmatrix}$$

It can be shown easily that

$$e^{Jt} = \mathcal{L}^{-1}[(sI - J)^{-1}] = \begin{bmatrix} e^{-t} & 0 & 0 \\ 0 & e^{-3t} & te^{-3t} \\ 0 & 0 & e^{-3t} \end{bmatrix}$$

The verification of this will be left as an exercise for the student.

DRILL PROBLEM 3.10

Repeat Drill Problem 3.9 by diagonalizing the A-matrix.

3.8 STATE TRANSITION EQUATIONS FOR A DISCRETE-TIME SYSTEM

Consider a linear system described by Equations (3.1) and (3.2). It is possible to convert it into a discrete-time system by introducing a sampler and hold device, so that the input is allowed to vary only at the sampling instants, $t = kT$, $k = 1, 2, \ldots$, and is held constant between the sampling instants. This type of discretization is necessary if a digital computer is used as a part of the control system. Computer-controlled systems are discussed in more detail in Chapter 11. Here, we shall be concerned only with finding the solution of the state equations for such systems. This can be done easily by making $t_0 = kT$ and $t = (k + 1)T$ in Equation (3.55) which now takes the form

$$x[(k+1)T] = e^{AT}x(kT) + \int_{kT}^{(k+1)T} e^{A(kT+T-\tau)} Bu(\tau)\,d\tau \qquad \text{...(3.68)}$$

Since $u(\tau) = u(kT)$ for $kT < \tau \le (k + 1)T$, we may simplify Equation (3.68) to

$$x[(k+1)T] = F(T)\,x(kT) + G(T)\,u(kT) \qquad \text{...(3.69)}$$

where $F(T) = e^{AT}$ and

$$G(T) = \int_0^T e^{At}\,dt\,B$$

The last result is obtained as follows. From Equations (3.68) and (3.69)

$$G(T) = \int_{kT}^{(k+1)T} e^{A(kT+T-\tau)}\, d\tau\, B$$

$$= -\int_{T}^{0} e^{A\zeta}\, d\zeta\, B \quad \text{where} \quad \zeta = kT + T - \tau$$

$$= \int_{0}^{T} e^{A\zeta}\, d\zeta\, B$$

It may be noted that both F and G are constant matrices if the sampling interval T is held constant. These matrices can be calculated conveniently on a digital computer using the power series expansions

$$F(T) = I + AT + \frac{(AT)^2}{2!} + \frac{(AT)^3}{3!} + \ldots \qquad \ldots(3.70)$$

$$G(T) = \left[I + \frac{AT}{2!} + \frac{(AT)^2}{3!} + \ldots \right] TB$$

where the infinite series are truncated after a certain number of terms to obtain a specified accuracy. It can be shown that if the sampling interval T is selected so that $|\lambda_m|T < 0.5$, where λ_m is the eigenvalue of A that has the largest magnitude (this is often called the *spectral radius* of A), one can truncate the series after only 12 terms with an error of less than 10^{-9} in each component of F and G. The disk available with this book contains a computer programme for calculating F and G for given A, B and T and also allows the user to specify the number of terms after which the series are to be truncated.

Equation (3.69) is called the *state transition equation* for the resulting discrete-time system. It shows how conveniently the response of the system can be calculated if the matrices $F(T)$ and $G(T)$ are known. This equation is suitable for use on a digital computer since it only requires matrix multiplication. We shall return to this equation in Chapter 12 when we study the control of discrete-time systems using the state-space formulation.

The following example illustrates how one can obtain the state transition equation for a sampled-data system.

EXAMPLE 3.10

Consider the discrete-time system shown in Fig. 3.10, where the transfer function

$$G(s) = \frac{4}{s(s+2)}$$

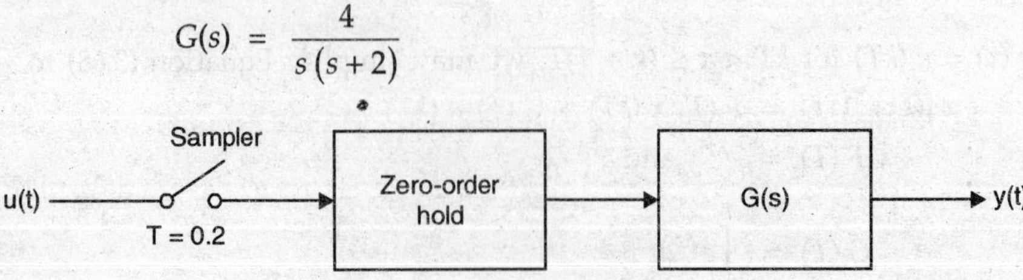

Figure 3.10. *A discrete-time system*

A state-space representation of the continuous-time system is obtained as

$$\begin{bmatrix} \dot{x}_1 \\ \dot{x}_2 \end{bmatrix} = \begin{bmatrix} -2 & 0 \\ 1 & 0 \end{bmatrix} \begin{bmatrix} x_1 \\ x_2 \end{bmatrix} + \begin{bmatrix} 1 \\ 0 \end{bmatrix} u(t)$$

$$y = [0 \quad 4] \begin{bmatrix} x_1 \\ x_2 \end{bmatrix}$$

Since this is a second-order system, we shall calculate the fundamental matrix using Laplace transforms.

$$e^{At} = \mathcal{L}^{-1}[(sI - A)^{-1}] = \mathcal{L}^{-1} \begin{bmatrix} \dfrac{1}{s+2} & 0 \\ \dfrac{1}{s(s+2)} & \dfrac{1}{s} \end{bmatrix} = \begin{bmatrix} e^{-2t} & 0 \\ 0.5\left(1 - e^{-2t}\right) & 1 \end{bmatrix}$$

We evaluate F and G as below:

$$F = e^{AT} = \begin{bmatrix} e^{-0.4} & 0 \\ 0.5\left(1 - e^{-0.4}\right) & 1 \end{bmatrix} = \begin{bmatrix} 0.67032 & 0 \\ 0.16484 & 1 \end{bmatrix}$$

and

$$G = \int_0^T e^{At} B \, dt = \int_0^{0.2} \begin{bmatrix} e^{-2t} \\ 0.5\left(1 - e^{-2t}\right) \end{bmatrix} dt = \begin{bmatrix} 0.16484 \\ 0.01758 \end{bmatrix}$$

We can now write the state transition equation for the system as

$$x[(k+1)T] = \begin{bmatrix} 0.67032 & 0 \\ 0.16484 & 1 \end{bmatrix} x(kT) + \begin{bmatrix} 0.16484 \\ 0.01758 \end{bmatrix} u(kT)$$

DRILL PROBLEM 3.11

The state equations of a helicopter near hover are as follows:

$$\begin{bmatrix} \dot{x}_1 \\ \dot{x}_2 \\ \dot{x}_3 \end{bmatrix} = \begin{bmatrix} -0.02 & -1.4 & 9.8 \\ -0.01 & -0.4 & 0 \\ 0 & 1 & 0 \end{bmatrix} \begin{bmatrix} x_1 \\ x_2 \\ x_3 \end{bmatrix} + \begin{bmatrix} 9.8 \\ 6.3 \\ 0 \end{bmatrix} u$$

$$y = [0 \quad 0 \quad 1] \begin{bmatrix} x_1 \\ x_2 \\ x_3 \end{bmatrix}$$

where x_1 = horizontal velocity, x_2 = pitch rate, x_3 = pitch angle, and u = rotor tilt angle.

Determine the state transition equation if the sampling period is 0.5 second and the input is held constant between sampling instants.

Ans.
$$F(T) = \begin{bmatrix} 0.989744 & 0.512262 & 4.87589 \\ -0.004509 & 0.818414 & -0.011435 \\ -0.001167 & 0.453201 & 0.998061 \end{bmatrix}$$

$$G(T) = \begin{bmatrix} 5.06819 \\ 2.84373 \\ 0.73565 \end{bmatrix}$$

SUMMARY

- In this chapter, we have studied the state-space representation of physical systems. The main idea is to rewrite the differential equation for the system as a set of first-order differential equations, called state equations. The state-space representation gives the complete description of the system, whereas the transfer function is only an input-output representation. It can also be used for modelling multivariable systems without any change in notation. For any given system, the state-variable description is not unique. In practice, it is desirable to use physical variables as state variables, which can then be measured and fed back. We have studied methods for relating state equations and transfer functions. Whereas we can determine a unique transfer function from state equations, numerous state variable realizations are possible for any given transfer function. Some canonical realizations have been studied. These equations can be readily solved using either analog or digital computers. The particular case of discrete-time systems, in which the input is held constant between sampling instants, can be solved very conveniently on a digital computer using the state transition equation.

Problems

1. Write a set of state equations for the dynamic vibration absorber shown in Fig. P3.1 and determine the transfer function relating $F(s)$ to $X_1(s)$.

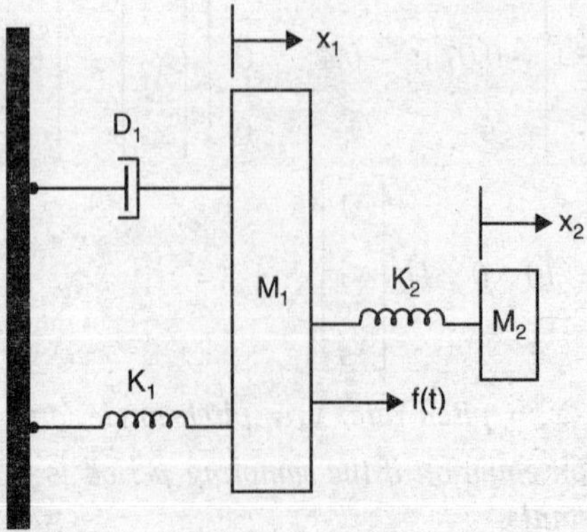

Figure P3.1. *A dynamic vibration absorber*

2. Write a set of state equations for the analog computer simulation diagram shown in Fig. P3.2, using x_1, x_2 and x_3 as the state variables, as labelled on the diagram. Use these equations to determine the transfer function relating C(s) to R(s). Verify the result by using Mason's rule.

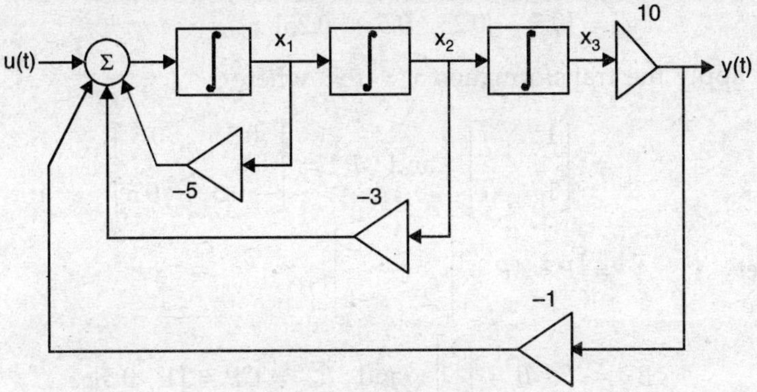

Figure P3.2. *An analog computer simulation diagram*

3. Write a set of state equations for the previous problem after interchanging the labels of the state variables x_1 and x_3. Can you identify a similarity transformation relating the state variables for the two cases?

4. The state equations of a linear system are as follows:

$$\dot{x} = \begin{bmatrix} -2 & 0 & 1 \\ 1 & -3 & 0 \\ 1 & 1 & -1 \end{bmatrix} x + \begin{bmatrix} 1 \\ 0 \\ 1 \end{bmatrix} u$$

$$y = [2 \quad 1 \quad -1]\, x$$

(*a*) Determine the transfer function relating $Y(s)$ to $U(s)$.

(*b*) Determine the state transition equation if the input is sampled at intervals of 0.5 second, and held constant between the sampling instants.

5. Draw an analog computer simulation diagram for the state equations for the helicopter given in Drill Problem 3.11. From this diagram, determine the transfer function relating $U(s)$ to $Y(s)$ by using Mason's rule.

6. The diagonal canonical form would result in some elements of the matrices A and C being complex numbers if the transfer function has complex poles. A simple transformation leads to a block diagonal structure for A, with each pair of complex conjugate eigenvalues replaced by 2 × 2 blocks of real numbers. For example, consider the transfer function.

$$G(s) = \frac{Y(s)}{U(s)} = \frac{s+2}{s^2+2s+5} = \frac{0.5-j0.25}{s+1-j2} + \frac{0.5+j0.25}{s+1+j2}$$

Let $$X_1(s) = \frac{U(s)}{s+1-j2} \quad \text{and} \quad X_2(s) = \frac{U(s)}{s+1+j2}$$

Then, we get the following state equations:

$$\dot{x} = \begin{bmatrix} -1+j2 & 0 \\ 0 & -1-j2 \end{bmatrix} x + \begin{bmatrix} 1 \\ 1 \end{bmatrix} u$$

$$y = [0.5 - j0.25 \quad 0.5 + j0.25]\ x$$

If we now apply the transformation $x = P\bar{x}$, where

$$P = \begin{bmatrix} 1 & j \\ 1 & -j \end{bmatrix} \quad \text{and} \quad P^{-1} = \begin{bmatrix} 0.5 & 0.5 \\ -j0.5 & j0.5 \end{bmatrix}$$

then we get $\quad \bar{A} = P^{-1} AP = \begin{bmatrix} -1 & -2 \\ 2 & -1 \end{bmatrix}$

$$\bar{B} = P^{-1} B = \begin{bmatrix} 1 \\ 0 \end{bmatrix} \quad \text{and} \quad \bar{C} = CP = [1 \quad 0.5]$$

(*a*) Draw a simulation diagram for the transformed set of state equations.

(*b*) Use the above procedure to show that the transfer function.

$$G(s) = \frac{Y(s)}{U(s)} = \frac{s+2}{s^2+2s+5} + \frac{1}{s} + \frac{2}{s+2}$$

can be realized using the following set of state equations:

$$\dot{x} = \begin{bmatrix} -1 & -2 & 0 & 0 \\ 2 & -1 & 0 & 0 \\ 0 & 0 & 0 & 0 \\ 0 & 0 & 0 & -2 \end{bmatrix} x + \begin{bmatrix} 1 \\ 0 \\ 1 \\ 1 \end{bmatrix} u$$

$$y = [1 \quad 0.5 \quad 1 \quad 2]\ x$$

and the transformation matrix is given by

$$P = \begin{bmatrix} 1 & j & 0 & 0 \\ 1 & -j & 0 & 0 \\ 0 & 0 & 1 & 0 \\ 0 & 0 & 0 & 1 \end{bmatrix}$$

7. The diagonal form can always be obtained if the poles of a transfer function are distinct. For the more general case, we can obtain the Jordan form. For example, consider

$$G(s) = \frac{Y(s)}{U(s)} = \frac{2s+8}{(s+2)^2} = \frac{2}{(s+2)^2} + \frac{4}{s+2}$$

Let $\quad X_1(s) = \dfrac{1}{s+2} X_2(s), \quad \text{or} \quad \dot{x}_1 = -2x_1 + x_2$

and $\qquad X_2(s) = \dfrac{1}{s+2}U(s), \quad \text{or} \quad \dot{x}_2 = -2x_2 + u$

Since $\qquad Y(s) = 2X_1(s) + 4X_2(s)$

we get the following state equations.

$$\begin{bmatrix} \dot{x}_1 \\ \dot{x}_2 \end{bmatrix} = \begin{bmatrix} -2 & 1 \\ 0 & -2 \end{bmatrix} x + \begin{bmatrix} 0 \\ 1 \end{bmatrix} u$$

$$y = [2 \quad 4] \begin{bmatrix} x_1 \\ x_2 \end{bmatrix}$$

Use the above procedure to show that the Jordan form for the state equations for the transfer function

$$G(s) = \frac{Y(s)}{U(s)} = \frac{2}{(s+2)^3} + \frac{4}{(s+2)^2} + \frac{5}{(s+2)} + \frac{3}{s+1}$$

is

$$\dot{x} = \begin{bmatrix} -2 & 1 & 0 & 0 \\ 0 & -2 & 1 & 0 \\ 0 & 0 & -2 & 0 \\ 0 & 0 & 0 & 1 \end{bmatrix} x + \begin{bmatrix} 0 \\ 0 \\ 1 \\ 1 \end{bmatrix} u$$

$$y = [2 \quad 4 \quad 5 \quad 3]\, x$$

8. (*a*) For the fourth-order system in the previous problem, determine the state equations in the controller and observer canonical forms.

(*b*) Draw an analog computer simulation diagram for each of the three canonical state-space representations (*i.e.*, Jordan, controller and observer).

9. Write a set of state equations for the electrical networks shown in Fig. P3.9. (*Hint*: See Appendix *D* for the procedure for writing these equations).

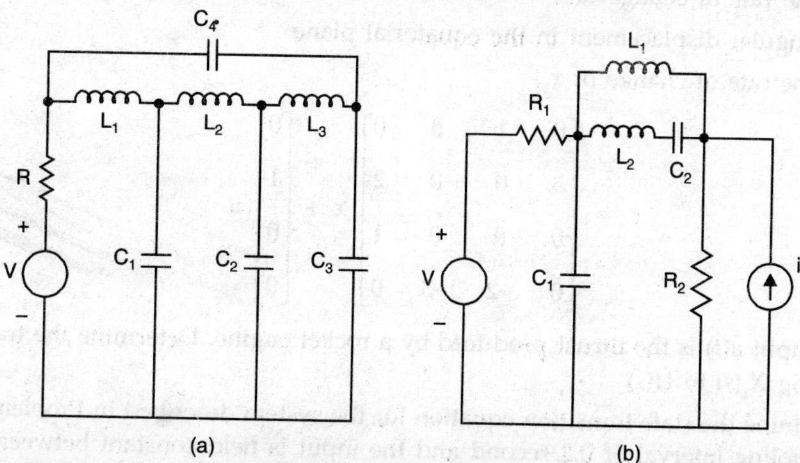

Figure P3.9. *Electrical networks*

10. Obtain a set of state equations for the Ward-Leonard speed control system shown in Fig. 2.40 with the parameters given in Chapter 2, Problem 7. Physical variables should be used as state variables.

11. The transfer function of a linear system is given below. Derive a set of state equations in (*a*) the controller canonical form, (*b*) the observer canonical form and (*c*) the diagonal form.

$$\frac{Y(s)}{U(s)} = \frac{3s^2 + 5s + 13}{(s+2)\left(s^2 + 4s + 8\right)}$$

12. Repeat Problem 10 if the transfer function is given by

$$\frac{Y(s)}{U(s)} = \frac{s^3 + 9s^2 + 18s + 25}{(s+2)\left(s^2 + 4s + 8\right)}$$

13. Determine e^{At} for each of the following *A*-matrices:

$$(a) \begin{bmatrix} 0 & 2 \\ -1 & -3 \end{bmatrix}, \quad (b) \begin{bmatrix} -1 & 1 \\ 0 & 1 \end{bmatrix}, \quad (c) \begin{bmatrix} -1 & 1 & 0 \\ 0 & -4 & 2 \\ 0 & 0 & -10 \end{bmatrix}$$

14. The following *A*-matrices are all in the Jordan form. Use the method of Laplace transform to obtain e^{At} for each of them. Can you think of a simple rule for finding e^{At} for such cases?

$$(a) \begin{bmatrix} 2 & 1 \\ 0 & 2 \end{bmatrix}, \quad (b) \begin{bmatrix} 2 & 1 & 0 \\ 0 & 2 & 1 \\ 0 & 0 & 2 \end{bmatrix}, \quad (c) \begin{bmatrix} 2 & 1 & 0 \\ 0 & 2 & 0 \\ 0 & 0 & 2 \end{bmatrix}$$

15. The linearized equations of a satellite in a circular orbit are given below in the normalized form, where the state variables are

x_1 = the distance from the center of the Earth

x_2 = the rate of change of x_1

x_3 = angular displacement in the equatorial plane

x_4 = the rate of change of x_3.

$$\dot{x} = \begin{bmatrix} 0 & 1 & 0 & 0 \\ 3 & 0 & 0 & 2 \\ 0 & 0 & 0 & 1 \\ 0 & -2 & -3 & 0 \end{bmatrix} x + \begin{bmatrix} 0 \\ 1 \\ 0 \\ 0 \end{bmatrix} u$$

The input $u(t)$ is the thrust produced by a rocket engine. Determine the transfer function relating $X_1(s)$ to $U(s)$.

16. Determine the state transition equation for the system described in Problem 14, assuming a sampling interval of 0.2 second and the input is held constant between the sampling instants.

17. Many control systems are of the 'multivariable' type, *i.e.*, they have many inputs and outputs. The state equations for such a system are given by Equations (3.1) and (3.2), with B being an $n \times m$ matrix and C a $p \times n$ matrix for a system with m inputs and p outputs. Such a system can also be described by a transfer function matrix given by

$$G(s) = C(sI - A)^{-1} B + D$$

Determine the transfer function matrix for a two-input two-output system described by the following state equations:

$$\dot{x} = \begin{bmatrix} -1 & 1 & 0 \\ 0 & -4 & 2 \\ 0 & 0 & -10 \end{bmatrix} x + \begin{bmatrix} 1 & -1 \\ 0 & 2 \\ -1 & 1 \end{bmatrix} u$$

$$y = \begin{bmatrix} 1 & 0 & 1 \\ 2 & 1 & -3 \end{bmatrix}$$

18. The linearized equations for an automatic ship steering system are given below:

$$\dot{x} = \begin{bmatrix} -0.05 & -6 & 0 & 0 \\ -0.001 & 0.16 & 0 & 0 \\ 1 & 0 & 0 & 12 \\ 0 & 1 & 0 & 0 \end{bmatrix} x + \begin{bmatrix} -0.5 \\ 0.9 \\ 0 \\ 0 \end{bmatrix} u$$

(*a*) Determine e^{At} using transformation to either a diagonal or Jordan form.

(*b*) Determine the state transition matrices if the sampling interval is 1 second.

19. The simplified transfer function for the longitudinal motion of an airplane is given by

$$G(s) = \frac{60s + 4}{(s + 10)(s^2 + 0.2s + 7.5)}$$

(*a*) Obtain a controller form realization for the system.

(*b*) Obtain an observer form realization for the system.

(*c*) Calculate the fundamental matrix e^{At} for the controller form realization obtained in part (*a*).

20. The transfer function relating the position of one joint of a robot to the voltage applied to the armature of the servomotor controlling it is given by

$$G(s) = \frac{\theta(s)}{V(s)} = \frac{5}{s(s + 0.2)(s + 5)}$$

(*a*) Obtain a state variable model in the observer form.

(*b*) Calculate the fundamental matrix e^{At} corresponding to this model.

(*c*) Determine the state transition equation if the system is discretized with a sampling period of 0.1 second and the input is held constant between sampling instants.

21. An important consideration in the design of the suspension system for an automobile is to increase the comfort of the passengers by absorbing the vibrations due to the terrain of the road. A model for the vertical suspension is shown in Fig. P3.21.

(*a*) Obtain an equivalent electrical network.

(*b*) Write a set of state equations for the system.

(*c*) Determine the transfer function matrix relating x_1 and x_2 to u_1 and u_2.

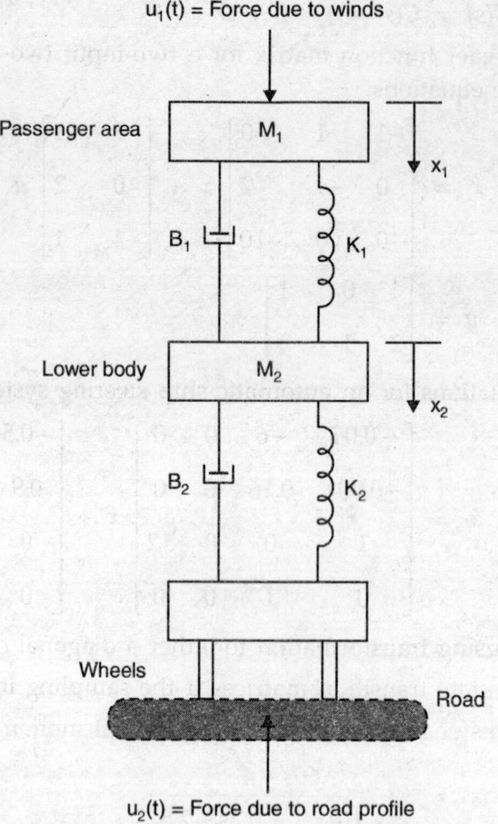

Figure P3.21. *Model for the vertical suspension of an automobile*

❑❑❑

CHAPTER 4

Characteristics of Closed-Loop Systems

4.1 INTRODUCTION

As mentioned in Chapter 1, all automatic control systems are of the closed-loop type. This is necessitated by the introduction of feedback for comparing the reference input, $r(t)$, with the controlled output, $c(t)$. A system without feedback is called an *open-loop* system. Both of them are shown in Fig. 4.1. In this chapter, we shall study some characteristics of closed-loop systems.

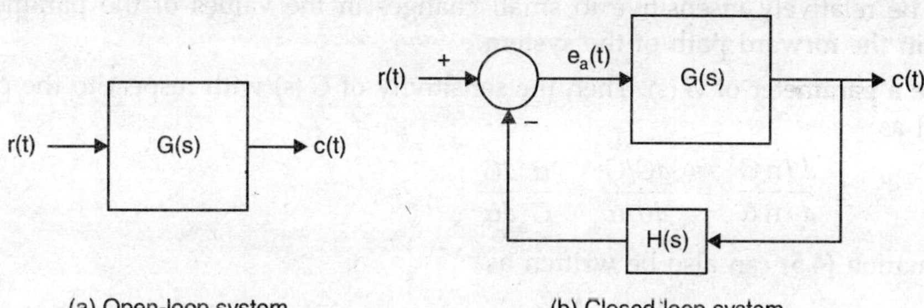

Figure 4.1. *Block diagrams of open- and closed-loop systems*

For the open-loop system,

$$C(s) = G(s)\,R(s) \qquad \text{...(4.1)}$$

For the closed-loop system,

$$C(s) = G(s)\,E_a(s) = G(s)\,[R(s) - H(s)\,C(s)]$$

Hence,

$$C(s) = \frac{G(s)}{1 + G(s)\,H(s)}\,R(s) \qquad \text{...(4.2)}$$

Also, the actuating error is given by

$$E_a(s) = \frac{1}{1 + G(s)\,H(s)}\,R(s) \qquad \text{...(4.3)}$$

To reduce the error, we must make the *loop gain*, $G(s)\,H(s)$ large over the range of frequencies of interest, that is,

$$|G(s)\,H(s)| >> 1 \qquad ...(4.4)$$

Please note that due to the low-pass nature of physical systems, this inequality will not be satisfied at high frequencies. Also, if $G(s)\,H(s)$ turns out to be a negative number close to one at some value of the frequency, then we may have a very large error.

4.2 SENSITIVITY TO PARAMETER VARIATIONS

One important property of negative feedback is the reduction in the sensitivity to variations in the parameters of the forward path.

Historically, this was the principal reason for the development of feedback amplifiers by engineers at the Bell Telephone Laboratories. With the expansion of the telephone system in North America over long distances, repeaters had to be installed. These were amplifiers employing vacuum tubes. It was soon noticed that with the mass-produced repeaters, the gain varied considerably between different elements, due primarily to the variation in the amplification factors of the vacuum tubes. For reasons of economy, one must always allow some tolerance in the manufacture of components, otherwise the products will be too expensive and should be unable to compete in the market. The primary object of introducing negative feedback was to make the gain of the amplifier relatively insensitive to variations in the parameters of the vacuum tube. Similarly, in the design of control systems, it is important that the transfer function of the closed-loop system be relatively insensitive to small changes in the values of the parameters of the components in the forward path of the system.

Let α be a parameter of $G(s)$. Then the sensitivity of $G(s)$ with respect to the parameter α is defined [1] as

$$S_\alpha^G \;\Delta\; \frac{d\,\ell n\,G}{d\,\ell n\,\alpha} = \frac{dG/G}{d\alpha/\alpha} = \frac{\alpha}{G}\frac{dG}{d\alpha} \qquad ...(4.5)$$

The Equation (4.5) can also be written as

$$S_\alpha^G = \lim_{\Delta\alpha \to 0} \frac{\dfrac{\Delta G}{G}}{\dfrac{\Delta\alpha}{\alpha}} \qquad ...(4.6)$$

which can be considered as the fractional change in G due to a very small fractional change in α. This is the sensitivity of the transfer function of the open-loop system to variations in α. The transfer function of the closed-loop system is given by

$$T(s) \;\Delta\; \frac{C(s)}{R(s)} = \frac{G(s)}{1+G(s)\,H(s)} \qquad ...(4.7)$$

Consequently, the sensitivity of the closed-loop system to variations in α is given by

$$S_\alpha^T = \frac{\alpha}{T}\cdot\frac{dT}{d\alpha} = \frac{\alpha(1+GH)}{G}\cdot\frac{dT}{dG}\cdot\frac{dG}{d\alpha}$$

$$= \frac{\alpha(1+GH)}{G}\cdot\frac{1}{(1+GH)^2}\cdot\frac{dG}{d\alpha} = \frac{\alpha}{G}\cdot\frac{dG}{d\alpha}\cdot\frac{1}{1+GH}$$

Hence, we get the following relationship between the sensitivities:

$$S_\alpha^T = \frac{S_\alpha^G}{1+G(s)H(s)} \qquad \text{...(4.8)}$$

Thus, feedback has reduced the sensitivity to variation in α by the factor $1/[1 + G(s) H(s)]$, which is small over the range of frequencies of interest. It may be emphasized that feedback amplifiers were developed by the telephone industry for this important advantage.

It is interesting to verify that feedback does not reduce the sensitivity to variations in the parameters in the feedback path. Let α be a parameter of $H(s)$. Then

$$S_\alpha^T = \frac{\alpha}{T}\cdot\frac{dT}{d\alpha} = \frac{\alpha}{H}\cdot\frac{dH}{d\alpha}\cdot\frac{H}{T}\cdot\frac{dT}{dH} = S_\alpha^H\cdot\frac{H}{T}\cdot\frac{(-G)\cdot G}{(1+GH)^2}$$

$$= -S_\alpha^H\cdot\frac{H\cdot(1+GH)}{G}\cdot\frac{G^2}{(1+GH)^2} = -S_\alpha^H\cdot\frac{GH}{1+GH} \qquad \text{...(4.9)}$$

It follows that the magnitudes of S_α^T and S_α^H are nearly equal if the loop gain $G(s)H(s)$ is much greater than one.

Example 4.1

Consider the simple closed-loop system shown in Fig. 4.2, which represents a position control system using a *dc* servomotor in the armature-control mode. We shall assume that the nominal value of the gain constant K is 10, that of α is 2, and the feedback parameter β is equal to 1.

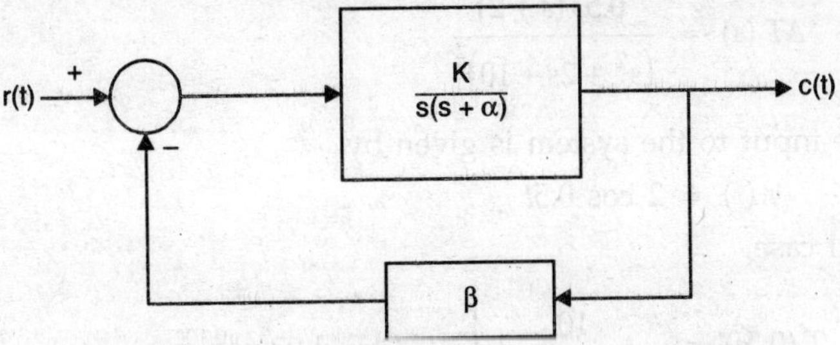

Figure 4.2. *Block diagram of a position-control system*

Hence, we have $G(s) = \dfrac{K}{s(s+\alpha)}$, $H(s) = \beta$

The sensitivity of G with respect to K is given by

$$S_K^G = \frac{K}{G}\cdot\frac{dG}{dK} = s(s+\alpha)\cdot\frac{1}{s(s+\alpha)} = 1$$

The sensitivity of G with respect to α is given by

$$S_\alpha^G = \frac{\alpha}{G}\cdot\frac{dG}{d\alpha} = \frac{\alpha s(s+\alpha)}{K}\cdot\frac{-K}{s(s+\alpha)^2} = -\frac{\alpha}{s+\alpha} = -\frac{2}{s+2}$$

The sensitivity of the feedback transfer function $H(s)$ with respect to β is given by

$$S_\beta^H = \frac{\beta}{H}\cdot\frac{dH}{d\beta} = 1$$

We are now ready to evaluate the sensitivities of the closed-loop system transfer function.

$$S_K^T = \frac{S_K^G}{1+G(s)H(s)} = \frac{1}{1+G(s)H(s)} = \frac{s(s+\alpha)}{s(s+\alpha)+K} = \frac{s^2+2s}{s^2+2s+10}$$

$$S_\alpha^T = \frac{S_\alpha^G}{1+G(s)H(s)} = -\frac{\alpha}{s+\alpha}\cdot\frac{s(s+\alpha)}{s(s+\alpha)+K} = -\frac{2s}{s^2+2s+10}$$

$$S_\beta^T = -S_\beta^H\cdot\frac{G(s)H(s)}{1+G(s)H(s)} = -\frac{K}{s^2+\alpha s+K} = -\frac{10}{s^2+2s+10}$$

The sensitivities of $T(s)$ with respect to K and α are seen to be small at low frequencies and are zero at zero frequency (*dc*). On the other hand, the sensitivity of $T(s)$ with respect to β has the magnitude of 1 at zero frequency that is equal to the sensitivity of $H(s)$ with respect to β. Thus, feedback has not reduced the sensitivity to variation of a parameter in the feedback path.

Now, consider the effect of a 5 per cent change in the parameter K. The resulting change in the transfer function of the closed-loop system is obtained as

$$\frac{\Delta T(s)}{T(s)} \approx S_K^{T(s)}\cdot\frac{\Delta K}{K} \approx \frac{s^2+2s}{s^2+2s+10}\cdot 0.05$$

Consequently, $\Delta T(s) \approx \dfrac{0.5s(s+2)}{(s^2+2s+10)^2}$

If the reference input to the system is given by

$$r(t) = 2\cos 0.5t$$

then for the nominal case,

$$T(0.5j) = \left.\frac{10}{s^2+2s+10}\right|_{s=0.5j} = 1.02e^{-j0.102}$$

Thus, the steady-state response is given by

$$c_{ss}(t) = 2.04\cos(0.5t - 0.102)$$

Also, for a 5 per cent change in K, the change in $T(s)$ is given by

$$\Delta T(0.5j) \approx \left.\frac{0.5s(s+2)}{(s^2+2s+10)^2}\right|_{s=0.5j} = 0.005e^{-j4.672}$$

and the resulting change in the steady-state response is

$$\Delta c_{ss}(t) = 0.01\cos(0.5t - 4.672)$$

Please note that this change is less than 0.5 per cent of the steady-state response in the nominal case.

DRILL PROBLEM 4.1

For the system described in Example 4.1 determine the change in T (s) for a 5 per cent change in the parameter α. Calculate the value of the change for ω = 0.5, and hence determine the change in the steady-state response to the input r (t) = 2 cos 0.5t.

Ans. $\Delta T(s) \approx \dfrac{-s}{\left(s^2+2s+10\right)^2}$, $\Delta T(0.5j) \approx 0.005e^{-j1.775}$

$\Delta c_{ss}(t) \approx 0.01 \cos (90.5t - 1.775)$

DRILL PROBLEM 4.2

The block diagram of the speed-control system of an internal combustion engine is shown in Fig. 4.3. The gain constant K of the system has a nominal value of 10, with a tolerance of ± 5 per cent. Determine the tolerance in the speed of the engine for step changes in reference speed setting.

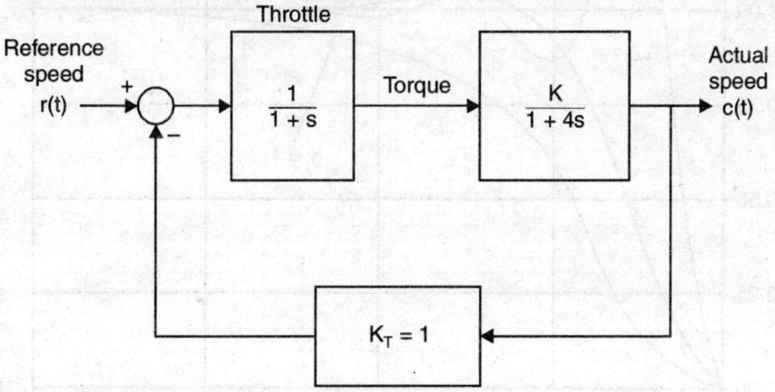

Figure 4.3. *Speed-control system for a gasoline engine*

Ans. ± 0.413 per cent.

4.3 TRANSIENT RESPONSE

The transient response of a closed-loop system is easily altered by changing the loop gain. For example, consider the closed-loop system shown in Fig. 4.4, which represents a position-control system. The transfer function of this system is given by

$$\frac{C(s)}{R(s)} = \frac{K}{s^2+2s+K} \qquad \text{...(4.10)}$$

where the value of K can be adjusted by changing the gain of an amplifier.

The response $c(t)$, for the case when $r(t)$ is a unit step, is shown in Fig. 4.5 for $K = 1$, $K = 2$ and $K = 10$. For these three cases, the poles of the closed-loop system transfer function are located at −1, −1, at $-1 \pm j1$, and at $-1 \pm j3$, respectively.

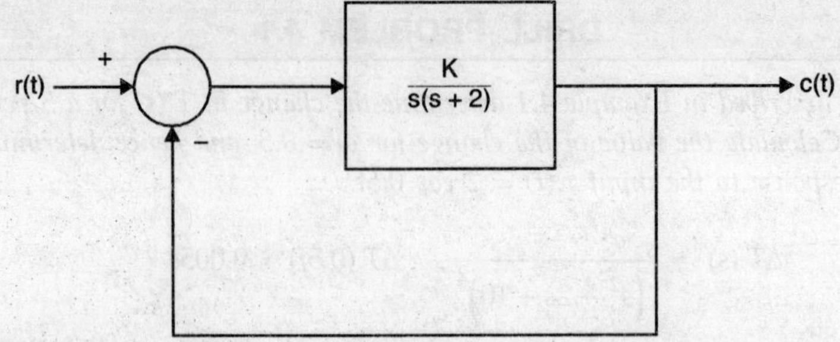

Figure 4.4. *A simple closed-loop system*

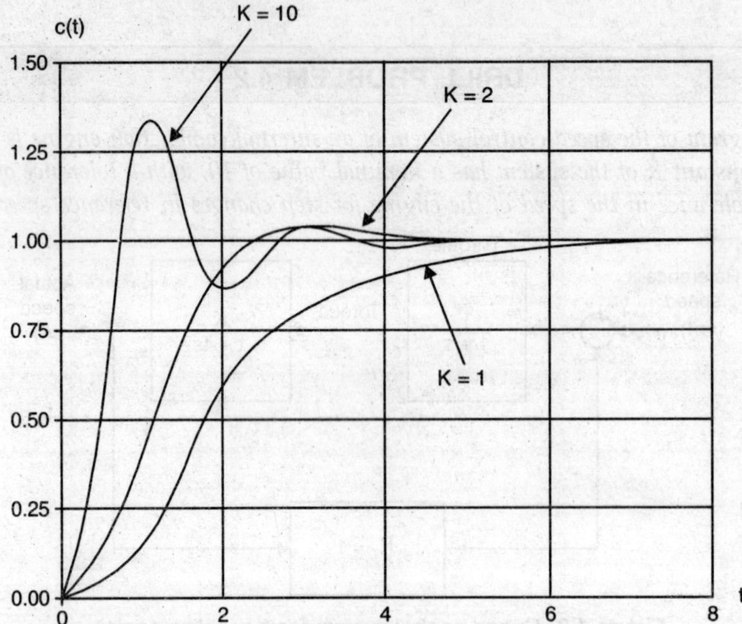

Figure 4.5. *Step responses for different values of K*

The damping ratio of a pair of complex poles is defined as

$$\zeta = \cos \phi \qquad ...(4.11)$$

where the angle ϕ is shown in Fig. 4.6. It will be seen that by varying K, one can adjust the damping ratio of the poles of the transfer function of the closed-loop system and thus alter the transient response. We shall return to the relationship between the transient response and the damping ratio of the complex poles in Chapter 5.

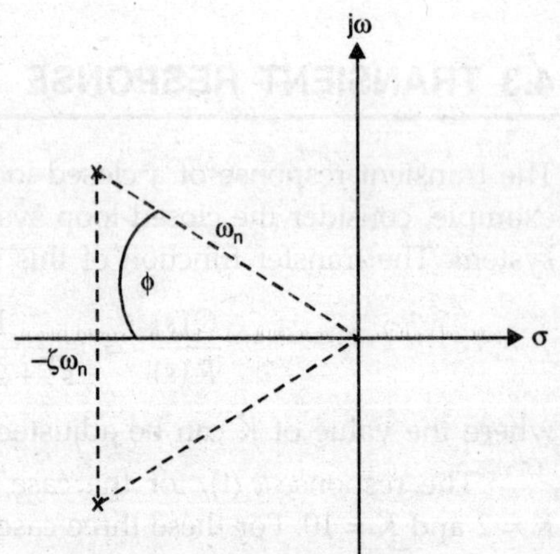

Figure 4.6. *Complex poles*

In Chapter 7, we shall study the root locus method, which is concerned with the movement of the poles of the transfer function of the closed-loop system as the open-loop gain K is varied. At this time, it is sufficient to note that the time constants as well as the damping ratios of the poles of the transfer function of the closed-loop system can be adjusted by changing K.

DRILL PROBLEM 4.3

The block diagram of a speed-control system using a dc motor with a time-constant of 4 seconds is shown in Fig. 4.7. Determine the time constant of the closed-loop system for (a) K = 3 and (b) K = 19. (Hint: The time constant of a first-order system is the reciprocal of the magnitude of the location of the pole.)

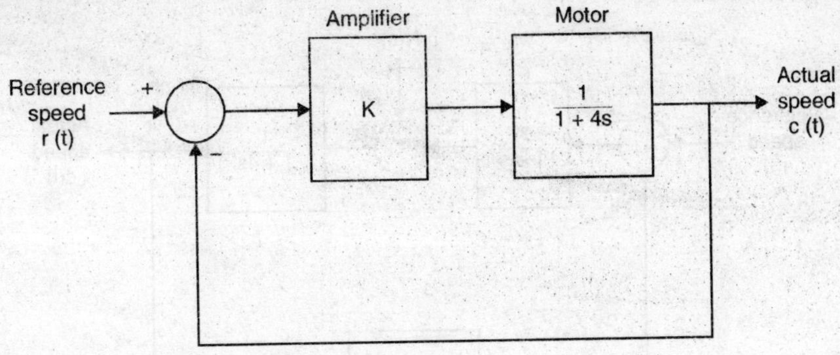

Figure 4.7. *Speed-control system*

Ans.

(*a*) 1 second, (*b*) 0.2 second. (Note that this assumes that the motor will be working in the linear mode for both the values of *K*!).

4.4 EFFECT OF DISTURBANCE SIGNALS

Most control systems are subject to unwanted disturbance signals. For example, we may have noise generated in electronic amplifiers, gusts of wind affecting radar antennas, load fluctuations, and so forth. The block diagram of such a system is shown in Fig. 4.8, where $N(s)$ is the disturbance (or noise) signal.

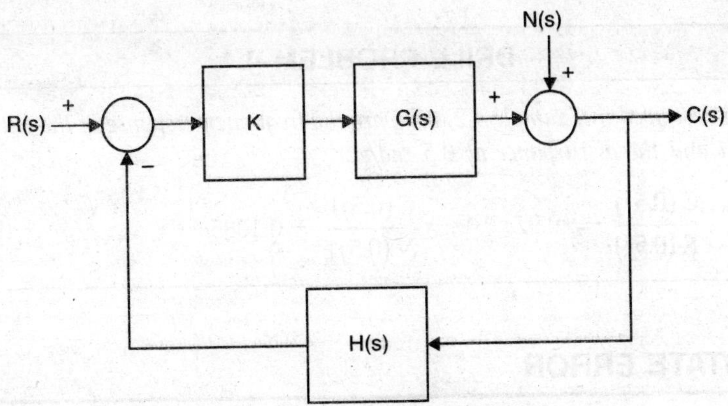

Figure 4.8. *System with disturbance*

Applying superposition, we get

$$C(s) = \frac{K\,G(s)}{1 + K\,G(s)\,H(s)} R(s) + \frac{1}{1 + K\,G(s)\,H(s)} N(s) \qquad \text{...(4.12)}$$

Hence, the effect of noise is reduced by the ratio $1/[1 + KG(s)H(s)]$ over the frequency range of interest. Thus, by increasing the loop-gain, we can alleviate the effect of noise occurring at the output level. Load disturbances fall under this category.

EXAMPLE 4.2

As an example, let us consider the speed control system for a gasoline engine with load disturbance, as shown in Fig. 4.9, where $K = 10$.

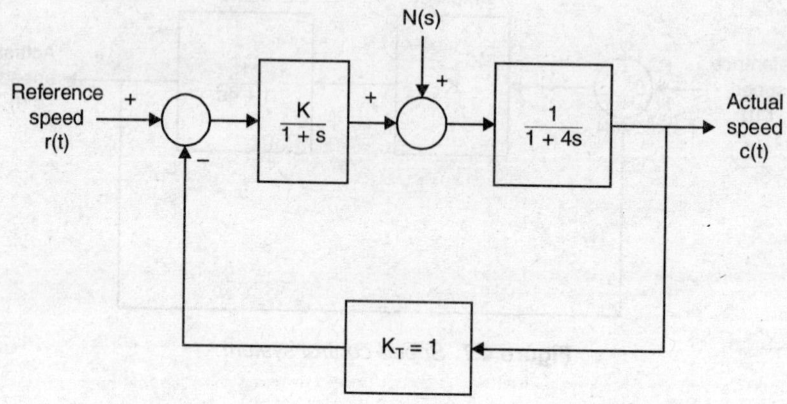

Figure 4.9. *Speed control system with load disturbance*

In this case, we have

$$C(s) = \frac{2.5}{s^2 + 1.25s + 2.75} R(s) + \frac{0.25(s+1)}{s^2 + 1.25s + 2.75} N(s) \quad \text{...(4.13)}$$

The reduction in the disturbance is evident by noting that the *dc* gain of $C(s)/R(s)$ is 10 times that of $C(s)/N(s)$, which is precisely the value of $KG(s)$ at $s = 0$.

It must be emphasized that this reduction is possible only when the disturbance occurs at the output level. If it had occurred at the input level, the effect of the reference input and disturbance would have been the same on the output.

DRILL PROBLEM 4.4

For the system described in Example 4.2, compare the frequency response of the output with respect to reference input and the disturbance at 0.5 rad/s.

Ans. $\frac{C(0.5j)}{R(0.5j)} = 0.97e^{-j0.245}$, $\frac{C(0.5j)}{N(0.5j)} = 0.1085e^{j0.2187}$

4.5 STEADY-STATE ERROR

Another effect of feedback is that it provided us with some control on the steady-state error to standard inputs by adjustment of the open-loop gain. As an example, we shall discuss the steady-state error to a unit step input.

Consider again the closed-loop system shown in Fig. 4.1(*b*). We shall now assume that $H(s) = 1$; that is, it is a unity-feedback system. For a unit step input,

$$C(s) = \frac{G(s)}{1+G(s)}R(s) = \frac{G(s)}{1+G(s)}\cdot\frac{1}{s} \quad ...(4.14)$$

and the error is given by

$$E(s) = R(s) - C(s) = \frac{1}{s}\left[1 - \frac{G(s)}{1+G(s)}\right] = \frac{1}{s}\cdot\frac{1}{1+G(s)} \quad ...(4.15)$$

Applying the final-value theorem the steady-state error is obtained as

$$e_{ss} = \lim_{s\to 0}[sE(s)] = \frac{1}{1+G(0)} \quad ...(4.16)$$

It is evident that the steady-stare error may be made as small as possible by increasing the *dc* gain, $G(0)$, of the system.

It is important to note that the above derivation assumes that the closed-loop system transfer function $T(s)$ has all its poles in the left-half of the *s*-plane; otherwise the final-value theorem cannot be applied. Furthermore, as will be seen in Chapters 6 and 7, in many cases a large increase in the loop-gain may lead to instability and thus violate the condition mentioned above. Hence, the possible decrease in the steady-state error is often limited.

DRILL PROBLEM 4.5

For the speed-control system described in Drill Problem 4.4 determine the steady-state error to a unit step input as well as the damping ratio of the poles of the transfer function of the closed-loop system for (a) K = 4 and (b) K = 20.

Ans. (*a*) $e_{ss} = 0.2$, damping ratio = 0.559.

(*b*) $e_{ss} = 0.048$, damping ratio = 0.273.

4.6 DISADVANTAGES OF FEEDBACK

Like most good things in life, the introduction of feedback in a system is not free of cost. The main disadvantages are given below:

1. Since feedback decreases the overall gain, this must be made up by an increase in the loop-gain of the system, requiring additional hardware and increased complexity.
2. The components in the feedback path must be made more accurate because feedback does not reduce the sensitivity to variations in these parameters. The result is a further increase in the overall cost.
3. The sensors required for the feedback path may introduce a small amount of noise in the system, thereby reducing the overall accuracy.
4. The introduction of feedback may lead to instability of the closed-loop system, even though the open-loop system may be stable. This is caused by inherent time lags in the system, with the result that what was intended as negative feedback may turn out to be positive feedback at some higher frequency. This was one of the first features of feedback

observed when automatic control was introduced. A thorough understanding of the Nyquist criterion of stability leads to the solution of this problem. It will be discussed in detail in Chapter 9.

SUMMARY

- In this chapter, we have studied the various advantages and disadvantages of the introduction of feedback necessary for automatic control.
- The main *advantages* are: (1) reduction in sensitivity to variations of parameters in the forward path, (2) control over the transient response as well as the steady-state accuracy by adjusting the loop-gain, and (3) reduction in the effect of noise or disturbance at the output level. The main *disadvantages* are: (1) the need for additional hardware with consequent increase in cost and (2) the possibility of instability due to phase lags in the feedback loop. The latter disadvantage is not serious and can be offset by proper design. The cost of additional hardware is often worthwhile due to the overall improvement in performance. Hence, it is fair to say that the advantages of feedback are much more significant than the disadvantages.

Problems

1. Figure P4.1 shows the block diagram for automatic control of the speed of an automobile. Determine (*a*) the sensitivity of the transfer function of the closed-loop system to changes in the engine gain K_e, (*b*) the effect of the load disturbance on the speed, and (*c*) the steady-state speed when the input setting $r(t)$ is 60 km/h if $K_e = 100$ and $d(t) = 0$.

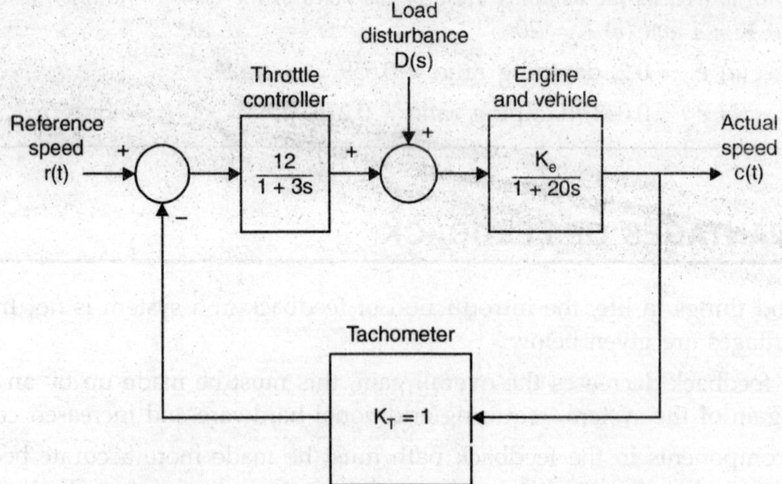

Figure P4.1. *A speed control system for an automobile*

2. Repeat part (*c*) of the previous problem if there is a 5 per cent variation in K_e and there is a 5 per cent variation in K_T.

3. The block diagram of a speed-control system is shown in Fig. P4.3.

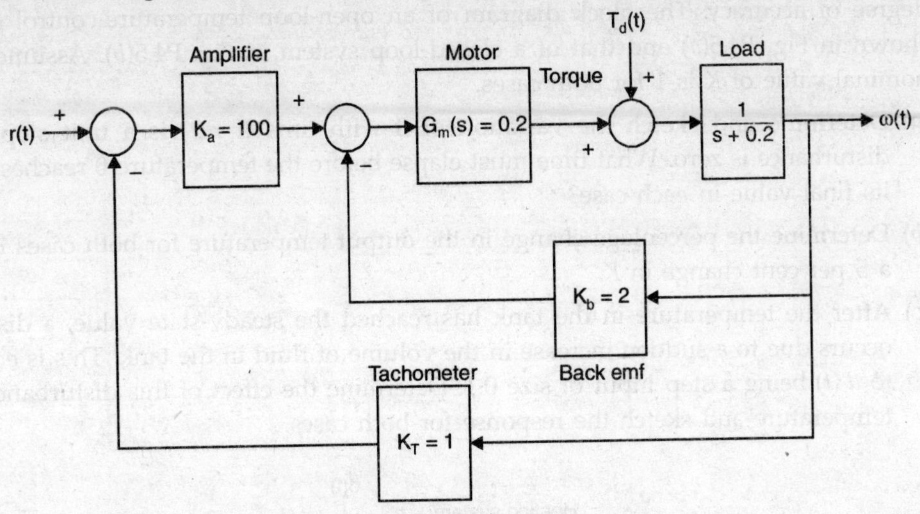

Figure P4.3. *A speed-control system*

(*a*) Compare the open- and closed-loop sensitivities of ω to small changes in the amplifier gain K_e.

(*b*) Determine the steady-state error in speed control if $r(t)$ is a unit step and if the disturbance $T_d(t) = 0$.

(*c*) Determine the error in the system if $T_d(t)$ is a unit step, and $r(t) = 0$.

4. The block diagram of an electronic pacemaker for controlling the rate of heartbeats is shown in Fig. P4.4.

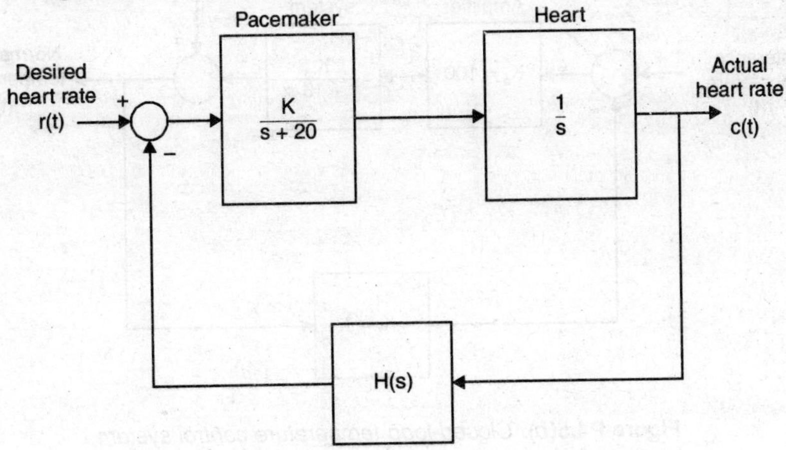

Figure P4.4. *A heart rate control system*

(*a*) Determine the sensitivity of the closed-loop system transfer function to small changes in K at the normal heart rate of 72 beats per minute if $H(s) = 1$ and the nominal value of $K = 400$.

(*b*) Repeat if $H(s) = 1/(1 + 0.1s)$.

5. In many chemical processes the temperature in a tank has to be maintained within a good degree of accuracy. The block diagram of an open-loop temperature-control system is shown in Fig. P4.5(*a*) and that of a closed-loop system in Fig. P4.5(*b*). Assume that the nominal value of K is 1 for both cases.

   (*a*) Determine and sketch the variation of θ with time if $r(t)$ is a unit step and the disturbance is zero. What time must elapse before the temperature θ reaches 63.2% of its final value in each case?

   (*b*) Determine the percentage change in the output temperature for both cases if there is a 5 per cent change in K.

   (*c*) After the temperature in the tank has reached the steady-state value, a disturbance occurs due to a sudden increase in the volume of fluid in the tank. This is equivalent to $d(t)$ being a step input of size 0.1. Determine the effect of this disturbance on the temperature and sketch the response for both cases.

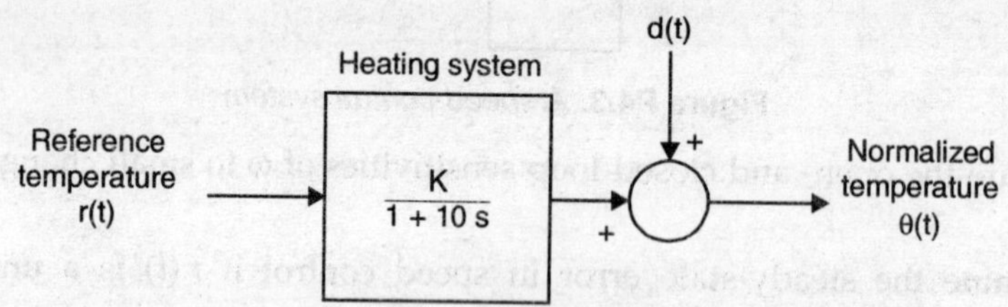

Figure P4.5(*a*). *Open-loop temperature control system*

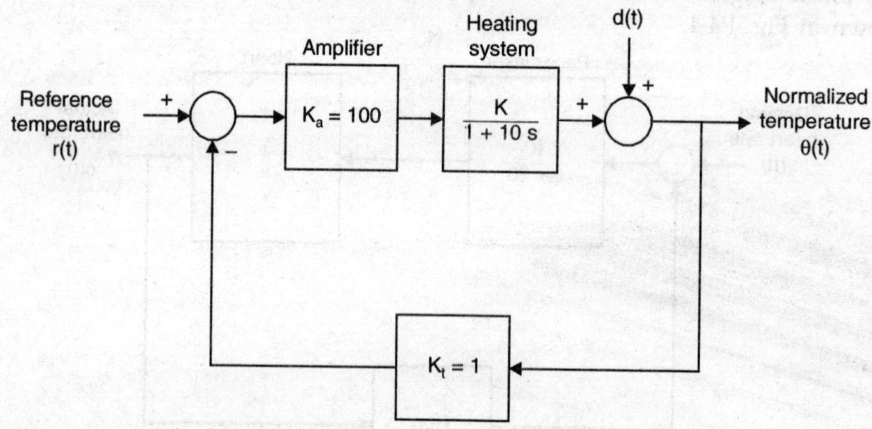

Figure P4.5(*b*). *Closed-loop temperature control system*

6. The block diagram of the cruise control system for an automobile is shown in Fig. P4.6, where $c(t)$ is the normalized speed.

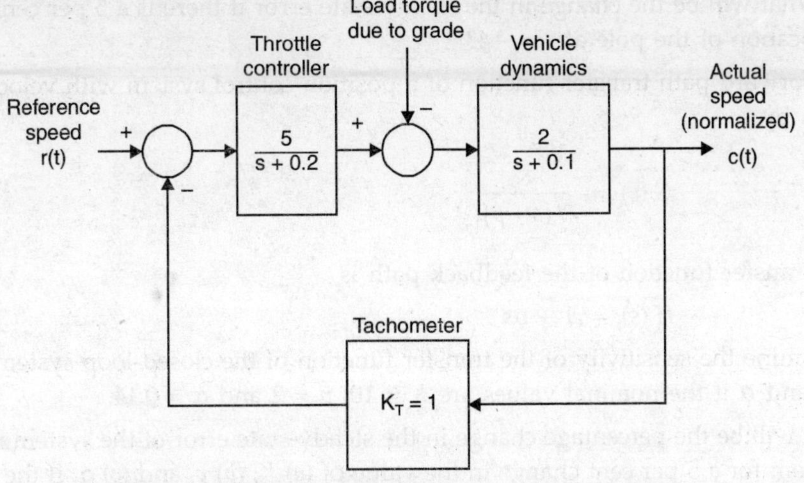

Figure P4.6. *Cruise control system for an automobile*

(*a*) Determine the steady-state error in the speed when the reference input is a unit step corresponding to the normalized value of the desired speed. Express this error as percentage relative to the desired speed.

(*b*) With the automobile moving at this steady speed along a level road, suddenly the grade changes. This corresponds to introducing a load torque disturbance of 0.1 times the unit step. What will be the new steady-state error in the speed as a percentage of the desired speed? (*Hint*: Use superposition.)

7. The Ward-Leonard scheme for controlling the speed of heavy loads is described in Chapter 2, Problem 7. Determine the percentage change in the speed if the load is increased by 5%.

8. The forward transfer function of a unity feedback system is given by

$$G(s) = \frac{K(s+2)}{s(s+1)(1+sT_0)}$$

Determine the relative change in the output $\Delta C/C$ for the relative change $\Delta T_0/T_0$ of the time constant.

9. Repeat the previous problem for the relative change $\Delta K/K$ in the forward gain of the system.

10. The transfer function of the forward path of a unity-feedback system is

$$G(s) = \frac{K(s+2)}{s^2(s+4)}$$

(*a*) For $K = 8$, determine the steady-state error when the input to the system is

$$r(t) = 4 - t - 2t^2$$

(*b*) What will be the change in the steady-state error if there is a 5 per cent change in the value of K?

(*c*) What will be the change in the steady-state error if there is a 5 per cent change in the location of the zero at $s = -2$?

(*d*) What will be the change in the steady-state error if there is a 5 per cent change in the location of the pole at $s = -4$?

11. The forward path transfer function of a position control system with velocity feedback is given by

$$G(s) = \frac{K}{s(s+p)}$$

The transfer function of the feedback path is

$$H(s) = 1 + \alpha s$$

Determine the sensitivity of the transfer function of the closed-loop system to changes in K, p and α if the nominal values are $K = 10$, $p = 2$ and $\alpha = 0.14$.

12. What will be the percentage change in the steady-state error of the system in the previous problem for a 5 per cent change in the value of (*a*) K, (*b*) p, and (*c*) α, if the input is a unit ramp function?

References

1. Bode, H.D., *Network Analysis and Feedback Amplifer Design*, Van Nostrand, Princeton, N.J., 1945.
2. Paul, M. Frank, *Introduction to System Sensitivity Theory*, Academic Press, New York, 1978.

CHAPTER 5

Performance of Control Systems

5.1 INTRODUCTION

As discussed in the previous chapter, it is possible to alter the transient as well as the steady-state behaviour of a closed-loop system by adjusting the open-loop gain. In practice, however, the two requirements are often contradictory, and in trying to improve the transient performance one may introduce a deterioration in the steady-state performance, and vice versa. This was observed even in the earliest days of automatic control when it was found that if one tried to improve the 'sensitivity' (or steady-state accuracy) of the Watt flyball governor, it led to 'hunting' or degradation of the transient performance. It was then felt that the engineer must make a compromise between the steady-state accuracy and the transient performance. Although it is true that such a compromise is necessary if the open-loop gain is the only adjustable parameter, in Chapter 10, we shall study methods that will enable us to obtain both high steady-state accuracy and desirable transient performance without sacrificing one for the other.

In general, we do not know the inputs to which a particular control system will be subjected while under operation. Hence, for the purpose of analysis and design, it is customary to consider the effect of application of certain standard inputs, which subject the system to sudden changes. The most commonly used inputs for this purpose are: (1) the unit step input, (2) the unit ramp input, and (3) the unit parabolic input. In terms of a position control system, these correspond to (1) a step change in position, (2) a step change in velocity, and (3) a step change in acceleration. Another standard test input is the sinusoid, which is of considerable importance since it provides a great deal of information about the system. Early work in the development of control theory was based primarily on the steady-state response of systems to sinusoidal inputs. This will be discussed in detail in Chapters 8 and 9.

The complete response of the system to any of these test inputs will consist of two components: (1) the *natural response* determined by the poles of the transfer function and the initial conditions, and (2) the *forced response* determined by the input signal. For a stable system, these are also the *transient* component and the *steady-state* component, respectively. The stability of a system will be studied in Chapter 6.

5.2 STANDARD TEST INPUTS

Some of the standard test inputs will be described in this section.

5.2.1 The Step Function

This is one of the most common and convenient test inputs because it can be implemented by a sudden change of the value of the input. This corresponds to switching on a constant voltage in an electrical circuit, applying a sudden change in the position of an input shaft, or suddenly opening or closing a valve.

Mathematically, the unit step function is described by the equation and is shown in Fig. 5.1.

$$r(t) = \begin{cases} 1 & \text{for} \quad t > 0 \\ 0 & \text{for} \quad t < 0 \end{cases} \qquad \text{...(5.1)}$$

Figure 5.1. *The unit step function*

Its Laplace transform is given by

$$R(s) = \frac{1}{s} \qquad \text{...(5.2)}$$

It should be noted that the function is undefined at $t = 0$, where its value suddenly changes from 0 to 1. In most practical cases, this can only be approximately correct, since the time for transition of the level can at best be made very small. For control system applications this approximation will not cause any problem as long as the time for transition is considerably smaller than the smallest time constant of the system.

5.2.2 The Ramp Function

The ramp function starts with the value of zero at $t = 0$ and increases linearly with time. The unit ramp function is shown in Figure 5.2 and is described by the equation.

$$r(t) = \begin{cases} t & \text{for} \quad t > 0 \\ 0 & \text{for} \quad t \le 0 \end{cases} \qquad \text{...(5.3)}$$

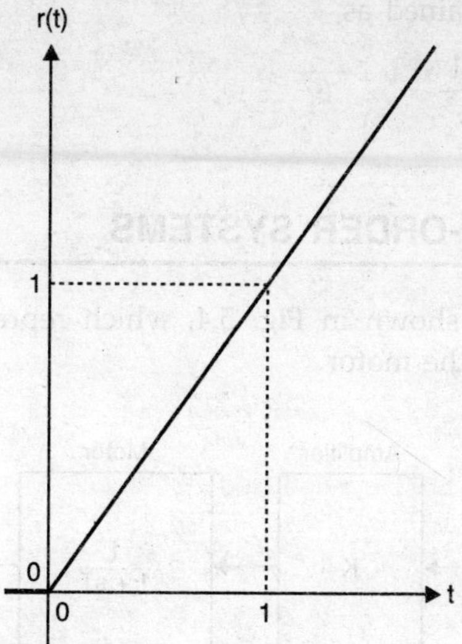

Figure 5.2. *The unit ramp function*

It is seen to be the integral of the unit step. Its Laplace transform is given by

$$R(s) = \frac{1}{s^2} \qquad ...(5.4)$$

5.2.3 The Parabolic Function

The unit parabolic function may be regarded as the integral of the unit ramp. This function is shown in Fig. 5.3 and is described by the equation.

$$r(t) = \begin{cases} \frac{1}{2}t^2 & \text{for} \quad t > 0 \\ 0 & \text{for} \quad t \leq 0 \end{cases} \qquad ...(5.5)$$

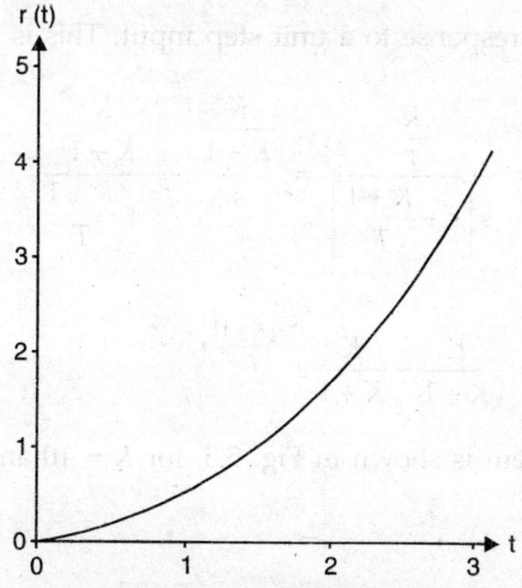

Figure 5.3. *The unit parabola*

Its Laplace transform is obtained as

$$R(s) = \frac{1}{s^3} \qquad \text{...(5.6)}$$

5.3 RESPONSE OF FIRST-ORDER SYSTEMS

Consider the first-order system shown in Fig. 5.4, which represents a speed-control system, where T is the time constant of the motor.

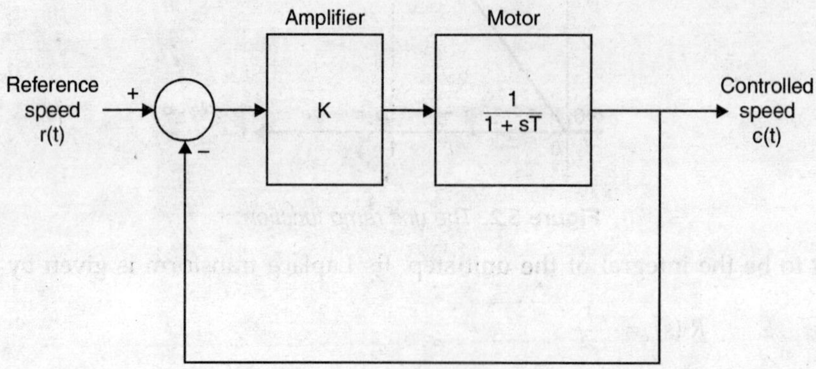

Figure 5.4. *A first-order system*

The transfer function of the system is given by

$$\frac{C(s)}{R(s)} = \frac{K}{1 + sT + K} = \frac{\frac{K}{T}}{s + \frac{K+1}{T}} \qquad \text{...(5.7)}$$

We shall first obtain the response to a unit step input. This is given by

$$C(s) = \frac{\frac{K}{T}}{s\left[s + \frac{K+1}{T}\right]} = \frac{\frac{K}{K+1}}{s} - \frac{\frac{K}{K+1}}{s + \frac{K+1}{T}} \qquad \text{...(5.8)}$$

and

$$c(t) = \frac{K}{K+1} - \frac{K}{K+1} e^{\frac{-(K+1)}{T}t} \qquad \text{...(5.9)}$$

The response of the system is shown in Fig. 5.5, for $K = 10$ and $T = 1$.

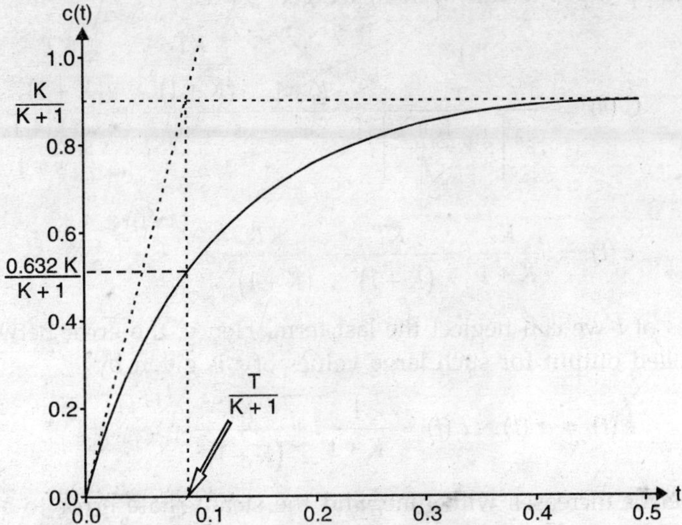

Figure 5.5. *Response of a first-order system to a unit step input*

It will be seen that the steady-state value of the response is given by

$$c_{ss} = \frac{K}{K+1} \quad \text{...(5.10)}$$

It is evidently less than one, the value of the input, but can be made closer to one by increasing *K*, the gain of the amplifier. In practice, *K* cannot be made infinitely large, since all practical amplifiers also generate internal noise, which increases with the gain. Furthermore, the linear model is valid only up to a certain range of inputs about the operating point. Hence, in such a system of this type, the steady-state error cannot be made zero, and is given by

$$e_{ss} = 1 - c_{ss} = \frac{1}{K+1} \quad \text{...(5.11)}$$

We shall now determine the time constant of the system, which can be defined as the time required by the response to reach 63.2 per cent of its steady-state value. Alternatively, the time constant may be defined as the time that the response would have taken to reach the steady-state if it had continued to increase at the initial rate. We have

$$\frac{dc}{dt} = \frac{K}{T} e^{\frac{-(K+1)t}{T}} \quad \text{...(5.12)}$$

so that the initial rate of increase is

$$\frac{dc}{dt}(0) = \frac{K}{T}, \quad \text{and} \quad \tau = \frac{K}{K+1} \div \frac{K}{T} = \frac{T}{K+1} \quad \text{...(5.13)}$$

Thus, we see that this system will track the unit step input with a steady-state error of $1/(K + 1)$ and will have a time constant $T/(K + 1)$, where *T* is the time constant of the open-loop system.

If we apply a ramp input to this system, we get

$$C(s) = \frac{\frac{K}{T}}{s^2\left[s+\frac{K+1}{T}\right]} = \frac{\frac{K}{K+1}}{s^2} - \frac{\frac{KT}{(K+1)^2}}{s} + \frac{\frac{KT}{(K+1)^2}}{s+\frac{K+1}{s+1}} \quad ...(5.14)$$

and
$$c(t) = \frac{K}{K+1}t - \frac{KT}{(K+1)^2} - \frac{KT}{(K+1)^2}e^{\frac{-(K+1)t}{T}} \quad ...(5.15)$$

For large values of t we can neglect the last term. Hence, the error between the reference input and the controlled output for such large values of t is given by

$$e(t) = r(t) - c(t) \approx \frac{1}{K+1}t + \frac{KT}{(K+1)^2} \quad ...(5.16)$$

Evidently, the error increases with time, and the steady-state error to a ramp input will approach infinity as t approaches infinity. This is an important property of such a system. We shall discuss this in more detail in Section 5.7.

DRILL PROBLEM 5.1

Determine the response to a unit step of the speed control system shown in Fig. 5.4 if K = 10 and T = 5. What are the values of the steady-state error and the time constant?

Ans. $c(t) = \frac{10}{11}\left(1 - e^{-\frac{5t}{11}}\right),\ e_{ss} = \frac{1}{11},\ \tau = \frac{5}{11}$ sec

5.4 RESPONSE OF A SECOND-ORDER SYSTEM

Consider the block diagram shown in Fig. 5.6. This represents a position-control system as discussed in Chapter 2 (Section 2.6).

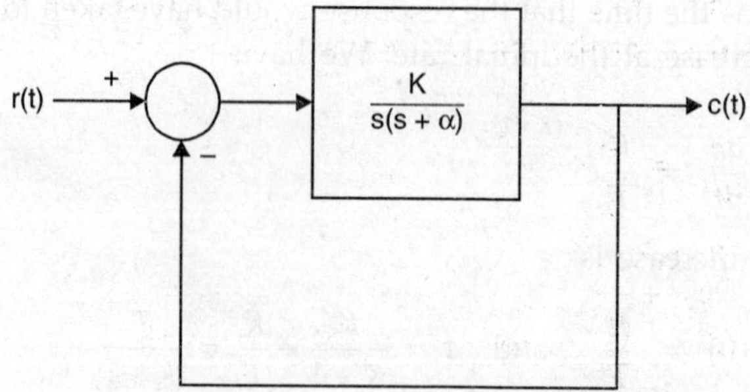

Figure 5.6. *A second-order system*

The transfer function is obtained as

$$\frac{C(s)}{R(s)} = \frac{K}{s^2 + \alpha s + K} \quad ...(5.17)$$

The denominator of the transfer function is a quadratic in *s*. Hence, the poles may either be real or complex. We shall first discuss the case of complex poles. Equation (5.17) may be rewritten as

$$\frac{C(s)}{R(s)} = \frac{\omega_n^2}{s^2 + 2\zeta\omega_n s + \omega_n^2} \quad ...(5.18)$$

where $\omega_n = \sqrt{K} \triangleq$ undamped natural frequency ...(5.19)

and $\zeta = \dfrac{\alpha}{2\omega_n} \triangleq$ damping ratio ...(5.20)

It should be noted that the poles will be complex only if $\zeta < 1$. Figure 5.7 shows the plot of the poles of the transfer function for this case.

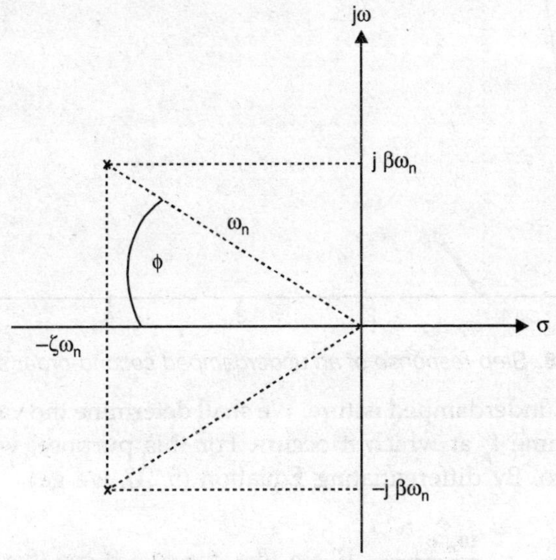

Figure 5.7. *Poles of the transfer function*

Define $\phi = \cos^{-1}\zeta$ ...(5.21)

and $\beta = \sin\phi = \sqrt{1-\zeta^2}$ ...(5.22)

as shown in Fig. 5.7.

We can now calculate the response of the closed-loop system to a unit step as below.

$$C(s) = \frac{\omega_n^2}{s\left(s^2 + 2\zeta\,\omega_n s + \omega_n^2\right)} = \frac{1}{s} - \frac{s + 2\zeta\,\omega_n}{s^2 + 2\zeta\,\omega_n s + \omega_n^2} \quad ...(5.23)$$

$$= \frac{1}{s} + \frac{(s+\zeta\omega_n) + \zeta\omega_n}{(s+\zeta\omega_n)^2 + (\beta\omega_n)^2}$$

Taking the inverse Laplace transform, we get

$$c(t) = 1 - e^{-\zeta\omega_n t}\left[\cos\beta\omega_n t + \frac{\zeta}{\beta}\sin\beta\omega_n t\right] \quad ...(5.24)$$

$$= 1 - \frac{1}{\beta}e^{-\zeta\omega_n t}\sin(\beta\omega_n t + \phi)$$

The response is shown in Fig. 5.8 for $K = 10$ and $\alpha = 2$, *i.e.*, $\omega_n = \sqrt{10}$ and $\zeta = \dfrac{1}{\sqrt{10}}$.

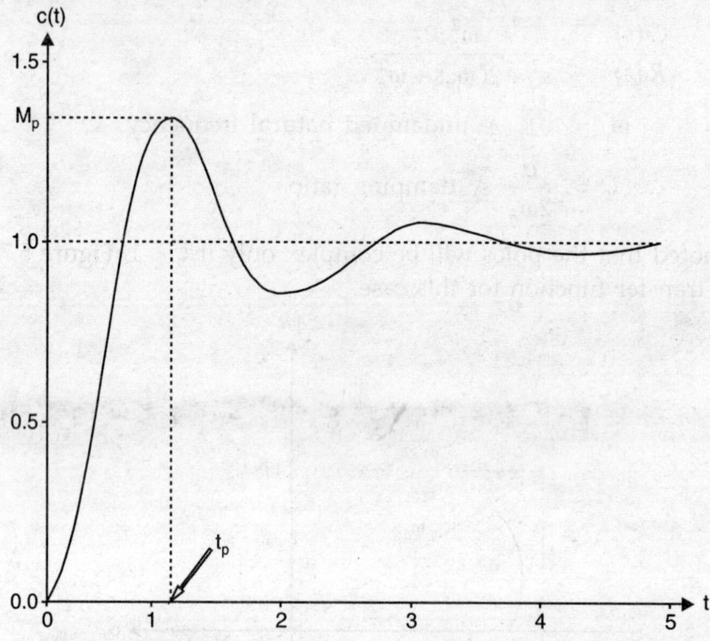

Figure 5.8. *Step response of an underdamped second-order system*

The response is of an underdamped nature. We shall determine the value M_p of the maximum response, as well as the time t_p at which it occurs. For this purpose, we must first determine dc/dt and equate it to zero. By differentiating Equation (5.24), we get

$$\frac{dc}{dt} = \frac{\omega_n e^{-\zeta\omega_n t}}{\beta} [\zeta \sin(\beta\omega_n t + \phi) - \beta \cos(\beta\omega_n t + \phi)] \qquad ...(5.25)$$

$$= \frac{\omega_n e^{-\zeta\omega_n t}}{\beta} \sin \beta\omega_n t$$

Alternatively, we may use the Laplace transform relationship

$$\frac{dc}{dt} = \mathcal{L}^{-1} [s\,C(s) - c(0)] = \mathcal{L}^{-1}\left[\frac{\omega_n^2}{s^2 + 2\zeta\omega_n s + \omega_n^2}\right] \qquad ...(5.26)$$

$$= \frac{\omega_n}{\beta} e^{-\zeta\omega_n t} \sin \beta\omega_n t$$

Equating dc/dt to zero, the smallest nonzero value of t gives us

$$t_p = \frac{\pi}{\beta\omega_n} = \frac{\pi}{\omega_n\sqrt{1-\zeta^2}} \qquad ...(5.27)$$

Substituting into Equation (5.24) we obtain the peak of the response as

$$M_p = 1 + e^{-\frac{\zeta\pi}{\beta}} = 1 + e^{-\frac{\pi}{\tan\phi}} = 1 + e^{-\frac{\pi\zeta}{\sqrt{1-\zeta^2}}} \qquad ...(5.28)$$

Expressing the difference $M_p - 1$ as a percentage, we get

$$\text{Maximum overshoot} = 100\, e^{\frac{-\pi\zeta}{\sqrt{1-\zeta^2}}} \text{ per cent} \qquad ...(5.29)$$

It follows that the maximum overshoot depends only on the value of the damping ratio ζ. We also note that the steady-state value of the response is given by

$$c_{ss} = \lim_{t\to\infty} c(t) = 1 \qquad ...(5.30)$$

so that the steady-state error is zero.

We could also have determined the steady-state error without inverse Laplace transformation by first obtaining.

$$E(s) = R(s) - C(s) = R(s)\cdot\left[1 - \frac{C(s)}{R(s)}\right] \qquad ...(5.31)$$

$$= \frac{1}{s}\left[1 - \frac{\omega_n^2}{s^2 + 2\zeta\omega_n s + \omega_n^2}\right] = \frac{s + 2\zeta\omega_n}{s^2 + 2\zeta\omega_n s + \omega_n^2}$$

Since $E(s)$ does not have any pole on the $j\omega$-axis or in the right half of the s-plane, we can apply the final value theorem to obtain

$$e_{ss} = \lim_{t\to\infty} c(t) = \lim_{s\to 0}\left[sE(s)\right] = 0 \qquad ...(5.32)$$

We shall now consider the effect of applying the unit ramp input to the system. Since $R(s) = \frac{1}{s^2}$, we get

$$C(s) = \frac{\omega_n^2}{s^2\left(s^2 + 2\zeta\omega_n s + \omega_n^2\right)} \qquad ...(5.33)$$

$$= \frac{1}{s^2} - \frac{2\zeta/\omega_n}{s} + \frac{(2\zeta s/\omega_n) - \left(1 - 4\zeta^2\right)}{s^2 + 2\zeta\omega_n s + w_n^2}$$

and

$$c(t) = t - \frac{2\zeta}{\omega_n} + e^{-\zeta\omega_n t}\left[\frac{2\zeta}{\omega_n}\cos\beta\omega_n t - \frac{1-2\zeta^2}{\beta\omega_n}\sin\beta\omega_n t\right] \qquad ...(5.34)$$

The response is shown in Fig. 5.9 for $K = 10$ and $\alpha = 2$, *i.e.*, $\omega_n = \sqrt{10}$ and $\zeta = \frac{1}{\sqrt{10}}$.

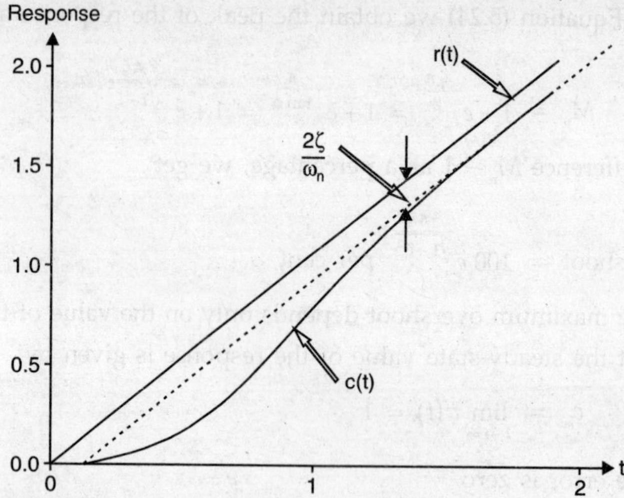

Figure 5.9. *Response of an underdamped second-order system to a unit ramp input*

It is seen that the output $c(t)$ follows the input $r(t)$ with the steady-state error

$$e_{ss} = \frac{2\zeta}{\omega_n} \quad \text{...(5.35)}$$

Again, the steady-state error could have been obtained directly from the transfer function using the final value theorem. This will be left as an exercise for the student.

The transfer function of a second-order system has real poles when ζ is greater than or equal to one. The system is said to be *critically damped* when ζ is equal to one, and *overdamped* when ζ is greater than one. For the critically damped case, the two real roots of the denominator are equal. The responses for the two cases are shown in Fig. 5.10. Where the poles are located at $s = -3$ for the former and at -2 and -4 for the latter. These responses are seen to be slower than for the underdamped case. The calculation of the responses will be left as an exercise for the student.

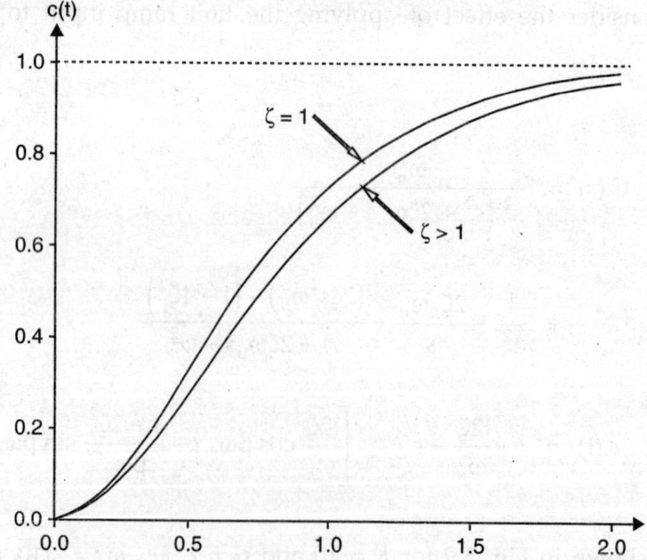

Figure 5.10. *Step responses of overdamped and critically damped second-order systems*

DRILL PROBLEM 5.2

Using the final value theorem, show that for a unit ramp input the steady-state error of the system described by the transfer function in Equation (5.18) is given by $2\zeta/\omega_n$.

DRILL PROBLEM 5.3

The transfer function of a second-order system is given by Equation (5.18), with $\zeta = 1$ *and* $\omega_n = 5$. *Determine its response to a unit step input.*

Ans. $c(t) = 1 - e^{-5t} - 5t\, e^{-5t}$

DRILL PROBLEM 5.4

Repeat the previous problem for $\zeta = 1.5$.

Ans. $c(t) = 1 - 1.17082\, e^{-13.0902t} + 0.17082\, e^{-1.90982t}$

DRILL PROBLEM 5.5

It is desired that the step response of a second-order system described by Equation (5.18) should have maximum overshoot of 10 per cent. What must be the value of the damping ratio in order to achieve this? [Hint: Use Equation (5.29)]

Ans. $\zeta = 0.5912$

5.5 PROPERTIES OF THE TRANSIENT RESPONSE

The nature of the transient response of a system depends upon the location of poles of the transfer function, which determine the natural frequencies. Consequently, the response of a system to a unit step is sufficient to bring out the main characteristics of the transient response. Various properties of the step response of a system are shown in Fig. 5.11. The following quantities are used for evaluating the transient response.

1. *Swiftness.* The speed with which the system responds to sudden changes is important. The common measures of the speed of response are: (*a*) the 10 to 90 per cent rise time, and (*b*) the time required to reach the first peak (for the case of underdamped systems).

   For an underdamped second-order system with no zeros, the time required to reach the first peak was derived in Equation (5.27) and is given by $t_p = \dfrac{\pi}{\omega_n\sqrt{1-\zeta^2}}$.

2. *Settling time.* This is defined as the time required for the response to stay within a specified tolerance of its final value. The tolerance is usually taken as either 2 or 5 per cent.

Considering only the exponentially decaying envelope of the response of the underdamped system, we get the following approximation for 2 per cent tolerance,

$$t_s \approx \frac{4}{\zeta \omega_n} \qquad \text{...(5.36)}$$

which is based on the relationship $e^{-4} \approx 0.02$.

3. *Maximum overshoot.* This is a measure of the oscillatory nature of the response of an underdamped system. For the second-order case (with no zeros), we had seen that the maximum overshoot in the step response depends only on the value of the damping ratio ζ. [Refer to Equation (5.29).]

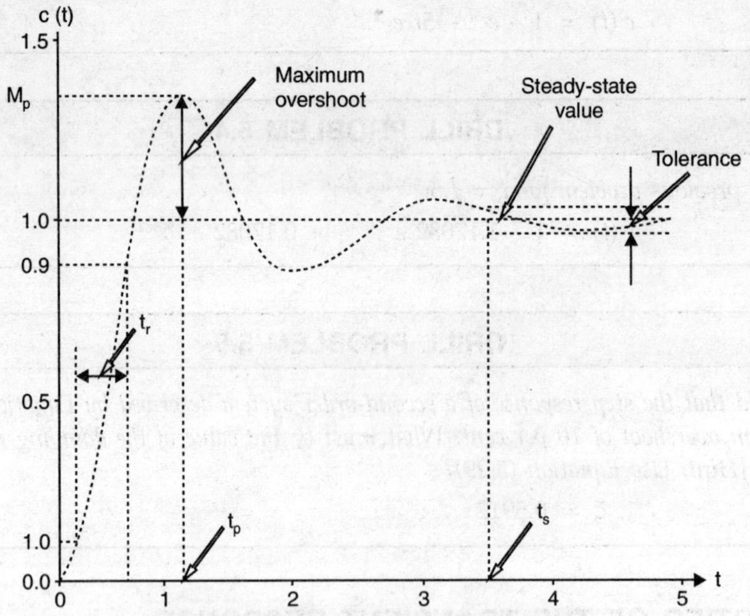

Figure 5.11. *Properties of the transient response*

In view of the relationship derived above, it is seen that for a second-order transfer function with no zeros, the various characteristics depend entirely upon the values of the parameters ω_n and ζ. Hence, we can expect only two of these to be independent.

For a system with more poles and zeros, the transient response is more difficult to calculate. Consequently, it is not possible to determine the various characteristics of the transient response directly. The only exception is the case when the system transfer function has a pair of complex poles much closer to the $j\omega$-axis in the s-plane than all the other poles and zeros. This is called the case of *dominant* poles, since the response is dominated by this pair of complex poles and the contributions of the other poles are negligible except for the initial part of the transient response. Such systems can be approximated by a second-order model consisting of these poles only. As we shall see later, in control system design it is often convenient to force the system to have a pair of dominant complex conjugate poles so that it may have desired transient response characteristics.

DRILL PROBLEM 5.6

It is desired that the system described in Equation (5.18) should have a maximum overshoot of 20 per cent, and the settling time must not exceed 2 seconds. Determine the values of ζ and ω_n that will just satisfy these specifications.

Ans. $\zeta = 0.4559, \quad \omega_n = 4.3864$

DRILL PROBLEM 5.7

A first-order system was described in Section 5.3. Show that the 10 to 90 per cent rise time t_r and the settling time t_s of this system are given by $t_r = 2.2\tau$ and $t_s = 4\tau$, where τ is the time constant of the system, given by T/(K + 1), as defined in Equation (5.13).

5.6 STEADY-STATE PERFORMANCE

Consider the closed-loop system shown in Fig. 5.12. The error is defined as

$$e(t) \triangleq r(t) - c(t) \qquad \text{...(5.37)}$$

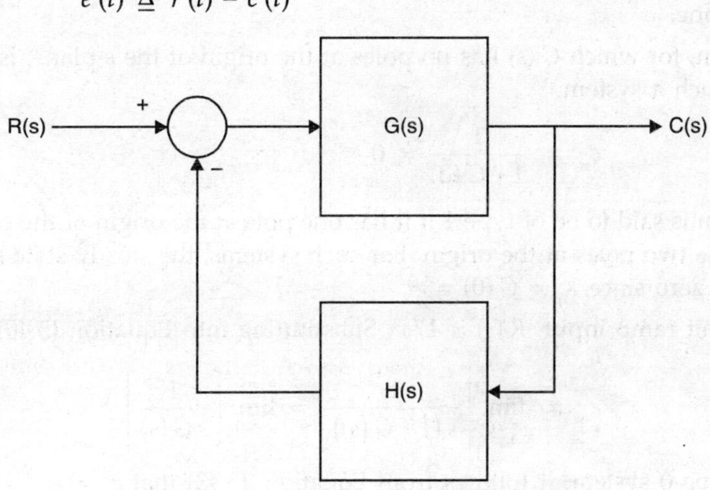

Figure 5.12. *A general closed-loop system*

Taking Laplace transforms of both sides

$$E(s) = R(s) - C(s) = R(s)\left[1 - \frac{C(s)}{R(s)}\right] \qquad \text{...(5.38)}$$

$$= R(s)\left[1 - \frac{G(s)}{1 + G(s)H(s)}\right] = \frac{1 + G(s)H(s) - G(s)}{1 + G(s)H(s)} R(s)$$

Applying the final-value theorem (assuming that we have a stable system) we get the following expression for the steady-state error.

$$e_{ss} = \lim_{t\to\infty} e(t) = \lim_{s\to 0}\left[s\,E(s)\right] \qquad \text{...(5.39)}$$

$$= \lim_{s\to 0}\left[\frac{1+G(s)\,H(s)-G(s)}{1+G(s)\,H(s)}\cdot s\,R(s)\right]$$

Equation (5.39) can be utilized for determining the steady-state error for the general case shown in Figure 5.12. We shall now consider the special case of unity feedback, that is, $H(s) = 1$, which is of considerable importance. This reduces Equation (5.39) to

$$e_{ss} = \lim_{s\to 0}\left[\frac{s\,R(s)}{1+G(s)}\right] \qquad \text{...(5.40)}$$

Let us now evaluate the steady-state error for some of the standard inputs for the unity-feedback case.

1. For a unit step input, $R(s) = 1/s$ and

$$e_{ss} = \lim_{s\to 0}\left[\frac{1}{1+G(s)}\right] = \frac{1}{1+G(0)} = \frac{1}{1+K_p} \qquad \text{...(5.41)}$$

   where $K_p = G(0)$ is called the position constant.

   Thus, $e_{ss} = 0$ if and only if $G(0) = \infty$, that is $G(s)$ has one or more poles at the origin of the *s*-plane.

   A system, for which $G(s)$ has no poles at the origin of the *s*-plane, is said to be of type 0. For such a system,

$$e_{ss} = \frac{1}{1+G(0)} \neq 0 \qquad \text{...(5.42)}$$

   A system is said to be of type 1 if it has one pole at the origin of the *s*-plane, and of type 2 if it has two poles at the origin. For such systems, the steady-state error to a unit step input is zero since $K_p = G(0) = \infty$.

2. For a unit ramp input, $R(s) = 1/s^2$. Substituting into Equation (5.40), we have

$$e_{ss} = \lim_{s\to 0}\left[\frac{1}{s\,(1+G(s))}\right] = \lim_{s\to 0}\left[\frac{1}{s\,G(s)}\right] \qquad \text{...(5.43)}$$

   For a type 0 system, it follows from Equation. (5.43) that $e_{ss} = \infty$.

   For a type 1 system, since $G(s)$ has a pole at the origin, $sG(s) \neq 0$. If we define

$$\lim_{s\to 0}\left[s\,G(s)\right] \;\Delta\; K_v,$$

   then we get $$e_{ss} = \frac{1}{K_v} \qquad \text{...(5.44)}$$

   It is customary to call K_v the *velocity constant* of the system since it is a measure of the steady-state error of a position-control system subject to a unit step change in velocity (equivalent to a unit ramp position input). The reason for this name is mainly historical, going back to the 1940s when the theory of control systems was applied to the development of gun control systems (or servo-mechanisms, as they were called in those days).

It follows that for a type 2 system, $K_v = \infty$. Consequently, the steady-state error to a ramp input will be zero for such a system.

3. For a unit parabolic input, $r(t) = \frac{1}{2}t^2$ and $R(s) = 1/s^3$. Hence, from Equation (5.40),

$$e_{ss} = \lim_{s \to 0} \left[\frac{1}{s^2\, G(s)} \right] \qquad \text{...(5.45)}$$

It is evident from Equation (5.45), the e_{ss} will be infinite unless $G(s)$ has two or more poles at the origin of the *s*-plane. For the latter case, we define

$$K_a = \lim_{s \to 0} \left[s^2\, G(s) \right] \qquad \text{...(5.46)}$$

so that the steady-state error is given by

$$e_{ss} = \frac{1}{K_a} \qquad \text{...(5.47)}$$

K_a is called the *acceleration constant* for the system.

The steady-state error to a parabolic input will be zero for systems in which $G(s)$ has more than two poles at the origin, since for such systems K_a will be infinite.

A simple interpretation of these results is obtained if we appreciate that every pole of $G(s)$ at the origin of the *s*-plane corresponds to an integration of the actuating error in the block diagram shown in Fig. 5.12. Hence, if we want a constant output with zero actuating error, we must have at least one integration in the forward path. Similarly, if we want a ramp output with zero actuating error, we must have at least two integrations in the forward path.

5.7 STEADY-STATE ERROR IN TERMS OF THE TRANSFER FUNCTION OF THE CLOSED-LOOP SYSTEM

The steady-state error is often more conveniently obtained from the transfer function of the closed-loop system. Define

$$T(s) \triangleq \frac{C(s)}{R(s)} \qquad \text{...(5.48)}$$

Consequently, the Laplace transform of the error is obtained as

$$E(s) \triangleq R(s) - C(s) = R(s)\,[1 - T(s)] \qquad \text{...(5.49)}$$

and the steady-state error is given by

$$e_{ss} = \lim_{s \to 0} \left[s\,E(s) \right] = \lim_{s \to 0} \left[s\,R(s)\left[1 - T(s)\right] \right] \qquad \text{...(5.50)}$$

We shall now consider the special inputs.

For a **unit step input,** we have

$$sR(s) = 1.$$

Hence, the steady-state error is given by

$$e_{ss} = \lim_{s \to 0} \left[1 - T(s) \right] = 1 - T(0) \qquad \text{...(5.51)}$$

Thus, $e_{ss} = 0$ if and only if $T(0) = 1$, that is the d.c. gain of the closed-loop system is one. This also follows from the physical interpretation of a constant as a *dc* or zero-frequency input so that the input and the steady-state output will be equal only if the *dc* gain is one.

For a unit **ramp input,** $sR(s) = 1/s$, so that the steady-state error is given by

$$e_{ss} = \lim_{s \to 0} \left[\frac{1 - T(s)}{s} \right] \qquad \text{...(5.52)}$$

It is evident from Equation (5.52), that $e_{ss} = \infty$, unless $T(0) = 1$. If $T(0) = 1$, we may use L'Hopital's rule to obtain

$$e_{ss} = \lim_{s \to 0} \left[-\frac{dT(s)}{ds} \right] \qquad \text{...(5.53)}$$

Let

$$T(s) = \frac{b_0 s^m + b_1 s^{m-1} + \cdots + b_{m-1} s + b_m}{s^n + a_1 s^{n-1} + \cdots + a_{n-1} s + a_n}$$

$$= \frac{K(s - z_1)(s - z_2) \cdots (s - z_m)}{(s - p_1)(s - p_2) \cdots (s - p_n)} \qquad \text{...(5.54)}$$

where $m \le n$.

Since differentiation of $T(s)$ is rather complicated, we shall use the natural logarithm of $T(s)$. First, we note that

$$\ln [T(s)] = \ln K + \sum_{i=1}^{m} \ln (s - z_i) - \sum_{i=1}^{n} \ln (s - p_i) \qquad \text{...(5.55)}$$

Consequently,

$$\frac{d \ln T(s)}{ds} = \sum_{i=1}^{m} \frac{1}{s - z_i} - \sum_{i=1}^{n} \frac{1}{s - p_i} \qquad \text{...(5.56)}$$

Since $T(0) = 1$, we obtain the following expression for the steady-state error.

$$e_{ss} = \lim_{s \to 0} \left[-\frac{dT(s)}{ds} \right] = \lim_{s \to 0} \left[-\frac{1}{T(s)} \cdot \frac{dT(s)}{ds} \right]$$

$$= \lim_{s \to 0} \left[-\frac{d \ln T(s)}{ds} \right] = \sum_{i=1}^{m} \frac{1}{z_i} - \sum_{i=1}^{n} \frac{1}{p_i} \qquad \text{...(5.57)}$$

Thus, we have obtained a very simple expression for the steady-state error for a unit ramp input which is valid only if $T(0) = 1$. This result is useful even if the locations of the poles and zeros are not known, that is, the transfer function is given in the form of the ratio of two polynomials. In that case, it can be shown that

$$\sum_{i=1}^{m} \frac{1}{z_i} = -\frac{b_{m-1}}{b_m} \qquad \text{...(5.58)}$$

and

$$\sum_{i=1}^{n} \frac{1}{p_i} = -\frac{a_{n-1}}{a_n} \qquad \text{...(5.59)}$$

Equations (5.58) and (5.59) can be proved using some basic properties of polynomials. The proof will be left as an exercise for the student. [Also see Problem 9 at the end of this chapter.]

For a unit **parabolic** input, $sR(s) = 1/s^2$ and the steady-state error is given by

$$e_{ss} = \lim_{s \to 0} \left[\frac{1 - T(s)}{s^2}\right] \qquad \text{...(5.60)}$$

Hence, as for the case of the ramp input, the steady-state error will be infinite unless

$$T(0) = 1$$

and
$$\lim_{s \to 0} \left[\frac{d \ln T(s)}{ds}\right] = 0$$

The latter condition is satisfied if

$$\sum_{i=1}^{m} \frac{1}{z_i} = \sum_{i=1}^{n} \frac{1}{p_i}$$

If both of these conditions are satisfied then by applying L'Hopital's rule twice, we get

$$e_{ss} = \lim_{s \to 0} \left[-\frac{1}{2} \frac{d^2 \ln T(s)}{ds^2}\right] \qquad \text{...(5.61)}$$

To evaluate this limit, we shall return to Equation (5.55) and differentiate it twice. Hence, we get

$$\frac{d^2 \ln T(s)}{ds^2} = -\sum_{i=0}^{m} \frac{1}{(s - z_i)^2} + \sum_{i=1}^{n} \frac{1}{(s + p_i)^2} \qquad \text{...(5.62)}$$

and
$$e_{ss} = \frac{1}{K_a} = \frac{1}{2}\left[\sum_{i=1}^{m} \frac{1}{z_i^2} - \sum_{i=1}^{n} \frac{1}{p_i^2}\right] \qquad \text{...(5.63)}$$

Again, it is not necessary to determine the location of all poles and zeros to evaluate the expression for e_{ss} in Equation (5.62). [See Problem 10 at the end of this chapter.]

EXAMPLE 5.1

Consider a second-order system for which

$$T(s) = \frac{\omega_n^2}{s^2 + 2\zeta\,\omega_n\, s + \omega_n^2}$$

Here, $T(0) = 1$. Hence, the steady-state error to a step input will be zero. We shall now calculate the steady-state error to a unit ramp input. First, we note that there are no zeros, but a pair of poles which will be complex if $\zeta < 1$, and real otherwise. Using the result expressed in Equations (5.57) to (5.59), we get

$$e_{ss} = \frac{1}{K_v} = -\sum_{i=1}^{2} \frac{1}{p_i} = \frac{2\zeta\omega_n}{\omega_n^2} = \frac{2\zeta}{\omega_n}$$

EXAMPLE 5.2

Consider a closed-loop system described by the transfer function

$$T(s) = \frac{10(s+5)}{(s+2)(s^2+5s+25)}$$

Here again, $T(0) = 1$, so that the steady-state error to a step input will be zero. The steady-state error to a unit ramp input will be given by

$$e_{ss} = \frac{1}{-5} - \left[\frac{1}{-2} - \frac{5}{25}\right] = \frac{1}{2}$$

DRILL PROBLEM 5.8

The transfer function of a closed-loop system used for speed control is given below, where the parameters K and z are adjustable.

$$T(s) = \frac{K(s+z)}{s^2+8s+65}$$

Determine the values of K and z in order that the steady-state error to a ramp input may be zero. What will be the steady-state error of this system to the unit parabolic input?

Ans. $K = 8, \quad z = \frac{65}{8}, \quad e_{ss} = \frac{1}{K_a} = \frac{1}{65}.$

5.8 STEADY-STATE ERROR FROM THE CLOSED-LOOP TRANSFER FUNCTION (AN ALTERNATIVE APPROACH)

In the previous section, we derived expressions for steady-state errors to various inputs in terms of the locations of the poles and zeros of the transfer function of the closed-loop system. These are often useful for design, as illustrated by Drill Problem 5.8. We shall now study another approach that provides some additional insight. Recall that the steady-state error is given by

$$e_{ss} = \lim_{s\to 0} \left[s R(s)[1-T(s)]\right] \quad ...(5.64)$$

Let $T(s)$ be as given below:

$$T(s) = \frac{b_0 s^n + b_1 s^{n-1} + \cdots + b_{n-1} s + b_n}{s^n + a_1 s^{n-1} + \cdots + a_{n-1} s + a_n} \quad ...(5.65)$$

where some of the coefficients in the numerator may be zero if the degree of the numerator is lower than that of the denominator. Consequently,

$$1 - T(s) = \frac{(1-b_0) s^n + (a_1 - b_1) s^{n-1} + \cdots + (a_{n-1} - b_{n-1}) s + (a_n - b_n)}{s^n + a_1 s^{n-1} + \cdots + a_{n-1} s + a_n} \quad ...(5.66)$$

For most practical cases, the transfer function of the closed-loop system, $T(s)$, will not have poles at the origin, otherwise the system will be unstable. Hence, we must have $a_n \neq 0$. Also, for a *type 0* system, $a_n \neq b_n$, so that the steady-state error to a *unit step input* will be

$$e_{ss} = \frac{a_n - b_n}{a_n} \qquad \text{...(5.67)}$$

For a *type* 1 system, $a_n = b_n$, but $a_{n-1} \neq b_{n-1}$. Therefore, the steady-state error to a *unit ramp input* will be given by

$$e_{ss} = \frac{a_{n-1} - b_{n-1}}{a_n} \qquad \text{...(5.68)}$$

Similarly, for a *type* 2 system, $a_n = b_n$, $a_{n-1} = b_{n-1}$, but $a_{n-2} \neq b_{n-2}$. Thus, the steady-state error to a *unit parabolic input* will be

$$e_{ss} = \frac{a_{n-2} - b_{n-2}}{a_n} \qquad \text{...(5.69)}$$

5.9 INTEGRAL PERFORMANCE CRITERIA

Often it is desirable to assess the quality of a control system by evaluating a performance index that can either be calculated or measured. An important application is in the area of adaptive control, where one can adjust certain parameters that will minimize the value of this index, also known as the *'cost function'*. A number of such cost functions are used by control engineers.

The most popular cost function is the integral of the square of the error (ISE) defined as

$$J_1 = \int_0^T e^2(t)\, dt \qquad \text{...(5.70)}$$

where $e(t) = r(t) - c(t)$ is the error between the desired output and the actual output. The upper limit T is often taken somewhat larger so that the settling time of the system is in order and it can be measured experimentally. In most cases, the computation of J_1 is convenient if the upper limit T is taken as infinity, since it is then possible to apply Parseval's theorem and calculate J_1 from the poles of the Laplace transform of $E(s)$, since

$$\int_0^\infty e^2(t)\, dt = \frac{1}{2\pi j} \int_{-j\infty}^{j\infty} E(s)\, E(-s)\, ds$$

$$= \text{sum of the residues of } E(s)\, E(-s) \text{ at all of its poles in the left half of the } s\text{-plane} \qquad \text{...(5.71)}$$

Another criterion that is quite popular is the integral of the absolute value of the error (IAE), defined as

$$J_2 = \int_0^T |e(t)|\, dt \qquad \text{...(5.72)}$$

The main advantage of this criterion is the ease with which it can be measured using an analog computer.

In many other cases, there is some advantage in providing a higher weighting to errors occurring later in the response than during the initial part. One criterion of this type is the integral of the time multiplied by the square of the error (ITSE), defined as

$$J_3 = \int_0^T t\, e^2(t)\, dt \qquad \text{...(5.73)}$$

Another is the time multiplied by the absolute value of the error (ITAE),

$$J_4 = \int_0^T t\, |e(t)|\, dt \qquad \text{...(5.74)}$$

As expected, the optimum response will be different depending upon the criterion used. For example, it was seen earlier that in the case of second-order transfer functions with no zeros, the nature of the step response depends entirely on the value of the damping ratio ζ. It can be shown theoretically that the ISE criterion gives the optimum value of the damping ratio as 0.5, while the ITSE cost function is minimized for $\zeta = 0.6$. On the other hand, the ITAE cost function is minimized for $\zeta = 0.7$. The values of the cost functions calculated for different values of ζ for step function inputs are shown in Fig. 5.13. It is seen that ITAE criterion is the most selective of the four criteria considered here.

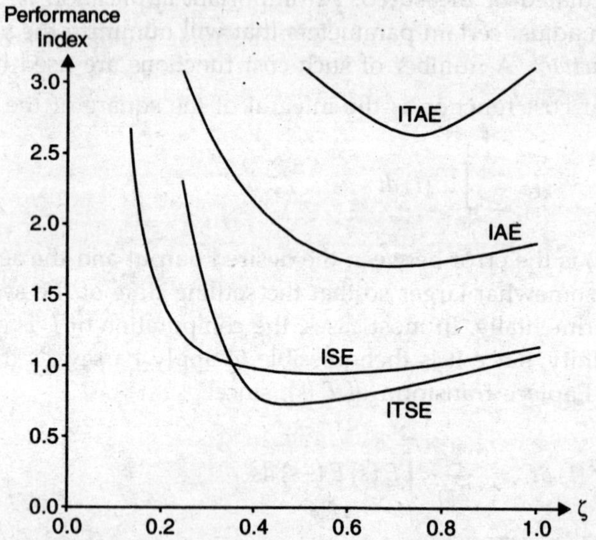

Figure 5.13. *Performance indices for a second-order system*

In practice, a performance index that is not very sensitive may be desirable for systems where there is some uncertainty about the values of the parameters of the model. Since the ISE criterion is mathematically the most convenient to calculate, it is often chosen for design of control systems although some of the other criteria may look more attractive in theory.

DRILL PROBLEM 5.9

Show that the ISE criterion applied to the step response of the second-order system with transfer function $\dfrac{\omega_n^2}{s^2 + 2\zeta\,\omega_n s + \omega_n^2}$ *is obtained as*

$$J_1 = \int_0^\infty e^2(t)\,dt = \frac{1}{2w_n}\left[\frac{1}{2\zeta} + 2\zeta\right] \qquad \text{...(5.75)}$$

Hence, show that, for a fixed value of ω_n, J_1 *is minimized by making* $\zeta = 0.5$.
*[**Hint:** Use the expression for c (t) in Equation (5.24) to obtain e (t) = 1 – c (t)].*

SUMMARY

- Some ideas for analyzing the performance of a control system have been discussed in this chapter. These relate to both the transient and the steady-state response to certain standard inputs. The test inputs most commonly used are the unit step, the unit ramp, and the unit parabola. For a position control system, these are equivalent to step changes in (1) position, (2) velocity, and (3) acceleration.
- The transient performance of a system can be quantified in terms of the following properties of the response to a unit step: (1) speed of response (either 10 to 90 per cent rise time, or the time to reach the first peak); (2) settling time, defined as the time after which the output remains within 2 per cent of its steady-state value; and (3) the maximum overshoot. Calculation of these parameters for a second-order system was studied and is relatively straightforward. For high-order systems it gets rather tedious. Often a high-order system can be approximated by a second-order model if it has a pair of complex poles much closer to the $j\omega$-axis than the other poles and zeros. In such cases, one may obtain an explicit relationship between the parameters and the pole locations.
- The steady-state error to the test inputs can be calculated if the transfer function of the forward and feedback paths are known. Expressions for the steady-state error to unit step, ramp and parabolic inputs were also derived for the case when the transfer function of the closed-loop system is given. These are useful for design purposes.
- Finally, some integral performance criteria were discussed. These are used as numerical measures of the suitability of the transient response. In each case, some function of the error in the response of the system to a unit step input, is integrated from $t = 0$ to $t = \infty$. The most popular among these is the integral of the square error (ISE) criterion mainly because it is conveniently calculated using Parseval's theorem.

Problems

1. The open-loop transfer function of a unity feedback system is

$$\frac{K}{s\,(s+2)}$$

It is specified that the response of the system to a unit step input should have a maximum overshoot of 10 per cent, and the settling time should be less than one second.

(*a*) Is it possible to satisfy both the specifications simultaneously by adjusting K?

(*b*) If not, determine the value of K that will satisfy the first specification. What will be the settling time and the time to reach the first peak for this case?

2. Find the steady-state error to (*i*) a unit step input, (*ii*) a unit ramp input, and (*iii*) a unit parabolic input ($r = ½t^2$) for unity-feedback systems that have the following forward transfer functions:

(*a*) $G(s) = \dfrac{20}{s(s+2)(s^2+2s+2)}$ (*b*) $G(s) = \dfrac{108}{s^2(s+4)(s^2+3s+12)}$

3. For each of the systems described in Problem 2, determine the steady-state error for each of the following inputs, (*i*) $r(t) = 8 + 3t$, and (*ii*) $r(t) = 2 + 5t + 2t^2$.

4. The block diagram of a position-control system with velocity feedback is shown in Fig. P5.4. Determine the value of α so that the step response has a maximum overshoot of 10 per cent. What is the steady-state error to a unit ramp input for this value of α?

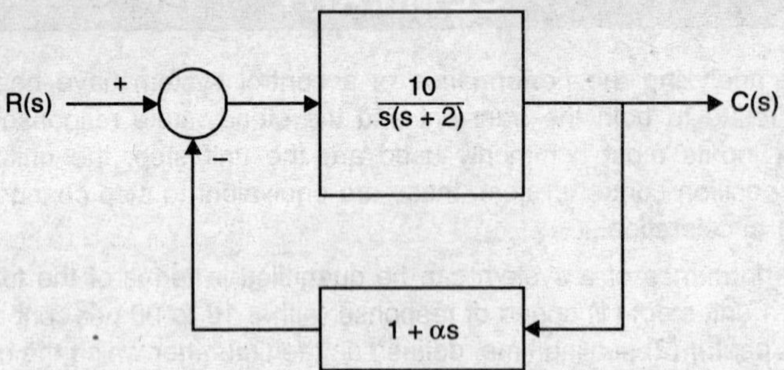

Figure P5.4. *Position-control system with velocity feedback*

5. A position-control system with velocity feedback damping is shown in Fig. P5.5.

(*a*) Determine the settling time and maximum overshoot in response to a unit step input as well as the steady-state error to a unit ramp input in the absence of velocity feedback, *i.e.*, for $\alpha = 0$.

(*b*) Repeat if $\alpha = 2$.

(*c*) Determine the value of α so that the damping ratio of the closed-loop system transfer function may be increased to 0.6. What are the values of the settling time, maximum overshoot, and the steady-state error to a unit ramp input in this case?

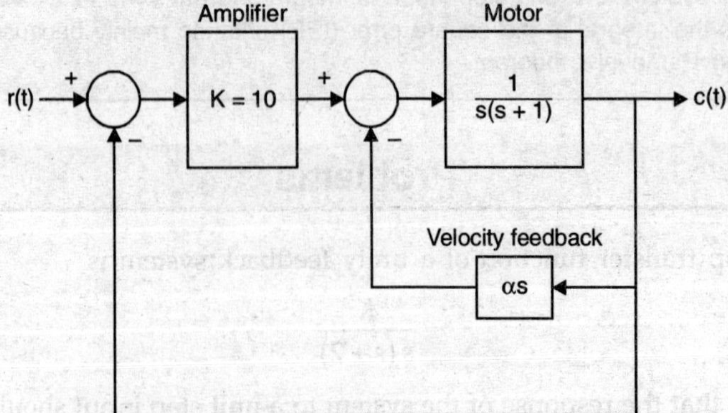

Figure P5.5. *Position-control system with velocity damping*

6. The block diagram of the attitude control system for a missile is shown in Fig. P5.6. Determine the values of K_1 and K_2 in order that the closed-loop system may have undamped natural frequency $\omega_n = 10$ rad/s and damping ratio $\zeta = 0.5$. What will be the settling time and maximum overshoot for a unit step input?

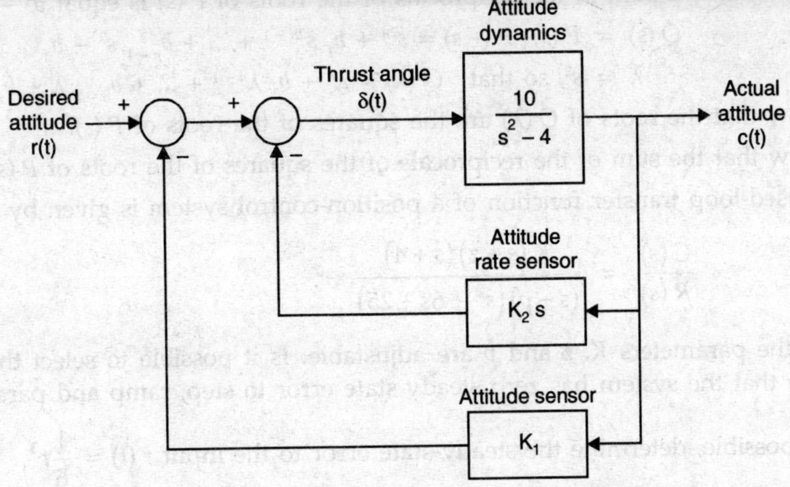

Figure P5.6. *Attitude control system for a missile*

7. The transfer function of a second-order closed-loop system with proportional and error-rate feedback is given by

$$\frac{C(s)}{R(s)} = \frac{K(s+z)}{s^2 + 4s + 8}$$

where the parameters K and z are adjustable.

(*a*) If $r(t) = t$, determine the values of K and z so that the steady-state error is zero.

(*b*) For these values of K and z, determine the steady-state error to the input $r(t) = \frac{1}{2}t^2$.

(*c*) Determine the integral of the square of the error if a unit step input is applied to the system, with the values of K and z determined in part (*a*).

8. Consider the position-control system with velocity feedback shown in Fig. P5.8. Find the values of K and α so that the resulting system satisfies the following conditions:

(*a*) The steady-state error to a unit ramp input is 0.1.

(*b*) The damping ratio of the poles of the closed-loop system is 0.5.

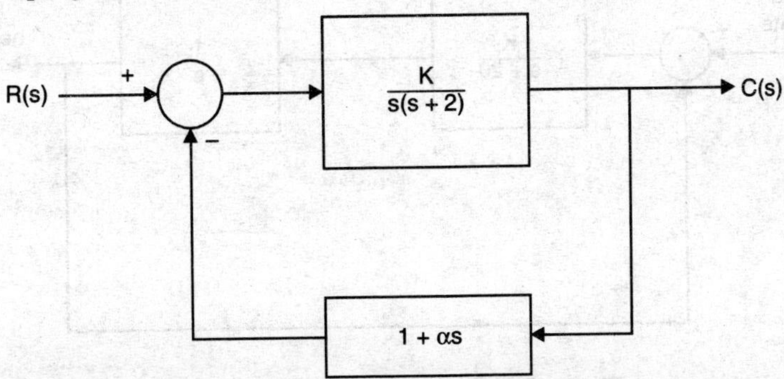

Figure P5.8. *A position-control system with velocity-feedback*

9. Consider the polynomial

$$P(s) = s^n + a_1 s^{n-1} + \ldots + a_{n-1} s + a_n$$

(*a*) Show that the sum of the roots of $P(s)$ is equal to $-a_1$.

(*b*) Show that the sum of the reciprocals of the roots of $P(s)$ is equal to $-a_{n-1}/a_n$.

10. Let $\quad Q(s) = P(s) \cdot P(-s) = s^{2n} + b_1 s^{2n-1} + \ldots + b_{n-1} s^2 + b_n$

Define $\quad \lambda = s^2$, so that $Q(\lambda) = \lambda^n + b_1 \lambda^{n-1} + \ldots + b_{n-1} \lambda + b_n$

(*a*) Show that the roots of $Q(\lambda)$ are the squares of the roots of $P(s)$.

(*b*) Show that the sum of the reciprocals of the squares of the roots of $P(s)$ is $-b_{n-1}/b_n$.

11. The closed-loop transfer function of a position-control system is given by

$$\frac{C(s)}{R(s)} = \frac{K(s+z)(s+4)}{(s+p)(s^2+6s+25)}$$

where the parameters K, *z* and *p* are adjustable. Is it possible to select them in such a manner that the system has zero steady-state error to step, ramp and parabolic inputs?

If it is possible, determine the steady-state error to the input $r(t) = \frac{1}{6}t^3$.

12. The linearized dynamic model relating changes in the angle of attack of a ship to changes in the angular position of the rudder is given by the transfer function.

$$\frac{K(1+2.5s)}{s(1+4s)(1+10s)}$$

Assume that this represents the forward path of a unity-feedback system and $K = 2$.

(*a*) Calculate the response to a unit step input.

(*b*) Calculate the integral of the square of the error for a unit step input.

(*c*) Determine the steady-state error to a unit ramp input.

13. Repeat Problem 12, with velocity feedback added so that the transfer function of the feedback path is changed from 1 to $(1 + 0.1s)$. In this case, the error is defined as the difference between the reference input and the actual output.

14. The block diagram of an electronic pacemaker is given below where $K = 400$.

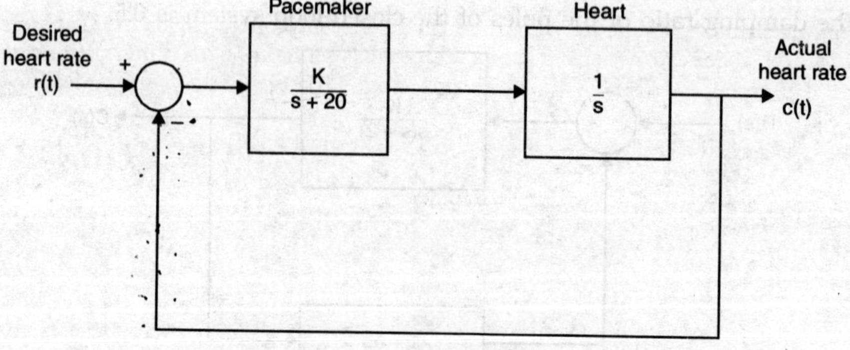

Figure P5.14. *An electronic pacemaker*

(*a*) Calculate the output $c(t)$ for a unit step input.

(*b*) Calculate the integral of the square of the error between the input and the output.

(*c*) Determine the steady-state error for a unit ramp input.

(*d*) Determine the value of K for which the steady-state error to a unit ramp will be 0.02.

15. Repeat Problem 14 if the transfer function of the feedback path is changed from $H(s) = 1$ to $H(s) = 1 + 0.12s$.

16. It is often possible to approximate a high-order system by a model of lower order. For example, consider a system described by the third-order transfer function.

$$G(s) = \frac{s+4}{s^3 + 6s^2 + 11s + 6}$$

Compare the step response of this system with that of the second-order approximation given by

$$G(s) = \frac{4/3}{s^2 + 3s + 2}.$$

CHAPTER 6

Stability of Linear Systems

6.1 INTRODUCTION

To be useful, a control system must be stable. A stable system may be defined as one that will have a bounded response for all possible bounded inputs. Here, we shall be concerned mainly with linear time-invariant systems.

The response of such a system to any input consists of two components: (*i*) the natural response and (2) the forced response. The natural response of the system will remain bounded if and only if all the poles of its transfer function are either in the left half of the *s*-plane or simple (i.e., not repeated) if on the $j\omega$-axis. On the other hand, for the forced response to be bounded for any input, it is necessary that all poles of the transfer function be strictly in the left half of the *s*-plane. This follows from the fact that if the transfer function has either a pole at the origin of the *s*-plane, or at $\pm j\omega_0$ on the imaginary axis, then application of a step input for the first case or the input $A \cos \omega_0 t$ for the second will cause the forced response to be unbounded. Thus, we may state that *a linear system will be stable if and only if all the poles of its transfer function are located on the left of the $j\omega$-axis.*

Some alternative definitions of stability, which are equivalent to the above, are given in Equations (6.1) and (6.2) in terms of the impulse response of the system. These can be easily proved due to the fact that the impulse response, $w(t)$, is the inverse Laplace transform of the transfer function, $G(s)$.

$$\lim_{t \to \infty} w(t) = 0 \qquad ...(6.1)$$

$$\int_0^\infty w^2(t)\, dt < \infty \qquad ...(6.2)$$

The most direct approach for investigating the stability of a linear system is to determine the location of the poles of its transfer function. Often, this is not very convenient as it requires evaluating the roots of a polynomial, the degree of which may be high. An analytical method for evaluating the roots of a polynomial of degree greater than five is not known. In the general case, therefore, one must use numerical search techniques to find the roots. These are often slow, and do not always give accurate results.

Fortunately, it is not necessary to determine the actual location of the poles of the transfer function for investigating the stability of a linear system. We only need to find out if the number of poles in the right half of the s-plane is zero or not. A procedure for determining the number of roots of a polynomial $P(s)$ in the right half of the s-plane without evaluating the roots will be discussed in the next section.[1]

6.2 THE ROUTH-HURWITZ CRITERION

The Routh-Hurwitz criterion was developed independently by A. Hurwitz (1895) in Germany and E.J. Routh (1892) in the United States. Let the characteristic polynomial (the denominator of the transfer function after cancellation of common factors with the numerator) be given by

$$\Delta(s) = a_0 s^n + a_1 s^{n-1} + \ldots + a_{n-1} s + a_n \qquad \text{...(6.3)}$$

Then the Routh table is obtained as follows:

$$\begin{array}{c|cccc} s^n & a_0 & a_2 & a_4 & \cdots \\ s^{n-1} & a_1 & a_3 & a_5 & \cdots \\ s^{n-2} & b_1 & b_3 & b_5 & \cdots \\ s^{n-3} & c_1 & c_3 & c_5 & \cdots \\ \cdot & & & & \\ \cdot & & & & \\ \cdot & & & & \\ s^0 & h_1 & & & \end{array}$$

where the first two rows are obtained from the coefficients of $\Delta(s)$. The elements of the following rows are obtained as shown below:

$$b_1 = \frac{a_1 a_2 - a_0 a_3}{a_1} \qquad \text{...(6.4)}$$

$$b_3 = \frac{a_1 a_4 - a_0 a_5}{a_1} \qquad \text{...(6.5)}$$

$$\vdots$$

$$c_1 = \frac{b_1 a_3 - a_1 b_3}{b_1} \qquad \text{...(6.6)}$$

and so on.

In preparing the Routh table for a given polynomial, as suggested above, some of the elements may not exist. In calculating the entries in the line that follows, these elements are considered to be zero. The procedure will be clearer from the examples that follow.

---

[1] It may be noted that if any coefficient of the polynomial is zero or negative, there must be one or more roots with zero or negative real parts.

The Routh-Hurwitz criterion states that the number or roots with positive real parts is equal to the number of changes in sign in the first column of the Routh table.

Consequently, the system is stable if and only if there are no sign changes in the first column of the table.

EXAMPLE 6.1

$$\Delta(s) = s^4 + 5s^3 + 20s^2 + 40s + 50 \quad ...(6.7)$$

The Routh table is as follows:

s^4	1	20	50
s^3	5	40	
s^2	12	50	
s^1	230/12		
s^0	50		

No sign changes in the first column indicate no root in the right half of the *s*-plane, and hence a stable system.

EXAMPLE 6.2

$$\Delta(s) = s^3 + s^2 + 2s + 24 \quad ...(6.8)$$

The Routh table is as follows:

s^3	1	2
s^2	1	24
s^1	−22	
s^0	24	

Two sign changes in the first column indicate two roots in the right half of the *s*-plane. Hence, this is the characteristic polynomial of an unstable system.

DRILL PROBLEM 6.1

Utilize the Routh table to determine the number of roots of each of the following polynomials in the right half of the s-plane.

(a) $s^3 + 2s^2 + 4s + 9$

(b) $s^4 + 6s^3 + 23s^2 + 40s + 50$

(c) $5s^6 + 8s^5 + 12s^4 + 20s^3 + 100s^2 + 150s + 200$

Ans. (*a*) Two, (*b*) none, (*c*) two.

6.3 SPECIAL CASES

Two special difficulties may arise while obtaining the Routh table for a given characteristic polynomial.

CASE I. If an element in the first column turns out to be zero, it should be replaced by a small positive number, ε, in order to complete the table. The sign of elements of the first column is then examined as ε approaches zero.

EXAMPLE 6.3

$$\Delta(s) = s^5 + 2s^4 + 3s^3 + 6s^2 + 10s + 15 \quad \text{...(6.9)}$$

The Routh table is shown below, where zero in the third row is replaced by ε.

s^5	1	3	10
s^4	2	6	15
s^3	ε	$5/2$	
s^2	$6 - 5/\varepsilon$	15	
s^1	$\dfrac{30\varepsilon - 25 - 30\varepsilon^2}{12\varepsilon - 10}$		
s^0	15		

As ε approaches zero the first column of the table may be simplified to obtain

s^5	1
s^4	2
s^3	ε
s^2	$-5/\varepsilon$
s^1	2.5
s^0	15

and two sign changes indicate two roots in the right half of the *s*-plane.

CASE II. Sometimes we may find that an entire row of the Routh table is zero. This indicates the presence of some roots that are negative of each other. In such cases, we should form an auxiliary polynomial from the row preceding the zero row. This auxiliary polynomial has only alternate powers of *s*, starting with the highest power indicated by the power of the leftmost column of the row. The auxiliary polynomial is a factor of the characteristic polynomial. The number of roots of the characteristic polynomial in the right half of the *s*-plane will be the sum of the number of right-half-plane roots of the auxiliary polynomial and the number determined from the Routh table of the lower-order polynomial obtained by dividing the characteristic polynomial by the auxiliary polynomial. The following example will elucidate the procedure.

EXAMPLE 6.4

$$\Delta(s) = s^5 + 6s^4 + 15s^3 + 30s^2 + 44s + 24 \quad \text{...(6.10)}$$

The Routh table is as follows:

s^5	1	15	44
s^4	6	30	24
s^3	10	40	
s^2	6	24	
s^1	0	← zero row	

The auxiliary row gives the equation

$$6s^2 + 24 = 0 \quad \text{...(6.11)}$$

Hence, $s^2 + 4$ is a factor of the characteristic polynomial, and we get

$$q(s) = \frac{\Delta(s)}{s^2 + 4} = s^3 + 6s^2 + 11s + 6 \quad \text{...(6.12)}$$

The Routh table for $q(s)$ is as follows:

s^3	1	11
s^2	6	6
s^1	10	
s^0	6	

Since this shows no sign changes in the first column, it follows that all roots of $\Delta(s)$ are in the left half of the s-plane, with one pair of roots $s = \pm j2$ on the imaginary axis. Consequently, $\Delta(s)$ is the characteristic polynomial of an oscillatory system. According to our definition, this is an unstable system.

EXAMPLE 6.5

$$\Delta(s) = s^6 + 4s^5 + 12s^4 + 16s^3 + 41s^2 + 36s + 72 \quad \text{...(6.13)}$$

The Routh table is as follows:

s^6	1	12	41	72
s^5	4	16	36	
s^4	8	32	72	
s^3	0	0		

The auxiliary polynomial is

$$p(s) = 8s^4 + 32s^2 + 72 = 8\left(s^2 + \sqrt{2}s + 3\right)\left(s^2 - \sqrt{2}s + 3\right) \quad \text{...(6.14)}$$

which has two roots in the right half of the s-plane. Also,

$$q(s) = \frac{\Delta(s)}{p(s)} = \frac{s^2 + 4s + 8}{8} \quad \text{...(6.15)}$$

The Routh table for $q(s)$ is as follows:

s^2	1	8
s^1	4	
s^0	8	

Hence, $\Delta(s)$ has two roots in the right half of the s-plane, and is the characteristic polynomial of an unstable system.

DRILL PROBLEM 6.2

How many roots does each of the following polynomials have in the right half of the s-plane?

(a) $s^4 + 2s^3 + 4s^2 + 8s + 15$

(b) $s^5 + 2s^4 + 6s^3 + 10s^2 + 8s + 12$

(c) $s^5 + 2s^4 + 4s^3 + 8s^2 + 16s + 32$

(d) $s^6 + 4s^5 + 11s^4 + 12s^3 + 26s^2 + 84s + 16$

Ans. (a) Two, (b) none (two on the $j\omega$-axis), (c) two, (d) two (two on the $j\omega$-axis).

6.4 RELATIVE STABILITY

Application of the Routh criterion, as discussed above, only tells us whether the system is stable or not. In many cases, we need more information. For example, if the Routh test shows that a system is stable, we often like to know how close it is to instability, *i.e.* how far from the $j\omega$-axis is the pole closest to it. This information can be obtained from the Routh criterion by shifting the vertical axis in the s-plane to obtain the p-plane, as shown in Figure 6.1.

Hence, in the polynomial $\Delta(s)$, if we replace s by $p - \alpha$, we get a new polynomial $\Delta(p)$. Applying the Routh test to this polynomial will tell us how many roots $\Delta(p)$ has in the right half of the p-plane. This is also the number of roots of $\Delta(s)$ located towards the right of the line $s = -\alpha$ in the s-plane.

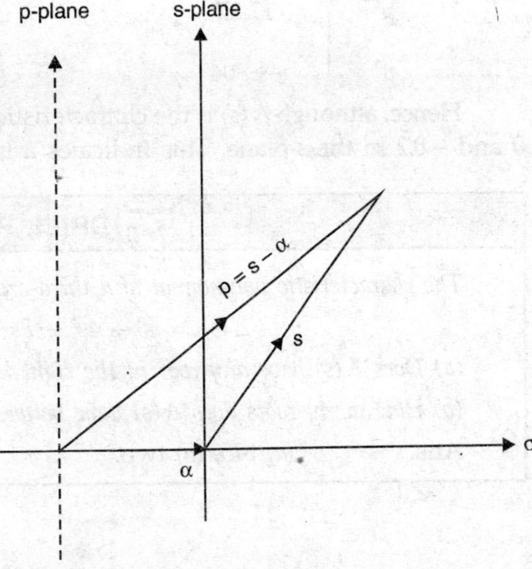

Figure 6.1. *Shift of the axis to the left by* α

The procedure is illustrated by the following example:

EXAMPLE 6.6

Consider the polynomial

$$\Delta(s) = s^3 + 10.1s^2 + 21s + 2 \quad \text{...(6.16)}$$

The Routh table shown below tells us that the system is stable.

s^3	1	21
s^2	10.1	2
s^1	20.802	
s^0	2	

Now, we shall shift the axis to the left by 0.2, that is,

$$p = s + 0.2 \quad \text{...(6.17)}$$

or

$$s = p - 0.2 \quad \text{...(6.18)}$$

Hence,

$$\Delta(p) = (p - 0.2)^3 + 10.1\,(p - 0.2)^2 + 21\,(p - 0.2) + 2$$
$$= p^3 + 9.5p^2 + 17.08p - 1.804 \quad \text{...(6.19)}$$

The Routh table shows that $\Delta(p)$ shows that it has one root in the right half of the p-plane. This implies one root between 0 and -0.2 in the s-plane. It follows that shifting by a smaller amount would enable us to get a better idea of the location of this root.

p^3	1	17.8
p^2	9.5	−1.804
p^1	17.27	
p^0	−1.804	

Hence, although $\Delta(s)$ is the characteristic polynomial of a stable system, it has a root between 0 and -0.2 in the s-plane. This indicates a large time constant of the order of 5 seconds.

DRILL PROBLEM 6.3

The characteristic polynomial of a third-order system is given by

$$\Delta(s) = s^3 + 4s^2 + 8s + 30$$

(a) Does $\Delta(s)$ have any root in the right half of the s-plane?

(b) How many roots does $\Delta(s)$ have between the lines $\sigma = 0$ and $\sigma = -0.05$?

Ans. *(a)* No, *(b)* two.

DRILL PROBLEM 6.4

The closed-loop transfer function of a position control system utilizing an ac servomotor is given by

$$T(s) = \frac{20}{s^3 + 2s^2 + 11s + 20}$$

It is desired that all poles of this transfer function lie to the left of the line $\sigma = -0.1$. Is this condition satisfied?

Ans. No, there are two poles between $\sigma = 0$ and $\sigma = -0.1$.

6.5 APPLICATION TO DESIGN

The Routh criterion may also be utilized to determine the range of the open-loop gain for which a feedback system may be stable. This is done by finding the limiting values of K that will cause changes in the signs of the elements in first column of the Routh table. It will be illustrated by means of an example.

EXAMPLE 6.7

Consider the feedback system shown in Fig. 6.2. Its characteristic equation is obtained as

$$\Delta(s) = s(s+3)(s+10) + K = s^3 + 13s^2 + 30s + K \qquad ...(6.20)$$

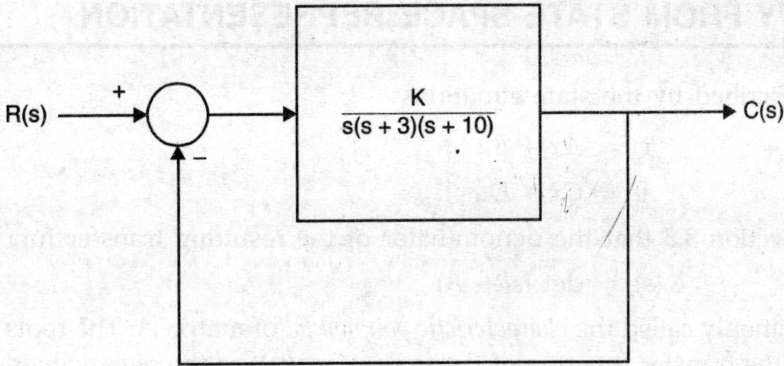

Figure 6.2. *A feedback control system*

The Routh table is shown below.

s^3	1	30
s^2	13	K
s^1	$\dfrac{390 - K}{13}$	
s^0	K	

From the Routh table it is obvious that the system will be stable if and only if

$$0 < K < 390 \qquad ...(6.21)$$

For $K = 390$, we get a zero row, and the auxiliary equation is

$$13s^2 + 390 = 0 \qquad ...(6.22)$$

Hence, for this case,

$$\Delta (s) = (s^2 + 30)(s + 13) \qquad ...(6.23)$$

with a pair of roots on the imaginary axis.

DRILL PROBLEM 6.5

The closed-loop transfer function of an antenna control system is given by

$$T(s) = \frac{K}{s^4 + 6s^3 + 30s^2 + 60s + K}$$

(a) Determine the range in which K must lie for the system to be stable.

(b) What should be the upper limit on K if all the poles of T (s) are required to be on the left of the line $\sigma = -1$?

*[**Hint:** Shift the axis suitably before applying the method discussed in this Section.]*

Ans. (*a*) $0 < K < 200$, (*b*) $K < 112$.

6.6 STABILITY FROM STATE-SPACE REPRESENTATION

For a system described by the state equations

$$\left.\begin{aligned} \dot{x} &= Ax + Bu \\ y &= Cx + Du \end{aligned}\right\} \qquad ...(6.24)$$

it was seen in Section 3.3 that the denominator of the resulting transfer function is given by

$$\Delta (s) = \det (sI - A) \qquad ...(6.25)$$

This is commonly called the *characteristic polynomial* of matrix A. The roots of this polynomial are the poles of the transfer function of the system as well as the eigenvalues of A, as seen from the denominator of Equation (3.25).

The system will be stable if and only if all the roots of $\Delta (s)$ are in the left half of the s-plane.

Alternatively, the system described by Equation (6.24) will be stable if and only if all the eigenvalues of A have negative real parts.

Instead of actually evaluating all the roots of $\Delta (s)$, we may use the Routh-Hurwitz criterion to determine the number of roots in the right half of the s-plane. We may also investigate relative stability by shifting the axis, as discussed in Section 6.4.

EXAMPLE 6.8

The state equations for a third-order system are given below. Use the Routh criterion to determine whether it is stable.

$$\dot{x} = \begin{bmatrix} 0 & 1 & -8 \\ 1 & 0 & -5 \\ 0 & 2 & -6 \end{bmatrix} x + \begin{bmatrix} 2 \\ 1 \\ 0 \end{bmatrix} \qquad ...(6.26)$$

$$y = [1 \;\; 2 \;\; 1]\, y$$

The characteristic polynomial is obtained as

$$\Delta(s) = \det(sI - A) = \begin{vmatrix} s & -1 & 8 \\ -1 & s & 5 \\ 0 & -2 & s+6 \end{vmatrix} \qquad ...(6.27)$$

$$= s(s^2 + 6s + 10) + 1(-s - 6 + 16) = s^3 + 6s^2 + 9s + 10$$

The Routh table is as follows:

s^3	1	9
s^2	6	10
s^1	44/6	
s^0	10	

Since every element in the first column is positive, the system is stable.

DRILL PROBLEM 6.6

The A-matrix in the state equation for a helicopter near hover is given below. Determine whether A has any eigenvalue with a positive real part.

$$A = \begin{bmatrix} -0.02 & -1.4 & 9.8 \\ -0.01 & -0.4 & 0 \\ 0 & 1 & 0 \end{bmatrix}$$

Ans. A has two eigenvalues with positive real parts.

6.7 STABILITY UNDER PARAMETER UNCERTAINTY

In our discussion so far, we have assumed that the characteristic polynomial $\Delta(s)$ for the system is known precisely. In practice, for a product to be economical, some tolerance must be allowed in the values of the components. As a result, for most systems, the coefficients a_i of the characteristic polynomial

$$\Delta(s) = a_0 s^n + a_1 s^{n-1} + \ldots + a_{n-1} s + a_n \qquad ...(6.28)$$

are unknown except for the bounds

$$\alpha_i \le a_i \le \beta_i, \quad i = 0, 1, \ldots n \qquad ...(6.29)$$

Consequently, the system will be stable if the roots of $\Delta(s)$ will be in the left half of the s-plane for each set of values of the coefficients within the ranges defined by Equation (6.29). This is called *robust parametric stability.*

According to an interesting result proved by Kharitonov [1], it is necessary to investigate the stability for only four polynomials formed from Equations (6.28) and (6.29). We shall now discuss how one can obtain these four *Kharitonov polynomials.* These are defined in the form of the polynomial

$$\Delta(s) = c_0 s^n + c_1 s^{n-1} + \ldots c_{n-1} s + c_n \qquad \ldots(6.30)$$

where the coefficients, c_i are defined in a pairwise fashion, c_{2k}, c_{2k+1}, $k = 0, 1, \ldots, m$, where

$$m = \begin{cases} \dfrac{n}{2} & \text{if } n \text{ is even,} \\[2ex] \dfrac{n-1}{2} & \text{if } n \text{ is odd.} \end{cases} \qquad \ldots(6.31)$$

Based on the pairwise assignment of the coefficients, the four polynomials are assigned a mnemonic {k = even, k = odd} name as shown below, for $(2k + 1) < n$.

$$\Delta_1(s) \text{ [max, max; min, min] } c_{2k} = \begin{cases} \beta_{2k} & \text{if } k \text{ is even,} \\ \alpha_{2k} & \text{if } k \text{ is odd.} \end{cases}$$

$$c_{2k+1} = \begin{cases} \beta_{2k+1} & \text{if } k \text{ is even,} \\ \alpha_{2k+1} & \text{if } k \text{ is odd.} \end{cases}$$

$$\Delta_2(s) \text{ [min, min; max, max] } c_{2k} = \begin{cases} \alpha_{2k} & \text{if } k \text{ is even,} \\ \beta_{2k} & \text{if } k \text{ is odd.} \end{cases}$$

$$c_{2k+1} = \begin{cases} \alpha_{2k+1} & \text{if } k \text{ is even,} \\ \beta_{2k+1} & \text{if } k \text{ is odd.} \end{cases}$$

$$\Delta_3(s) \text{ [min, max; max, min] } c_{2k} = \begin{cases} \alpha_{2k} & \text{if } k \text{ is even,} \\ \beta_{2k} & \text{if } k \text{ is odd.} \end{cases}$$

$$c_{2k+1} = \begin{cases} \beta_{2k+1} & \text{if } k \text{ is even,} \\ \alpha_{2k+1} & \text{if } k \text{ is odd.} \end{cases}$$

$$\Delta_4(s) \text{ [max, min; min, max] } c_{2k} = \begin{cases} \beta_{2k} & \text{if } k \text{ is even,} \\ \alpha_{2k} & \text{if } k \text{ is odd.} \end{cases}$$

$$c_{2k+1} = \begin{cases} \alpha_{2k+1} & \text{if } k \text{ is even,} \\ \beta_{2k+1} & \text{if } k \text{ is odd.} \end{cases}$$

Proofs of this remarkable result can be found in references [1], [2] and [3]. We shall illustrate its usefulness by a simple example.

EXAMPLE 6.9

Consider the feedback control system discussed in Example 6.7. Here, we shall assume that the transfer function of the forward path is given by

$$G(s) = \frac{K}{s(s+a)(s+b)} \quad \text{...(6.32)}$$

where, the uncertainty ranges of the parameters K, a and b are as given below:

$$K = 10 \pm 2, \; a = 3 \pm 0.5 \text{ and } b = 4 \pm 0.2 \quad \text{...(6.33)}$$

We shall determine the stability of the closed-loop system using Kharitonov's four polynomials. First we determine the characteristic polynomial

$$\begin{aligned} \Delta(s) &= a_0 s^3 + a_1 s^2 + a_2 s + a_3 \\ &= s^3 + (a+b)s^2 + abs + K \end{aligned} \quad \text{...(6.34)}$$

From the given ranges on the parameters, the coefficients of the characteristic polynomial are

$$\begin{aligned} a_0 &= 1 \\ 6.3 \le a_1 &\le 7.7 \\ 9.5 \le a_2 &\le 14.7 \\ 8 \le a_3 &\le 12 \end{aligned} \quad \text{...(6.35)}$$

We now obtain the four polynomials as

$$\begin{aligned} \Delta_1(s) &= s^3 + 7.7s^2 + 9.5s + 8 \\ \Delta_2(s) &= s^3 + 6.3s^2 + 14.7s + 12 \\ \Delta_3(s) &= s^3 + 7.7s^2 + 14.7s + 8 \\ \Delta_4(s) &= s^3 + 6.3s^2 + 9.5s + 12 \end{aligned}$$

Application of the Routh-Hurwitz criterion to each of these four polynomials show that the first column of Routh table is positive for all of them. This shows that the system in this example will be stable for all possible sets of parameters within the ranges given by Equation (6.35).

SUMMARY

- A linear system will be stable if and only if all the poles of its transfer function have negative real parts, that is they are in the left half of the *s*-plane. This can be determined conveniently by using the Routh-Hurwitz criterion, which determines the number of roots of the characteristic polynomial with positive real parts without actually requiring the evaluation of the roots. By shifting the axis, it is also possible to determine whether the system is on the verge of instability. The criterion may be utilized for determining the range of the open-loop gain over which a feedback system may be stable. When a system is described by state equations, the Routh criterion may be applied to determine the number of eigenvalues of the *A*-matrix with positive real parts. If the parameters of the system are not known precisely, but their ranges are known, one can investigate robust parametric stability by using the Routh-Hurwitz criterion on the four polynomials formed using Kharitonov's theorem.

Problems

1. Using the Routh-Hurwitz criterion, determine if the following characteristic polynomials represent stable systems:

   (a) $s^3 + 5s^2 + 10s + 20$

   (b) $s^4 + 3s^3 + 10s^2 + 20s + 100$

   (c) $s^4 + 2s^3 + s^2 + 2s + 3$

   (d) $s^5 + 2s^4 + 3s^3 + 4s^2 + 6s + 12$

2. Determine the range of K over which the following characteristic polynomials belong to stable systems.

   (a) $s^3 + 6s^2 + 11s + K$

   (b) $s^4 + 5s^3 + 9s^2 + 20s + K$

   (c) $s^5 + 8s^4 + 15s^3 + 32s^2 + 60s + K$

3. For Problem 2, determine the value of K in each case so that all the roots of the polynomials are to the left of the line $\sigma = -0.5$.

4. The block diagram of a type 2 system stabilized by a lead compensator is shown in Fig. P6.4 where $\alpha = 1$.

   (a) Determine the range of values of K for which the system will be stable.

   (b) Determine the maximum value of K for which all the poles of the transfer function of the closed-loop system will to the left of the line $\sigma = -0.2$.

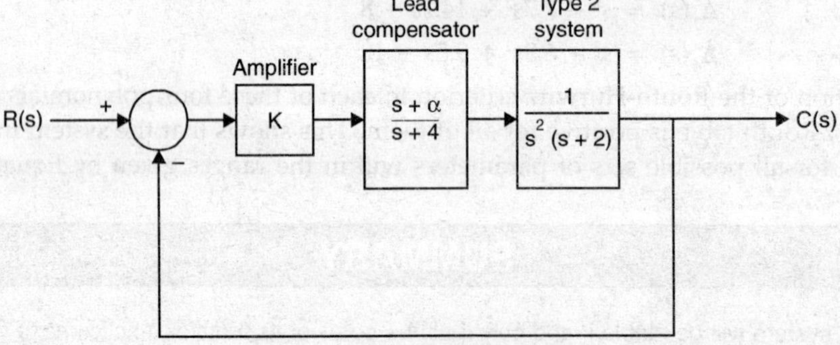

Figure P6.4. *A type 2 system*

5. Suppose that in the previous problem the gain K is fixed at 12 but the location of the zero, that is, the value of α, is adjustable.

   (a) Determine the range of values of α so that the system may be stable.

   (b) Determine the range of values of α so that all the poles of the transfer function of the closed-loop system are to the left of the line $\sigma = -0.2$.

6. The forward transfer function of a unity-feedback system is given by

$$G(s) = \frac{K(s+1)}{s(s+2)(s+3)}$$

   (a) Determine the range of values of K so that the system may be stable.

(*b*) Determine the maximum value of *K* for which all the poles of the transfer function of the closed-loop system are to the left of the line $\sigma = -0.5$.

7. The state equations of a linear system are

$$\dot{x} = \begin{bmatrix} 0 & 1 & -6 \\ 1 & 0 & 5 \\ 0 & -1 & 2 \end{bmatrix} x + \begin{bmatrix} 1 \\ 1 \\ 0 \end{bmatrix} u$$

$$y = \begin{bmatrix} 0 & 1 & 2 \end{bmatrix} y$$

Is this a stable system?

8. Determine the number of eigenvalues with positive real parts for each of the matrices given below.

(*a*) $\begin{bmatrix} -12 & -5 & -20 \\ 10 & -8 & 0 \\ 0 & 1 & 0 \end{bmatrix}$ (*b*) $\begin{bmatrix} -1 & 2 & 0 \\ 0 & -5 & -3 \\ 2 & 1 & -6 \end{bmatrix}$

[**Hint:** Leverrier's algorithm, described in Section 3.3 can be used for calculating the characteristic polynomial of a matrix, if desired.]

9. The forward transfer function of a unity-feedback system is given by

$$G(s) = \frac{K(s+12)}{(s+5)(s+3)(s+6)}$$

Determine the range of values of *K* for which the closed-loop system will be stable.

10. For Problem 9 determine the range of values of *K* for which of the poles of the transfer function of the closed-loop system will be to the left of the line $\sigma = -1$.

11. The forward transfer function of a position control system with velocity feedback is given by

$$G(s) = \frac{K}{s(s+4)}$$

The transfer function of the feedback path is

$$H(s) = 1 + 0.14s$$

Determine the range of values of *K* for which all the poles of the transfer function of the closed-loop system will be to the left of the line $\sigma = -1$.

12. The characteristic polynomial of a closed-loop system is given by

$$\Delta(s) = s^4 + 5s^3 + (K + 12)s^2 + (2K + 8)s + 2K$$

(*a*) Determine the range of values of *K* for which all the roots of $\Delta(s)$ will be in the left half of the *s*-plane.

(*b*) Determine the range of values of *K* for which all the roots of $\Delta(s)$ will be to the left of the line $\sigma = -1$.

13. The steady-state response of a stable system to a step input can be obtained directly from the state equations. This follows from the fact that, for such an input, all the states must be constant in the steady state; that is, $\dot{x}$ must approach zero as t approaches infinity. Hence, the steady-state value of x is obtained as $-A^{-1}Bu$, where u is the constant input. Use this approach to determine the steady-state response to a unit step of the system described by the state equations.

$$\dot{x} = \begin{bmatrix} -2 & 0 & 1 \\ 1 & -2 & 0 \\ 1 & 1 & -1 \end{bmatrix} x + \begin{bmatrix} 1 \\ 0 \\ 1 \end{bmatrix} u$$

$$y = [2 \quad 1 \quad -2]\, x$$

Verify your answer by calculating the transfer function of the system. (Note that this requires that all the eigenvalues of A have negative real parts, which also implies the nonsingularity of the matrix A.)

14. The result of Problem 13 can be easily extended to the case when the input is a ramp function. For such an input the state x in the steady state must be of the form

$$x = \alpha t + \beta$$

where α and β are constant vectors. Substituting this expression into the state equation one can solve for α and β by equating the corresponding coefficients of t.

Use this approach to determine the steady-state output of the system described by the state equations in Problem 13, when the input is a unit ramp.

15. The transfer function of the forward path of a unity-feedback system is given by

$$G(s) = \frac{K}{s(s+p_1)(s+p_2)}$$

where the ranges of the parameters K, p_1 and p_2 are as given below.

$$K = 200 \pm 20,\ p_1 = 4 \pm 0.5 \text{ and } p_2 = 6 \pm 1.$$

Use Kharitonov's technique to determine if the system will be stable for the given ranges of parameter values.

16. Reconsider Problem 15 if the ranges of p_1 and p_2 are as specified, but it is possible to adjust K precisely to a desired value. Use Kharitonov's technique to determine the range of values of K for which the system will be stable.

17. The transfer function of the forward path of a unity-feedback system is given by

$$\frac{K(s+z)}{s(s+p_1)(s+p_2)}$$

where the ranges of the parameters z, p_1 and p_2 are as given below.

$$z = 2 \pm 0.2,\ p_1 = 5 \pm 0.4 \text{ and } p_2 = 8 \pm 1.$$

Use Kharitonov's technique to determine the range of values of K for which the system will be stable.

References

1. Bose, N.K., ''A System-Theoretic Approach to Stability of Sets of Polynomials'', *Contemporary Mathematics*, Vol. 47, 1985, pp. 25-34.
2. Kharitonov, V. L., ''Asymptotic stability of an equilibrium position of a family of systems of linear differential equations,'' *Differential'nye Uraveniya*, Vol. 14, pp. 1483-1485, 1978.
3. Minnichelli, R. J., J. J. Anagnost and C. A. Desoer, ''An elementary proof of Kharitonov's stability theorem with extensions,'' *IEEE Trans. Automat. Contr.*, Vol. AC-34, pp. 995-998, 1989.

❑❑❑

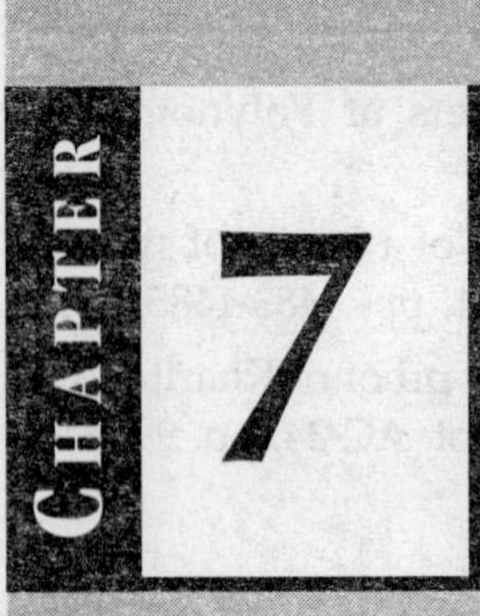

The Root Locus Method

7.1 INTRODUCTION

In the previous chapter, we saw that it is possible to adjust the location of the poles of the transfer function of a closed-loop system by varying the loop gain. Since the nature of the transient response is closely related to the location of the poles, it is very important to know how they move in the s-plane, as the gain, or some other parameter, varies. Furthermore, since for economy in manufacture it is necessary to allow some tolerance in the values of the parameters, it is also desirable to determine the movement of poles with small changes in these parameters.

The root locus method, introduced by W.R. Evans in 1948, is a graphical procedure for sketching the movement in the s-plane of poles of the transfer function of the closed-loop system as a parameter varies from zero to infinity. The method was developed as an alternative to repeatedly determine the roots of the characteristic polynomial using numerical methods. Although originally the method was intended for the case when the open-loop gain was the variable parameter, it can be generalized as follows. Starting with the characteristic polynomial of the transfer function of the closed-loop system, rewrite it as

$$KP(s) + Q(s) = 0 \qquad ...(7.1)$$

where K is the variable parameter. This can be done by separating all terms that contain K from those that do not. (Note that both $P(s)$ and $Q(s)$ are monic polynomials.)

Our objective is to plot the locus of the roots of Equation (7.1) as K is varied from 0 to ∞. We shall first consider the example of a second-order system to understand the various implications and the power of the method.

7.2 ROOT LOCI FOR A SECOND-ORDER SYSTEM

Consider the second-order system shown in Fig. 7.1, which represents a typical position control system with velocity feedback. By applying Mason's rule, we obtain the transfer function of the closed-loop system as

$$T(s) = \frac{C(s)}{R(s)} = \frac{K}{s^2 + (K\alpha + 2)s + K} \quad ...(7.2)$$

so that the characteristic polynomial is

$$\Delta(s) = s^2 + (K\alpha + 2)s + K \quad ...(7.3)$$

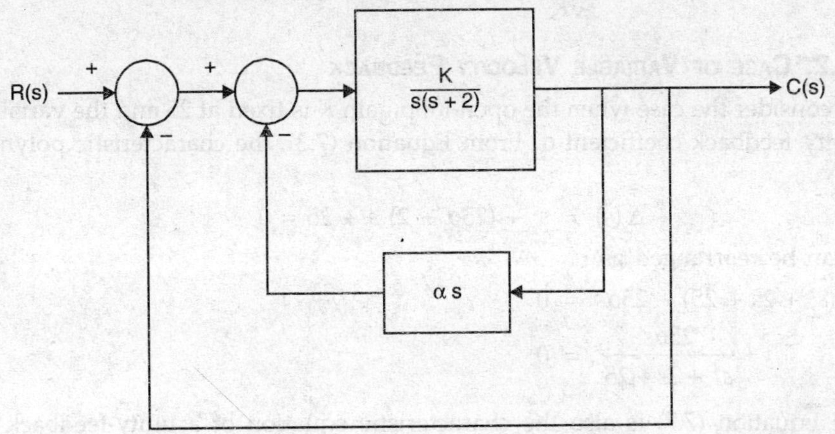

Figure 7.1. *Second-order system*

We shall consider two cases. First, we shall assume that α is zero, that is, there is no velocity feedback, and investigate the effect of varying the gain K from zero to infinity. Next, we shall assume that K is equal to 25 and examine the effect of varying α from zero to infinity. In both cases, we want to determine how the poles of $T(s)$ will move in the s-plane.

EXAMPLE 7.1

We shall consider the case of variable K, with $\alpha = 0$. In this case, we have to determine the roots of the characteristic polynomial.

$$\Delta(s) = s^2 + 2s + K = 0 \quad ...(7.4)$$

First consider the case when $K = 0$. From Equation (7.4), the roots are obtained as $s = 0$ and $s = -2$, which are also the poles of the open-loop transfer function with $\alpha = 0$.

As K is increased from 0 to 1, we have real roots, but for K greater than one, we get a pair of complex conjugate roots with real part equal to –1. This movement of the roots of Equation (7.4) is shown in Fig. 7.2.

It will be seen that the root locus consists of two straight lines, AB and DE, intersecting at C. One may visualize the loci starting from the two open-loop poles (A and B). The locus starting from A moves to the right, while the locus starting from B moves to the left on the real axis until they meet at C for $K = 1$. For larger values of K, the loci move along the lines CD and CE, as shown in Fig. 7.2.

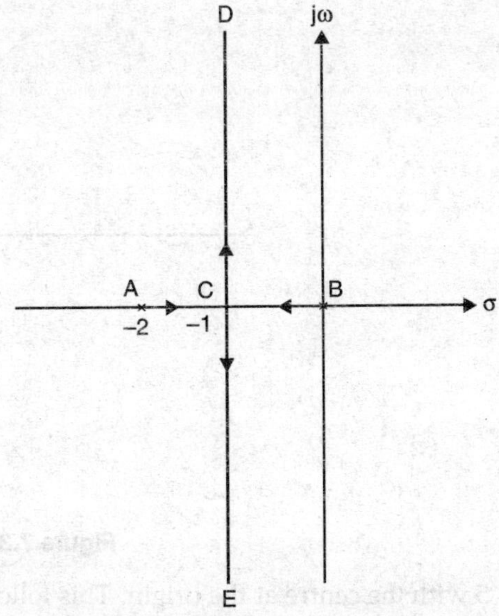

Figure 7.2. *Root locus for Equation (7.4)*

From the plot of the root locus, we can conclude that this system will be stable for all values of K, and the poles of the closed-loop system transfer function will be complex only if K is greater than one. Furthermore, for larger values of K, although the system will be stable, it will be very lightly damped, the damping ratio being given by

$$\zeta = \frac{1}{\sqrt{K}} \quad ...(7.5)$$

EXAMPLE 7.2: CASE OF VARIABLE VELOCITY FEEDBACK

Let us now consider the case when the open-loop gain K is fixed at 25 and the variable parameter is the velocity feedback coefficient α. From Equation (7.3), the characteristic polynomial can be written as

$$\Delta(s) = s^2 + (25\alpha + 2)\, s + 25 = 0 \quad ...(7.6)$$

This can be rearranged as

or
$$(s^2 + 2s + 25) + 25\alpha s = 0 \quad ...(7.7)$$

$$1 + \frac{25\alpha s}{s^2 + 2s + 25} = 0 \quad ...(7.8)$$

Hence Equation (7.7) is also the characteristic equation of a unity-feedback system with forward transfer function

$$G(s) = \frac{Ks}{s^2 + 2s + 25} \quad ...(7.9)$$

which has a zero at the origin and a pair of complex conjugate poles at $s = -1 \pm j\sqrt{24}$ with $K = 25\alpha$. The pole-zero plot of the open-loop transfer function is shown in Fig. 7.3, along with the root locus. We note that for $\alpha = 0$ the roots of $\Delta(s)$ are located at $s = -1 \pm j\sqrt{24}$, labelled A and B in Fig. 7.3. As α is increased from zero to 0.32, the roots move along a circular arc of radius

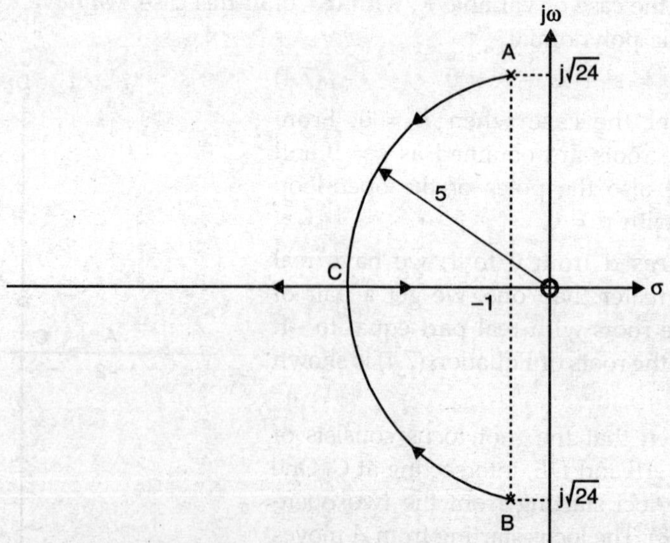

Figure 7.3. *Root locus for Equation (7.7)*

5 with the centre at the origin. This follows from the fact that Equation (7.6) has a pair of complex conjugate roots for these values of α and all of these roots have magnitude equal to 5. For α equal

to 0.32, we get two equal real roots located at $s = -5$. As α is further increased, there will again be two real (but unequal) roots, with one moving towards the left and the other towards the right.

From the root locus plot we can conclude that the effect of velocity feedback has been to increase the damping ratio of the poles of the closed-loop system, without altering the undamped natural frequency, as long as $\alpha \leq 0.32$. For larger values of α, we get an over-damped system.

DRILL PROBLEM 7.1

For the problem considered in Example 7.2 determine the value of α so that the poles of the closed-loop system transfer function have a damping ratio of 0.6.

Ans. 0.16.

7.3 BASIC PRINCIPLES OF THE ROOT LOCUS

In the previous section, we considered two examples of a second-order system where the root locus can be obtained from analytical considerations. In most other cases, it is not so straightforward. For example, if we had assumed in Example 7.1 that α was fixed at 0.1 and K was to be varied from zero to infinity, we would have obtained the characteristic polynomial.

$$\Delta(s) = s^2 + (0.1K + 2)s + K \qquad ...(7.10)$$

In this case, the root locus is not obtained as simply as before, even though we have a polynomial of the second degree. For higher order systems the problem gets more involved.

Hence, we shall look at Equation (7.1) again and try to develop some basic conditions that apply to the root locus. These will then be utilized for deriving certain properties of the root locus that will be useful in the construction of the root locus plot.

Equation (7.1) may be written as

$$1 + \frac{K P(s)}{Q(s)} = 0 \qquad ...(7.11)$$

which may be rearranged as

$$K\frac{P(s)}{Q(s)} = -1 = 1e^{j\pi} = 1\angle\pi \qquad ...(7.12)$$

Consequently, we derive the following basic conditions that must be satisfied for all points on the root locus.

CONDITION 1 (ANGLE CONDITION). For some value of K, any point in the s-plane will be a root of Equation (7.12) if and only if the angle condition is satisfied, *i.e.*, at this point the algebraic sum of the angles (arguments) of the vectors drawn to it from the open-loop poles and zeros is an odd multiple of π (or 180°). [Note that the roots of $P(s)$ are the zeros and the roots of $Q(s)$ are the poles of the open-loop transfer function $KP(s)/Q(s)$.]

CONDITION 2 (MAGNITUDE CONDITION). At any point on the root locus (with the angle condition satisfied) the value of K (for this point to be a root) is obtained as

$$K = \frac{\text{Product of lengths of vectors from poles}}{\text{Product of lengths of vectors from zeros}} \quad \text{...(7.13)}$$

Thus, we first apply the angle condition to determine whether a point in the s-plane lies on the root locus. If this condition is satisfied, we can use the magnitude condition to obtain the value of K for which this will be a root of the characteristic equation.

EXAMPLE 7.3

We shall now use these principles to obtain the root locus for Equation (7.10). Since we can rearrange it as

$$\Delta(s) = (s + 1 + 0.05K)^2 + K - (1 + 0.05K)^2 = 0 \quad \text{...(7.14)}$$

it follows that the roots will be complex if and only if

$$K - (1 + 0.05K)^2 > 0 \quad \text{...(7.15)}$$

To determine these values of K, let us solve (7.15) as an equality. This gives two values of K, as follows

$$K = 1.1146 \quad \text{or} \quad 358.8854 \quad \text{...(7.16)}$$

It can be verified that the inequality (7.15) is satisfied if

$$1.1146 < K < 358.8854 \quad \text{...(7.17)}$$

For the locus of the complex roots, we shall rearrange Equation (7.10) in the form of Equation (7.11) so that we have

$$1 + \frac{0.1K(s+10)}{s(s+2)} = 0 \quad \text{...(7.18)}$$

The pole-zero plot of the corresponding forward-path transfer function $G'(s)$ is shown in Fig. 7.4, where

$$G'(s) = \frac{K'(s+10)}{s(s+2)} \quad \text{...(7.19)}$$

and

$$K' = 0.1K \quad \text{...(7.20)}$$

Consider any point $\tilde{P}(s)$ in the s-plane where $s = \sigma + j\omega$. In order that $\tilde{p}$ may be a point on the root locus, the angle condition must be satisfied, that is,

$$\tan^{-1}\frac{\omega}{\sigma} + \tan^{-1}\frac{\omega}{\sigma+2} - \tan^{-1}\frac{\omega}{\sigma+10} = \pm\pi \quad \text{...(7.21)}$$

It can be rearranged as

$$\tan^{-1}\frac{\omega}{\sigma+2} - \tan^{-1}\frac{\omega}{\sigma+10} = \pm\pi - \tan^{-1}\frac{\omega}{\sigma} \quad \text{...(7.22)}$$

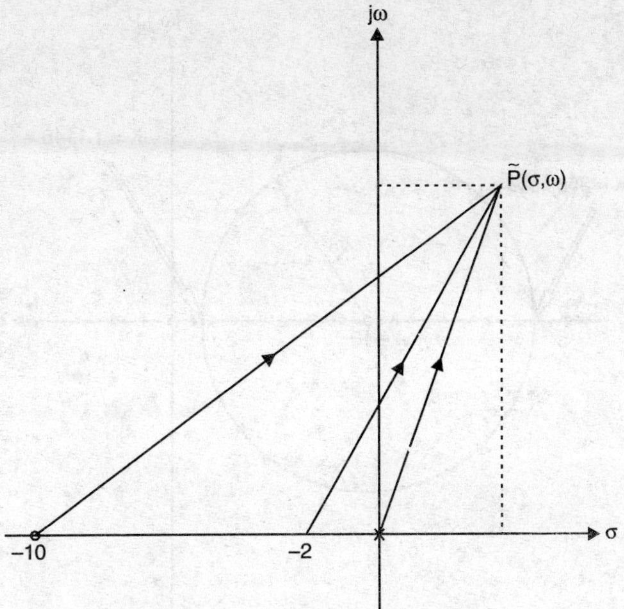

Figure 7.4. *Pole-zero plot for G (s)*

The two terms on the left-hand side of Equation (7.22) may be combined to give

$$\tan^{-1}\frac{\dfrac{\omega}{\sigma+2}-\dfrac{\omega}{\sigma+10}}{1+\dfrac{\omega^2}{(\sigma+2)(\sigma+10)}} = \pm\pi-\tan^{-1}\frac{\sigma}{\omega} \quad ...(7.23)$$

Taking the tangent of both sides of Equation (7.23), we get

$$\frac{\dfrac{\omega}{\sigma+2}-\dfrac{\omega}{\sigma+10}}{1+\dfrac{\omega^2}{(\sigma+2)(\sigma+10)}} = \tan\left(\pm\pi-\tan^{-1}\frac{\omega}{\sigma}\right) = -\frac{\omega}{\sigma} \quad ...(7.24)$$

or
$$\frac{8\omega}{\sigma^2+12\sigma+20+\omega^2} = -\frac{\omega}{\sigma} \quad ...(7.25)$$

Equation (7.25) can be simplified further to obtain

$$(\sigma + 10)^2 + \omega^2 = 80 \quad ...(7.26)$$

which is recognized as the equation of a circle with radius equal to $\sqrt{80}$, and center at $\sigma = -10$ and $\omega = 0$. The resulting plot is shown in Fig. 7.5.

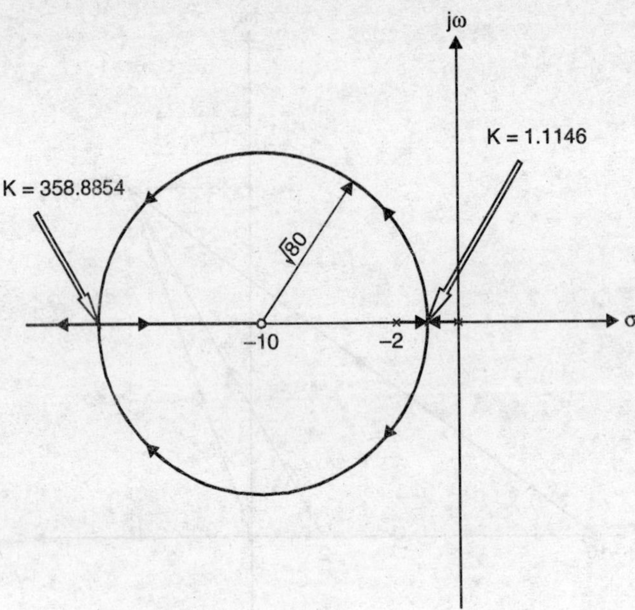

Figure 7.5. *Root locus plot for Equation (7.26)*

7.4 SOME PROPERTIES OF THE ROOT LOCUS

In the previous section, we were able to obtain the root locus using the basic principles for determining whether a point in the *s*-plane lies on the root locus. This will not be suitable if we have polynomials of higher order. Furthermore, it is not very practical to use a trial-and-error procedure to determine the set of all points satisfying the angle criterion.

Hence, we shall now discuss some properties of the root locus that will enable us to obtain an approximate plot with much less effort. These are given in the following paragraphs:

1. The root locus is symmetrical about the real axis of the *s*-plane.

   This follows since the coefficients of the polynomials $P(s)$ and $Q(s)$ are real with the result that the polynomial $KP(s) + Q(s)$ can have complex roots only in conjugate pairs.

2. A branch of the root locus starts from each pole and terminates at each zero of the open-loop transfer function or at infinity. The number of branches terminating at infinity is equal to the number of poles minus the number of zeros of the open-loop transfer function.

   This property is evident if we examine Equation (7.1), which is also written as Equation (7.11), by dividing both sides by $Q(s)$. For $K = 0$, the equation takes the form

   $$Q(s) = 0 \qquad \text{...(7.27)}$$

   That is, for this case the roots of the characteristic polynomial are the roots of $Q(s)$, which are also the poles of the open-loop transfer function $G(s) = KP(s)/Q(s)$.

   If we divide both sides of Equation (7.1) by K, we obtain

   $$P(s) + \frac{Q(s)}{K} = 0 \qquad \text{...(7.28)}$$

It follows that for K equal to infinity, the roots of the characteristic polynomial will coincide with the roots of

$$P(s) = 0 \quad \text{...(7.29)}$$

which are the zeros of $G(s)$.

3. Any point on the real axis is a part of the root locus if and only if the number of poles and zeros to its right is odd.

   This property is proved by applying the angle condition at any point on the real axis of the s-plane. Since the poles and zeros of the open-loop transfer function are symmetrical about the real axis (*i.e.*, all poles and zeros occur in conjugate pairs, if complex), the angles of the vectors drawn to any point on the real axis from each pair of complex poles and zeros will add up to zero. Thus, we need to only consider the poles and zeros on the real axis. Furthermore, the angle subtended at any point on the real axis by a pole or zero to its left is also zero. Hence, only the poles and zeros on the real axis that are to the right of point need be considered. Since each of these subtends an angle equal to π, only an odd number will satisfy the angle condition.

4. If the number of finite zeros, m, is less than the number of finite poles, n, then $n - m$ branches of the root locus must end at zeros at infinity. The asymptotes to these branches intersect at a common point on the real axis, given by

$$\sigma_A = \frac{\text{Sum of poles} - \text{Sum of zeros}}{n - m} \quad \text{...(7.30)}$$

and are inclined to the real axis at angle ϕ_A, given by

$$\phi_A = \frac{2q+1}{n-m}\pi, \quad q = 0, 1, 2, ..., n - m - 1 \quad \text{...(7.31)}$$

Equations (7.30) and (7.31) can be proved by rewriting Equation (7.1) as

$$-K = \frac{Q(s)}{P(s)} = \frac{(s-p_1)(s-p_2)\cdots(s-p_n)}{(s-z_1)(s-z_2)\cdots(s-z_m)} \quad \text{...(7.32)}$$

$$\approx \frac{s^n - \sum_{i=1}^{n} p_i \cdot s^{n-1}}{s^m - \sum_{i=1}^{m} z_i \cdot s^{m-1}} \quad \text{(for large } s\text{)}$$

or

$$K e^{j\pi} \approx s^{n-m} - \left(\sum_{i=1}^{n} p_i - \sum_{i=1}^{m} z_i\right) \text{ by long division} \quad \text{...(7.33)}$$

Taking the $(n - m)$th root of both sides of Equation (7.33), we get

$$K^{\frac{1}{n-m}} \cdot e^{\frac{(2q+1)\pi}{n-m}} = s - \frac{\sum_{i=1}^{n} p_i - \sum_{i=1}^{m} z_i}{n-m} \quad \text{...(7.34)}$$

From Equation (7.34) it is seen that all the asymptotes intersect at σ_A, as given by Equation (7.30). Also, the angles ϕ_A that the asymptotes make with the positive real axis are given by Equation (7.31).

5. If the locus crosses the $j\omega$-axis for some value of K, this is readily determined through the Routh-Hurwitz criterion.

   From the Routh table we can determine the value of K that causes the system to be just unstable, as well as the location of the resulting roots on the $j\omega$-axis. This was discussed in Section 6.5, and is illustrated by means of an example.

EXAMPLE 7.4

Consider a unity-feedback system with open-loop transfer function given by

$$GH(s) = \frac{K}{s(s+4)(s+16)} \qquad \text{...(7.35)}$$

The characteristic polynomial for the system is given by

$$\Delta(s) = s^3 + 20s^3 + 64s + K \qquad \text{...(7.36)}$$

and the Routh table is shown below.

s^3	1	64
s^2	20	K
s^1	$\frac{1280-K}{20}$	
s^0	K	

Hence, we get a zero row if $K = 1280$. Substituting this value in the auxiliary equation obtained from the row above the zero row, we get

$$20s^2 + 1280 = 0 \qquad \text{...(7.37)}$$

giving roots at

$$s = \pm j8 \qquad \text{...(7.38)}$$

DRILL PROBLEM 7.2

Determine the $j\omega$-axis crossover points of the root locus for a unity-feedback system with open-loop transfer function

$$G(s) = \frac{K}{s(s^2+6s+16)} \qquad \text{...(7.39)}$$

where K is the variable parameter. What is the value of K for this crossover?

Ans. $s = \pm j4, \quad K = 96.$

6. The break-away and re-entry points on the root locus are determined from the roots of $dK/ds = 0$. If r branches meet at a point, they break away at an angle of $\pm \pi/r$.

In the examples of root locus that we have studied so far, we saw instances where two branches of the root locus met at a point on the real axis, and then moved away from the real axis at right angles to it. These are called 'break-away' points and are characterized by the fact that the characteristic polynomial has multiple roots at these points. The dual situation occurs when the branches of the root loci enter the real axis. These are called 'break-in' or 're-entry' points. Normally, a break-away point occurs on a segment of the root locus on the real axis connecting two poles and a re-entry point occurs between two zeros, as shown in Fig. 7.6. Later, we shall see some cases where loci can enter and leave the real axis even on a segment of the root locus connecting a pole and a zero.

There may be points on the real axis where more than two branches of the root locus meet. The angles at which the branches will leave the real axis are then given by

$$\phi = \pm\frac{\pi}{r} \qquad \text{...(7.40)}$$

where r is the multiplicity of the roots. Equation (7.40) follows from the fact that the angle condition must be satisfied at this point in all the branches of the root locus. Two such cases are shown in Fig. 7.7.

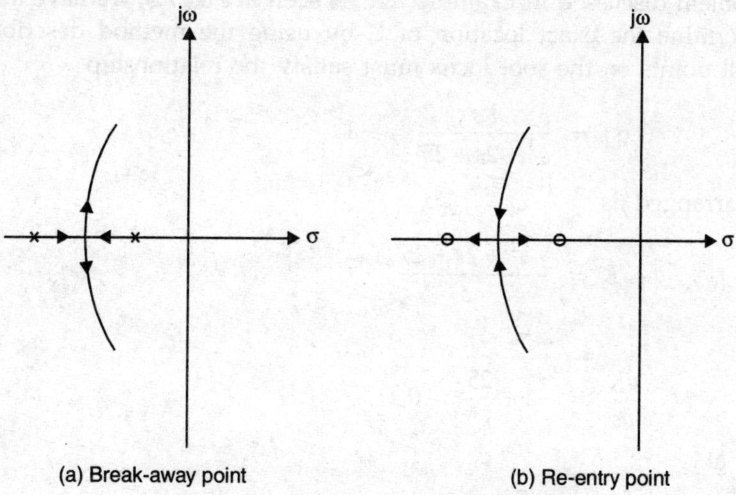

Figure 7.6. *Break-away and re-entry points*

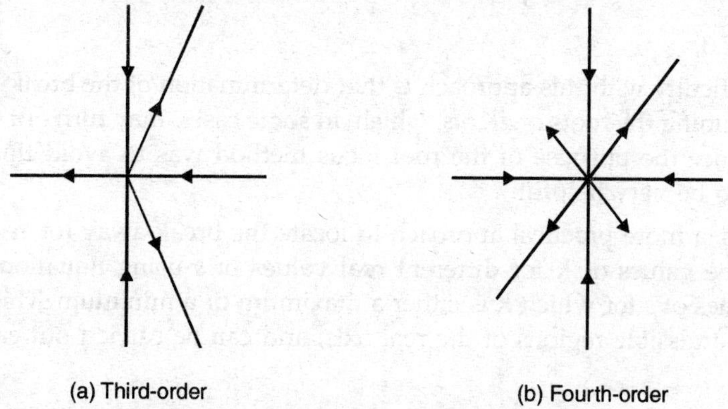

Figure 7.7. *Higher-order break-away points*

Several methods can be utilized for determining the location of break-away points. The simplest of these is based on the fact that a break-away point represents the maximum value of K for which the roots can be real. Similarly, a re-entry point represents the minimum value of K for which the roots can be real. Hence, the real roots of

$$\frac{dK}{ds} = 0 \qquad \text{...(7.41)}$$

where $$K = -\frac{Q(s)}{P(s)} \qquad \text{...(7.42)}$$

obtained by rearranging Equation (7.1) are the break-away or re-entry points, provided that they also satisfy the angle condition (*i.e.*, the number of poles and zeros to the right of the point under consideration is odd). The following example will illustrate this method.

EXAMPLE 7.5

Consider the problem discussed in Example 7.2. As seen in Fig. 7.3, we have a re-entry point at C. We shall determine the exact location of C by using the method described above. From Equation (7.9), all points on the root locus must satisfy the relationship

$$G(s) = \frac{Ks}{s^2 + 2s + 25} = -1 \qquad \text{...(7.43)}$$

which can be rearranged as

$$K = -\frac{s^2 + 2s + 25}{s} \qquad \text{...(7.44)}$$

Hence

$$\frac{dK}{ds} = -1 + \frac{25}{s^2} = 0 \qquad \text{...(7.45)}$$

which has roots at

$$s = \pm 5 \qquad \text{...(7.46)}$$

Since the point $s = +5$ does not satisfy the angle condition, the only admissible solution is

$$s = -5 \qquad \text{...(7.47)}$$

as shown in Fig. 7.3.

The main difficulty with this approach is that determination of the break-away (or re-entry) point requires obtaining the roots of dK/ds, which, in some cases, may turn out to be a polynomial of high degree. Since the purpose of the root locus method was to avoid this very problem, it does not appear to be very helpful.

In such cases, a more practical approach to locate the break-away (or re-entry) points is to make a table of the values of K for different real values of s using Equation (7.42), and hence determine the values of s for which K is either a maximum or a minimum. This search need only be made for the admissible regions of the real axis, and can be carried out easily with the help of a calculator.

DRILL PROBLEM 7.3

Determine the break-away and re-entry points in the root locus shown in Fig. 7.5 by solving for the roots of dK/ds obtained from Equation (7.18).

Ans. $s = -10 \pm \sqrt{80}$

7. The angle of departure of the root locus from a complex pole is given by

$$\phi_d = \pi - \text{sum of angles of vectors drawn to this pole from other poles} + \text{sum of angles of vectors drawn to this pole from the zeros} \quad ...(7.48)$$

The angle of arrival at a complex zero is obtained in a similar manner,

i.e.,
$$\phi_a = \pi - \text{sum of angles of vectors drawn to this zero from other zeros} + \text{sum of angles of vectors drawn to this zero from the poles} \quad ...(7.49)$$

Both of these properties follow directly from the application of the angle condition at a point on the root locus very close to the pole or zero under consideration.

The properties of the root locus are summarized in Table 7.1.

We shall now discuss several examples of sketching the root locus using these properties. For the sake of clarity, the procedure will be described in a step-by-step fashion.

EXAMPLE 7.6

Consider a unity feedback system with forward-path transfer function.

$$G(s) = \frac{K}{s(s+4)(s+5)} \quad ...(7.50)$$

It is desired to sketch the locus of the poles of the transfer function of the closed-loop system as K varies from zero to infinity.

SOLUTION

(*i*) The first step is to show the plot of the poles and zeros of the open-loop transfer function. Here, we have no zeros ($m = 0$), but three poles ($n = 3$) at $s = 0, -4$ and -5.

(*ii*) The second step is to mark the real axis segments of the root locus. These are between 0 and -4, and to the left of -5.

(*iii*) There are $n - m = 3$ branches going to infinity. The asymptotes intersect at

$$\sigma_A = \frac{0-4-5}{3} = -3$$

and the angles ϕ_A are $\pi/3$, π and $5\pi/3$. These are then drawn in the s-plane.

(*iv*) For $G(s)$ given in Equation (7.50) to be equal to -1, we have

$$K = -s(s+4)(s+5) = -s^3 - 9s^2 - 20s$$

TABLE 7.1: Properties of the Root Locus

Property Number	*Property*
1.	The root locus is symmetrical about the real axis.
2.	Each branch of the root locus starts from a pole (for $K = 0$) and terminates at a zero (for $K = \infty$) of the open-loop transfer function or at infinity. The number of branches terminating at infinity is equal to $n - m$, for a system with n poles and m zeros.
3.	A segment of the real axis is part of the root locus if and only if the total number of real poles and zeros to its right is odd.
4.	The $n - m$ branches terminating at infinity are asymptotic to straight lines intersecting at the real axis at the point $\sigma_A = \dfrac{\text{Sum of poles} - \text{Sum of zeros}}{n-m}$ and inclined to the real axis at angles $\phi_A = \left(\dfrac{2q+1}{n-m}\right)\pi, \quad q = 0, 1, 2, ..., n-m-1$
5.	The points at which the root locus intersects the imaginary axis can be determined by using the Routh-Hurwitz criterion.
6.	The break-away and re-entry points on the root locus are determined from the roots of $dK/ds = 0$. If r branches meet at a point, they break away at angles given by $\pm \pi/r$.
7.	The angle of departure from a complex pole of the open-loop transfer function is $\phi_d = \pi$ – sum of angle of vectors drawn from other poles to this pole + sum of angles of vectors drawn from the zeros to this pole. The angle of arrival at a complex zero of the open-loop transfer function is $\phi_a = \pi$ – sum of angles of vectors drawn from other zeros to this zero + sum of angles of vectors drawn from the poles to this zero.

Setting the derivative dK/ds to zero, we get

$$-3s^2 - 18s - 20 = 0$$

which has roots at $s = -1.4725$ and -4.5275. Of these two roots, only the first is admissible as a break-away point since the segment between -4 and -5 is not part of the root locus.

(*v*) We now determine the $j\omega$-axis crossover using the Routh table. We have

$$\Delta(s) = s^3 + 9s^2 + 20s + K$$

and the Routh table is as follows:

$$\begin{array}{c|cc} s^3 & 1 & 20 \\ s^2 & 9 & K \\ s^1 & \dfrac{180-K}{9} & \\ s^0 & K & \end{array}$$

Hence, for $K = 180$, we get roots at $s = \pm j\sqrt{20}$.

(*vi*) A sketch of the complete root locus is shown in Fig. 7.8.

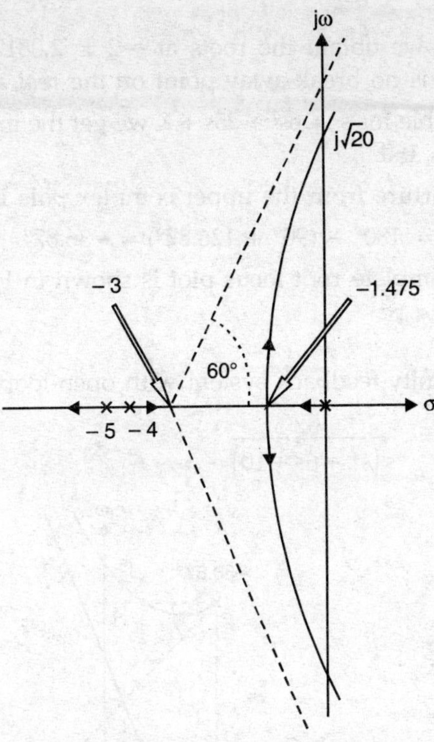

Figure 7.8. *Root locus for Equation (7.50)*

EXAMPLE 7.7

Obtain the root locus for a unity feedback system with open-loop transfer function

$$G(s) = \frac{K}{s\left(s^2 + 6s + 25\right)} \quad ...(7.51)$$

SOLUTION

We shall obtain the root locus for this case in the same way as for the previous example.

(*i*) We have no zeros and three poles, so that $m = 0$ and $n = 3$. The poles are located at $s = 0$ and $-3 \pm j4$. These are plotted in the *s*-plane.

(*ii*) Since there is only one real pole at $s = 0$, the entire left half of the real axis is part of the root locus.

(*iii*) Since $n - m = 3$, we have three branches going to infinity. The asymptotes intersect at

$$\sigma_A = \frac{0 - 3 - 3}{3} = -2$$

and the angles ϕ_A are 60°, 180° and 300°.

(*iv*) From the open-loop transfer function, we get

$$K = -s\,(s^2 + 6s + 25) = -s^3 - 6s^2 - 25s$$

and $$\frac{dK}{ds} = -3s^2 - 12s - 25$$

Setting it to zero, we obtain the roots at $-2 \pm j2.0817$. Since these are complex, it follows that there is no break-away point on the real axis.

(*v*) From the Routh table for $s^3 + 6s^2 + 25s + K$ we get the imaginary axis crossover points at $s = \pm j5$ for $K = 150$.

(*vi*) The angle of departure from the upper complex pole is calculated as

$$\phi_d = 180° - (90° + 126.87°) = -36.87°$$

(*vii*) A sketch of the complete root locus plot is shown in Fig. 7.9.

Example 7.8

Obtain the root locus for a unity-feedback system with open-loop transfer function

$$G(s) = \frac{K}{s\left(s^2 + 6s + 10\right)} \quad \text{...(7.52)}$$

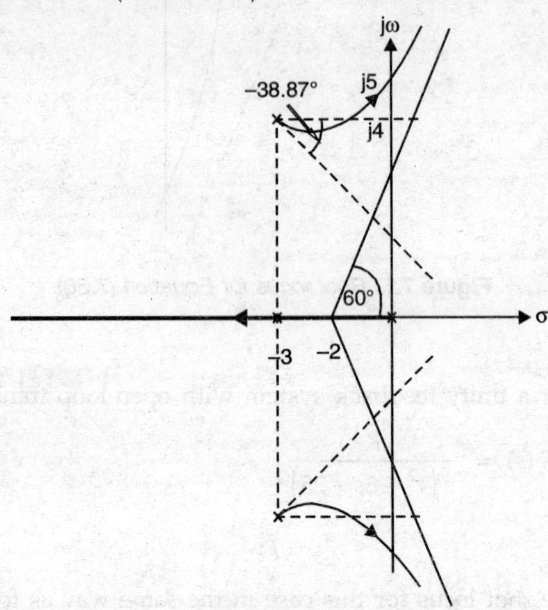

Figure 7.9. *Root locus for Equation (7.51)*

Solution

We shall see that although this transfer function is similar to Equation (7.51), the root locus plot for $0 \le K \le \infty$ is quite different.

(*i*) As before, we have $m = 0$ and $n = 3$. The poles are at $s = 0$ and $-3 \pm j1$.

(*ii*) Since there is only one real pole at $s = 0$, the entire left half of the real axis is part of the root locus.

(*iii*) Since $n - m = 3$, we have three branches going to infinity. The asymptotes intersect at

$$\sigma_A = \frac{0-3-3}{3} = -2$$

and the angles ϕ_A are 60°, 180° and 300°.

(*iv*) From the open-loop transfer function, we get

$$K = -s(s^2 + 6s + 10) = -s^3 - 6s^2 - 10s$$

and $$\frac{dK}{ds} = -3s^2 - 12s - 10$$

with roots at $s = -1.1835$ and -2.8165. As both of these roots are admissible, we have a break-away and a re-entry point on the negative real axis.

(*v*) From the Routh table for $s^3 + 6s^2 + 10s + K$, we get the imaginary axis crossover at $s = \pm j\sqrt{10}$ for $K = 60$.

(*vi*) The angle of departure from the upper complex pole is calculated as

$$\phi_d = 180° - (90° + 161.57°) = -71.5°.$$

(*vii*) The complete root locus is sketched in Fig. 7.10. Note that there is a break-away point and also a re-entry point on the negative real axis. This follows since d^2K/ds^2 is negative at $s = -1.1835$ (making it a break-away point), and positive at $s = -2.8165$ (making it a re-entry point).

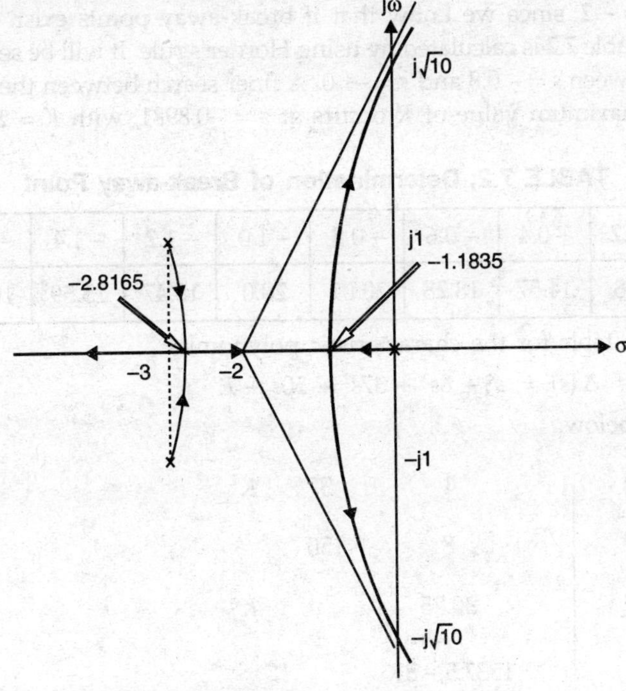

Figure 7.10. *Root locus for Equation (7.52)*

EXAMPLE 7.9

The open-loop transfer function of a unity-feedback system is given by

$$G(s) = \frac{K}{s(s+2)(s^2 + 6s + 25)} \quad ...(7.53)$$

Sketch the root locus for $0 \le K \le \infty$.

SOLUTION

(*i*) We have $m = 0$ and $n = 4$. The poles are located at $s = 0, -2$ and $-3 \pm j4$.

(*ii*) The segment of the real axis between $s = 0$ and $s = -2$ is part of the root locus.

(*iii*) Since $n - m = 4$, four branches of the root locus will go to infinity. The asymptotes intersect at

$$\sigma_A = \frac{0-2-6}{4} = -2$$

and the angle ϕ_A are 45°, 135°, 225° and 315°.

(*iv*) From the open-loop transfer function, we get

$$K = -s(s+2)(s^2+6s+25) = -(s^4+8s^3+37s^2+50s)$$

and
$$\frac{dK}{ds} = -(4s^3+24s^2+74s+50)$$

To determine the break-away points, we would have to solve for the roots of a cubic. To avoid doing this, we shall calculate the values of K for different values of s in the range 0 to -2, since we know that if break-away points exist, they must lie in this interval. Table 7.2 is calculated by using Horner's rule. It will be seen that the maximum occurs between $s = -0.8$ and $s = -1.0$. A finer search between these two points reveals that the maximum value of K occurs at $s = -0.8981$, with $K = 20.206$.

TABLE 7.2: Determination of Break-away Point

s	−0.2	−0.4	−0.6	−0.8	−1.0	−1.2	−1.4	−1.6	−1.7	−1.8
K	8.58	14.57	18.28	20.01	20.0	18.47	15.59	11.49	9.02	6.28

(*v*) The Routh table for the characteristic polynomial

$$\Delta(s) = s^4 + 8s^3 + 37s^2 + 50s + K$$

is shown below.

$$\begin{array}{c|ccc} s^4 & 1 & 37 & K \\ s^3 & 8 & 50 & \\ s^2 & 30.75 & & K \\ s^1 & \dfrac{1537.5-8K}{30.75} & & \\ s^0 & K & & \end{array}$$

Hence, for $K = 192.1875$, we get roots at $s = \pm j2.5$.

(*vi*) The angle of departure from the upper complex pole is given by

$$\phi_d = 180° - (90° + 126.87° + 104.04°) = -140.91°$$

The root locus is shown in Figure 7.11.

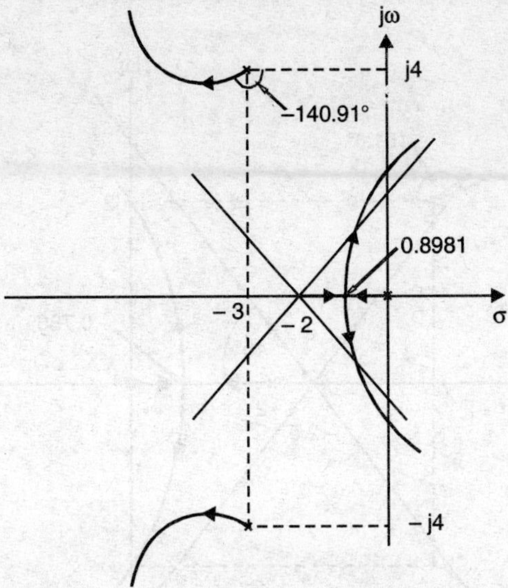

Figure 7.11. *Root locus for Equation (7.53)*

DRILL PROBLEM 7.4

Sketch the root locus for each of the following open-loop transfer functions as K varies from zero to infinity. Be sure to label all the important points.

(a) $\dfrac{K}{s\left(s^2+6s+12\right)}$, (b) $\dfrac{K}{s\left(s+2\right)\left(s^2+8s+20\right)}$, (c) $\dfrac{K\left(s^2+6s+25\right)}{s\left(s+1\right)\left(s+2\right)}$

Ans.

(a)

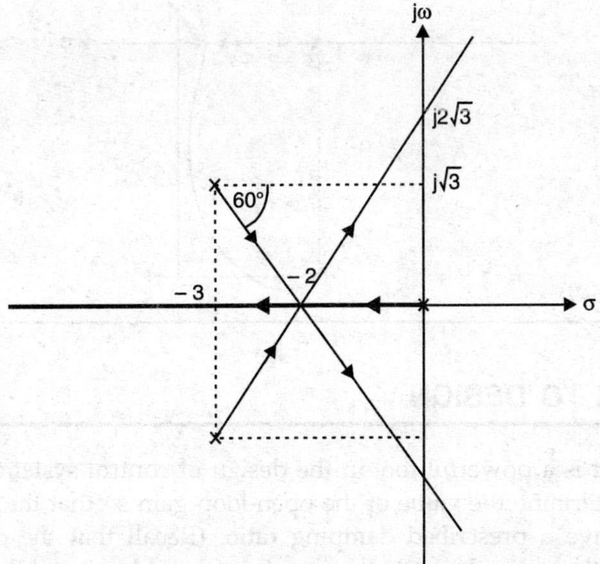

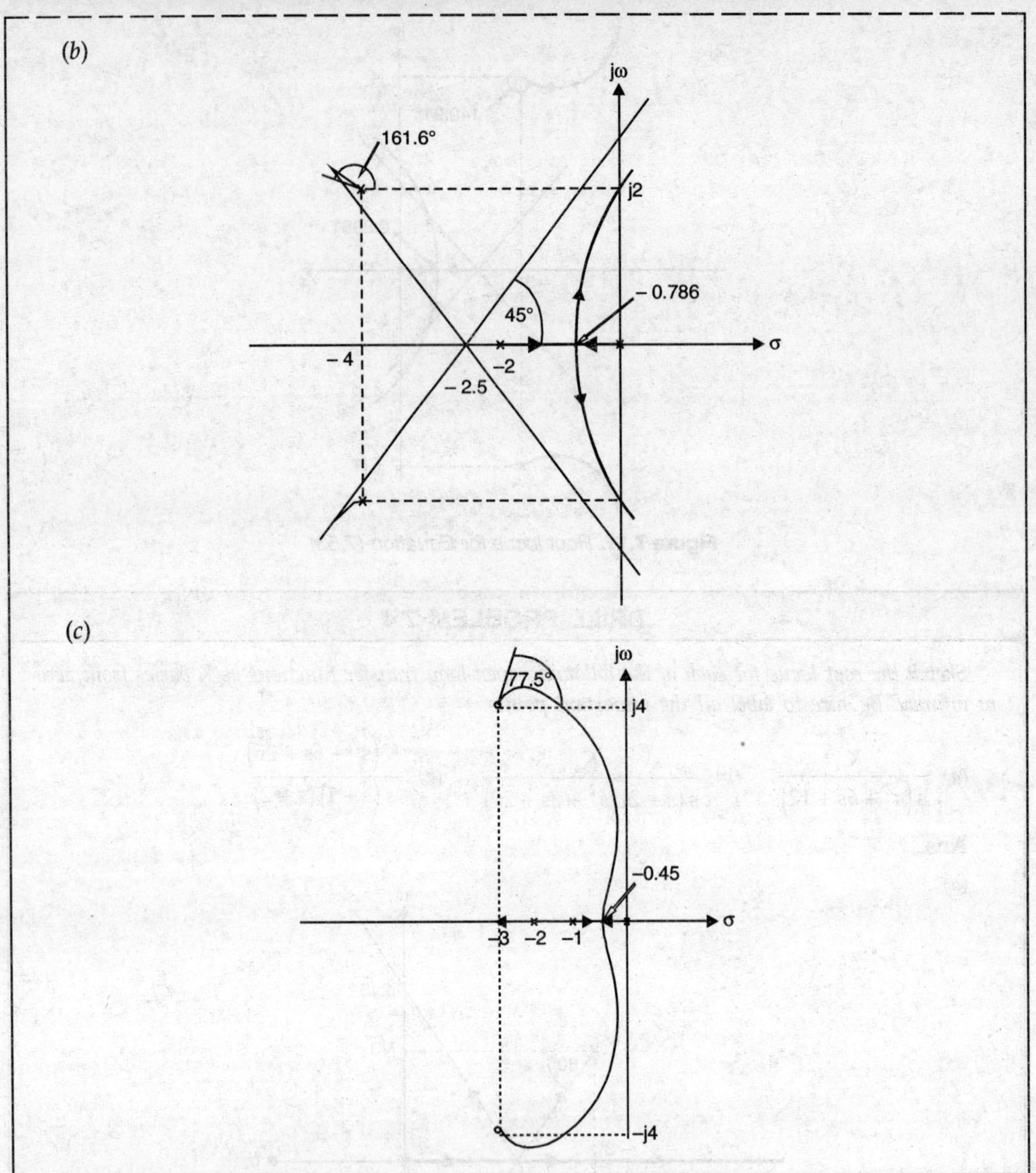

7.5 APPLICATION TO DESIGN

The root locus method is a powerful tool in the design of control system. For example, we can use this method to determine the value of the open-loop gain so that the dominant poles of the closed-loop system have a prescribed damping ratio. (Recall that the dominant poles of the system are defined as the poles closest to the $j\omega$-axis as they 'dominate' the transient response of the system if all other poles are much farther to the left.) Alternatively, in the design of compensators (as will be seen in Chapter 10), a pole or zero of the open-loop transfer function may be a variable parameter. In such cases, we can use the root locus method to determine the value of the parameter that will cause the dominant poles to have the desired damping ratio.

This may be followed by using the root locus to determine the location of all the poles of the transfer function of the closed-loop system for this value of the adjustable parameter. It is usually helpful to note that if the characteristic polynomial is given by

$$\Delta(s) = s^n + a_1 s^{n-1} + \dots + a_{n-1} s + a_n \qquad \text{...(7.54)}$$

then $\qquad$ Sum of the roots of $\Delta(s) = -a_1$ $\qquad$ ...(7.55)

and $\qquad$ Product of the roots of $\Delta(s) = (-1)^{n}a_n$ $\qquad$ ...(7.56)

The following examples will illustrate some applications of the root locus method to the design of control systems.

Example 7.10

Consider the third-order position-control system with velocity feedback shown in Fig. 7.12. Determine the value of the open-loop gain K so that the dominant poles of the transfer function of the closed-loop system have damping ratio of 0.5. What will be the response of the system to a unit step input for this value of K?

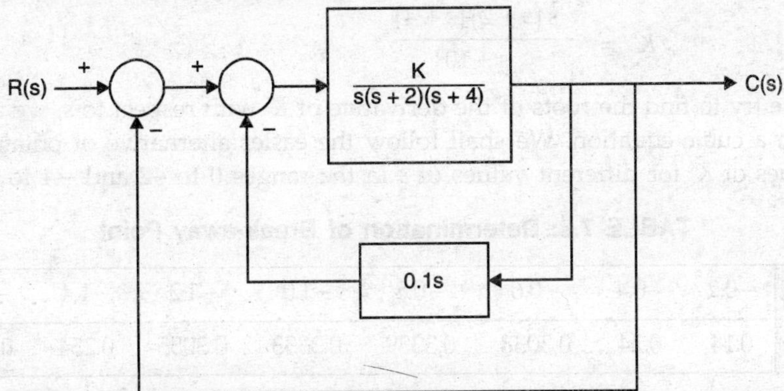

Figure 7.12. *A third-order position-control system with velocity feedback*

Solution

Applying Mason's rule, the transfer function of the closed-loop system is obtained as

$$\frac{C(s)}{R(s)} = \frac{K}{s(s+2)(s+4) + 0.1\,Ks + K} \qquad \text{...(7.57)}$$

and the characteristic polynomial is given by

$$\Delta(s) = s^3 + 6s^2 + 8s + 0.1K(s + 10) = 0 \qquad \text{...(7.58)}$$

which can be arranged as

$$1 + \frac{K'(s+10)}{s(s+2)(s+4)} = 0 \qquad \text{...(7.59)}$$

where $K' = 0.1K$.

Therefore, we shall sketch the root locus of the auxiliary system

$$G'(s) = \frac{K'(s+10)}{s(s+2)(s+4)} \quad ...(7.60)$$

which has the same characteristic polynomial as our original system. We shall proceed as in the previous section.

(*i*) The open-loop transfer function has three poles located at $s = 0, -2$ and -4, and one zero at $s = -10$. Thus, $n = 3$ and $m = 1$.

(*ii*) The real-axis segments of the root locus are obtained between $s = 0$ and -2 and also between $s = -4$ and -10.

(*iii*) As $n - m = 2$, two branches of the locus go to the infinity. The asymptotes intersect at

$$\sigma_A = \frac{0-2-4-(-10)}{2} = 2$$

and the angles ϕ_A are $\pm 90°$.

(*iv*) From Equation (7.60), we have

$$K' = -\frac{s(s+2)(s+4)}{s+10}$$

If we try to find the roots of the derivative of K' with respect to s, we are again faced with a cubic equation. We shall follow the easier alternative of obtaining a table of values of K' for different values of s in the ranges 0 to -2 and -4 to -10.

TABLE 7.3: Determination of Break-away Point

s	−0.2	−0.4	−0.6	−0.8	−1.0	−1.2	−1.4	−1.6	−1.8
K'	0.14	0.24	0.3038	0.3339	0.3333	0.3055	0.254	0.183	0.097

s	−4.5	−5.0	−5.5	−6.0	−6.5	−7.0	−7.5	−8.0	−8.5	−9.0	−9.5
K'	1.023	3.0	6.42	12.0	20.9	35.0	57.8	97.0	167.8	315.0	783.7

From the first part of the table we locate a maximum between – 0.8 and – 1.0. A finer search can be used to determine the break-away point at $s = -0.8951$. The second part of the table show K' increasing monotonically from 1.02 to 783.7, indicating that there is no break-away or re-entry point in this segment.

(*v*) The Routh table for the characteristic polynomial indicates that the imaginary-axis crossover occurs at $s = \pm j\sqrt{20}$ for $K' = 12$. The table is shown on next page.

s^3	1	$8 + K'$
s^2	6	$10K'$
s^1	$\frac{48 - 4K'}{6}$	
s^0	$10K'$	

(*vi*) The complete root locus is shown in Fig. 7.13.

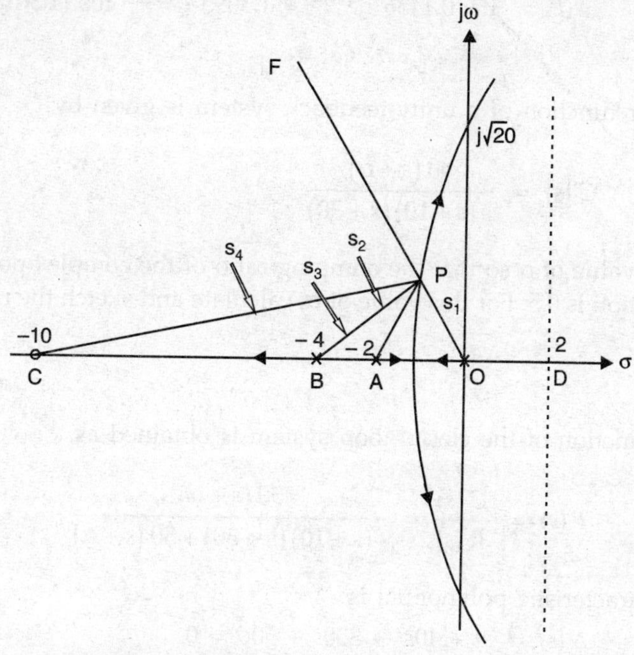

Figure 7.13. *Root locus for Equation (7.59)*

We can now draw the line *OF* making an angle of 60° = $\cos^{-1} \zeta$ with the negative real axis, intersecting the root locus at *P*. It is important to note that since we have sketched the root locus only approximately, with the knowledge of the break-away point, the *j*ω-axis crossover and the asymptotes, we might not have obtained the root precisely. This may be verified by measuring arguments of the various vectors and checking that

$$\angle POD + \angle PAD + \angle PBD + \angle PCD = 180°$$

If necessary, one should move along the line *OF* by a small amount to locate *P* in such a way that the angle condition is satisfied. This may be done either graphically, using a ruler and protractor, or one may use a computer program (see Appendix *E*). Using the latter, the point *P* was located at $s = -0.7513 + j1.3013$. Then the value of *K′* was obtained as

$$K' = \frac{s_1 s_2 s_3}{s_4} = 1.01539$$

where, again for convenience the computer program was utilized.

Hence, $$K = 10K' = 10.1539$$

and
$$\frac{C(s)}{R(s)} = \frac{10.1539}{(s+0.7513+j1.3013)(s+0.7513-j1.3013)(s+4.4974)} \quad ...(7.61)$$

Note that from the root locus plot we already know the location of the pair of complex poles at $s = -0.7513 \pm j1.3013$. The third pole is obtained by using the fact that the sum of the three poles is equal to -6, from Equations (7.55) and (7.56).

Finally, the response to a unit step input is obtained from the transfer function by substituting $1/s$ for $R(s)$ and taking the inverse Laplace transform of $C(s)$. It is given by

$$c(t) = 1 - 0.1436\, e^{-4.4974t} + 1.3095\, e^{-0.7513t} \cos(1.3013t - 3.9995) \quad ...(7.62)$$

EXAMPLE 7.11

The forward transfer function of a unity-feedback system is given by

$$G(s) = \frac{50(s+\alpha)}{s(s+10)(s+30)} \quad ...(7.63)$$

Determine the value of α so that the damping ratio of the complex poles of the closed-loop system transfer function is 0.5. For this value of α, calculate and sketch the response of the system to a unit step.

SOLUTION

The transfer function of the closed-loop system is obtained as

$$T(s) = \frac{C(s)}{R(s)} = \frac{50(s+\alpha)}{s(s+10)(s+30)+50(s+\alpha)} \quad ...(7.64)$$

Hence, the characteristic polynomial is

$$\Delta(s) = s^3 + 40s^2 + 350s + 50\alpha = 0$$

which can be rearranged as

$$1 + \frac{50\alpha}{s^3 + 40s^2 + 350s} = 0$$

Consequently, the poles of $T(s)$ can be obtained from the root locus of the auxiliary transfer function, $G'(s)$,

where

$$G'(s) = \frac{50\alpha}{s(s^2+40s+350)} = \frac{K}{s(s+12.93)(s+27.07)}, \quad \text{where } K = 50\alpha$$

We have $n = 3$ and $m = 0$. Hence, for the three asymptotes, we have

$$\sigma_A = -\frac{40}{3} = -13.33, \quad \text{and} \quad \phi_A = 60°, 180° \text{ and } 300°.$$

There is a break-away point between 0 and -12.93. By finding the roots of the derivative of K with respect to s, this is located at -5.516.

Using the Routh table, the $j\omega$-axis crossover is determined at $\pm j18.7$ for $K = 14000$.

The root locus is shown in Fig. 7.14.

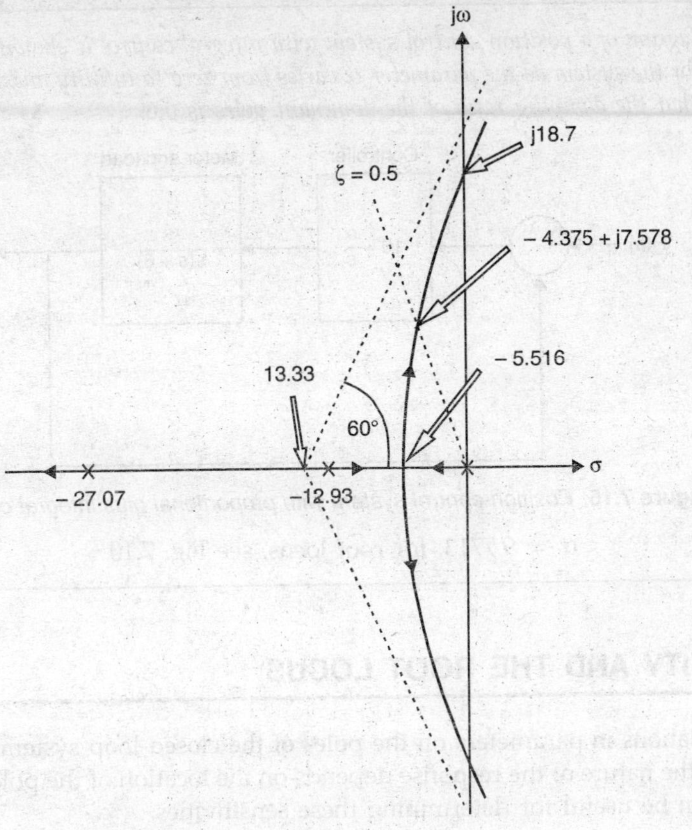

Figure 7.14. *Root locus for Equation (7.63)*

For $\zeta = 0.5$, we draw a line from the origin of the *s*-plane, ($\phi = 60°$), as shown. From the intersection of this line with the root locus, we get the roots at $s = -4.375 \pm j7.578$, with $K = 2392.56$ so that $\alpha = K/50 = 47.8521$.

For the original system, using this value of α, we get

$$T(s) = \frac{50\,(s + 47.8521)}{(s + 31.25)\,(s + 4.375 + j7.578)\,(s + 4.375 - j7.578)} \qquad ...(7.65)$$

$$= \frac{50\,(s + 47.5821)}{(s + 31.25)\left(s^2 + 8.75s + 75.763\right)}$$

(**Note:** The location of the third pole is obtained by utilizing the relationship that the sum of the poles is – 40.)

The response to a unit step is now obtained as

$$c(t) = 1 - 0.0341\, e^{-31.25t} + 1.1918\, e^{-4.375t} \cos(7.578t - 3.767)$$

DRILL PROBLEM 7.5

The block diagram of a position control system with integral control is shown in Fig. 7.15. Draw the root locus for the system as the parameter α varies from zero to infinity and hence determine the value of α so that the damping ratio of the dominant poles is 0.6.

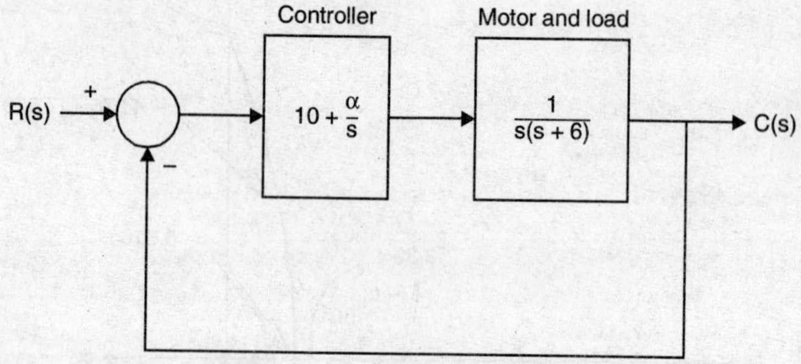

Figure 7.15. *Position-control system with proportional plus integral control*

Ans. α = 9.7713; for root locus, see Fig. 7.10

7.6 SENSITIVITY AND THE ROOT LOCUS

The effect of variations in parameters on the poles of the closed-loop system transfer function is important, since the nature of the response depends on the location of the poles. We shall see that the root locus can be useful for determining these sensitivities.

Let us rearrange Equation (7.1) as

$$K(s - z_1)(s - z_2) \ldots (s - z_m) + (s - p_1)(s - p_2) \ldots (s - p_n) = (s - q_1)(s - q_2) \ldots (s - q_n) \quad \text{...(7.66)}$$

where K is the variable parameter, z_i are the open-loop zeros, p_i are the open-loop poles, and q_i are the roots (*i.e.*, the poles of the closed-loop system transfer function).

It follows that the transfer function of the closed-loop system is given by

$$T(s) = \frac{K(s - z_1)(s - z_2) \cdots (s - z_m)}{(s - q_1)(s - q_2)(s - q_n)} \quad \text{...(7.67)}$$

The sensitivities ('root sensitivities') are defined below.

$$S_K^i = \frac{\partial q_i}{\partial K/K} \triangleq \text{the sensitivity of the ith root, } q_i\text{, to } K \quad \text{...(7.68)}$$

$$S_{p_j}^i = \frac{\partial q_i}{\partial p_j} \triangleq \text{the sensitivity of the ith root, } q_i \text{ to the open-loop pole } p_j \quad \text{...(7.69)}$$

$$S_{z_j}^i = \frac{\partial q_i}{\partial z_j} \triangleq \text{the sensitivity of the ith root, } q_i \text{ to the open-loop zero } z_j \quad \text{...(7.70)}$$

The total change Δq_i in the root q_i due to small variations in the parameter K and the locations of the zeros and poles of the open-loop transfer function is given by

$$\Delta q_i \simeq \frac{\partial q_i}{\partial K}\Delta K + \sum_{j=1}^{m}\frac{\partial q_i}{\partial z_j}\Delta z_j + \sum_{j=1}^{n}\frac{\partial q_i}{\partial p_i}\Delta p_i \quad ...(7.71)$$

$$= S_K^i \frac{\Delta K}{K} + \sum_{j=1}^{m} S_{z_j}^i \Delta z_i + \sum_{j=1}^{n} S_{p_j}^i \Delta p_j$$

which is valid for small changes.

It can be shown [by differentiation of Equation (7.66) and simplification] that

$$S_K^i = [-(s-q_i)\,T(s)]_{s=q_i} \quad ...(7.72)$$

$$S_{z_j}^i = \left[\frac{(s-q_i)\,T(s)}{s-z_j}\right]_{s=q_j} = \frac{S_K^i}{z_j - q_i} \quad ...(7.73)$$

$$S_{p_j}^i = \left[\frac{(s-q_i)\,T(s)}{s-p_j}\right]_{s=q_i} = \frac{S_K^i}{q_i - p_j} \quad ...(7.74)$$

Hence, we see that the sensitivities of q_i with respect to z_j and p_j are easily determined after S_K^i is calculated. The latter is obtained from the root locus plot and can be interpreted graphically as

$$S_K^i = -\frac{K \times \text{product of distances from the zeros } z_j \text{ to } q_i}{\text{Product of distances from the other roots } q_j \text{ to } q_i} \quad ...(7.75)$$

Since these distances are complex numbers, S_K^i will also be a complex number, denoting the direction in the s-plane in which the root will move if the parameter K is varied by a small amount.

The following example will illustrate the procedure:

EXAMPLE 7.12

Determine the sensitivity of the complex poles to α in Example 7.11.

SOLUTION

It was seen that for $\zeta = 0.5$, $\alpha = 47.8521$ and $K = 50\alpha$. Consequently,

$$T'(s) = \frac{50\alpha}{(s-q_1)(s-q_2)(s-q_3)} \quad ...(7.76)$$

$$= \frac{2392.6}{(s+31.24)(s+4.375+j7.578)(s+4.375-j7.5780)}$$

The sensitivity of the pole q_3 to K is obtained as

$$S_K^3 = -\left[(s-q_3)\,T(s)\right]\Big|_{s=q_3} = -\frac{2392.6}{(-4.375+j7.578+31.23)\times j7.578\times 2} = 5.654\, e^{j1.296} \quad ...(7.77)$$

Also, since $K = 50\alpha$, it follows that

$$S_\alpha^3 = \alpha \frac{\partial q_3}{\partial \alpha} = \frac{K}{50}\frac{\partial q_3}{\partial K}\frac{\partial K}{\partial \alpha} = K\frac{\partial q_3}{\partial K} = S_K^3. \qquad \text{...(7.78)}$$

DRILL PROBLEM 7.6

Determine the sensitivity of the dominant poles in Drill Problem 7.5 to small changes in α about its normal value α = 9.7713.

Ans. $S_K' = 1.15\, e^{j0.196}$

SUMMARY

- In this chapter, we have studied the root locus method for determining the movement of the poles of the transfer function of a closed-loop system when a parameter of the open-loop transfer function varies from zero to infinity. This is very important information since the transient response of the system depends on the location of these poles. It is especially useful in the design of control system where one may adjust either the open-loop gain or the location of one pole or zero in a compensator in order to obtain desirable response. These design procedures will be discussed in Chapter 10.
- Another important application of the root locus method is in determining the sensitivity of the poles of the closed-loop transfer function to variations in certain parameters of the open-loop transfer function. This is of great practical value, since in actual design we have to allow for some tolerance in component values in order to make the product economical.
- The procedure discussed in this chapter enables us to obtain a rough sketch of the root locus. In most practical situations we require a more accurate plot. This can be obtained by searching for points in the vicinity of the rough sketch where the angle condition is satisfied precisely. In the earlier days, this was done by means of a ruler and a protractor or a spirule. At the present time, one can conveniently use an electronic calculator, or a short computer program for the same purpose.
- Although the root locus method was developed in a period when computers were not readily accessible to all engineers, in the present age one may use computer programs for plotting the root locus and then use these plots for analysis and design. Nevertheless, the graphical construction techniques discussed in this chapter provide the insight needed to obtain the general shape of the root locus rapidly. The main advantage of the plot is that it presents a great deal of information in one figure, which would, otherwise, require pages of numerical data. Computer programs for this purpose, on the disk available with this book, are described in Appendix *E*.

Problems

1. The transfer function of the forward path of a unity-feedback system is given by

$$G(s) = \frac{K}{s(s+1)(s+9)}$$

Determine if the following points will be on the locus of roots of the characteristic polynomial of the closed-loop system. If, for any point, the answer is 'yes', determine the value of K for this point to be a root.

(*a*) $s = -0.5$, (*b*) $s = -2$, (*c*) $s = j3$, (*d*) $s = -0.5 + j0.5$.

2. Sketch the root locus diagrams for each of the following open-loop transfer functions as K is varied from zero to infinity. Be sure to label all the critical points, and in each case, determine the maximum value of K for which the closed-loop system will be stable.

(*a*) $\dfrac{K}{s(s+5)(s+7)}$ (*b*) $\dfrac{K}{s(s^2+10s+34)}$

(*c*) $\dfrac{K}{s(s+1)(s^2+4s+20)}$ (*d*) $\dfrac{K}{s(s+4)(s^2+4s+20)}$

(*e*) $\dfrac{K}{s^2(s^2+6s+25)(s+10)}$

3. The block diagram of a position-control system with velocity feedback is shown in Fig. P7.3. Use the root locus method to determine the value of α so that the damping ratio of the complex poles may be 0.6. What is the sensitivity of these poles to α in this case?

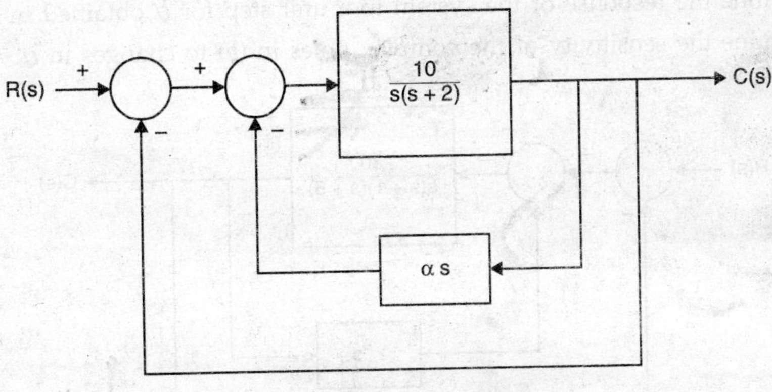

Figure P7.3. *Block diagram*

4. Consider the block diagram shown in Fig. P7.4.
   (*a*) Draw the locus of the poles of the overall system as z is varied from zero to infinity.
   (*b*) Determine the value of z so that the damping ratio of the dominant poles may be 0.4. (*Hint*: There may be two answers!)
   (*c*) Determine the response of the system to a unit step input for the value of z calculated in part (*b*). What is the maximum overshoot?

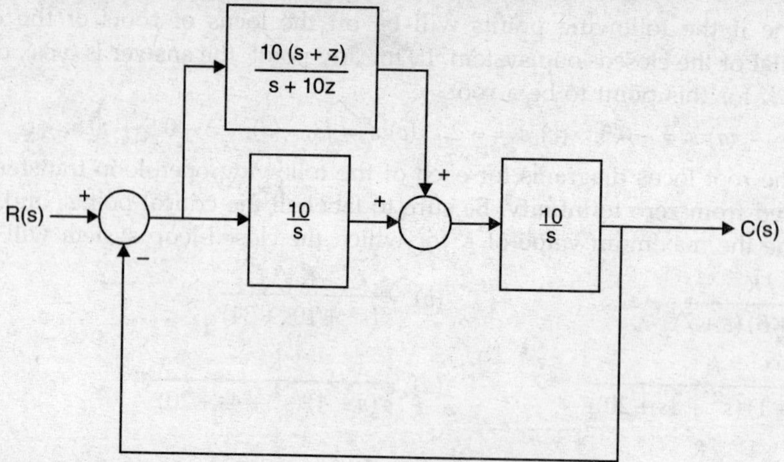

Figure P7.4. *Block diagram*

5. A third-order position-control system with velocity feedback is shown in Fig. 7.1.
   (*a*) What value of α will give the maximum damping ratio for the complex poles?
   (*b*) What value of α will give a damping ratio of 0.5 for the complex poles?
   (*c*) Determine the response of the system to a unit step for α obtained in (*b*).
   (*d*) Determine the sensitivity of the complex poles in (*b*) to changes in α.

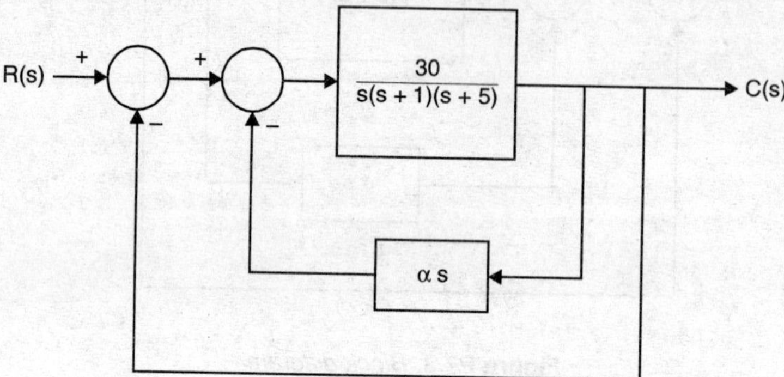

Figure P7.5. *Position-control system with velocity feedback*

6. The block diagram of a system is shown in Fig. P7.6. Determine the value of α so that the damping ratio of the complex poles may be 0.5. What will be the steady-state error to a unit ramp input?

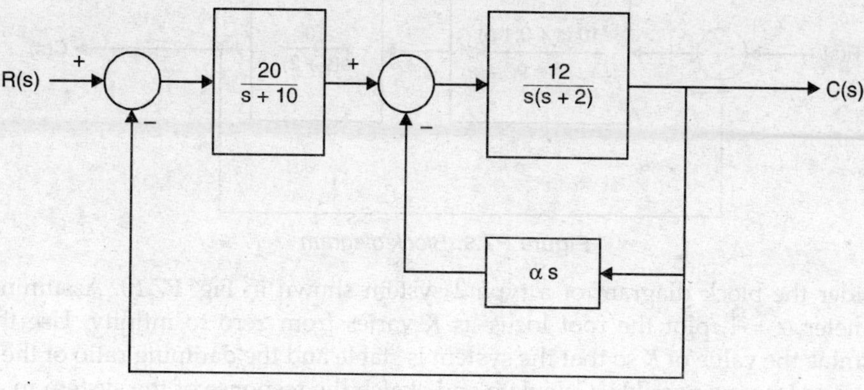

Figure P7.6. *Block diagram*

7. The open-loop transfer function of a control system with unity feedback is given by

$$G(s) = \frac{25(s+\alpha)}{s^2(s+6)}$$

Sketch the root locus as α varies from zero to infinity, and determine the value of α so that the damping ratio of the complex poles may be 0.5. What is the sensitivity of these poles to small changes in α?

8. The simplified block diagram of the pitch control system of a supersonic airplane, flying at an altitude of 60,000 feet, is shown in Fig. P7.8. Sketch the root locus as *K* varies from zero to infinity. For what value of *K* will the damping ratio of the complex poles be 0.5? Determine the response to a unit step input for this case.

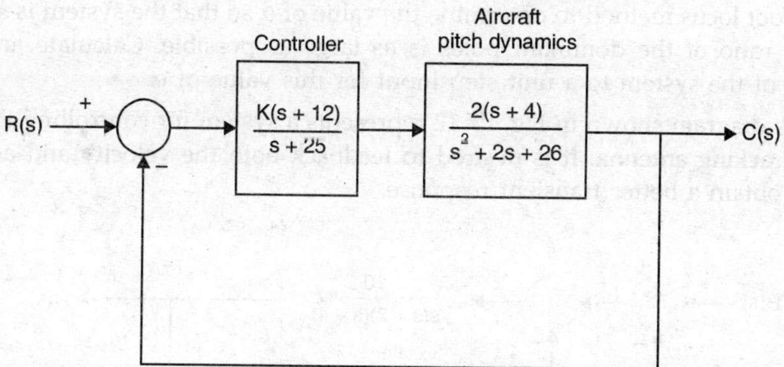

Figure P7.8. *Block diagram*

9. The block diagram of a position-control system with a lead compensator is shown in Fig. P7.9. Sketch the root locus as *p* varies from zero to infinity, and determine the value of *p* so that the damping ratio of the complex poles may be 0.5. What will be the response to a unit step for this value of *p*?

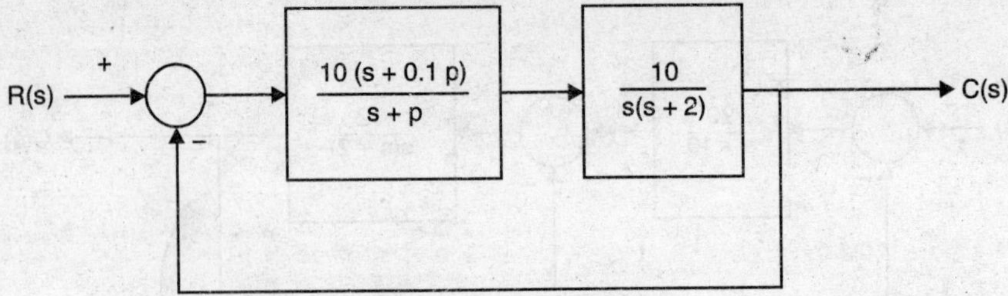

Figure P7.9. *Block diagram*

10. Consider the block diagram of a type 2 system shown in Fig. P7.10. Assuming that the parameter $\alpha = 1$, plot the root locus as K varies from zero to infinity. Use this plot to determine the value of K so that the system is stable and the damping ratio of the dominant poles is as large as possible. Calculate and sketch the response of the system to a unit step input for this value of K.

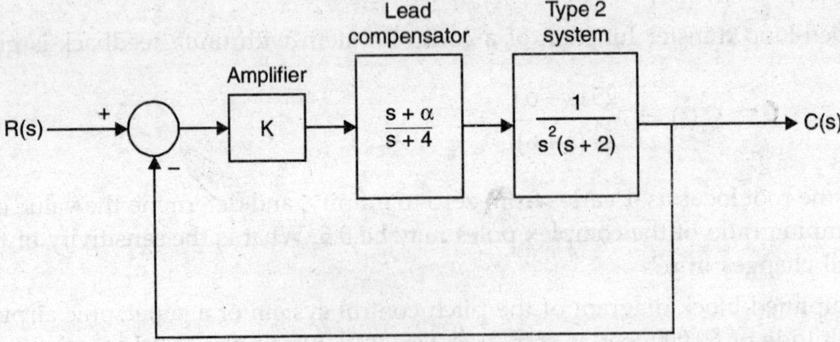

Figure P7.10. *Block diagram*

11. Reconsider the previous problem but with $K = 12$ and α adjustable in the range 0 to 10. Use the root locus method to determine the value of α so that the system is stable and the damping ratio of the dominant poles is as large as possible. Calculate and sketch the response of the system to a unit step input for this value of α.

12. The block diagram shown in Fig. P7.12 represents a system for controlling the position of a radar tracking antenna. It is desired to feedback both the velocity and acceleration in order to obtain a better transient response.

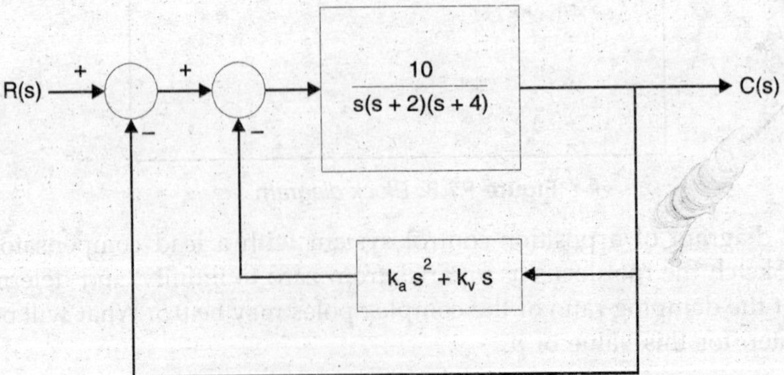

Figure P7.12. *System with velocity and acceleration feedback*

Assuming that the value of k_a is fixed at 1, use the root locus method to determine the value of k_v in such a way that the damping ratio of the complex poles will be 0.45. What will be the steady-state error of the system to a unit ramp input for this case? Also, calculate the maximum overshoot in response to a unit step input.

13. Plot the root locus for Problem 12 if the value of k_v is fixed at 2, but k_a is adjustable. Determine the value of k_a so that the damping ratio of the complex poles may be 0.5. Calculate and sketch the response of this system to a unit step input.

14. The block diagram in Fig. P7.14 represents a more realistic position control system with velocity feedback, including the time constant of the tachometer, and also a compensator. Assuming that the forward gain K has been fixed at 10, use the root locus plot to determine the value of k_v so that the damping ratio of the dominant poles of the closed-loop system transfer function is 0.5. What will be the steady-state error of this system to a unit ramp input?

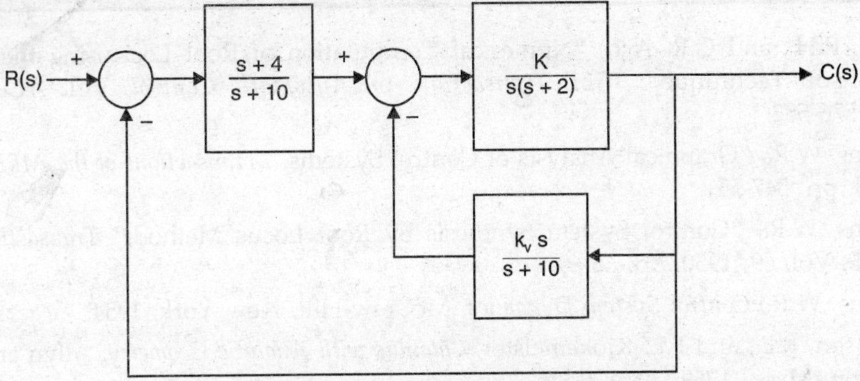

Figure P7.14. *System with tachometer feedback*

15. Plot the root locus for Problem 14 if the value of k_v is fixed at 2, but the gain K can be adjusted. Determine the value of K for which the damping ratio of the dominant poles of the transfer function of the closed-loop system will be 0.707.

16. Determine the sensitivity of the dominant poles in the previous problem to changes in the value of K from the nominal value calculated.

17. The transfer function of the forward path of a unity-feedback system is given by

$$G(s) = \frac{10(s+0.7)}{s(s+p)(s+4)}$$

(*a*) Sketch the locus of the poles of the closed-loop system, as p varies from 0 to infinity.

(*b*) Use this plot to determine the location of poles of the closed-loop system transfer function if the nominal value of $p = 2$.

(*c*) What is the sensitivity of the complex poles to small variations in p for this case?

18. The transfer function of the forward path of a unity-feedback system is given by

$$\frac{K(s+5)}{s(s+2)(s+3)}$$

(*a*) Sketch the locus of the poles of the closed-loop system as K varies from zero to infinity and determine these poles if $K = 8$.

(*b*) Determine the sensitivity of the complex poles to variations in the location of the open-loop pole at $s = -2$ (for $K = 8$).

19. In the previous problem, determine the sensitivity of the complex poles to variations in the location of the open-loop zero.

20. The block diagram of the pitch-control system for a supersonic airoplane is shown in Fig. P7.8 in connection with Problem 8.

(*a*) Determine the value of K so that the damping ratio of the dominant poles will be 0.6.

(*b*) How will these poles move in the *s*-plane for a small variation in the value of K?

(*c*) How will these poles move in the *s*-plane for a small variation in the location of the zero at $s = -4$?

References

1. Ash, R.H., and G.R. Ash, "Numerical Computation of Root Loci using the Newton-Raphson Technique," *IEEE Transactions on Automatic Control,* Vol. AC-13, 1968, pp. 576-582.
2. Evans, W.R., "Graphical Analysis of Control Systems," *Transactions of the AIEE,* Vol. 67, 1948, pp. 547-551.
3. Evans, W.R., "Control System Synthesis by Root Locus Method," *Transactions of the AIEE,* Vol. 69, 1950, pp. 66-69.
4. Evans, W.R., *Control System Dynamics,* McGraw-Hill, New York, 1954.
5. Johnson, R.E., and F.L. Kiokemeister, *Calculus with Analytic Geometry,* Allyn and Bacon, Boston, Mass., 1969.
6. Phillips, C.L., and R.D. Harbor, *Feedback Control Systems,* Prentice-Hall, Englewood Cliffs, New Jersey, 1988.
7. Sinha, N.K., "The Impact of the Electronic Pocket Calculator on Electrical Engineering Education," *IEEE Transactions on Education,* Vol. E-20, 1977, pp. 6-9.
8. Thaler, G.J., *Design of Feedback Systems,* Dowden, Hutchinson and Ross, Inc., Stroudsburg, PA., 1973.

CHAPTER

8 Frequency Response Plots

8.1 INTRODUCTION

In the previous chapters, we discussed several methods for studying the stability of a closed-loop system. The Routh-Hurwitz criterion enables us to determine the stability from the characteristic polynomial without actually determining the roots, whereas the root locus method shows how roots of the characteristic polynomial move in the s-plane, as one parameter varie, from zero to infinity. Both these methods are very helpful in providing an insight into the transient performance of the closed-loop system. However, they are based on the assumption that the transfer function of the system is known precisely. In many practical situations, this is not the case.

An alternative procedure is to utilize the frequency response of the open-loop system to determine the stability of the closed-loop system. As we shall see later, this may also provide information about the transient performance of the closed-loop system. The main advantage of this approach is that the frequency response of a stable open-loop system can be obtained experimentally and it is not necessary to know the transfer function. We simply apply a sine-wave input to the system, and after waiting until steady-state conditions are reached, we measure the ratio of the magnitudes of the output and the input as well as the phase difference between them. These measurements are easily carried out using digital gain-phase meters currently available in the market, although it is possible to utilize more sophisticated (and expensive) transfer-function analyzers. The measurements are repeated at different frequencies until the entire range of frequencies of interest is covered.

Another advantage of frequency response analysis is that it can be applied to systems that do not have rational transfer functions. Systems with pure delay belong to this category because for these the transfer function is the ratio of two polynomials multiplied by the factor $e^{-\beta s}$, where β is the transport lag (in seconds) of the system. For such cases it is not possible to utilize the Routh-Hurwitz criterion. The root locus method can be applied only after the irrational transfer function $e^{-\beta s}$ is replaced by an approximate rational function.

In this chapter we shall first consider the relationship between the transfer function and the frequency response of a linear system. This will be followed by the study of different types of frequency response plots that are used by control engineers. Finally, we shall discuss the estimation of the transfer function of a linear system from its frequency response.

8.2 TRANSFER FUNCTION AND FREQUENCY RESPONSE

If we apply a sine-wave input to a stable linear system, the steady-state output will also be a sine-wave of the same frequency. The output will, in general, differ from the input in both magnitude and phase. As shown in Appendix *A*, the frequency response can be calculated by simply replacing *s* in the transfer function by $j\omega$, that is

$$G(j\omega) = G(s)\big|_{s=j\omega} = M(\omega)\, e^{j\phi(\omega)} = M\angle\phi \qquad ...(8.1)$$

In Equation (8.1), *M* is the ratio of the amplitudes of the output and input sinusoids and is called the magnitude ratio or gain. Also, ϕ is the angle by which the output leads the input. Both *M* and ϕ are functions of the angular frequency ω.

The function $G(j\omega)$ is called the frequency response of the system and is given by the same graphical interpretation as for the calculation of residues. Thus, if

$$G(s) = \frac{K(s-z_1)(s-z_2)\cdots(s-z_m)}{(s-p_1)(s-p_2)\cdots(s-p_n)} \qquad ...(8.2)$$

then for $s = j\omega_1$, we get

$$G(j\omega_1) = \frac{K(j\omega_1-z_1)(j\omega_1-z_2)\cdots(j\omega_1-z_m)}{(j\omega_1-p_1)(j\omega_1-p_2)\cdots(j\omega_n-p_n)} \qquad ...(8.3)$$

$$= K\cdot\frac{\text{product of directed distances from each zero to } s = j\omega_1}{\text{product of directed distances from each pole to } s = j\omega_1}$$

If each of the directed distances in Equation (8.3) is expressed in the polar form, then we can separate the moduli and arguments to obtain

$$M = K\cdot\frac{\text{product of lengths of vectors from the zeros}}{\text{product of lengths of vectors from the poles}} \qquad ...(8.4)$$

and ϕ = sum of arguments of vectors from the zeros

– sum of arguments of vectors from the poles ...(8.5)

where it is understood that all of these "vectors" (or directed distances) are drawn to the point $s = j\omega_1$ on the imaginary axis of the *s*-plane.

EXAMPLE 8.1

Consider a linear system described by the transfer function

$$G(s) = \frac{10(s+1)}{s^2+4s+13} = \frac{10(s+1)}{(s+2+j3)(s+2-j3)} \qquad ...(8.6)$$

The pole-zero plot of the transfer function is shown in Fig. 8.1, which also illustrates graphical evaluation of $G(j2)$.

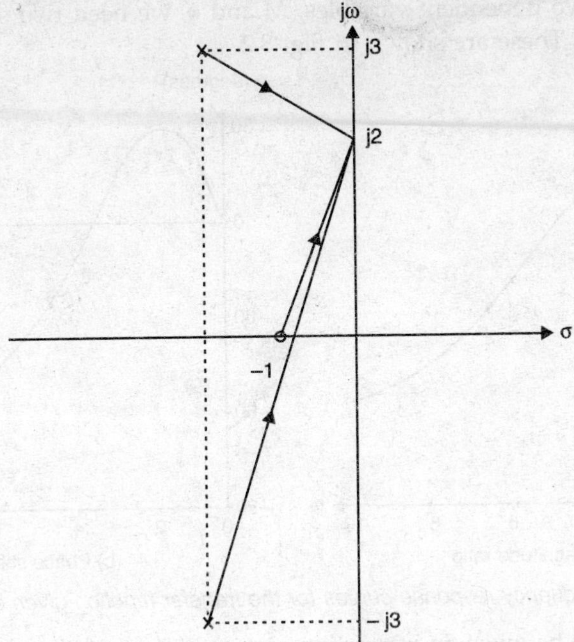

Figure 8.1. *Frequency response calculation for the transfer function given by Equation (8.6)*

From Fig. 8.1, it follows that

$$G(j2) = \frac{10\,(j2+1)}{(j2+2+j3)\,1\,(j2+2-j3)} \qquad ...(8.7)$$

$$= \frac{10 \times \sqrt{5}\ \angle 63.4^\circ}{\sqrt{29}\ \angle 68.2 \times \sqrt{5}\ \angle -26.6^\circ} = 1.857\angle 21.8^\circ$$

so that $M\,(2) = 1.857$ and $\phi\,(2) = 21.8°$.

This calculation can be repeated for different values of ω. Table 8.1 shows the values of M and ϕ for different values of ω for this example.

TABLE 8.1: Frequency Response for the Transfer Function given by Equation (8.6)

ω	M	ϕ	ω	M	ϕ
0	0.769	0	5	2.186	–42.3°
1	1.118	26.6°	6	1.830	–53.2°
2	1.857	21.8°	7	1.550	–60.3°
3	2.5	0°	8	1.339	–65.0°
3.5	2.596	–12.8°	10	1.050	–71.0°
4	2.533	–24.7°			

Since we have two dependent variables, M and ϕ, we need two separate plots to indicate their variation with ω. These are shown in Fig. 8.2.

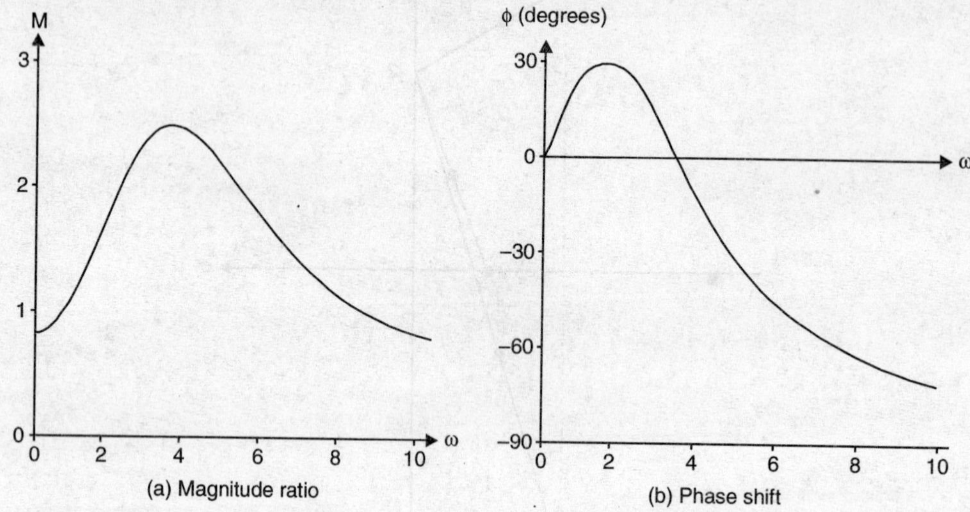

Figure 8.2. *Frequency response curves for the transfer function given by Equation (8.6)*

Equation (8.4) can be given an interesting pictorial interpretation. Imagine a flexible rubber sheet suspended over the complex frequency plane. At each pole location, the sheet is picked up by a thin rod of infinite height, and, at the location of each zero, the sheet is tacked down to the plane. The height of the rubber sheet, at any point in the s-plane, will then represent the magnitude of $G(s)$ for that value of s, and the magnitude ratio for different values of ω is represented by the cross-section of the sheet along the imaginary axis of the s-plane.

DRILL PROBLEM 8.1

Sketch the frequency response curves for the following transfer functions:

(a) $G(s) = \dfrac{4(s+1)}{s+4}$ (b) $G(s) = \dfrac{10}{s^2 + 2s + 10}$

Ans.

(a)

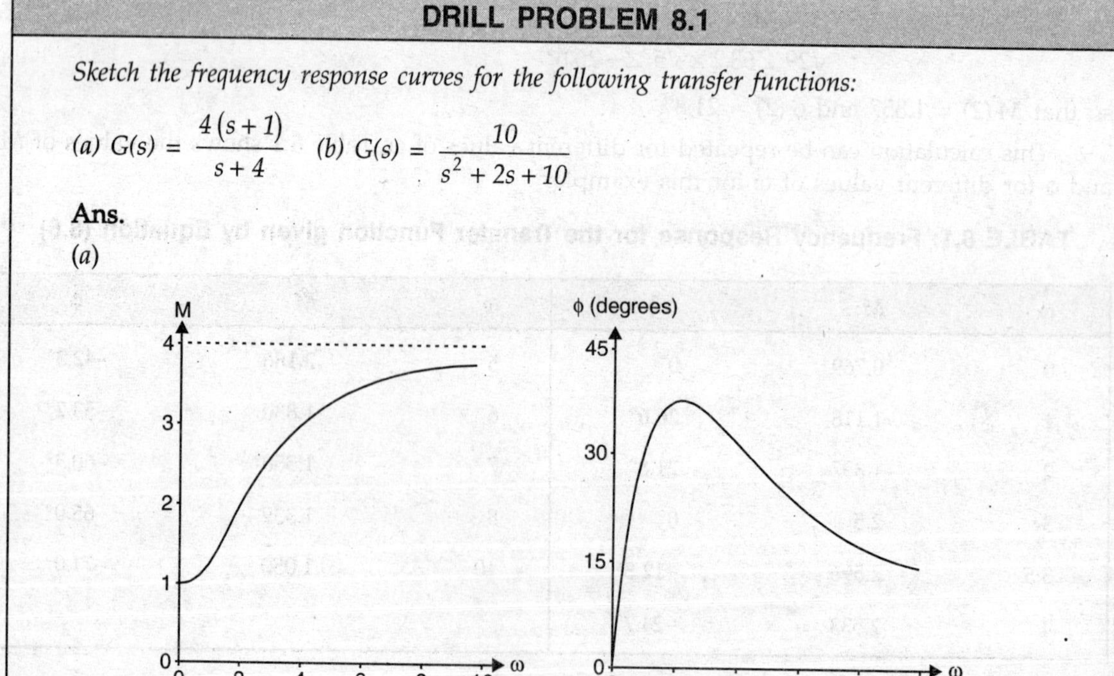

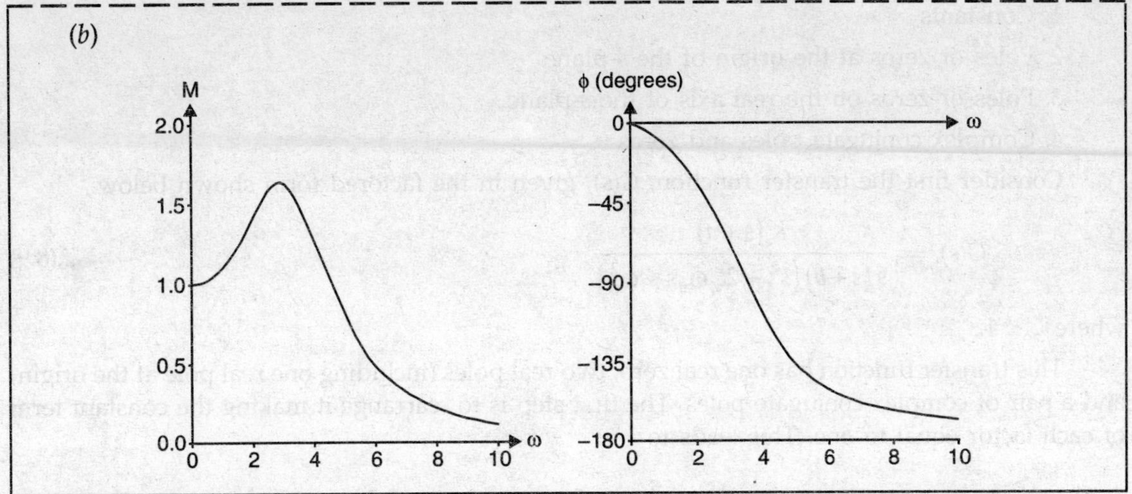

8.3 BODE PLOTS

The procedure described in the previous section gets quite tedious when the transfer function has several poles and zeros. The drudgery in computation is considerably reduced by using logarithmic coordinates. The resulting plots are called Bode plots, honoring H.W. Bode, who used them in the study of feedback amplifiers. These plots require semilog graph paper, where the logarithmic scale is used for the ω-axis. Two graphs are required: one for the gain in decibels and the other for the phase shift in degrees plotted against frequency on the log scale. The simplification in Bode plots results partly due to the basic advantage of logarithmic representation that multiplication and division are replaced by addition and subtraction respectively. There are several other advantages, as we shall see later.

8.3.1 Logarithmic Scales

The logarithmic scale, used for frequency in Bode plots, has some interesting properties. First, we observe that the logarithmic scale is non linear, that is, the distance between 1 and 2 is greater than the distance between 2 and 3, and so on. As a result, use of this scale enables us to cover a greater range of frequencies. It is interesting to note that on the log scale, the distance between 1 and 10 is equal to the distance between 10 and 100, or between 100 and 1000. This distance is called a *decade*. In fact, the distance between k and $10k$, where k is any positive (nonzero) number, is equal to the decade. This follows because

$$\log 10k - \log k = \log 10 = 1 \qquad \text{...(8.8)}$$

Similarly, the distance between k and $2k$ is equal to the constant log 2, and is called an *octave*.

Semilog paper comes in one, two, three, or four cycles, indicating the range of coverage in decades. It may be noted that we cannot locate the point $\omega = 0$ on the log scale since $\log 0 = -\infty$.

8.3.2 Magnitude Plot: Straight-Line Approximation

We shall now discuss how the magnitude of the frequency response can be sketched with very little effort by using logarithmic coordinates. In general, the transfer function can be factored into four types of terms:

1. Constants
2. Poles or zeros at the origin of the s-plane
3. Poles or zeros on the real axis of the s-plane
4. Complex conjugate poles and zeros

Consider first the transfer function, $G(s)$, given in the factored form shown below.

$$G(s) = \frac{K(s+a)}{s(s+b)\left(s^2 + 2\zeta\,\omega_n s + \omega_n^2\right)} \qquad \text{...(8.9)}$$

where $\zeta < 1$.

This transfer function has one real zero, two real poles (including one real pole at the origin) and a pair of complex conjugate poles. The first step is to rearrange it making the constant term of each factor equal to one. This leads to

$$G(s) = \frac{Ka}{b\omega_n^2} \frac{1 + \frac{s}{a}}{s\left(1 + \frac{s}{b}\right)\left(1 + 2\zeta \frac{s}{\omega_n} + \frac{s^2}{\omega_n^2}\right)} \qquad \text{...(8.10)}$$

If we now replace s by $j\omega$ and take the magnitude of both sides, we have

$$M = |G(j\omega)| = \frac{Ka}{b\omega_n^2} \frac{\left|1 + \frac{j\omega}{a}\right|}{|j\omega| \cdot \left|1 + j\frac{\omega}{b}\right| \cdot \left|1 + j2\zeta\frac{\omega}{\omega_n} + \left(j\frac{\omega}{\omega_n}\right)^2\right|} \qquad \text{...(8.11)}$$

Taking the logarithm of both sides and multiplying by 20, we get

$$\begin{aligned} 20 \log M &= 20 \log\left(\frac{Ka}{b\omega_n^2}\right) + 20 \log\left|1 + j\frac{\omega}{a}\right| - 20 \log|j\omega| - 20 \log\left|1 + j2\zeta\frac{\omega}{\omega_n} + \left(j\frac{\omega}{\omega_n}\right)^2\right| \\ &= 20 \log M_1 + 20 \log M_2 - 20 \log M_3 - 20 \log M_4 - 20 \log M_5 \end{aligned} \qquad \text{...(8.12)}$$

In the left-hand side of Equation (8.12), the gain in decibels is seen to be the algebraic sum of the gains obtained from each factor of the transfer function given in Equation (8.10). The first term on the right-hand side of Equation (8.12) is a constant and easily evaluated. Let us now examine the second term on the right-hand side.

In particular, we shall study the asymptotic behaviour of the second term for both small and large ω. First consider the case when

$$\frac{\omega}{a} << 1 \qquad \text{...(8.13)}$$

It is easily seen that for this case

$$M_2 = \left|1 + j\frac{\omega}{a}\right| = \sqrt{1 + \left(\frac{\omega}{a}\right)^2} \approx 1 \qquad \text{...(8.14)}$$

and

$$20 \log M_2 = 20 \log\left|1 + j\frac{\omega}{a}\right| \approx 20 \log 1 \approx 0 \qquad \text{...(8.15)}$$

Similarly, for $\frac{\omega}{a} >> 1$ ...(8.16)

we have $$M_2 = \left|1 + j\frac{\omega}{a}\right| \approx \frac{\omega}{a} \quad ...(8.17)$$

and $$20 \log M_2 \approx 20 \log \frac{\omega}{a} \quad ...(8.18)$$

For the intermediate case, when $\omega/a = 1$, we have

$$\left|1 + j\frac{\omega}{a}\right| = |1 + j1| = \sqrt{2} \quad ...(8.19)$$

and $$20 \log M_2 = 20 \log \sqrt{2} \approx 3 \text{ dB} \quad ...(8.20)$$

A plot of the asymptotes is shown in Fig. 8.3, where the actual curve is shown dotted.

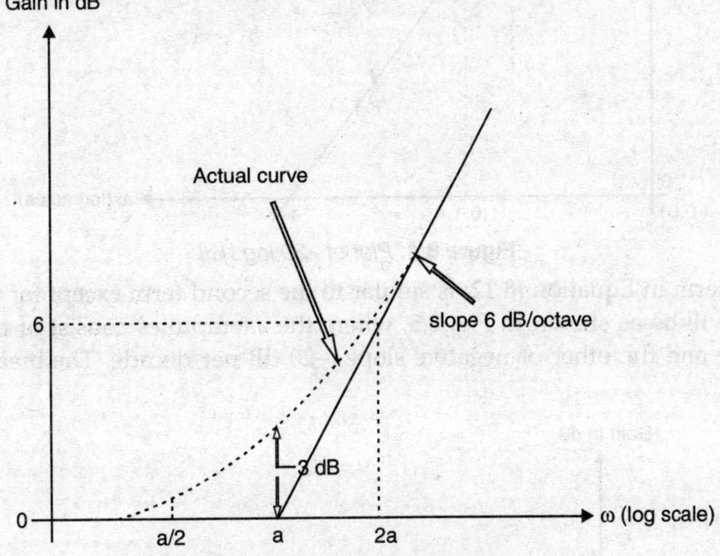

Figure 8.3. *Plot of 20 log |1 + jω/a|*

It will be seen that the asymptotes consist of a pair of straight lines intersecting at $\omega = a$, which is called the *break* or *corner* frequency. For $\omega < a$, the asymptote is a straight line of zero slope and zero height, represented by Equation (8.15). For $\omega > a$, the asymptote is a straight line of slope 20 dB/decade (or 6 dB per octave), as given by Equation (8.18). The maximum difference between the exact curve and the asymptotic approximation occurs at the break frequency, and is equal to 3 dB. If we move one octave away from the break frequency the error is approximately 1 dB, and one decade away from the break frequency the error is less than 0.05 dB. Hence, for most applications, the straight-line approximations are often of sufficient accuracy. However, if a better approximation is desired, one may introduce a correction of 3 dB at the break frequency and corrections of 1 dB at points one octave above and below the break frequency.

Let us now consider the third term in Equation (8.12). Here,

$$M_3 = |j\omega| = \omega \quad ...(8.21)$$

and $$-20 \log M_3 = -20 \log \omega \quad ...(8.21a)$$

The resulting plot is shown in Fig. 8.4. This is simply a straight line of slope –20 dB/decade, crossing the zero-decibel line at $\omega = 1$. Note that this is exact.

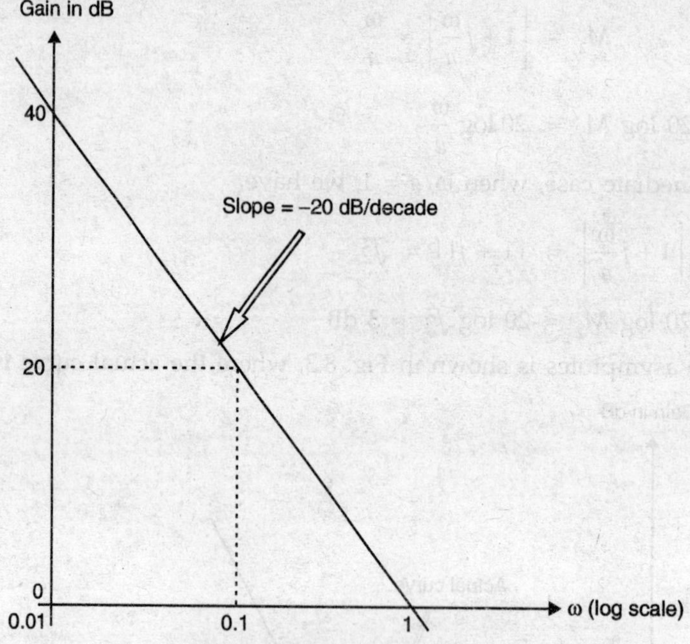

Figure 8.4. *Plot of –20 log |jω|*

The fourth term in Equation (8.12) is similar to the second term except for the negative sign. Hence, the plot will be as shown in Fig. 8.5, where the asymptotes consist of two straight lines: one of zero slope and the other of negative slope, –20 dB per decade. The break frequency is at $\omega = b$.

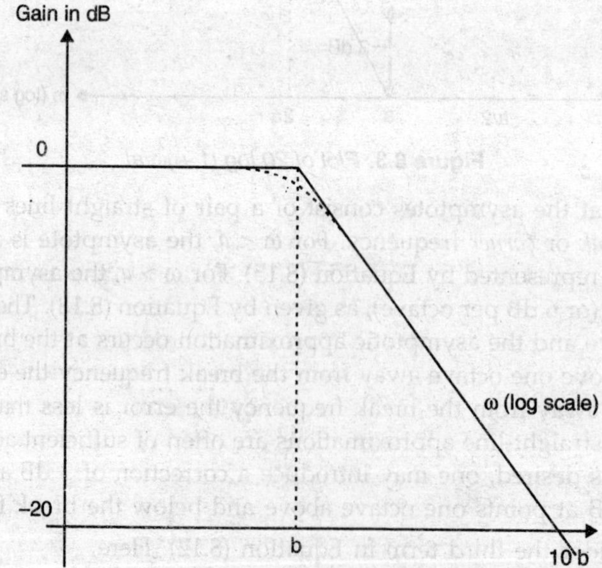

Figure 8.5. *Plot of –20 log |1 + jω/b|*

The last term in Equation (8.12) is different from all others because it involves a quadratic factor with complex roots. Here, we have

$$M_5 = \left| 1 + 2\zeta\left(j\frac{\omega}{\omega_n}\right) + \left(j\frac{\omega}{\omega_n}\right)^2 \right| \qquad ...(8.22)$$

$$= \left| \left(1 - \frac{\omega^2}{\omega_n^2}\right) + j2\zeta\frac{\omega}{\omega_n} \right|$$

We shall again study the asymptotic behaviour for the two cases: $\omega/\omega_n << 1$, and $\omega/\omega_n >> 1$. For the former we get

$$M_5 \approx 1 \qquad ...(8.23)$$

and $$-20 \log M_5 \approx 0 \qquad ...(8.24)$$

which is a straight line of zero slope, coinciding with the zero-decibel line. For $\omega/\omega_n >> 1$, we have

$$M_5 \approx \left(\frac{\omega}{\omega_n}\right)^2 \qquad ...(8.25)$$

and $$-20 \log M_5 \approx -20 \log \left(\frac{\omega}{\omega_n}\right)^2 \qquad ...(8.26)$$

which is a straight line of slope –40 dB/decade, crossing the zero-decibel line at $\omega = \omega_n$.

Hence, it is seen that the asymptotes consist of a pair of straight lines, intersecting at $\omega = \omega_n$, the undamped natural frequency. For $\omega < \omega_n$, the asymptote is the zero-decibel line of slope zero, whereas for $\omega > \omega_n$, the asymptote is a line of slope – 40 dB/decade. At the intermediate point $\omega = \omega_n$, we get

$$M_5 = 2\zeta \qquad ...(8.27)$$

and $$-20 \log M_5 = -20 \log 2\zeta \qquad ...(8.28)$$

Consequently, the exact curve will differ from the asymptotes near the break frequency ω_n, and the deviation will depend upon the value of the damping ratio ζ. In particular, for ζ less than 0.5, the gain for $\omega = \omega_n$ will be positive. The plot for the gain is shown in Fig. 8.6, where the exact curves are shown dotted. It will be seen that actual gain at the break frequency is quite high for small values of ζ.

From the derivative of M_5 with respect to ω, it can be shown that the exact curve has a peak value for

$$\omega_{max} = \omega_n\sqrt{1 - 2\zeta^2} \qquad ...(8.29)$$

This indicates that a maximum occurs only if $\zeta < \sqrt{0.5}$. If this condition is satisfied, then the maximum value of M_5 is obtained by substituting ω_{max} for ω in Equation (8.22).

$$M_5(\omega_{max}) = 2\zeta\sqrt{1 - \zeta^2} \quad \text{for} \quad \zeta \leq \sqrt{\frac{1}{2}} \qquad ...(8.30)$$

Hence, $$-20 \log M_5 = 20 \log \frac{1}{2\zeta\sqrt{1 - \zeta^2}} \qquad ...(8.31)$$

is the peak gain in decibels.

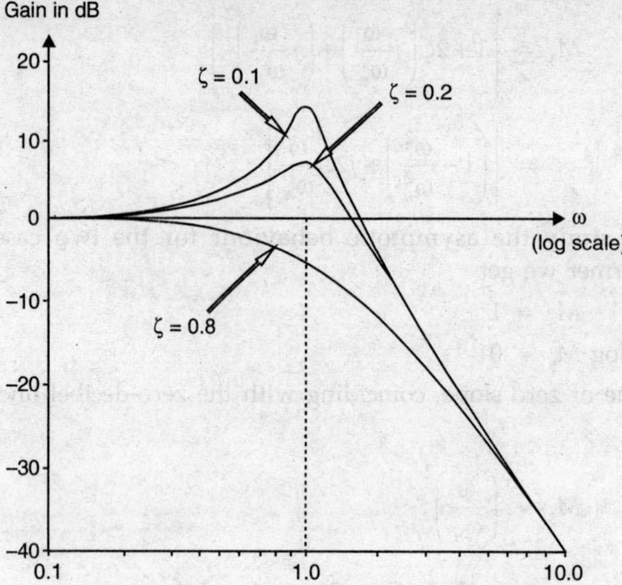

Figure 8.6. *Gain plot for quadratic factor*

The overall gain is now obtained as the algebraic sum of the gain for the different factors. For the asymptotic plot, the summation can be carried out very easily if we realize that for factor any contribution to the gain is made only for frequencies above the break frequency for that factor. Hence, if we start the plot for ω less than the lowest break frequency all we have to do is to account for the change in slope at each of the break frequencies. For the transfer function given in Equation (8.10), the overall gain is shown in Fig. 8.6, where it is assumed that $a < \omega < b$.

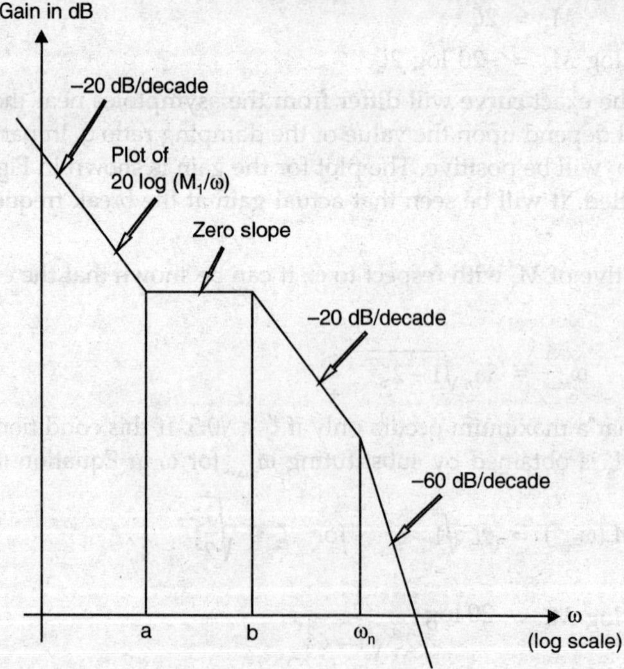

Figure 8.7. *Asymptotic plot of gain for the transfer function given in Equation (8.10)*

For $\omega < a$, only the first and the third terms on the right-hand side of Equation (8.12) are significant and they give us a straight line of slope –20 dB/decade, corresponding to the expression 20 log M_1/ω, where $M_1 = Ka/b\omega_n^2$.

As the frequency is increased, we first get the effect of the zero at $s = -a$ when ω reaches the value of a. The linear factor causes the slope to be increased by 20 dB/decade, resulting in a straight line of zero slope. At the next break frequency, $\omega = b$, the effect of the pole is to reduce the slope to –20 dB/decade. Finally, the break frequency at ω_n, the effect of the complex poles is to further reduce the slope by 40 dB/decade, that is to –60 dB/decade.

EXAMPLE 8.2

To get a clear idea of the procedure we shall go over each step for the transfer function.

$$G(s) = \frac{200\,(s+2)}{s\left(s^2+10s+100\right)} \qquad \text{...(8.32)}$$

SOLUTION

(*i*) First we rewrite the transfer function as

$$G(s) = \frac{4\,(1+s/2)}{s\left[1+s/10+(s/10)^2\right]} \qquad \text{...(8.33)}$$

(*ii*) Next, we order the poles and zeros in sequence as frequency increases, to get

(*a*) a pole at the origin (with multiplicity one),

(*b*) a zero at $s = -2$, giving a break frequency at $\omega = 2$,

(*c*) a pair of complex conjugate poles with damping ratio $\zeta = 0.5$ and undamped natural frequency $\omega_n = 10$.

(*iii*) Start with the factor $4/s$.

(*iv*) The plot for $4/s$ is a straight line of slope –20 dB/decade. To draw this line, we need one point on it, since the slope is known. This can be done by noting that when ω = 2, (or $s = j2$), we get a gain of 6 dB from the factor $4/s$.

(*v*) Since $\omega = 2$ is a break frequency corresponding to a zero, at this point the slope is increased by 20 dB/decade. Hence, the slope will be zero until the next break frequency is reached.

(*vi*) Finally, at the break frequency $\omega = 10$, corresponding to the pair of complex poles with undamped natural frequency $\omega_n = 10$, the slope must decrease by 60 dB/decade.

The resulting asymptotic plot is shown in Fig. 8.8.

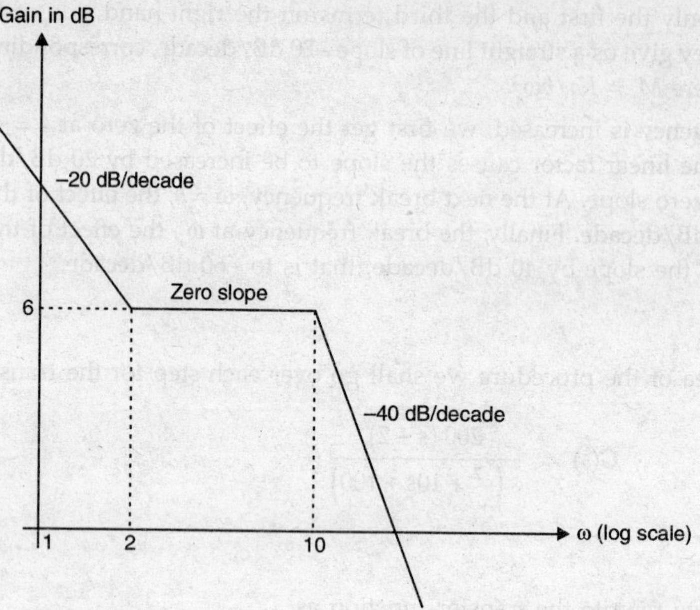

Figure 8.8. *Asymptotic plot of the gain for Equation (8.32)*

DRILL PROBLEM 8.2

Sketch the asymptotic gain for the following transfer functions:

(a) $\dfrac{20s}{(s+1)(s+10)}$ (b) $\dfrac{300(s^2+2s+4)}{s(s+10)(s+20)}$

Ans.

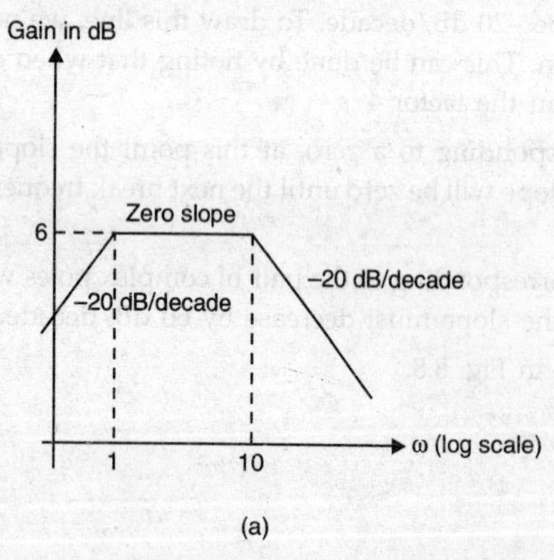

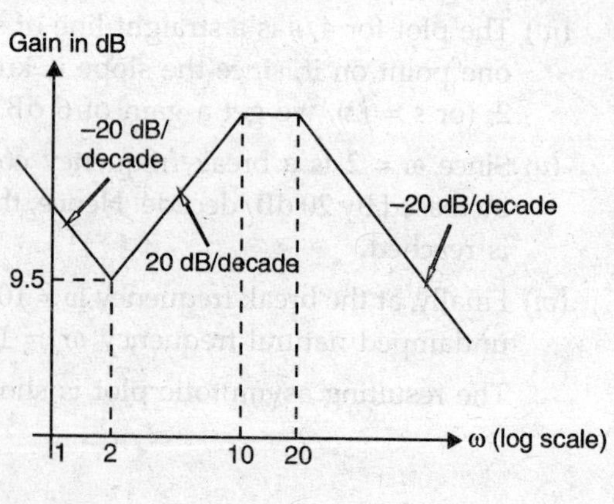

8.3.3 Phase Plot

The phase shift at any frequency can be obtained as the algebraic sum of the phase shifts due to the various factors in the transfer function, in a manner similar to that used for obtaining magnitude plots. To illustrate the procedure in detail, let us again consider the transfer function given in Equation (8.9). The argument of $G(s)$ for $s = j\omega$ is given by

$$\phi = \arg G(j\omega)$$

$$= \arg\left(\frac{Ka}{b\omega_n^2}\right) + \arg\left(1 + j\frac{\omega}{a}\right) - \arg(j\omega) - \arg\left(1 + j\frac{\omega}{b}\right) - \arg\left(1 - \frac{\omega^2}{\omega_n^2} + j2\zeta\frac{\omega}{\omega_n}\right)$$

$$= \phi_1 + \phi_2 - \phi_3 - \phi_4 - \phi_5 \quad \text{...(8.34)}$$

The first term on the right-hand side, which is the phase shift due to a real constant, will be identically zero. Hence, we shall now determine the second term, given by

$$\phi_2 = \arg\left(1 + j\frac{\omega}{a}\right) = \tan^{-1}\frac{\omega}{a} \quad \text{...(8.35)}$$

We can obtain asymptotic approximations by considering three cases: (*i*) $\omega/a << 0.1$, (*ii*) $0.1 \le \omega/a \le 10$ and (*iii*) $\omega/a >> 10$. For the first, we have

$$\phi_2 \approx 0 \quad \text{...(8.36)}$$

Similarly, for $\omega/a >> 10$, we get

$$\phi \approx 90° \quad \text{...(8.37)}$$

For the intermediate case, $0.1 \le \omega/a \le 10$, we get

$$\phi_2 = \log\left(\frac{10\omega}{a}\right)\frac{\pi}{4} \quad \text{...(8.38)}$$

The straight line approximation and the exact curve are shown in Fig. 8.9 for $a = 1$.

The asymptotic plot consists of three straight-line segments. For ω up to one-tenth of the break frequency, the phase shift is approximated as 0°, and for ω above ten times the break frequency, the phase-shift is approximately 90°. From one-tenth of the break frequency to ten times the break frequency, we have a straight-line of slope 45° per decade. The maximum error in the approximation is $\tan^{-1} 0.1$, or about 5.7°, and occurs at the two corners, $0.1a$ and $10a$. Note that there is no error at the break frequency, $\omega = a$.

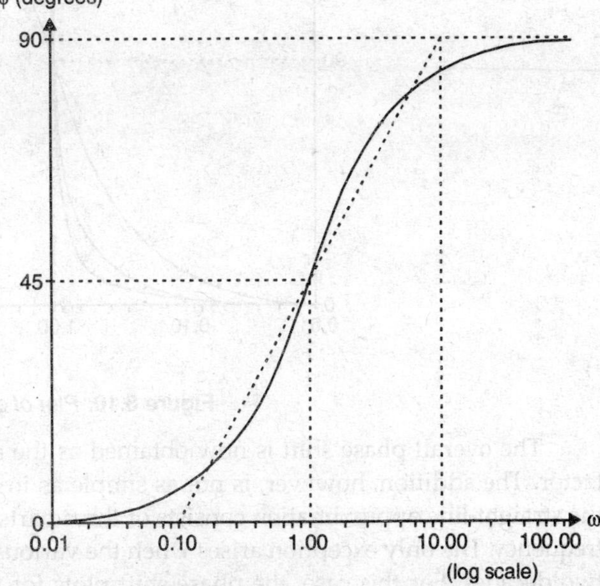

Figure 8.9. *Plot of arg (1 + jω)*

The third term on the right-hand side is the phase shift due to a purely imaginary expression and is identically equal to 90°.

The fourth term in Equation (8.34) is, again, similar to the second term, except for the negative sign. Hence, it can be approximated by phase shift of 0° up to $\omega = 0.1b$, a straight-line of slope –45°/decade from $0.1b$ to $10b$, and –90° for ω greater than $10b$.

We shall now look at the fifth term of Equation (8.34), given by

$$\phi_5 = \tan^{-1} \frac{2\zeta \dfrac{\omega}{\omega_n}}{1 - \left(\dfrac{\omega}{\omega_n}\right)^2} \qquad ...(8.39)$$

For $\omega/\omega_n << 0.1$, we have

$$\phi_5 \approx 0 \qquad ...(8.40)$$

and for $\omega/\omega_n >> 10$, we get

$$\phi_5 \approx 180° \qquad ...(8.41)$$

At the intermediate point, $\omega/\omega_n = 1$, we have

$$\phi_5 = 90° \qquad ...(8.42)$$

The plots of the asymptotic approximation and the actual curve for several values of ζ are shown in Fig. 8.10, where $\omega_n = 1$. The straight-line approximation has a slope of 90° per decade between 0.1ω and 10ω. It is seen that for small values of ζ, the actual phase shift changes very rapidly for ω near the undamped natural frequency ω_n.

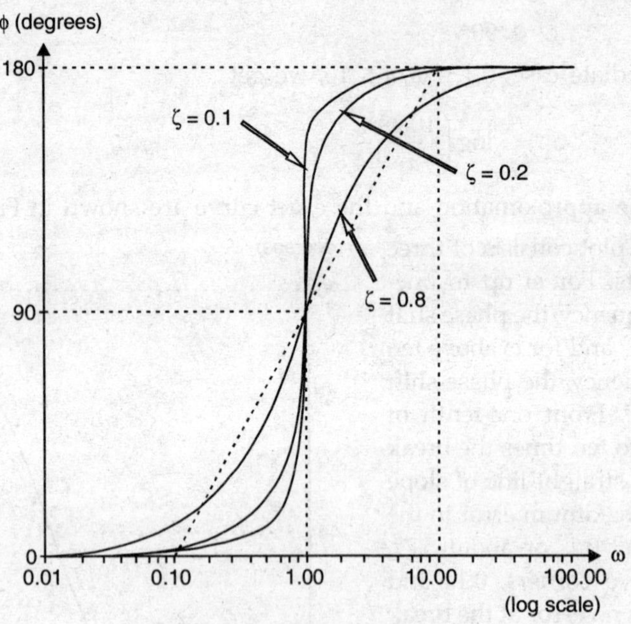

Figure 8.10. *Plot of ϕ_5 for $\omega_n = 1$*

The overall phase shift is now obtained as the algebraic sum of the phase shift due to each factor. The addition, however, is not as simple as in the case of the gain plot due to the fact that the straight-line approximation consists of three parts over four decades centred around the break frequency. The only exception arises when the various poles and zeros are separated by more than two decades. For this case, the phase-shift plots for the various factors can be superimposed on each other to obtain the overall phase-shift.

For the more general case, it is usually advisable to make a table of phase shift against frequency for each factor and then obtain the total phase shift as the algebraic sum of these. Better accuracy can be obtained by using the exact values in the table instead of those obtained through the asymptotic approximation.

EXAMPLE 8.3

Consider the transfer function

$$G(s) = \frac{20s}{s+5} = \frac{4s}{1+s/5} \qquad \text{...(8.43)}$$

In this case, we have one zero at the origin and one real pole at $s = -5$. The two break frequencies are more than two decades apart. Hence, we can obtain the overall plot by adding the straight-line approximations of the phase shift due to the real pole to the 90° phase lead caused by the zero at the origin. The resulting phase-shift plot is shown in Fig. 8.11.

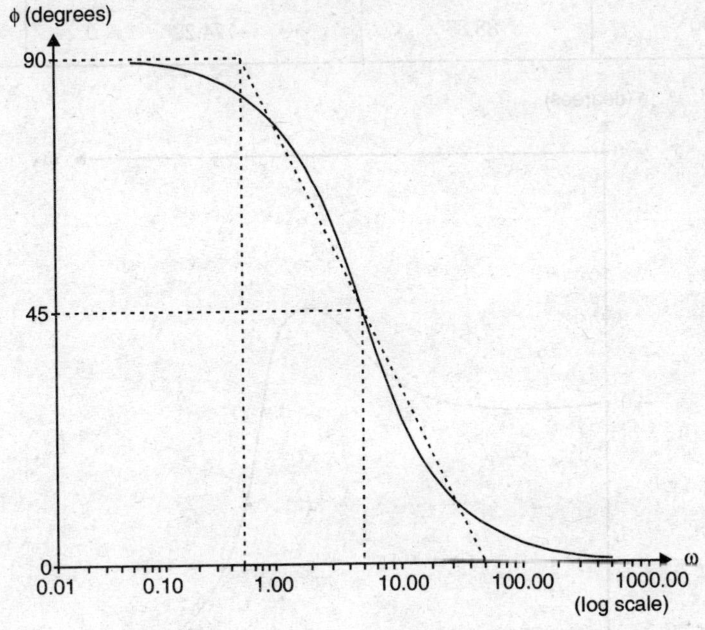

Figure 8.11. *Phase shift plot for the transfer function given by Equation (8.43)*

EXAMPLE 8.4

Consider the transfer function given by Equation (8.32), for which the gain curve was obtained in Fig. 8.8. We shall now plot the phase-shift curve.

This transfer function has a pole at the origin, a zero at $s = -2$, and a pair of complex conjugate poles with undamped natural frequency $\omega_n = 10$ and damping ratio $\zeta = 0.5$. The phase shift due to each factor for different values of ω was calculated and is given in Table 8.2. The plot of the total phase-shift is shown in Fig. 8.12.

TABLE 8.2: Phase Shift Due to Different Factors of the Transfer Function given by Equation (8.32)

ω	*Pole at the origin*	*Zero at s = –2*	*Complex conjugate poles*	*Total phase-shift*
0.1	–90°	2.86°	–0.57°	–87.71°
0.2	–90	5.71°	–1.15°	–85.44°
0.5	–90°	14.04°	–2.87°	–78.33°
1	–90°	26.57°	–5.77°	–69.20°
2	–90°	45.00°	–11.77°	–56.77°
5	–90°	68.20°	–33.69°	–55.49°
10	–90°	78.69°	–90.00°	–101.32°
20	–90°	84.29°	–146.31°	–152.02°
50	–90°	87.71°	–168.23°	–170.52°
100	–90°	88.85°	–174.22°	–175.68°

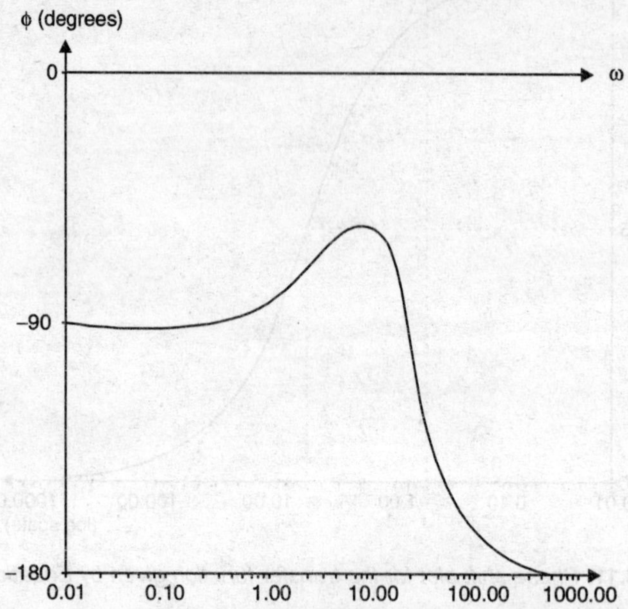

Figure 8.12. *Phase shift plot for the transfer function given by Equation (8.32)*

DRILL PROBLEM 8.3

Sketch the phase-shift curves for the following transfer functions:

(a) $\dfrac{20s}{(s+1)(s+10)}$ (b) $\dfrac{300\left(s^2+2s+4\right)}{s(s+10)(s+20)}$

Ans.

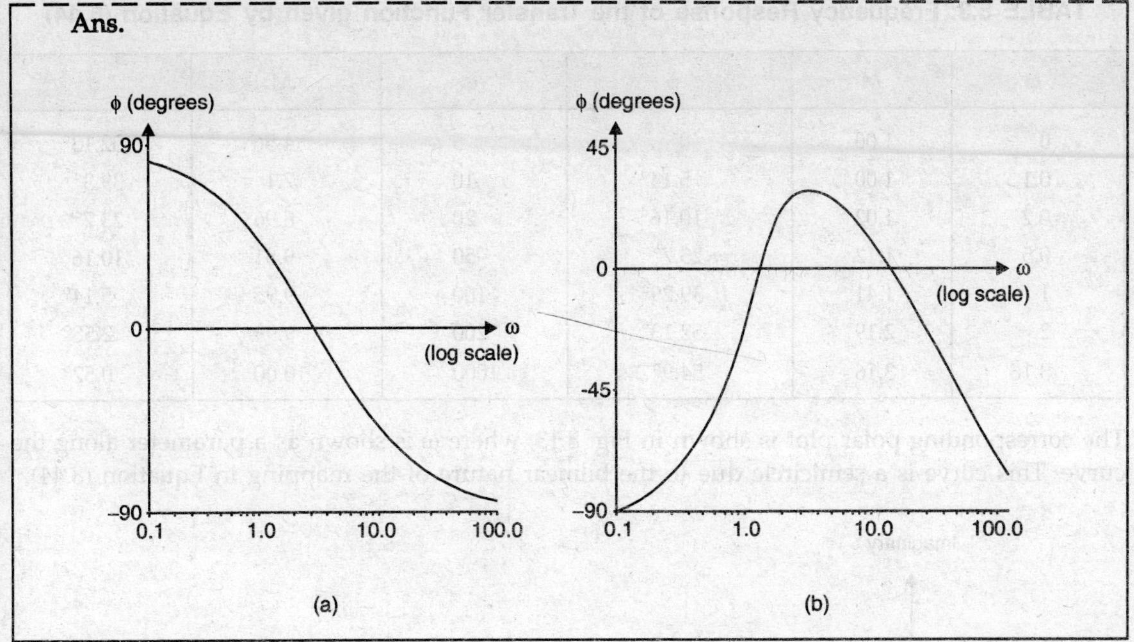

(a) (b)

8.4 POLAR PLOTS

One disadvantage of Bode plots is that we have two separate curves showing the variation of the gain and the phase shift with frequency. It is possible to combine the information contained in these two plots into one curve in two ways. One of these will be discussed in this section and the other in the next. If we express the frequency response function $G(j\omega)$ in the polar form $M(\omega)\, e^{j\phi(\omega)}$, and plot the tip of the vector $M(\omega)\, e^{j\phi(\omega)}$ in the G-plane as ω varies from zero to infinity, we obtain the *polar* plot. The value of ω is labelled at selected points on the plot. Mathematically, the polar plot may be regarded as the mapping of the positive part of the $j\omega$-axis of the s-plane into a curve in the G-plane.

This type of plot is very useful for determining the stability of a closed-loop system from its open-loop frequency response, as will be seen in Chapter 9 when we study the Nyquist criterion of stability.

EXAMPLE 8.5

Consider the transfer function

$$G(s) = \frac{10\,(s+1)}{s+10} \qquad ...(8.44)$$

which represents a lead compensator commonly used in control systems (Chapter 10 describes it in detail). We shall obtain the polar plot of the frequency response of this transfer function. The first step is to calculate the magnitude ratio M and the phase shift ϕ for different values of ω. These are given in Table 8.3.

TABLE 8.3: Frequency Response of the Transfer Function given by Equation (8.44)

ω	*M*	φ	ω	*M*	φ
0	1.00	0°	5	4.56	52.13°
0.1	1.00	5.14°	10	7.1	39.3°
0.2	1.02	10.16°	20	8.96	23.7°
0.5	1.12	23.7°	50	9.81	10.16°
1	1.41	39.29°	100	9.95	5.14°
2	2.19	52.13°	200	9.99	2.58°
3.16	3.16	54.9°	1000	10.00	0.52°

The corresponding polar plot is shown in Fig. 8.13, where ω is shown as a parameter along the curve. This curve is a semicircle due to the bilinear nature of the mapping in Equation (8.44).

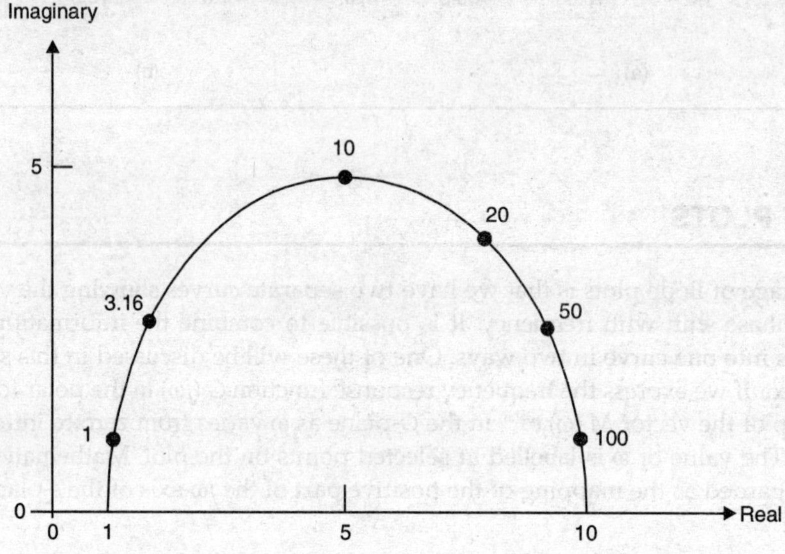

Figure 8.13. *Polar plot of frequency response for G(s) in Equation (8.44)*

EXAMPLE 8.6

Consider the transfer function

$$G(s) = \frac{10}{s\,(s+1)} \qquad \text{...(8.45)}$$

which represents the forward path of a position control system. The frequency response of this transfer function is given in Table 8.4 and its polar plot is shown in Fig. 8.14.

TABLE 8.4: Frequency Response of the Transfer Function given by Equation (8.45)

ω	M	φ	ω	M	φ
0	∞	−90°	5	0.39	−168.69°
0.1	99.5	−95.71°	7	0.202	−171.81°
0.2	49.03	−101.3°	10	0.100	−174.29°
0.5	17.89	−116.57°	20	0.025	−177.14°
1	7.07	−135.0°	50	0.004	−178.85°
2	2.24	−151.43°	70	0.002	−179.18°

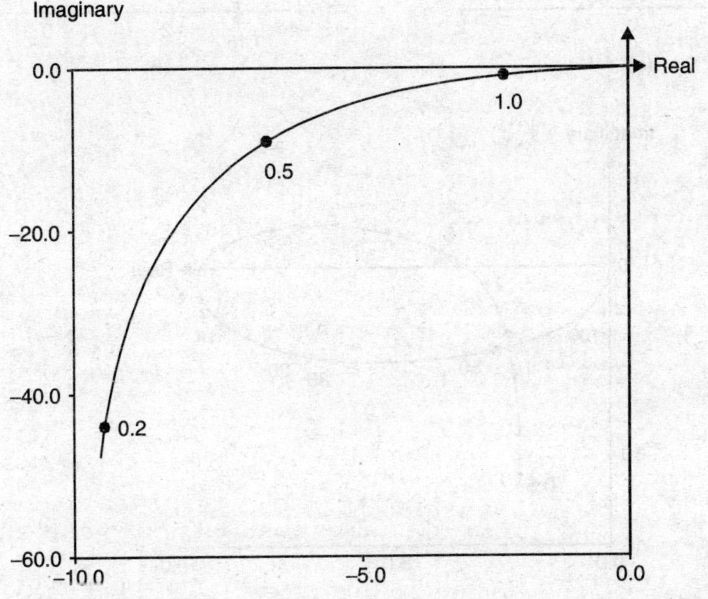

Figure 8.14. *Polar plot of the frequency response for G(s) given by Equation (8.45)*

DRILL PROBLEM 8.4

Sketch the polar plot of the frequency response for each of the following transfer functions.

(a) $\dfrac{20s}{(s+1)(s+10)}$ (b) $\dfrac{10}{s(s+1)(s+4)}$ (c) $\dfrac{300(s^2+2s+4)}{s(s+10)(s+20)}$

Ans.

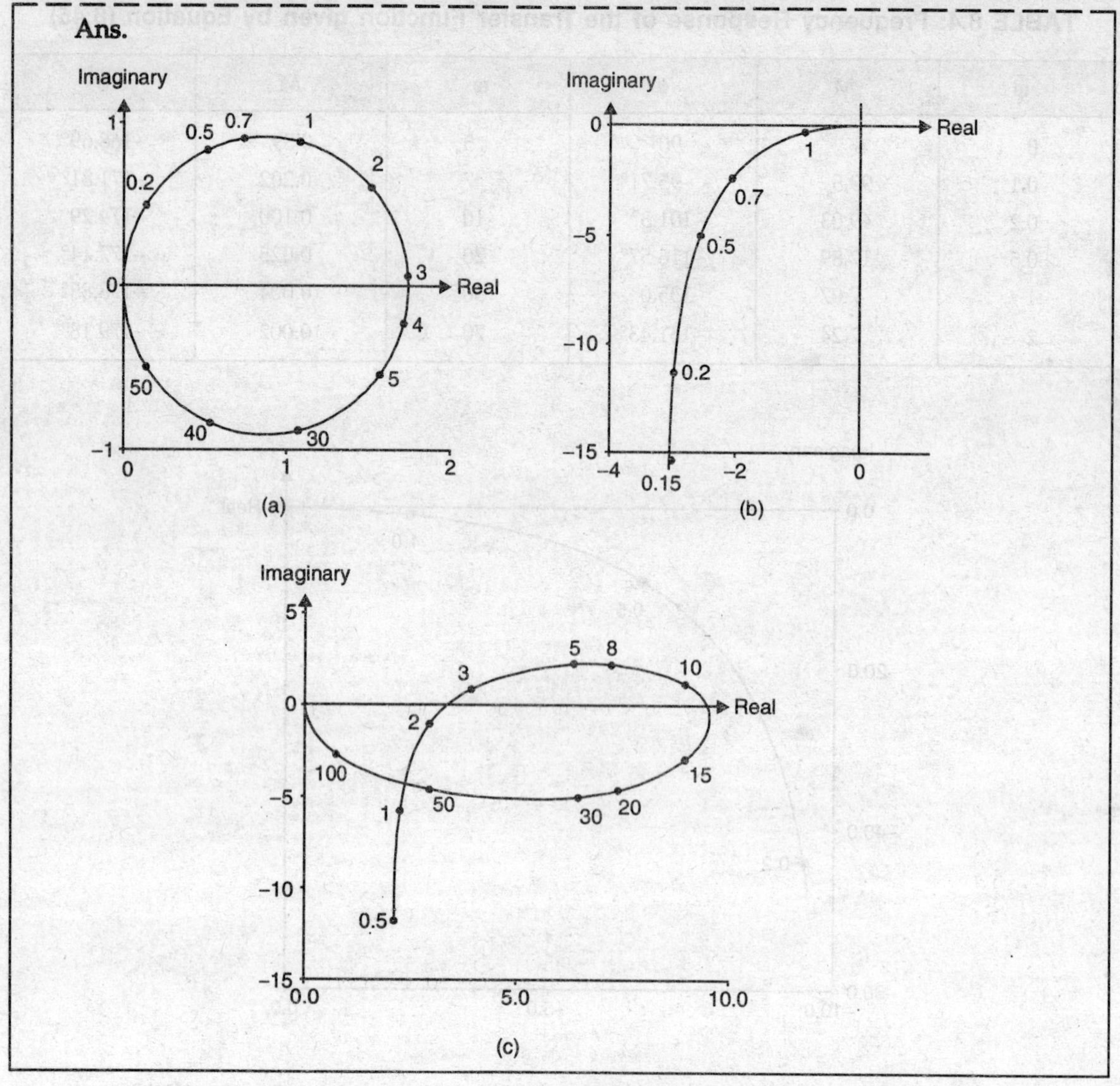

8.5 LOG-MAGNITUDE AND PHASE DIAGRAMS

The polar plot usually requires more computation than Bode plots but has the advantage of simultaneously providing information about gain as well as phase shift. An alternative is to plot logarithmic gain against phase shift for different values of ω. The main advantage is that the points can be obtained from the Bode plots of gain and phase shift for each value of ω. Such plots are especially useful for investigating the relative stability of closed-loop systems.

Example 8.7

We shall again consider the transfer function of the lead compensator given in Equation (8.44) for Example 8.5. The frequency response for this system is given in Table 8.5, where the gain is in decibels and the resulting log-magnitude/phase diagram is shown in Fig. 8.15.

TABLE 8.5: Frequency Response for the Transfer Function given by Equation (8.44)

ω	M (dB)	φ	ω	M (dB)	φ
0.1	0.0	5.14°	5	13.18	52.13°
0.2	0.17	10.16°	10	17.03	39.3°
0.5	0.96	23.7°	20	19.04	23.7°
1	2.79	39.29°	50	19.83	10.16°
2	6.82	52.13°	100	19.96	5.14°
3.16	9.99	54.9°	200	19.99	2.58°

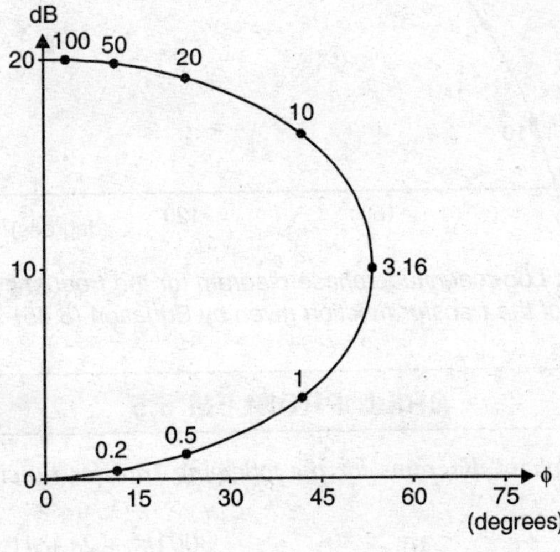

Figure 8.15. *Log-magnitude/phase plot for frequency response of the transfer function given by Equation (8.44)*

EXAMPLE 8.8

We shall now consider the transfer function of the position control system given in Equation (8.45) and discussed in Example 8.6. The frequency response for this system is given in Table 8.6, with the gain in decibels. The resulting log-magnitude/phase diagram is shown in Fig. 8.16.

TABLE 8.6: Frequency Response for the Transfer Function given by Equation (8.45)

ω	M (dB)	φ	ω	M (dB)	φ
0.1	39.97	–95.7°	5	–8.13	–168.69°
0.2	33.8	–101.3°	7	–13.89	–171.87°
0.5	25.05	–116.57°	10	–20.04	–174.29°
1	16.99	–115.0°	20	–32.05	–177.14°
2	6.99	–153.4°	50	–47.96	–178.85°

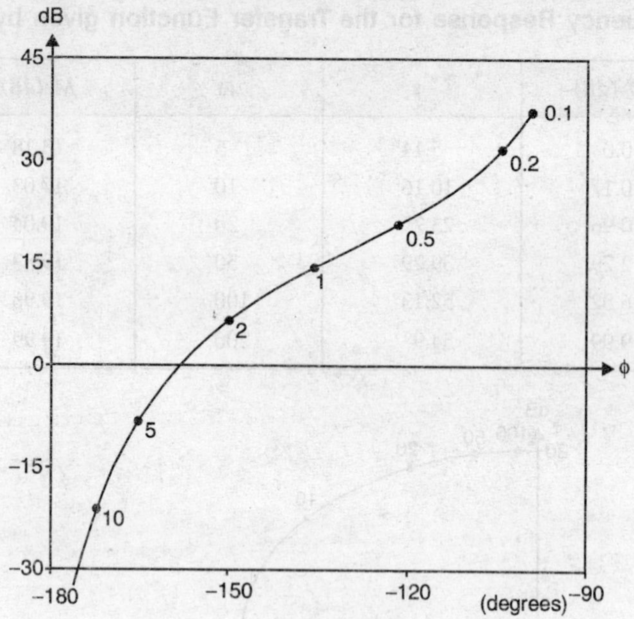

Figure 8.16. *Log-magnitude/phase diagram for the frequency response of the transfer function given by Equation (8.45)*

DRILL PROBLEM 8.5

Plot the log-magnitude/phase diagrams for the following transfer functions:

(a) $\dfrac{20s}{(s+1)(s+10)}$ (b) $\dfrac{10}{s(s+1)(s+4)}$ (c) $\dfrac{300(s^2+2s+4)}{s(s+10)(s+20)}$

Ans.

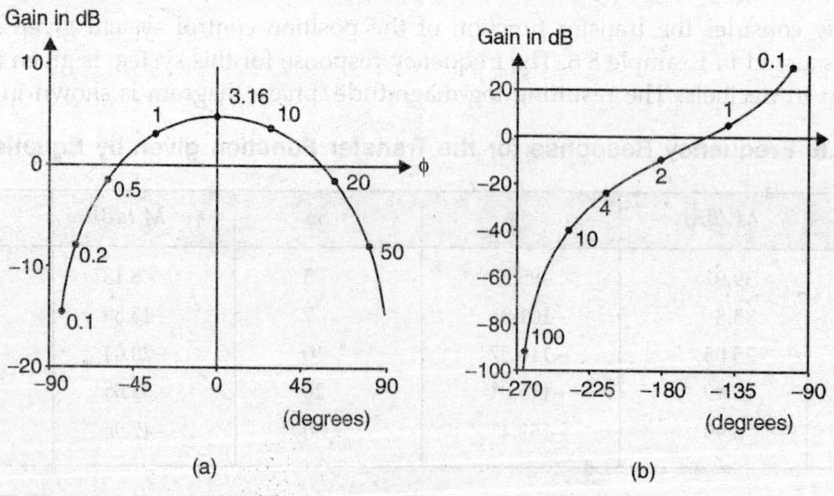

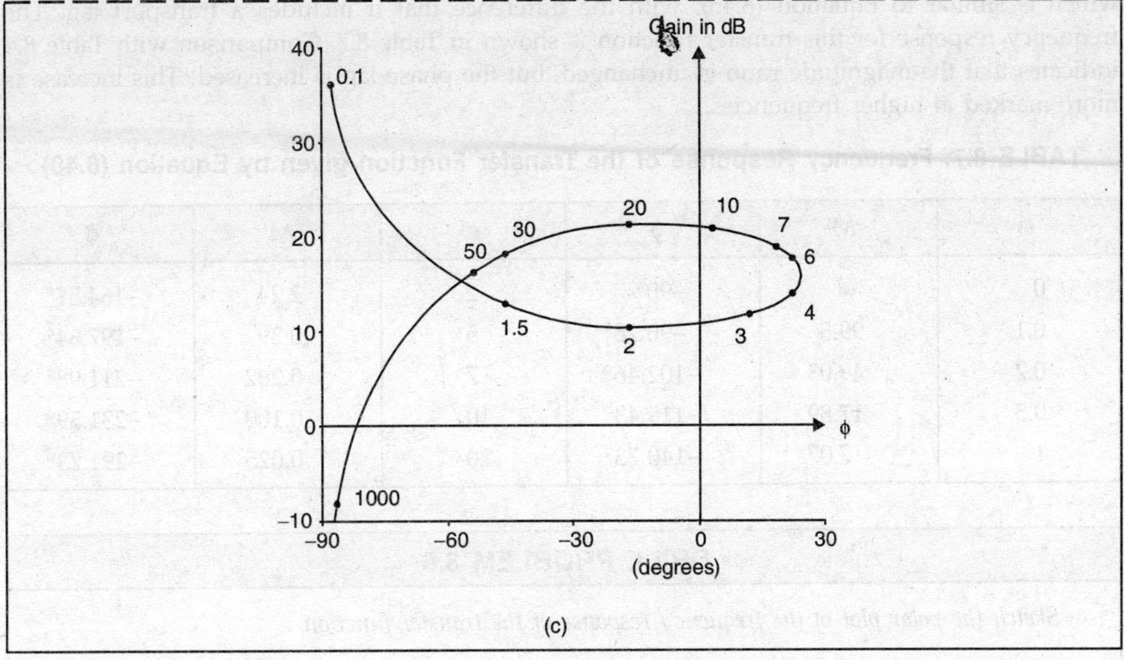

(c)

8.6 SYSTEMS WITH TRANSPORT LAG

The systems that we have considered so far have transfer functions in the form of the ratio of two polynomials of the complex variable s. In many systems, pure delays occur. Typical examples are systems with hydraulic, pneumatic, or mechanical components. An electrical transmission line is another example. In these systems, there is a time delay between the application of the input and its effect on the output. This is often called "transport lag" or "pure delay". The input and the output in such cases are related through the transfer function e^{-Ts}, where T represents the delay in seconds.

The frequency response of this transfer function is obtained easily by replacing s by $j\omega$. Consequently, we have

$$G(j\omega) = e^{-j\omega T} = Me^{j\phi} \quad ...(8.46)$$

It is evident that the gain $M = 1$ for all values of ω and the phase shift

$$\phi = -\omega T \quad ...(8.47)$$

Hence, transport lag does not affect the magnitude ratio, but introduces a phase lag that is proportional to the frequency. It should be noted that Equation (8.47) gives the phase lag in radians (and not degrees) if ω is in radians per second and T is in seconds.

EXAMPLE 8.9

Consider the transfer function

$$G(s) = \frac{10e^{-0.1s}}{s(s+1)} \quad ...(8.48)$$

which is similar to Equation (8.45), with the difference that it includes a transport lag. The frequency response for this transfer function is shown in Table 8.7. Comparison with Table 8.4 indicates that the magnitude ratio is unchanged, but the phase lag is increased. This increase is more marked at higher frequencies.

TABLE 8.7: Frequency Response of the Transfer Function given by Equation (8.48)

ω	M	ϕ	ω	M	ϕ
0	∞	–90°	2	2.24	–164.51°
0.1	99.5	–96.28°	5	0.39	–197.34°
0.2	49.03	–102.46°	7	0.202	–211.98°
0.5	17.89	–119.43°	10	0.100	–231.59°
1	7.07	–140.73°	20	0.025	–291.73°

DRILL PROBLEM 8.6

Sketch the polar plot of the frequency response of the transfer function

$$G(s) = \frac{10e^{-0.01s}}{s(s+1)(s+4)}$$

Ans.

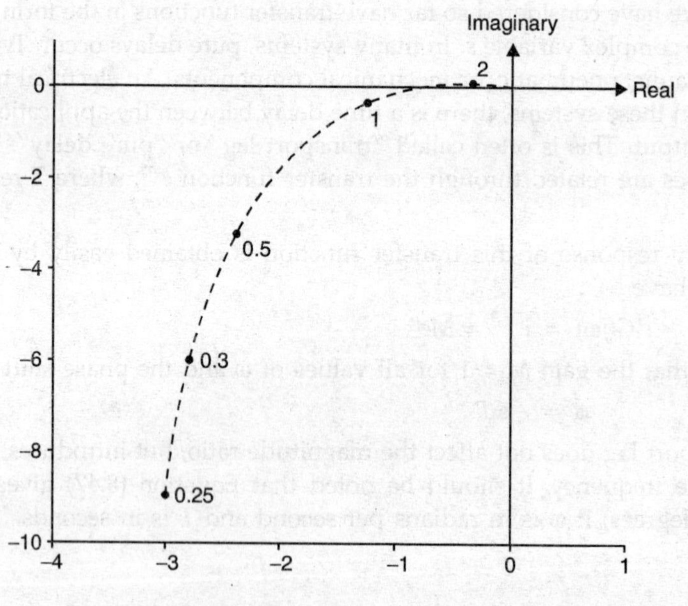

8.7 ESTIMATION OF TRANSFER FUNCTION FROM FREQUENCY RESPONSE PLOTS

It is often possible to estimate the transfer function from the frequency response. In particular, Bode plots are quite attractive for this purpose, since one can estimate the location of the poles and zeros of the transfer function by fitting straight-line segments of slope $6n$ dB per octave, where n is an integer. The points of intersection of these straight lines are the "corner" or "break" frequencies of the transfer function, and can be either poles or zeros. A particular corner frequency will be a simple pole if the slope decreases by 6 dB/octave at that point, and it will be a simple zero if the slope increases by 6 dB/octave. The constant term in the transfer function can be determined from the straight-line asymptote of the left-most part of the frequency response curve, that is, the part to the left of the lowest corner frequency. If the slope at a corner frequency changes by 12 dB/octave, this indicates the existence of a pair of complex poles or zeros ($0 \le \zeta \le 1$) at that point. The damping ratio, ζ, can be estimated either from the value of the actual gain at the corner frequency, or from the value of ω_m, the frequency where the gain is maximum, using Equation (8.29).

From the above it may appear that the transfer function can be estimated from the Bode plot of the gain curve alone. In fact this is not so. It is important to note that except for minimum-phase systems (*i.e.*, systems with stable transfer functions that have all zeros in the left half of the *s*-plane), the transfer function cannot be determined uniquely from the gain curve alone. This follows from the fact that if we multiply any transfer function by all pass functions of the form

$$\frac{s-a}{s+a} \quad \text{or} \quad \frac{s^2 - 2\zeta\,\omega_n s + \omega_n^2}{s^2 + 2\zeta\,\omega_n s + \omega_n^2}$$

or introduce an ideal delay of the form $e^{-\beta s}$, the phase shift is changed, but the gain remains unaltered. Since we do not always know that the system is of the minimum-phase type, it is necessary to examine the measured phase shift curve and compare it with that obtained for the transfer function obtained from the gain curve alone.

The following properties of the phase shift are useful in estimation of the transfer function:

(*i*) If the phase shift does not approach a constant value for high frequencies, this indicates the presence of an ideal delay.

(*ii*) If the transfer function of the system has n finite poles and m finite zeros, the phase shift at very high frequencies will approach $(n - m)\pi/2$ radians or $(n - m)90°$. Also, the slope of the gain curve for these frequencies should be $6(m - n)$ decibels per octave.

(*iii*) If the transfer function has r poles at the origin, the phase shift will approach $-r\pi/2$ radians for low frequencies. Similarly, for r zeros at the origin, the phase shift will approach $\pi r/2$ radians or $90r$ degrees at low frequencies. Also, the slope of the gain curve should be $6r$ decibels per octave for these frequencies, positive for zeros and negative for poles.

This last property is particularly useful in applying the Nyquist criterion for stability, which requires information about the number of poles at the origin of the *s*-plane to be determined from the polar plot of the frequency response. We shall return to this topic in Chapter 9 while studying the Nyquist criterion.

The following examples will illustrate the procedure discussed above.

EXAMPLE 8.10

The Bode plots of the transfer function of a dc servomotor-amplifier combination (the output is the angular velocity of the shaft and the input is the applied voltage) are shown in Fig. 8.17(*a*) and (*b*). The straight-line approximation to the gain curve is shown by dashed lines. It is seen to consist of two straight lines, one with zero slope and the other with slope –20 dB/decade, intersecting at $\omega = 2$ and $M = 18$ dB (for a dc gain of 8). Hence, the transfer function can be approximated as

$$G(s) = \frac{8}{1+s/2} = \frac{16}{s+2}$$

This agrees with the phase-shift curve which shows a phase lag of 45° at $\omega = 2$, in addition to approaching 0° for small ω and –90° for large ω.

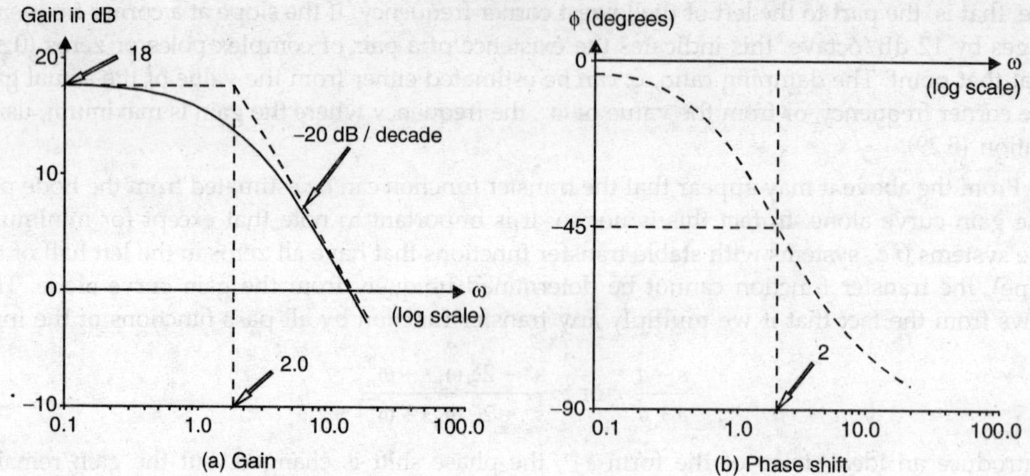

Figure 8.17. *Bode plots for a dc servomotor-amplifier combination*

EXAMPLE 8.11

The Bode plots for the frequency response of a closed-loop position-control system are shown in Fig. 8.18.

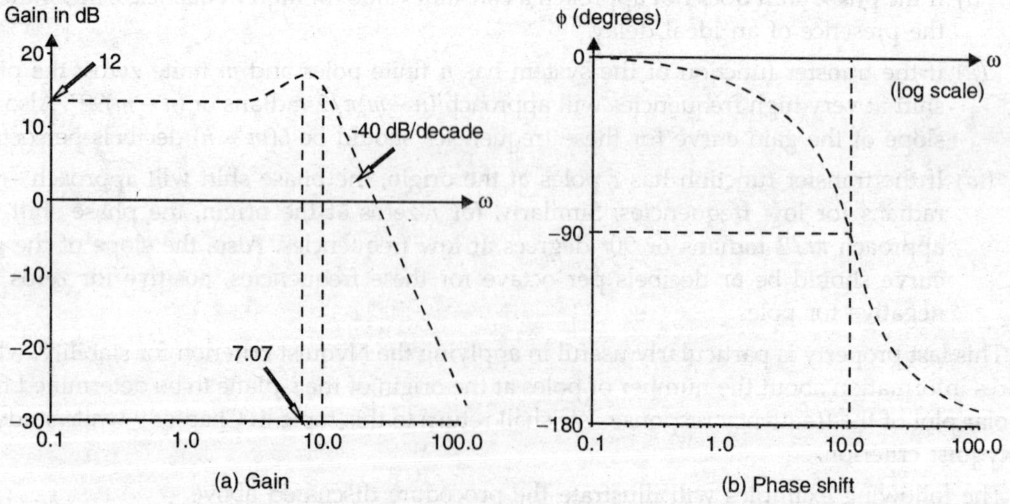

Figure 8.18 *Bode plots of the frequency response of a closed-loop position control system*

From the straight-line approximation of the gain curve, it is seen that there is a corner frequency at $\omega = 10$, where the slope changes from 0 to –40 dB/decade. This implies a pair of complex conjugate poles with undamped natural frequency $\omega_n = 10$. This is verified from the phase shift curve, which shows zero phase shift for small ω and –180° for large ω. Also, the phase shift is –90° at $\omega = 10$, confirming that this is the undamped natural frequency of the complex poles. The value of the damping ratio ζ is determined by noting that the peak of the gain curve occurs at $\omega = 7.07$. Using Equation (8.29), it is seen that

$$7.07 = 10\sqrt{1-2\zeta^2}$$

which is easily solved to give $\zeta = 0.5$.

Consequently, the transfer function of the system is estimated as

$$G(s) = \frac{4}{1+2\zeta\dfrac{s}{\omega_n}+\dfrac{s^2}{\omega_n^2}} = \frac{400}{s^2+10s+100}$$

DRILL PROBLEM 8.7

Estimate the transfer function from the Bode plots shown in Fig. 8.19(a) and (b), assuming that they are minimum-phase.

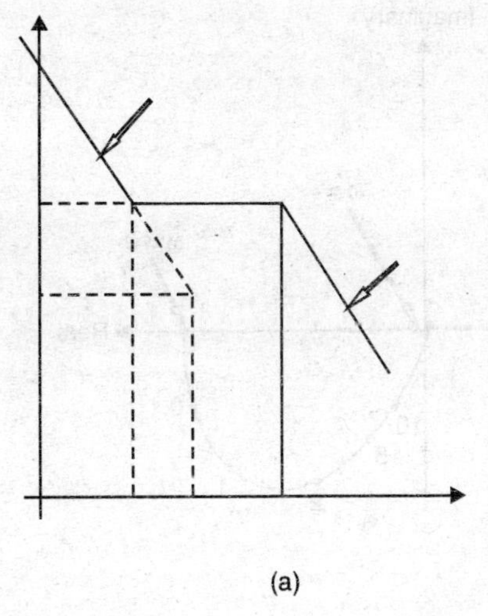
(a)

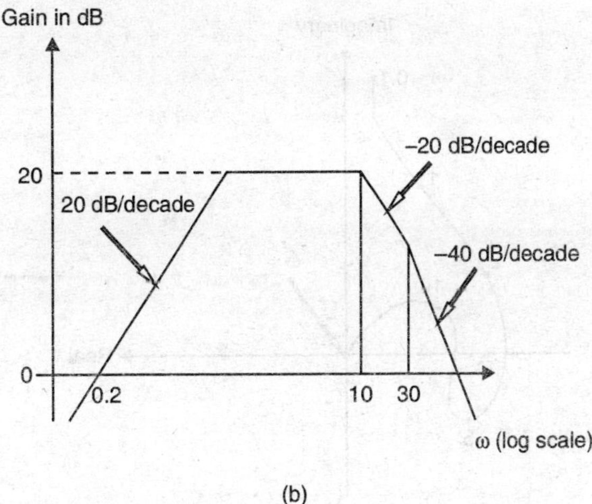

(b)

Figure 8.19. *Bode plots of gain*

Ans. (*a*) $\dfrac{4(1+s/2)}{s(1+s/10)}$, (*b*) $\dfrac{5s}{(1+s/2)(1+s/10)(1+s/30)}$

DRILL PROBLEM 8.8

The polar plots of the frequency response of some linear systems are shown in Figure 8.20. For each case, determine the number of poles the corresponding transfer function has at the origin of the s-plane. It is known that none of the transfer functions has more than three poles at the origin.

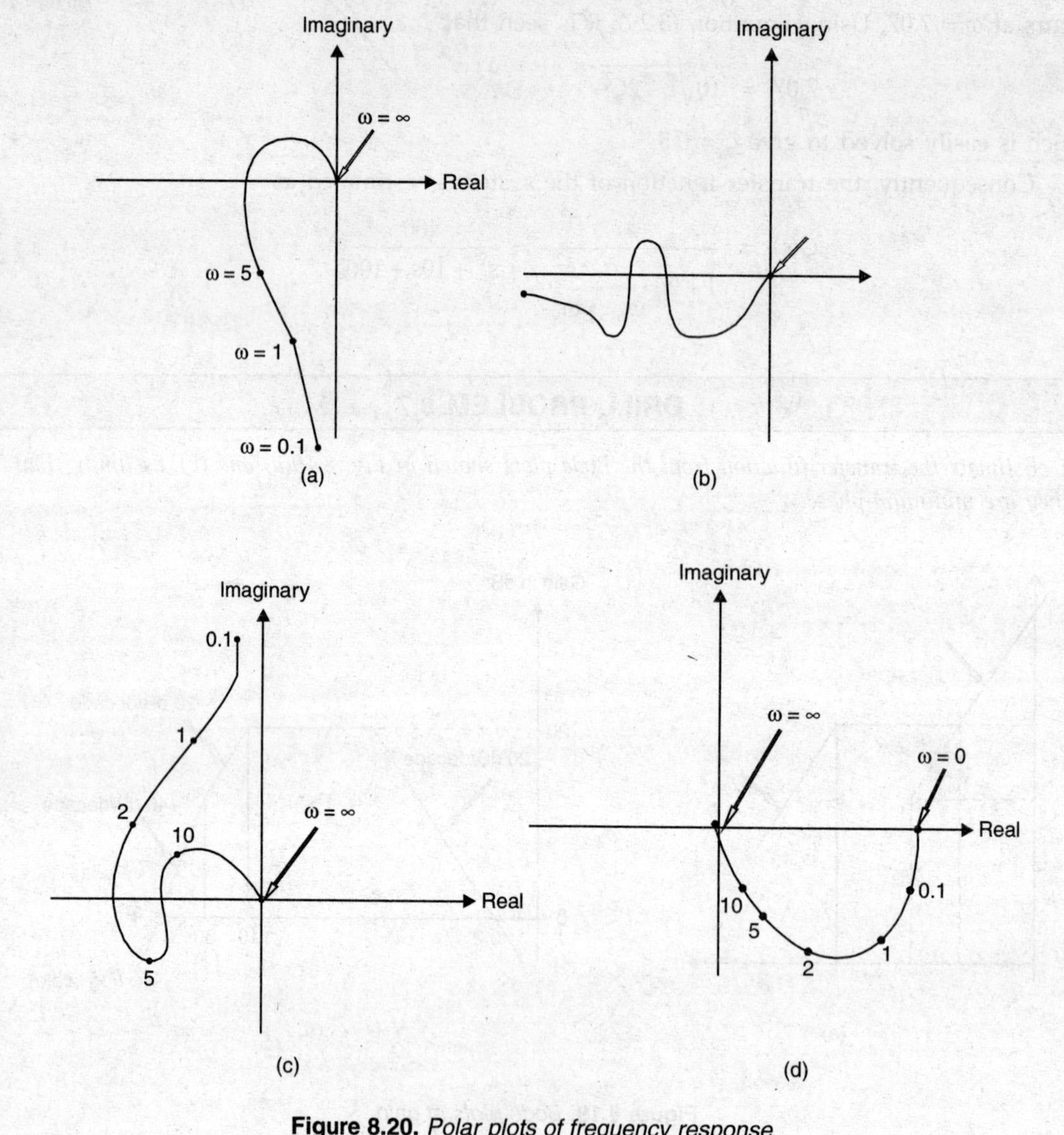

Figure 8.20. *Polar plots of frequency response*

Ans. (*a*) one, (*b*) two, (*c*) three, (*d*) none

SUMMARY

- In this chapter, we have discussed three different types of frequency response plots. These are: (*a*) Bode plots, (*b*) Polar plots, and (*c*) Log-magnitude/phase diagrams. The first and the last using logarithmic coordinates are utilized considerably in the design of compensators for closed-loop systems. Polar plots are most commonly used for stability analysis with the Nyquist criterion of stability. Bode plots have been very popular in the past since they can be obtained without much computation. They can also be used for estimating the transfer function of a system. Computation of frequency response can be carried out conveniently using either a programmable calculator or the computer program discussed in Appendix *E*.

Problems

1. Sketch the Bode plots for each of the following transfer functions:

(*a*) $\dfrac{300}{s(s+4)(s+8)}$ (*b*) $\dfrac{100(s+2)}{s(s+4)\left(s^2+8s+25\right)}$ (*c*) $\dfrac{100(s+2)\,e^{-0.2s}}{s\left(s^2+10s+100\right)}$

2. The frequency response of a control system was obtained experimentally and is given in Table P8.2. Draw the Bode plots and hence estimate the transfer function of the system.

TABLE P8.2: Frequency Response for the Control System

ω	*M (dB)*	φ	ω	*M (dB)*	φ
0.1	20.0	–90.3°	10	–14.0	–180°
0.2	14.0	–90.6°	15	–26.8	–239°
0.4	8.0	–91.2°	20	–36.0	–252°
0.6	4.46	–91.7°	30	–47.8	–259°
1	0.08	–92.9°	40	–55.6	–262°
2	–5.71	–95.9°	50	–61.6	–264°
4	–10.77	–103.4°	60	–66.5	–265°
6	–12.55	–115.1°	80	–74.2	–266°
8	–12.7	–138.0°	100	–79.9	–267°

3. Sketch the polar plot and the log-magnitude/phase diagram for the frequency response in Problem 2.

4. Sketch the polar plot and the log-magnitude/phase diagram for the frequency response of each of the transfer functions given in Problem 1.

5. Estimate the transfer function from the Bode plot of the gain shown in Fig. P8.5. It may be assumed that it is of the minimum-phase type.

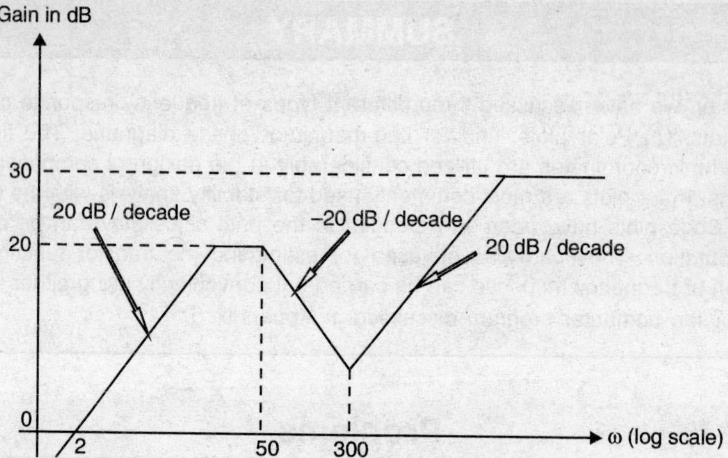

Figure P8.5. *Bode plot of gain*

6. Repeat Problem 5 for the plot shown in Fig. P8.6.

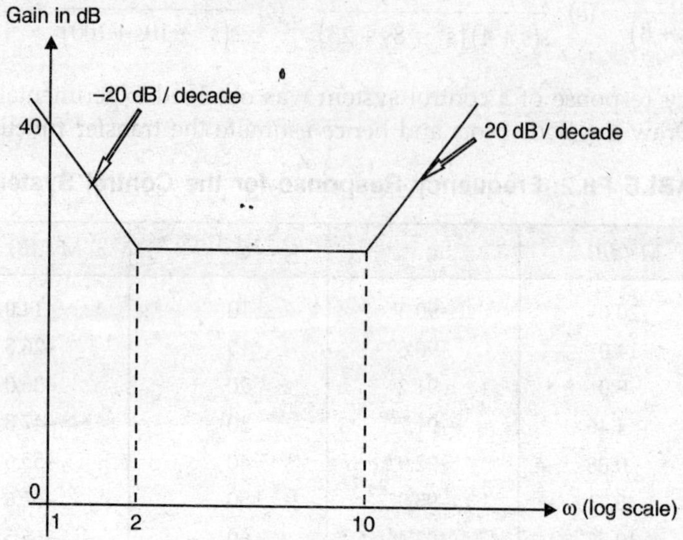

Figure P8.6. *Bode plot of gain*

7. The block diagram for the pitch control system for a supersonic airplane is shown in Fig. P7.8 (see Chapter 7, Problem 8). Draw the log-magnitude/phase plot for the frequency response of the open-loop as well as the closed-loop system if $K = 5$. [*Hint*: If the open-loop frequency response of a unity-feedback system for some value of ω is given by $Me^{j\phi}$, then frequency response of the closed-loop system is $Me^{j\phi}/(1 + Me^{j\phi})$].

8. The transfer function of the forward path of a unity-feedback system representing a chemical reactor is given by

$$G_p(s) = \frac{300e^{-0.05s}}{s(s+5)(s+10)}$$

Sketch the Bode plots of the frequency response of the open-loop system as well as the closed-loop system. What is the maximum magnitude of the frequency response of the closed-loop system?

9. Draw the log-magnitude/phase plot of the frequency response of the system described in the previous problem with the loop open as well as closed.

10. The block diagram of the Otolith linear model of a human vestibular system is shown in Fig. P8.10. Draw the log-magnitude/phase plot of the frequency response of this system assuming that the gain K is set at 1.5.

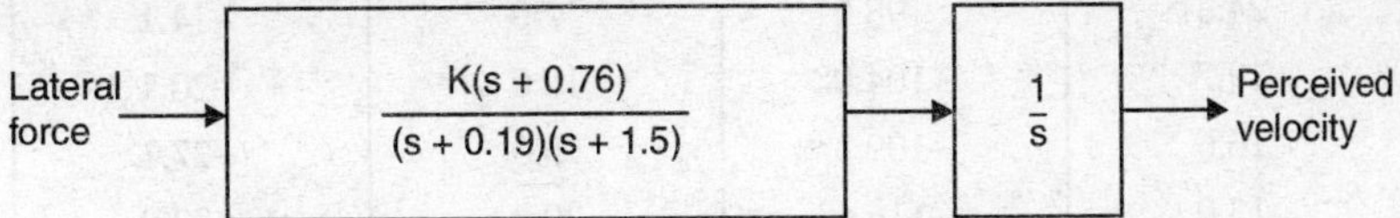

Figure P8.10. *Otolith linear model of a human vestibular system*

11. The transfer function relating the exciter voltage to the bucket-wheel current in a large coal dredger is given by

$$G_p(s) = \frac{K(1+0.613s)}{(1+4.42s)(1+0.23s)(1+0.66s)(1+0.4s)\left(s^2+0.41s+1.5\right)}$$

Draw the log-magnitude/phase plot of the frequency response if K is set at 0.5.

12. Draw the polar plot of the frequency response of the system described in the previous problem.

13. The transfer function of a normalized third-order Butterworth filter is given below. Draw the Bode plots of the frequency response for ω varying from 0.1 to 10 rad/s.

$$G(s) = \frac{1}{s^3 + 2s^2 + 2s + 1}$$

14. The block diagram in Fig. P8.14 shows the speed-control system for an ac induction motor commonly used in industry. Assuming that $K = 80$, draw the Bode plots of the frequency response of the open-loop system as well as the closed-loop system.

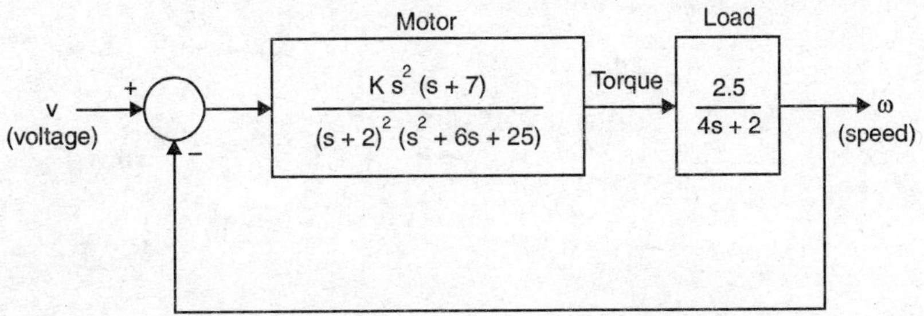

Figure P8.14. *Speed control of an induction motor*

15. Table P8.15 gives the frequency response of a minimum-phase system. Draw the Bode plots and hence estimate the transfer function of the system.

TABLE P8.15: Frequency Response for the Control System

ω	M (dB)	φ	ω	M (dB)	φ
0.1	34.0	–93°	3.0	–0.68	–146.3°
0.2	28.9	–96°	5.0	–8.6	–158.2°
0.3	24.3	–98.5°	7.0	–14.1	–164.0°
0.5	19.7	–104.0°	10.0	–20.1	–168.7°
0.7	16.6	–109.3°	15.0	–27.1	–172.0°
1.0	13.0	–116.6°	20.0	–32.0	–176.6°
2.0	4.95	–135.0°	50.0	–48.0	–177.7°

16. Repeat for the data in Table P8.16.

TABLE P8.16: Frequency Response for the Control System

ω	M (dB)	φ	ω	M (dB)	φ
0.5	0.04	–11.6°	4.0	–4.7	–110.6°
1.0	0.17	–24.0°	4.5	–6.3	–120.0°
1.5	0.2	–37.8°	5.0	–8.0	–127.0°
2.0	0.0	–53.1°	6.0	–12.3	–131.0°
2.5	–0.57	–69.4°	8.0	–16.8	–149.6°
3.0	–1.6	–85.2°	10.0	–19.9	–156.0°
3.5	–3.0	–99.0°	20.0	–32.0	–168.4°

❑❑❑

CHAPTER

9 Stability from Frequency Response

9.1 INTRODUCTION

In order to be useful, as stated earlier a control system must be stable. In Chapter 6, we discussed the Routh-Hurwitz criterion, which enables us to determine whether a system is stable by examining the characteristic polynomial, corresponding to the denominator of the transfer function of the closed-loop system. The root locus method, described in Chapter 7, enables us to determine the relative stability in terms of the damping ratio of the dominant poles. Both these methods require the knowledge of the transfer function of the open-loop system, which must be a rational function of the complex frequency variable *s*, that is a ratio of two finite-degree polynomials of *s*. These methods cannot be used when the system contains an ideal delay of the form e^{-Ts}, although it is possible to obtain approximate analysis by replacing e^{-Ts} with a truncated power series or a rational function such as a Padé approximant.

In this chapter, we shall study the use of frequency response methods described in Chapter 8, for the determination of the stability of closed-loop systems. Our objective is to utilize the frequency response of the open-loop system not only to determine the absolute stability of the closed-loop system but also to evaluate some measures of relative stability. The transfer function need not be known since the frequency response can be obtained experimentally.

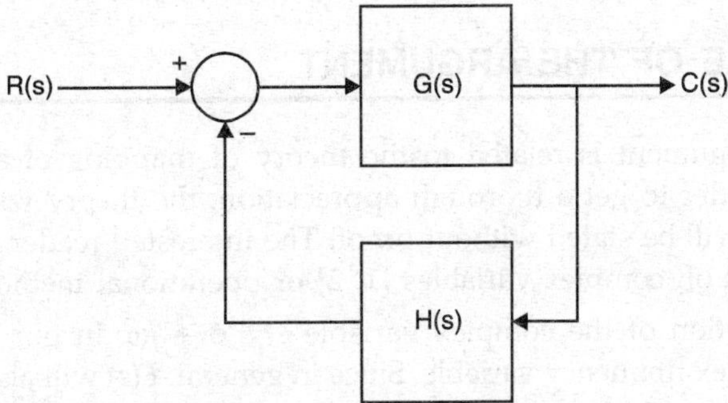

Figure 9.1. *A closed-loop system*

Consider the closed-loop system shown in Fig. 9.1. The main reason for possible instability of this system is that what is designed as negative feedback may turn out to be positive feedback if at some frequency a phase lag of 180° occurs in the loop. If the feedback at this frequency is sufficient to sustain oscillations, the system will act as an oscillator. The frequency response of the open-loop system can therefore be utilized for investigating the stability of the closed-loop system.

From superficial considerations, it will appear that the closed-loop system shown in Fig. 9.1 will be stable, provided that the open-loop frequency response $GH(j\omega)$ does not have gain of 1 or more at the frequency where the phase shift is 180°. However, this is not always true. For example, consider the polar plot of the frequency response, shown in Fig. 9.2. For two values of ω, the phase shift is 180° and the gain is more than one (at points A and B), and yet this system will be stable when the loop is closed. A thorough understanding of the problem is possible by applying the criterion of stability developed by H. Nyquist in 1932. It is based on a theorem in complex variable theory, due to Cauchy, called the principle of the argument, which is related to mapping of a closed path (or contour) in the s-plane for a function $F(s)$.

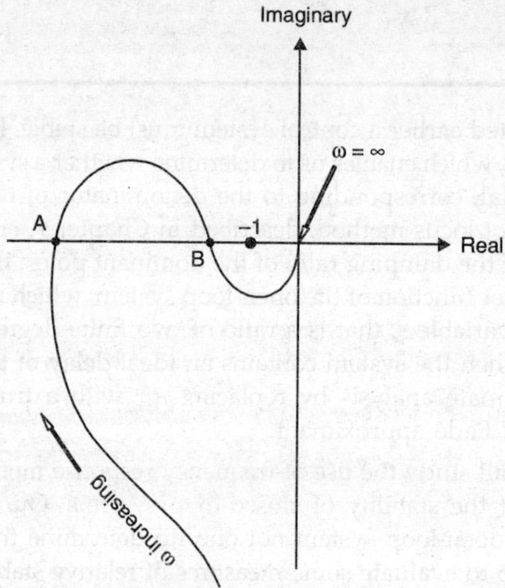

Figure 9.2. *Polar plot of the frequency response of a conditionally stable system*

9.2 THE PRINCIPLE OF THE ARGUMENT

The principle of the argument is related to the theory of mapping of analytic functions of a complex variable. In order to get a thorough appreciation, the theory will be reviewed briefly, and then the principle will be stated without proof. The interested reader may follow that up by reading a suitable book on complex variables [1, 2] or operational methods [3].

Let $F(s)$ be a function of the complex variable $s = \sigma + j\omega$. In our case, s has a physical meaning; it is the complex frequency variable. Since in general, $F(s)$ will also be complex, we may write

$$F(s) = U(\sigma, \omega) + jV(\sigma, \omega) \qquad \text{...(9.1)}$$

where U and V are real functions of the variables σ and ω. Thus, U is the real part of F, and V is the imaginary part.

The function $F(s)$ is said to be a real function of s if $F(s)$ is real for real s. For example,

$$F(s) = \frac{10s + 48}{s(s^2 + 5s + 21)} \qquad \text{...(9.2)}$$

is a real function of s, while

$$F(s) = \frac{3s + (2 + 3j)}{s(s^2 + 3s + 4 + 2j)} \qquad \text{...(9.3)}$$

is not a real function of s.

The transfer function of a physical system is always a real function of the complex variable s.

As $F(s)$, defined in a domain D in the s-plane, is said to be *analytic* in D if and only if the derivative dF/ds is continuous in D. This is possible if and only if the Cauchy-Riemann equations, given below, are satisfied in the domain D.

$$\frac{\partial U}{\partial \sigma} = \frac{\partial V}{\partial \omega} \qquad \text{...(9.4)}$$

$$\frac{\partial U}{\partial \omega} = -\frac{\partial V}{\partial \sigma}$$

It can be proved that transfer functions of physical systems are analytic everywhere in the s-plane, except at their poles.

Just as the complex variable s is shown in a plane with real axis σ and the imaginary axis ω, it is possible to illustrate F in a plane with real axis U and imaginary axis V.

Any point in the s-plane will be 'mapped' into the F-plane by locating the values of U and V for the given value of s. For example, consider the function

$$F(s) = \frac{2s + 3}{s + 5} \qquad \text{...(9.5)}$$

For $s_1 = 1 + j2$, we get

$$F(1 + j2) = \frac{2(1 + j2) + 3}{1 + j2 + 5} = 0.95 + j0.35 \qquad \text{...(9.6)}$$

This mapping is shown in Fig. 9.3.

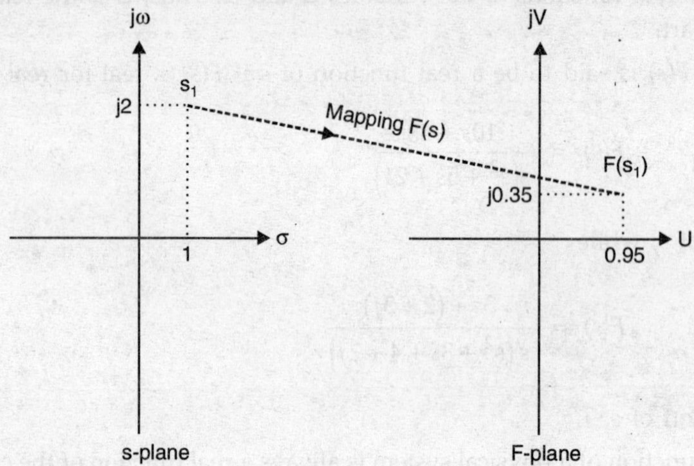

Figure 9.3. *Mapping of a function F(s)*

This correspondence between points in the two planes is called a *mapping* or *transformation.* For an analytic function, every point in the *s*-plane maps into a unique point in the *F*-plane. This concept can be extended to mapping a line or curve in the *s*-plane to the *F*-plane. In particular, a smooth curve in the *s*-plane will map into a smooth curve in the *F*-plane if *F*(*s*) is analytic at every point on the curve.

The polar plot of the frequency response of a transfer function *G*(*s*) is the map of the positive part of the *j*ω-axis of the *s*-plane.

We shall now consider mapping a closed curve in the *s*-plane. Such a curve will be called a contour, denoted as Γ_s. Its map in the *F*-plane will be denoted by Γ_F. The principle of the argument is stated below.

> If *F*(*s*) is an analytic function, except for a finite number of poles within Γ_s, then with a complete traversal along Γ_s in the clockwise direction, the corresponding contour Γ_F in the *F*-plane will encircle the origin of *F*-plane *N* times in the same direction, where $N = Z - P$.

In the above, *Z* is the number of zeros and *P* is the number of poles of the function *F*(*s*) inside the contour Γ_s.

Note that it is understood that *F*(*s*) does not have any pole or zero on the contour Γ_s.

EXAMPLE 9.1

Consider the function

$$F(s) = s + 2 \qquad ...(9.7)$$

which has a zero at $s = -2$ and no pole. We shall determine the maps of two contours, Γ_1 and Γ_2 which are circles with centre at the origin of the *s*-plane, and radius 1 and 3, respectively, as shown in the Fig. 9.4. Note that the traversal in both is in the clockwise direction. The corresponding contours in the *F*-plane have also been labelled as Γ_1 and Γ_2. These are also circles, obtained easily since the mapping represents a simple translation.

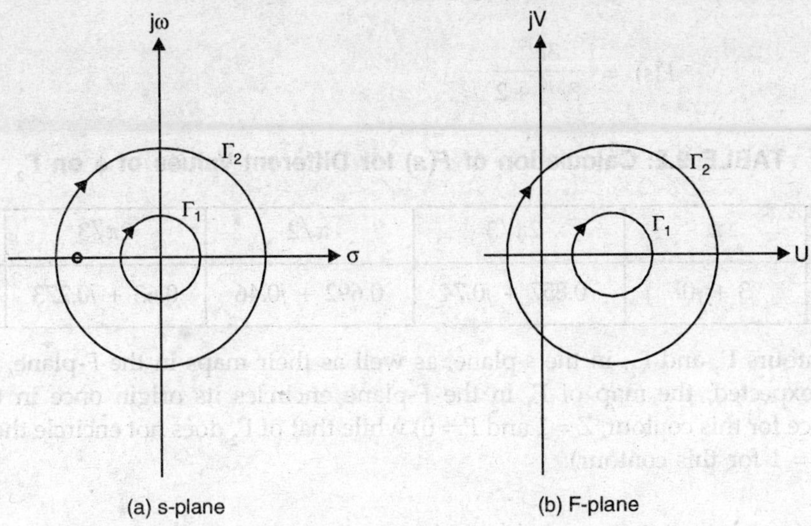

Figure 9.4. *Mapping F(s) = s + 2*

It will be seen that Γ_1 does not enclose any pole or zero in the *s*-plane. Therefore, the map of Γ_1 in the *F*-plane does not encircle the origin. On the other hand, Γ_2 encloses one zero and no pole in the *s*-plane. Consequently, its map in the *F*-plane encircles the origin once in the clockwise direction.

Both these observations agree with the principle of the argument.

EXAMPLE 9.2

Consider the function

$$F(s) = \frac{s}{s+2} \qquad \text{...(9.8)}$$

which has a zero at the origin of the *s*-plane and a pole at $s = -2$.

We shall determine the maps of two clockwise circular contours, with centre at the origin, Γ_1 has radius 1 and Γ_2 has radius 3. Since *F*(*s*) is not a simple translation in this case, we shall have to calculate this function along the contours. On Γ_1, we have $s = e^{j\phi}$, and which can be evaluated

$$F(s) = \frac{e^{j\phi}}{e^{j\phi} + 2} \qquad \text{...(9.9)}$$

for different values of ϕ to obtain the map of Γ_1 in the *F*-plane. Some values are given in the following table.

TABLE 9.1: Calculation of *F*(*s*) for Different Values of ϕ on Γ_1

ϕ	π	$2\pi/3$	$\pi/2$	$2\pi/3$	0
F(*s*)	$-1 + j0$	$0 + j0.289$	$0.2 + j0.4$	$0.286 + j0.25$	$0.33 + j0$

Similarly, on Γ_2, we have $s = 3e^{j\phi}$, so that

$$F(s) = \frac{3e^{j\phi}}{3e^{j\phi} + 2} \quad ...(9.10)$$

TABLE 9.2: Calculation of *F(s)* for Different Values of ϕ on Γ_2

ϕ	π	$2\pi/3$	$\pi/2$	$2\pi/3$	0
$F(s)$	$3 + j0$	$0.857 + j0.74$	$0.692 + j0.46$	$0.63 + j0.273$	$0.6 + j0$

The contours Γ_1 and Γ_2, in the *s*-plane, as well as their maps in the *F*-plane, are shown in Fig. 9.5. As expected, the map of Γ_1 in the *F*-plane encircles its origin once in the clockwise direction (since for this contour, $Z = 1$ and $P = 0$) while that of Γ_2 does not encircle the origin (since $Z = 1$ and $P = 1$ for this contour).

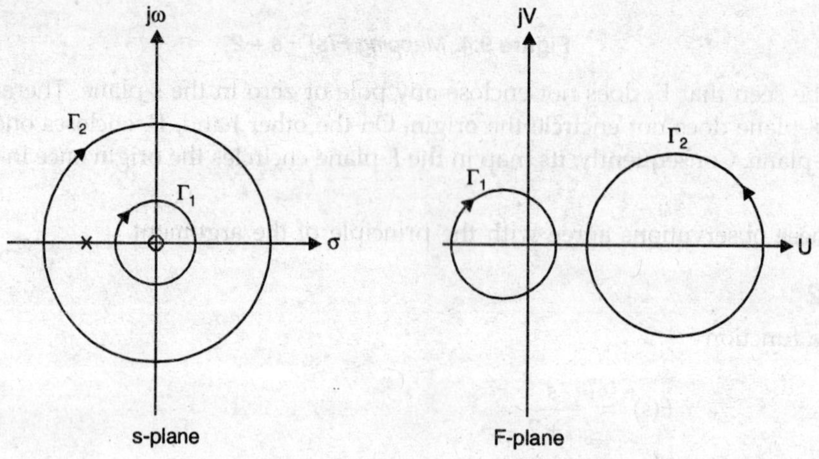

Figure 9.5. *Mapping F(s) in Equation (9.9)*

DRILL PROBLEM 9.1

The transfer function of a servomotor is given by

$$G(s) = \frac{10}{s(s+1)} \quad ...(9.11)$$

Determine the number of times the origin of the G-plane will be encircled by the map of each of the following contours in the s-plane: (a) a circle of radius 0.5 with centre located at the origin, (b) a circle of radius 3, with centre at the origin. Assume that the contours are traversed in the clockwise direction.

Ans. (*a*) –1, (*b*) –2. (*Note*: The negative sign implies encirclement in the opposite or counterclockwise direction.)

9.3 NYQUIST CRITERION

The overall transfer function of the system shown in Fig. 9.1 is given by

$$T(s) = \frac{G(s)}{1+G(s)\,H(s)} \qquad \text{...(9.12)}$$

To find out if the closed-loop system will be stable, we must determine whether $T(s)$ has any pole in the right-half of the s-plane (including the $j\omega$-axis), that is whether $F(s) = 1 + G(s)\,H(s)$ has any root in this part of the s-plane. For this purpose, we must take a contour in s-plane that encloses the entire right half plane as shown in Fig. 9.6(*a*).

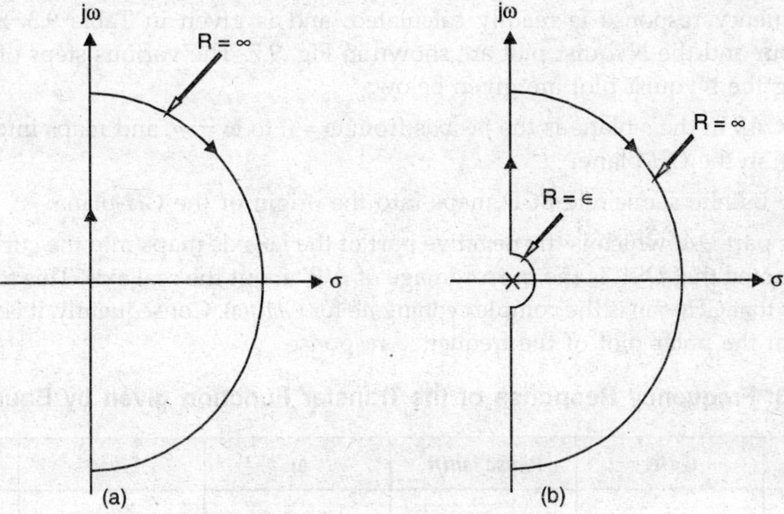

Figure 9.6. *The Nyquist contour*

If $F(s)$ has a pole or zero at the origin of the s-plane or at some points on the $j\omega$-axis, we must make a detour along an infinitesimal semicircle, as shown in Fig. 9.6(*b*). This is called the Nyquist contour, and our object is to determine the number of zeros of $F(s)$ inside this contour.

Let $F(s)$ have P poles and Z zeros within the Nyquist contour. Note that the poles of $F(s)$ are also the poles of $GH(s)$, but the zeros of $F(s)$ are different from those of $GH(s)$, and not known. For the system to be stable, we must have $Z = 0$, that is the characteristic polynomial must not have any root within the Nyquist contour.

From the principle of the argument, a map of the Nyquist contour in the F-plane will encircle the origin of the F-plane N times in the clockwise direction, where

$$N = Z - P \qquad \text{...(9.13)}$$

Thus, the system is stable if and only if $N = -P$, so that Z will be zero. Further, note that the origin of the F-plane is the point $-1 + j0$ in the GH-plane. Hence, we get the following criterion in terms of the loop transfer function $GH(s)$.

> A feedback system will be stable if and only if the number of counter-clockwise encirclements of the point $-1 + j0$ by the map of the Nyquist contour in the GH-plane is equal to the number of poles of $GH(s)$ inside the Nyquist contour in the s-plane.

The map of the Nyquist contour in the *GH*-plane is called the Nyquist plot of *GH*(*s*). The polar plot of the frequency response of *GH*(*s*), which is the map of the positive part of the *j*ω-axis, is an important part of the Nyquist plot and can be obtained experimentally or by computation if the transfer function is known. The procedure of completing the rest of the plot will be illustrated by a number of examples.

EXAMPLE 9.3

Consider the transfer function

$$GH(s) = \frac{60}{(s+1)(s+2)(s+5)} \qquad ...(9.14)$$

The frequency response is readily calculated, and is given in Table 9.3. Sketches of the Nyquist contour and the Nyquist plot are shown in Fig. 9.7. The various steps of the procedure for completing the Nyquist plot are given below.

(*i*) Part *AB* in the *s*-plane is the *j*ω-axis from ω = 0 to ω = ∞, and maps into the polar plot *A′B′* in the *GH*-plane.

(*ii*) The infinite semicircle *BCD* maps into the origin of the *GH*-plane.

(*iii*) The part *DA*, which is the negative part of the *j*ω-axis maps into the curve *D′A′*. It may be noted that *D′A′* is the mirror image of *A′B′* about the real axis. This follows from the fact that *GH*(–*j*ω) is the complex conjugate for *GH*(*j*ω). Consequently, it is easily sketched from the polar plot of the frequency response.

TABLE 9.3: Frequency Response of the Transfer Function given by Equation (9.14)

ω	*Gain*	*Phase shift*	ω	*Gain*	*Phase shift*
0	0	0°	3.0	0.90	–158.8°
0.25	5.77	–24.0°	4.123	0.476	–180°
0.6	5.18	–46.3°	5.0	0.309	–191.9°
1.0	3.72	–82.9°	10.0	0.052	–226.4°
2.0	1.76	–130.2°	20.0	0.007	–247.4°

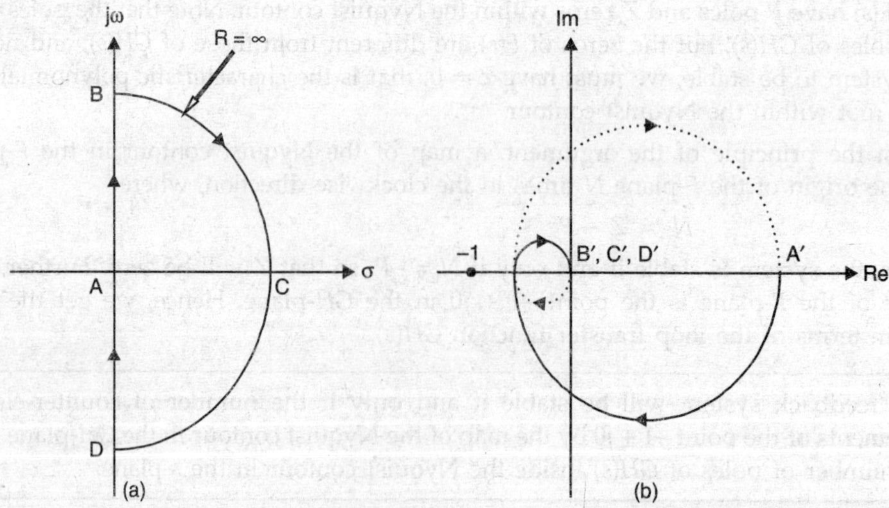

Figure 9.7. *Nyquist contour and plot for the transfer function given by Equation (9.14)*

The Nyquist plot does not encircle the point $-1 + j0$ in the GH-plane. Hence, we have $N = 0$. Also, the function $GH(s)$ does not have any pole inside the Nyquist contour. This makes the system stable since $Z = N + P = 0$.

If we increase the open-loop gain of this system by a factor of 2.15 or more, the resulting Nyquist plot will encircle the point $-1 + j0$ twice in the GH-plane, as shown in Fig. 9.8. Consequently, for this case, $N = 2$. Since P is still zero, the closed-loop system is unstable, with two poles in the right half of the s-plane.

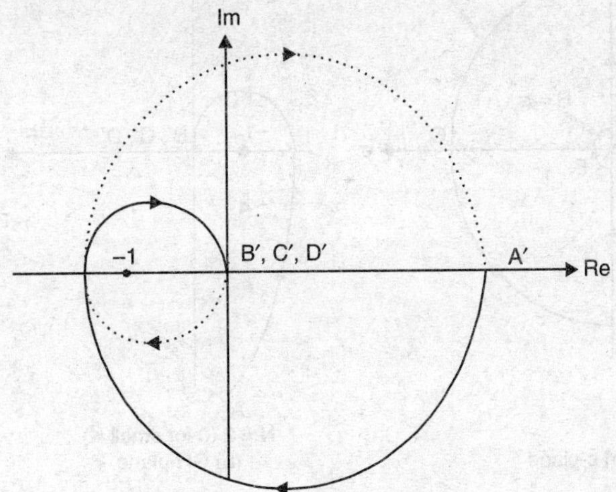

Figure 9.8. *Nyquist plot for the transfer function given by Equation (9.15) if the open-loop gain is increased by a factor of 2.15 or more*

EXAMPLE 9.4

Consider the open-loop transfer function

$$G(s) = \frac{K}{s(s+2)(s+6)} \qquad ...(9.15)$$

Since this time we have a pole at the origin of the s-plane, the Nyquist contour must take a small detour around this pole while still attempting to enclose the entire right half of the s-plane, including the $j\omega$-axis. Hence, we get the semicircle EFA of infinitesimal radius, $\in$. The resulting Nyquist contour and plot are shown in Fig. 9.9. The various steps in obtaining the Nyquist plot are described below.

(*i*) Part AB of the s-plane is the positive part of the $j\omega$-axis from $j\in$ to $j\infty$ and maps into the polar plot of the frequency response $A'B'$ in the GH-plane (shown as a solid curve).

(*ii*) The infinite semicircle BCD in the s-plane maps into the origin of the GH-plane.

(*iii*) The part DE (negative $j\omega$-axis) maps into the image $D'E'$ of the frequency response $A'B'$.

(*iv*) The infinitesimal semicircle EFA is the plot of the equation $s = \in e^{j\theta}$, with θ increasing from $-90°$ to $90°$. It maps into the infinite semicircle, $E'F'A'$, since $GH(s)$ can be approximated as $\dfrac{K}{12 \in e^{j\theta}} = \dfrac{Ke^{-j\theta}}{12 \in}$ for small s. Note that $s = \in e^{j\theta}$ in the denominator will cause the infinitesimal semicircle to map into an infinite semicircle with traversal in the opposite direction due to the change in the sign of θ as it is brought into the numerator from the denominator. For the plot shown, $N = 2$ and $P = 0$; consequently, the system will be unstable, with $Z = 2$. If the gain is sufficiently reduced, the point $-1 + j0$ will not be encircled, with the result that the system stable, since $N = 0$ and $Z = 0$.

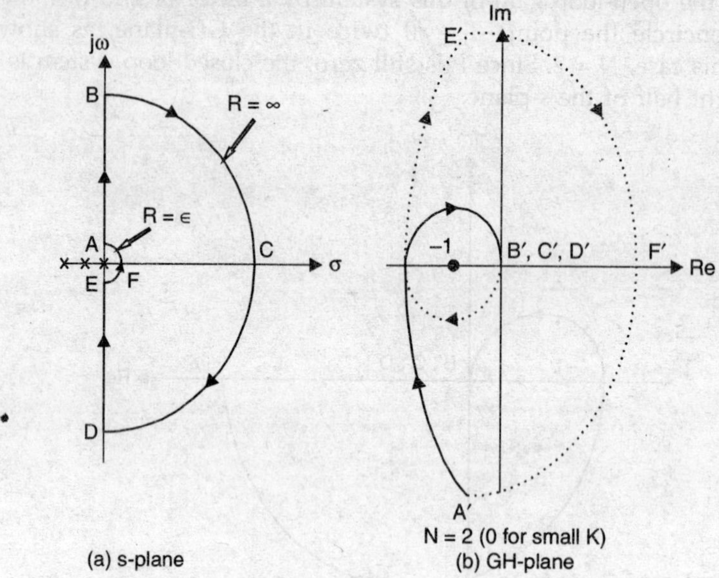

Figure 9.9. *Nyquist plot for the transfer function given by Equation (9.15)*

EXAMPLE 9.5

We shall reconsider the transfer function of the preceding example but this time we shall take a different Nyquist contour. The detour around the pole at the origin will be taken from its left, as shown in Fig. 9.10(*a*). The resulting Nyquist plot is shown in Fig. 9.10(*b*).

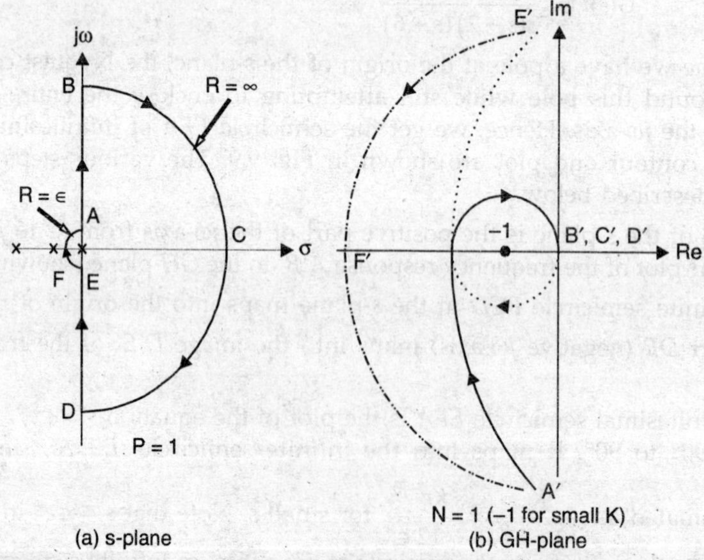

Figure 9.10. *Nyquist plot for the transfer function given by Equation (9.15) with the Nyquist contour enclosing the pole at the origin*

The essential difference with Fig. 9.9(*b*) is in the map of the infinitesimal semicircle in the Nyquist contour, which is an infinite semicircle in the counterclockwise direction. Consequently, we have $P = 1$. Also, the Nyquist plot encircles the point $-1 + j0$ once in the clockwise direction, giving us $N = 1$, so that $Z = N + P = 2$, and the system will be unstable. Also, reducing the gain sufficiently will cause the Nyquist plot to encircle the point $-1 + j0$ once in the counter-clockwise direction, making $N = -1$, with the result that now we shall have $Z = N + P = 0$. Thus, the system will be stable for small values of K. Both these results agree with those obtained in the previous example, as expected.

EXAMPLE 9.6

The polar plot of the open-loop frequency response of a closed-loop system is shown in Fig. 9.11(*b*) by the solid curve for positive ω, indicating two poles at the origin of the *s*-plane. We shall complete the Nyquist plot in the *GH*-plane to determine stability. The system is known to be open-loop stable, that is $GH(s)$ has no pole in the right half of the *s*-plane.

The Nyquist contour is shown in Fig. 9.11(*a*), and the corresponding Nyquist plot is shown as the dotted curve in Fig. 9.11(*b*) with the exception of the polar plot (map of the positive part of the *j*ω-axis) which is the solid part of the plot.

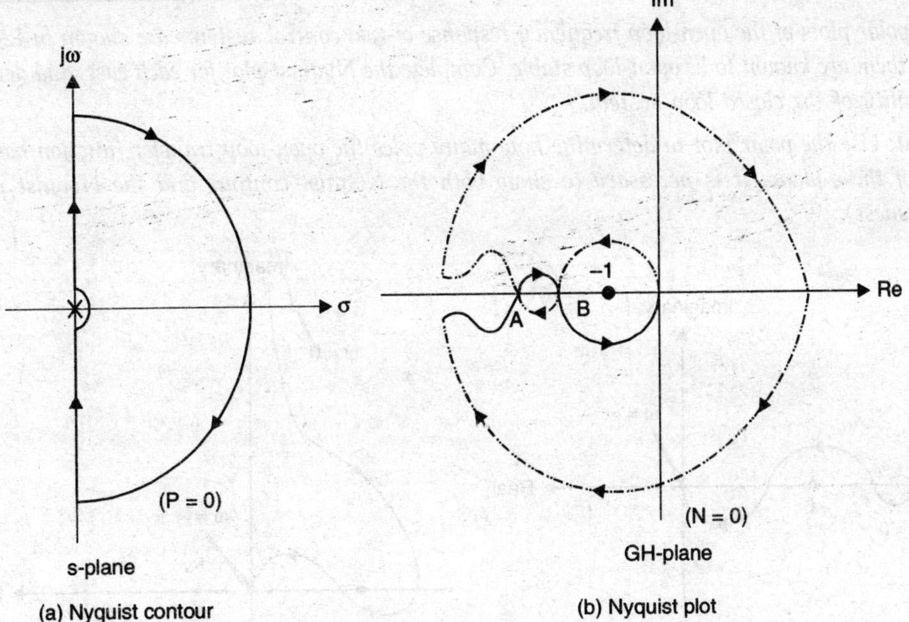

(a) Nyquist contour (b) Nyquist plot

Figure 9.11. *Complete Nyquist plot from the polar plot of the frequency response*

From the plot, we have $N = 0$, and $P = 0$, so that $Z = 0$. Hence the closed-loop system will be stable. It is interesting to note that although the polar plot of the frequency response crosses the negative real axis two points on the left of the point $-1 + j0$, the system is stable. This system will be stable for low and high gains but unstable for the medium range of gains, when the point $-1 + j0$ lies between the points A and B on the polar plot.

DRILL PROBLEM 9.2

For the open-loop transfer function given by

$$GH(s) = \frac{10(s+1)}{s^2(s+4)} \qquad ...(9.15)$$

obtain the Nyquist plot and determine whether the closed-loop system will be stable.

Ans. Stable

DRILL PROBLEM 9.3

Complete the Nyquist plot for the polar plot in Example 9.6, with Nyquist contour changed so that the detour around the pole at the origin of the s-plane is taken from the left.

DRILL PROBLEM 9.4

The polar plots of the open-loop frequency response of two control systems are shown in Fig. 9.12. Both of them are known to be open-loop stable. Complete the Nyquist plot for each case, and determine the stability of the closed-loop system.

(***Hint:*** *Use the polar plot to determine how many poles the open-loop transfer function has at the origin of the s-plane. It is necessary to show both the Nyquist contour and the Nyquist plot for completeness*).

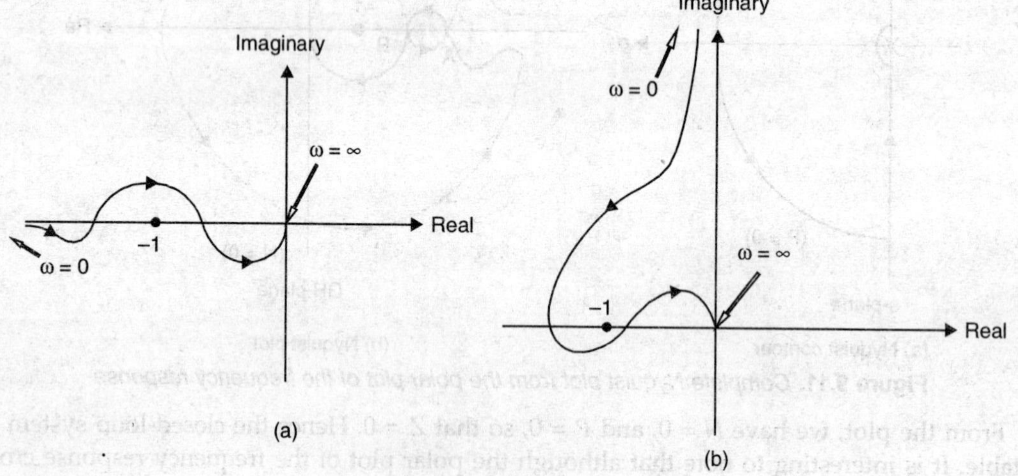

Figure 9.12. *Polar plots*

Ans. (*a*) Unstable, (*b*) Stable.

9.4 RELATIVE STABILITY

In most of the practical situations, it addition to finding out whether a closed-loop system is stable, it is also desirable to determine how close it is to instability. This information can be readily determined from the open-loop frequency response.

The proximity of the open-loop frequency response to the point $-1 + j0$ in the *GH*-plane provides a quantitative measure of the relative stability of a closed-loop system.

Two commonly used measures of relative stability are the gain margin and the phase margin. These are defined below.

The *gain margin* is defined as the additional gain required to make the system just unstable. It may be expressed either as a factor or in decibels.

The *phase margin* is defined as the additional phase lag required to make the system just unstable.

In the plot shown in Fig. 9.13, from the location of the point *A* where the phase lag is 180° we see that the gain must be increased by the factor of 2 to make the system just unstable. Consequently, the gain margin is 2. Alternatively, we may say that the gain margin is 20 log (2) or 6 dB.

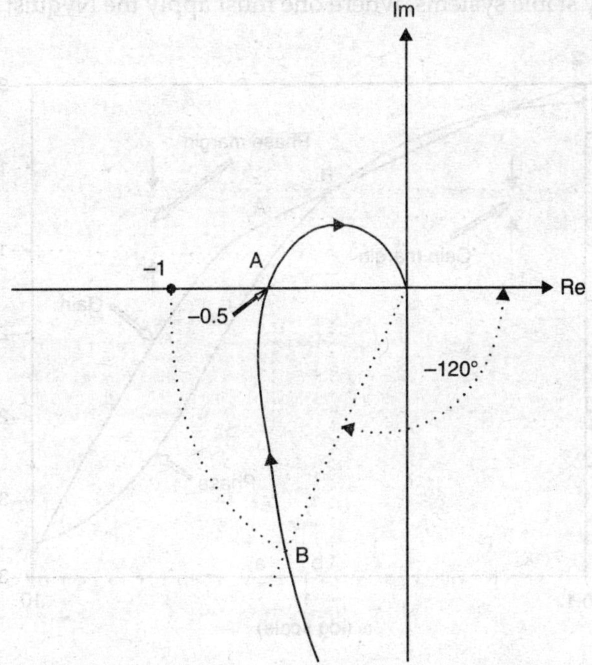

Figure 9.13. *Gain margin and phase margin*

Similarly, from point *B*, where the gain is one, the phase margin is 180° – 120°, or 60°.

The gain margin as well as the phase margin can also be determined either from the Bode plots or the log-magnitude/phase diagram. The gain margin is determined from the gain at the *phase crossover frequency, i.e.*, the frequency at which the phase lag is 180°. Similarly, the phase margin is obtained from the phase lag at the *gain crossover frequency, i.e.*, the frequency at which the gain is 0 dB. This is illustrated in the following example.

EXAMPLE 9.7

Consider the following forward-path transfer function of a unity-feedback system.

$$G(s) = \frac{5}{s(s+1)\left(s^2+2s+5\right)} \qquad ...(9.16)$$

The Bode plots of the gain and phase-shift are shown in Fig. 9.15. To obtain the gain margin, we first determine the phase-crossover frequency, denoted by *a*, from the point *A* on the phase curve, where the phase lag is 180°. We then read the gain at this frequency. Subtracting the value of this gain from 0 dB gives the gain margin in dB. Similarly, to obtain the phase margin, we first determine the gain-crossover frequency, *b*, from the point B on the gain curve, where the gain is 0 dB. We then read the phase lag at this frequency and subtract it from 180° to obtain the phase margin. Both the gain margin and the phase margin are labelled in Fig. 9.15, where the two cross-over frequencies are also shown. It will be seen from the plots that the gain margin is approximately 5 dB and the phase margin is about 30° for this system. Note that the gain margin in decibels will be negative if the gain at the phase-crossover frequency is greater than one. Similarly, a negative phase margin implies that the phase lag at the gain-crossover frequency is more than 180°. Negative values of these margins normally imply that the system is unstable. Exceptions arise in the case of conditionally stable systems, where one must apply the Nyquist criterion to investigate stability.

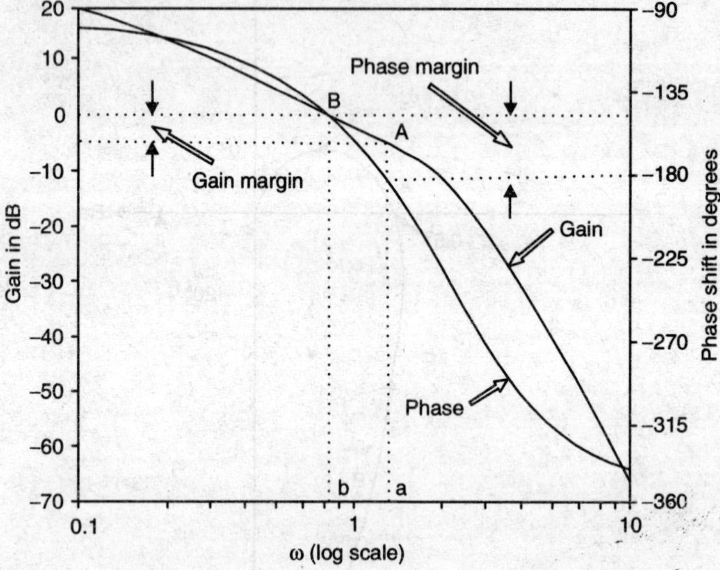

Figure 9.14. *Bode plots illustrating gain margin and phase margin*

The gain margin and phase margin are determined more directly from the log-magnitude/ phase diagram, as shown in Fig. 9.16, but the two crossover frequencies are not shown right away.

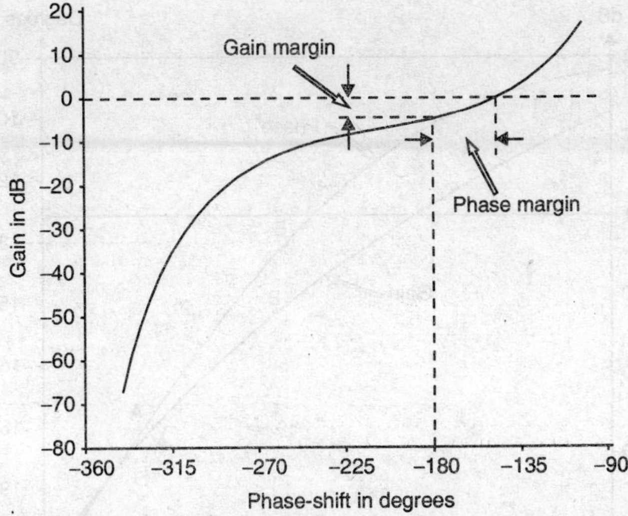

Figure 9.15. *Log-magnitude/phase plot showing gain margin and phase margin*

DRILL PROBLEM 9.5

How should the gain be adjusted in the system in Example 9.7 so that (a) the gain margin may be 10 dB, (b) the phase margin may be 45°?

Ans. (*a*) It should be reduced by 5 dB.

(*b*) It should be reduced by 3.5 dB.

EXAMPLE 9.8

The open-loop transfer function of control system is given by

$$G(s) = \frac{K}{s(s+2)(s+10)} \qquad \text{...(9.17)}$$

Determine the value of K so that the system may be stable with (*a*) gain margin equal to 6 dB, (*b*) phase margin equal to 45°.

Solution. We start by assuming a suitable value of K and drawing the Bode plots. Here we shall take $K = 10$. The calculated values of the frequency response for this case are shown in Table 9.4, and the Bode plots are shown in Fig. 9.16.

TABLE 9.4: Frequency Response of the Transfer Function given by Equation (9.17)

ω	*Gain*	*Phase shift*	ω	*Gain*	*Phase shift*
0.1	14	−93.4°	3.0	−21.0	−163.0°
0.2	7.9	−96.9°	4.0	−25.7	−175.2°
0.5	−0.25	−106.9°	5.0	−29.7	−184.8°
0.7	−3.43	−113.3°	6.0	−34.7	−192.5°
1.0	−7.01	−122.3°	10.0	−37.2	−213.7°
2.0	−3.2	−146.3°			

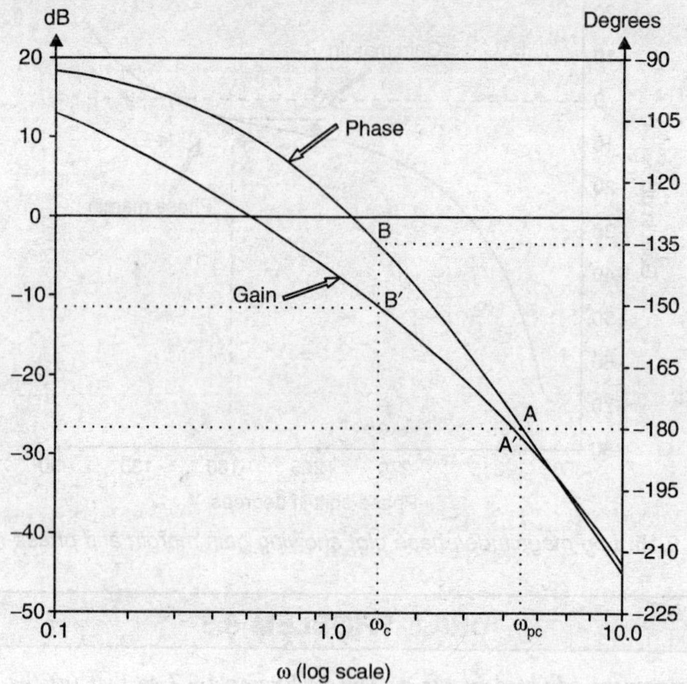

Figure 9.16. *Bode plots of the frequency response for transfer function given by Equation (9.17)*

(*a*) To obtain gain margin of 6 dB, first we determine the phase-crossover frequency. This is indicated by the point *A* on the phase curve with $\omega_{pc} = 4.47$. The gain at this frequency is found as –27.58 dB as indicated by the point *A*′. Hence, for the desired gain margin, we must increase the gain by 21.58 dB. The resulting value of *K* is obtained as

$$K = 10 \times 10^{21.58/20} = 120.28 \qquad \text{...(9.18)}$$

(*b*) To obtain phase margin of 45°, we must first determine the frequency at which the phase shift is –135° and then adjust the value of *K* so that the gain at this frequency is zero. From the Bode plot we find that for $\omega_c = 1.5$, we get this phase shift and a gain of –11.5 dB, as indicated by the points *B* and *B*′. Hence, the gain must be set at

$$K = 10 \times 10^{11.5/20} = 37.67 \qquad \text{...(9.19)}$$

In many cases, the phase margin is a better measure of relative stability than the gain margin. This will be illustrated by the following example.

Example 9.9

Consider a unity-feedback system with forward-path transfer function

$$G(s) = \frac{K}{s(s+2)} \qquad \text{...(9.20)}$$

The polar plots of the open-loop frequency response for $K = 10$ and $K = 50$ are shown in Fig. 9.17. It is evident that for both cases the gain margin is infinity since the plot never crosses the negative real axis. On the other hand, the phase margins for the two cases are 35° for $K = 10$ and 16° for $K = 50$. This shows that the phase margin is a better criterion for stability. A phase margin between 45° to 60° is usually considered satisfactory.

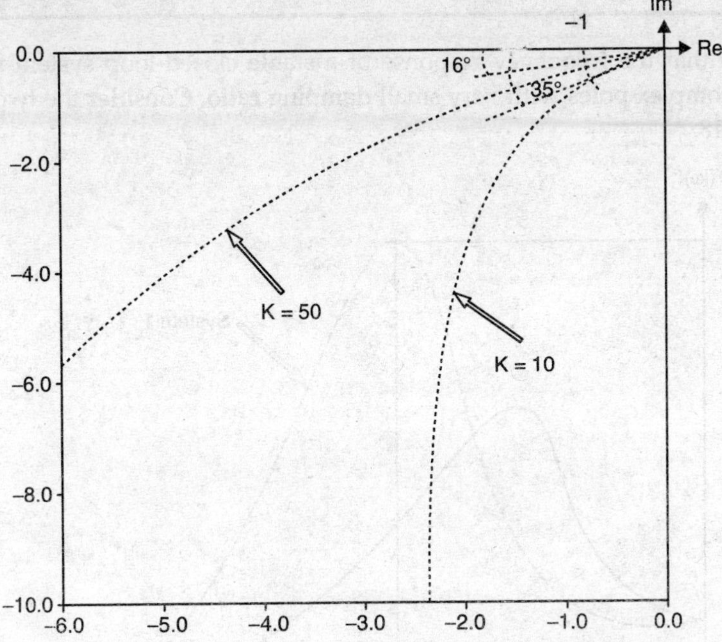

Figure 9.17. *Gain and phase margins for a second-order system*

DRILL PROBLEM 9.6

The transfer function of the forward path of a unity-feedback system is given by

$$G(s) = \frac{40}{s(s+2)(s+8)} \qquad ...(9.21)$$

Determine the gain margin, the phase margin, the gain crossover frequency and the phase crossover frequency.

Ans. Phase crossover frequency = 4 rad/s, Gain margin = 12 dB,
Gain crossover frequency = 1.81 rad/s, Phase margin = 35°.

DRILL PROBLEM 9.7

The open-loop transfer function of a unity feedback system is given by

$$G(s) = \frac{K}{s\left(s^2+2s+5\right)} \qquad ...(9.22)$$

What should be the value of K to obtain (a) a gain margin of 6 dB, (b) a phase margin of 50°?

Ans. (*a*) $K = 5$, (*b*) $K = 5.6$

9.5 FREQUENCY RESPONSE OF THE CLOSED-LOOP SYSTEM

It is well known that the frequency response of a stable closed-loop system will have a sharp peak if $T(s)$ has complex poles with very small damping ratio. Consider the two magnitude plots shown in Fig. 9.18.

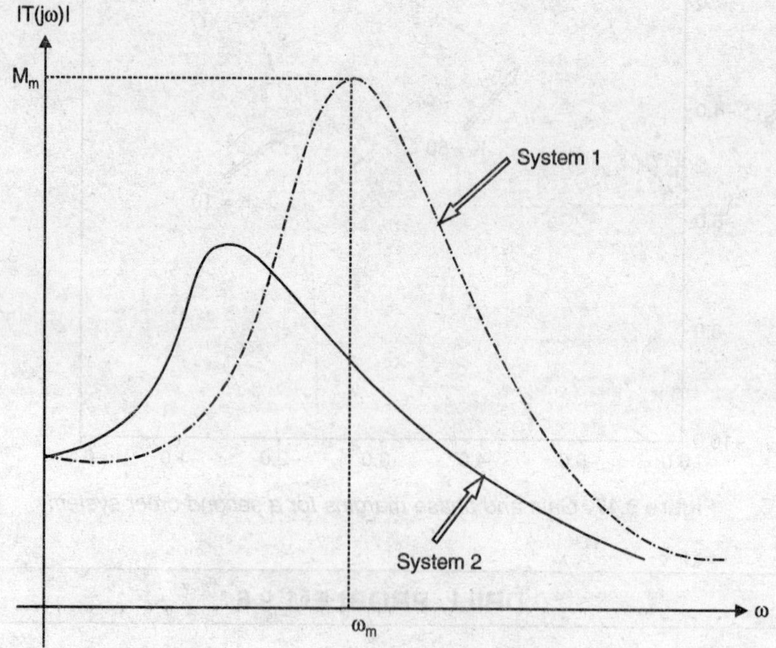

Figure 9.18. *Closed-loop frequency response of two systems*

It is evident that system 2 will have a much better transient response than the system 1, as the damping ratio of the dominant poles of the former is greater than the damping ratio of the dominant poles of the former. Consequently, M_m, defined as the *maximum magnitude of the closed-loop system frequency response,* is a very good measure of the relative stability of the system, in view of the fact that smaller values of this quantity will lead to a better transient response.

For a unity feedback system, the frequency response of the closed-loop system can be determined conveniently from the open-loop frequency, since

$$T(j\omega) = \frac{G(j\omega)}{1+G(j\omega)} = M(\omega)e^{j\phi(\omega)} \qquad ...(9.23)$$

This is shown graphically in Fig. 9.19.

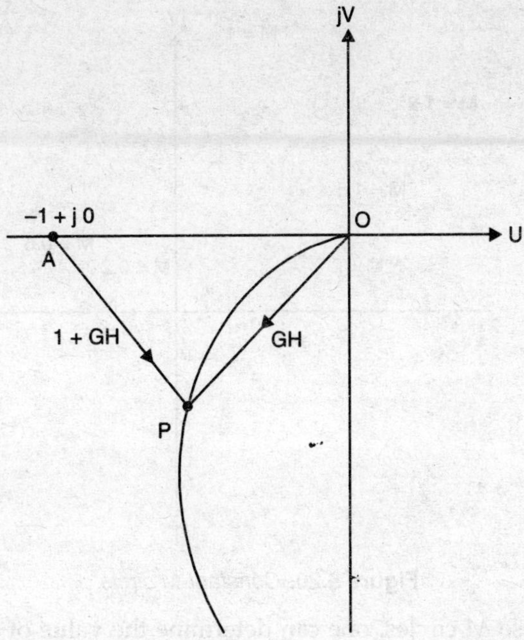

Figure 9.19. *Relation between open-loop and closed-loop frequency response*

Let $$G(j\omega) = U(\omega) + jV(\omega) \quad \text{...(9.24)}$$

Then, we have

$$T(j\omega) = \frac{U + jV}{1 + U + jV} \quad \text{...(9.25)}$$

It follows that the magnitude of the gain will be given by

$$M = |T(j\omega)| = \frac{\sqrt{U^2 + V^2}}{\sqrt{(1+U)^2 + V^2}} \quad \text{...(9.26)}$$

For constant M, we can rewrite Equation (9.26) as

$$(1 - M^2)\,U^2 + (1 - M^2)\,V^2 - 2M^2U = M^2 \quad \text{...(9.27)}$$

Equation (9.27) can be rearranged to get the equation of a circle in the G-plane.

$$\left(U - \frac{M^2}{1 - M^2}\right)^2 + V^2 = \left(\frac{M}{1 - M^2}\right)^2 \quad \text{...(9.28)}$$

It is seen from Equation (9.28) that the centre of the circle is at $U = -M^2/(M^2 - 1)$, $V = 0$, and its radius is $|M/(M^2 - 1)|$. This family of constant-M circles can be drawn and labelled on the graph paper for different values of M. Consequently, the value of M_m can be determined from the polar plot of the frequency response of $G(s)$. For example, Fig. 9.20 depicts several M-curves.

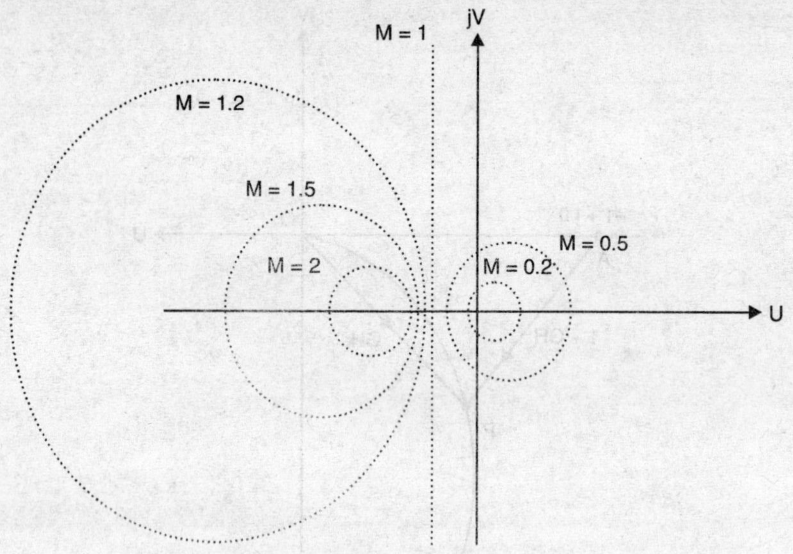

Figure 9.20. *Constant-M circle*

With the help of these M circles, one can determine the value of M_m from the polar plot of the open-loop frequency response, as the largest value of M for the circle tangent to the polar plot and satisfying Equation (9.28). For example, from the polar plot of the frequency response shown in Fig. 9.21, it is evident that for this case, $M_m = 1.5$.

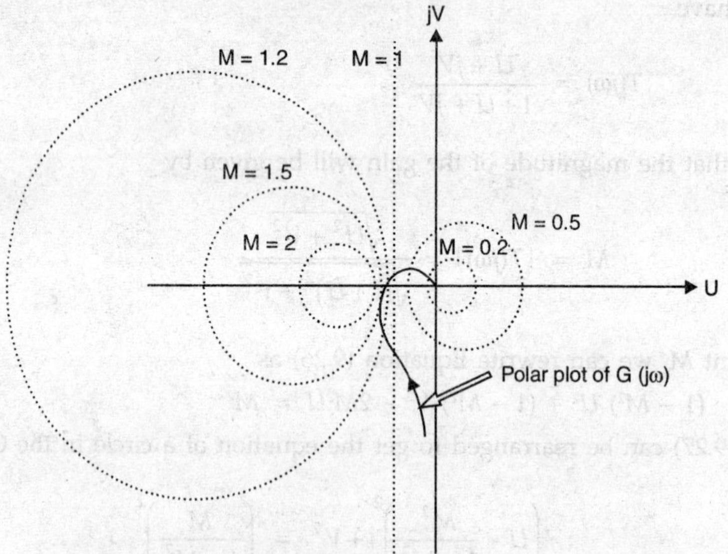

Figure 9.21. *Determination of M_m from the family of M-circles*

It is also possible to use a simple geometric construction to determine the value of K that will provide a desired M_m. The various steps are given below (see Fig. 9.22).

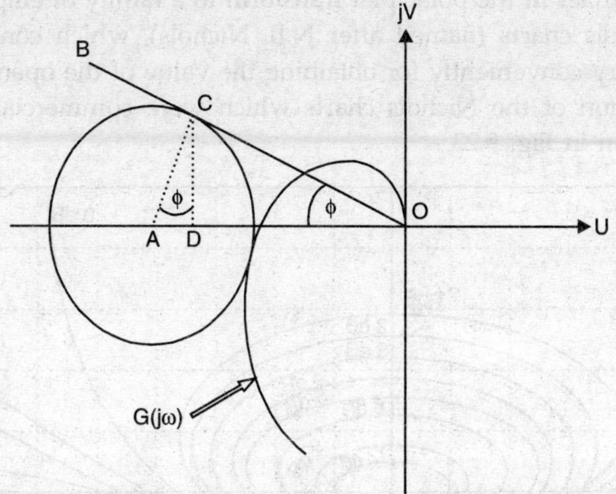

Figure 9.22. *Construction for determining the open-loop gain for a specified M_m*

1. Draw the polar plot of $G(j\omega)$ for some assumed value of K.
2. Calculate $\phi = \sin^{-1}(1/M_m)$, and draw a radial line OB from the origin of the G-plane making an angle ϕ with the negative real axis.
3. Draw a circle with centre on the negative real axis which is tangent to OB at C and also tangential to the polar plot of $G(j\omega)$. Let this centre be at A.
4. From C draw a line CD perpendicular to the negative real axis.
5. The desired gain is obtained by dividing the assumed value of K by the length of the line OD.

The proof for this construction follows from the expressions for the radius and the centre of the circle in Equation (9.28), which leads immediately to the relationships

$$\sin\phi = \frac{AC}{OA} = \frac{\dfrac{M}{M^2-1}}{\dfrac{M^2}{M^2-1}} = \frac{1}{M} \qquad \text{...(9.29)}$$

and
$$OD = OA - AC\sin\phi = \frac{M^2}{M^2-1} - \frac{M}{M^2-1}\cdot\frac{1}{M} = 1 \qquad \text{...(9.30)}$$

Thus, the location of the point D represents the point $-1 + j0$ in the GH-plane and gives the scale factor by which the loop gain must be changed to obtain the desired M_m.

DRILL PROBLEM 9.8

The transfer function of the forward path of a unity-feedback system is given by

$$G(s) = \frac{K}{s(s+2)(s+4)} \qquad \text{...(9.31)}$$

Determine the value of K in such a way that $M_m = 2$.

Ans. $K = 17.3$

The constant-M circles in the polar plot transform to a family of ellipses in the logarithmic gain phase plot. Nichols charts (named after N.B. Nichols), which contain these constant-M curves, can be used very conveniently for obtaining the value of the open-loop gain for a given M_m. A simplified version of the Nichols charts which were commercially available until the middle 1970's, is shown in Fig. 9.23.

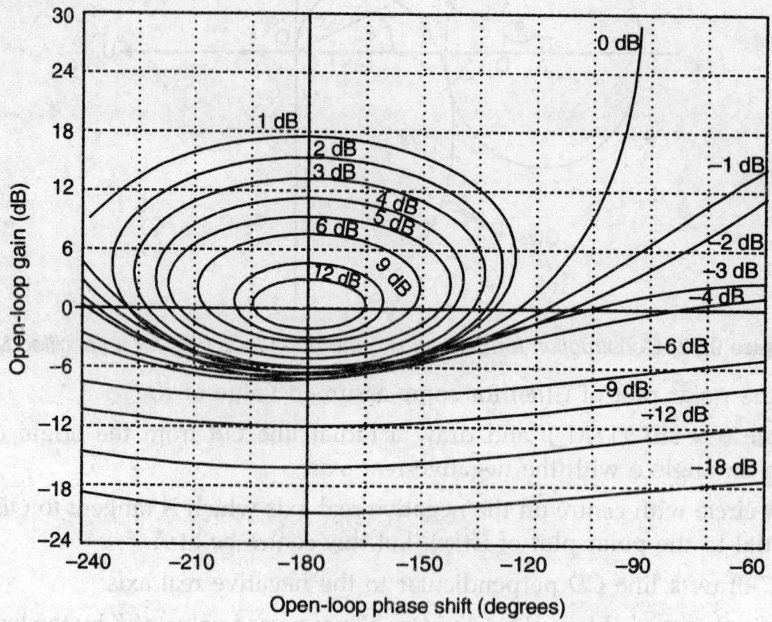

Figure 9.23. *Modified Nichols chart showing M-curves*

Although these charts cannot be purchased any more, they can be easily plotted with the help of a computer. The computer program Nichols.EXE on the disk that can be obtained with this book, can be used efficiently for this purpose. The procedure for obtaining the value of the loop-gain K that will provide a specified M_m is as follows.

For an assumed value of K, plot the frequency response for the given system on the Nichols chart. Also, locate plot the M-curve corresponding to the specified value of M_m. Next examine by how many decibels the frequency response curve should be raised or lowered so that it is tangential to the M-curve plotted earlier. This gives the gain adjustment necessary to obtain the desired M_m.

EXAMPLE 9.10

The forward transfer function of a unity feedback system is given by

$$G(s) = \frac{K}{s\left(s^2 + 2s + 5\right)}$$

Determine the value of K so that M_m = 3 dB.

Solution. Assuming K = 10, the frequency response was calculated, and is given in the Table 9.5.

TABLE 9.5: Frequency Response of the Transfer Function given by Equation (9.17)

ω	*Gain (dB)*	*Phase shift*	ω	*Gain (dB)*	*Phase shift*
0.1	26	−92.3°	2.0	1.67	−166°
0.2	20	−94.6°	2.2	0.28	−177.9°
0.5	12.3	−101.9°	2.3	−0.5	−183.6°
1.0	7	−116.6°	2.4	−1.34	−189.0°
1.5	4.3	−137.5°	2.6	−3.09	−198.7°
1.8	2.8	−153.9°	2.8	−4.9	−206.9°

This is plotted in the Nichols chart and is shown in Fig. 9.24. The plot indicates that the system is unstable for the assumed value of K, with M_m approximately infinity. To obtain $M_m = 3$ dB, we must lower the frequency response curve by approximately 6 dB. This implies that the gain must be reduced by 6 dB. Consequently, the desired value of K should be 5 instead of the assumed value of 10.

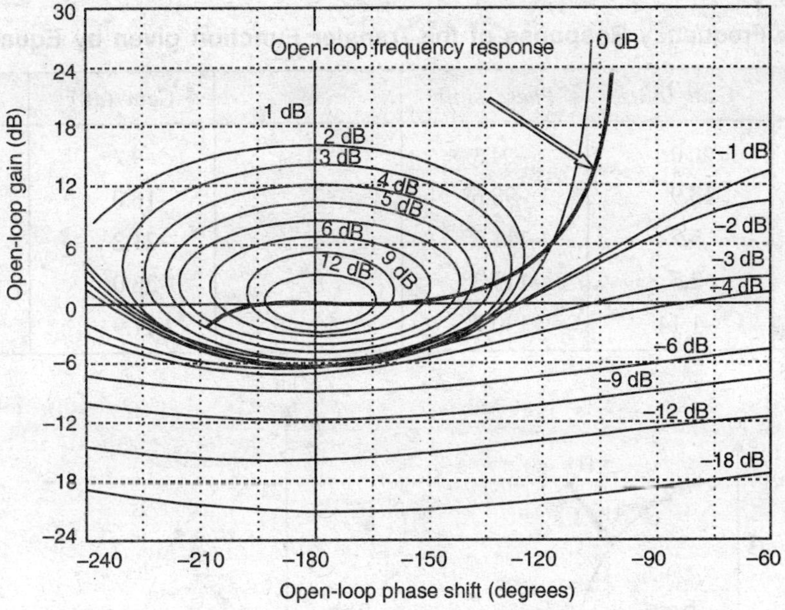

Figure 9.24. *Nichols chart with plot of open-loop frequency response*

DRILL PROBLEM 9.9

Use the Nichols chart procedure to verify the solution to Drill Problem 9.8.

9.6 SYSTEMS WITH PURE DELAY

In many practical situations the open-loop transfer function contains a pure delay (or transport lag) term of the form $e^{-\beta s}$. The Nyquist criterion remains valid for such cases, and we may use the frequency response curve as before. We may also use the various frequency response curves to determine relative stability, with the gain margin, phase margin, and the maximum magnitude of the closed-loop frequency response, as in the previous section. It is worth noting that neither the Routh criterion nor the root locus method is applicable in this case.

EXAMPLE 9.11

The transfer function of a chemical plant that is in the forward path of a unity-feedback system can be represented as

$$G(s) = \frac{K e^{-0.05s}}{s(s+2)(s+5)} \qquad ...(9.32)$$

We shall first use the Nyquist criterion to determine whether the system will be stable if the gain K is set to 10. After that, we shall use the Nichols chart to adjust K so that the system may be stable with (*a*) phase margin = 45°, and (*b*) M_m = 3 dB.

The frequency response for K = 10 for the open-loop transfer function is given in Table 9.6. The Nyquist contour and plot are shown in Fig. 9.25. These indicate that the system is stable with large gain and phase margins.

TABLE 9.6: Frequency Response of the Transfer Function given by Equation (9.17)

ω	*Gain (dB)*	*Phase shift*	ω	*Gain (dB)*	*Phase shift*
0.1	20.0	–94.3°	2	–9.7	–162.5°
0.2	13.9	–98.6°	3	–16.0	–185.9°
0.5	5.7	–111.2°	4	–21.2	–203.6°
0.7	2.5	–119.3°	5	–26.0	–217.5°
1.0	–1.14	–130.7°	10	–41.9	–260.8°

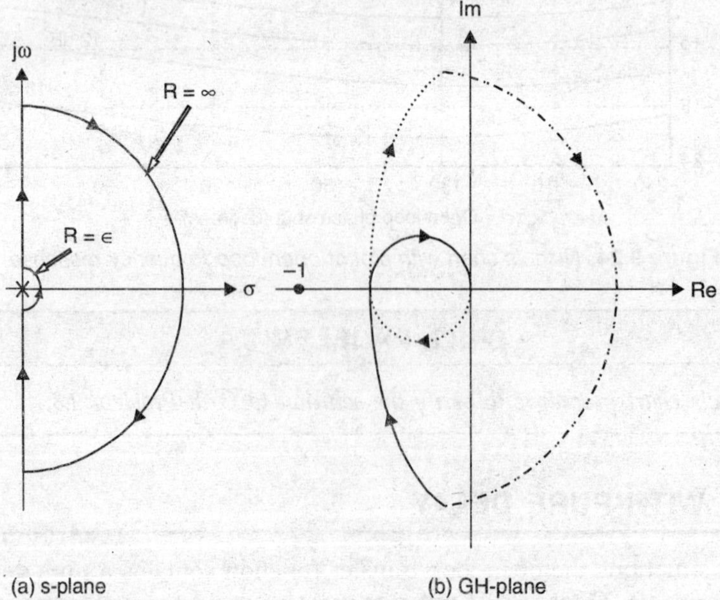

Figure 9.25. *Nyquist plot for system represented by Equation (9.32)*

The plot on the Nichols chart is shown in Fig. 9.26.

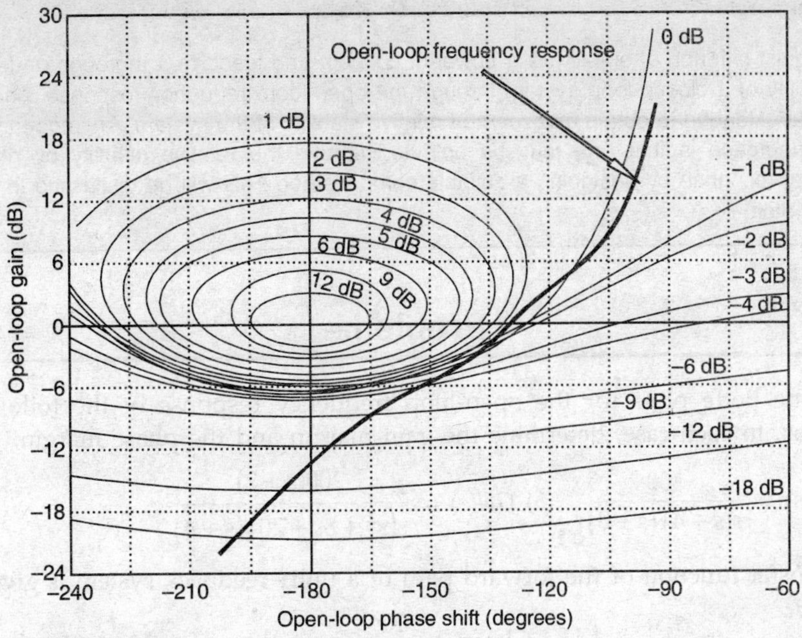

Figure 9.26. *Nichols chart with polar plot of the open-loop frequency response*

(*a*) From the plot, it is seen that the phase margin is 53°. To obtain phase margin of 45°, the plot should be raised by 2 dB. Hence, the desired value of the gain is obtained as

$$K = 10 \times 10^{2/20} = 12.59 \qquad ...(9.33)$$

(*b*) From the plot we get M_m = 1 dB. To make M_m = 3 dB, the plot should be raised by 3 dB. Consequently, the desired value of the gain is obtained as

$$K = 10 \times 10^{3/20} = 14.13 \qquad ...(9.34)$$

DRILL PROBLEM 9.10

The block diagram of a satellite control system is shown in Fig. 9.27. The time required for the signals sent from the earth to reach the satellite is represented by the ideal delay in the block diagram.

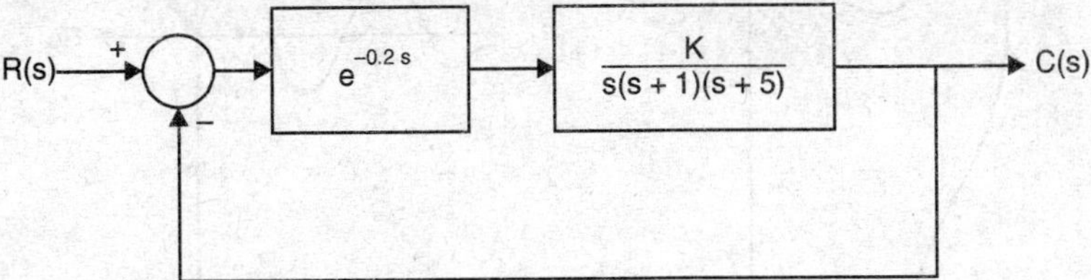

Figure 9.27. *Block diagram of a satellite control system*

(*a*) *Determine the value of the gain constant K so that the closed-loop system may have phase margin equal to 45°.*

(*b*) *Determine the value of K so that M_m may be 3 dB. What is the value of ω_m, the frequency at which the maximum gain is obtained for this case?*

Ans. (*a*) K = 3.56; (*b*) K = 3.9, ω_m = 0.7.

SUMMARY

- The Nyquist criterion of stability is a powerful concept and leads to a thorough understanding of the stability of a closed-loop system through the open-loop frequency response. The ideas may be further extended to obtain measures of relative stability of the system. An important advantage of this approach is that one may be able to improve the relative stability by reshaping the frequency response by designing a suitable compensator. This will be discussed in detail in the next chapter.

Problems

1. Draw the Bode plots for the open-loop frequency response of the following transfer functions. In each case, determine the gain margin and the phase margin.

$$(a)\ GH(s) = \frac{300}{s(s+4)(s+8)} \qquad (b)\ GH(s) = \frac{200(s+2)}{s(s^2+8s+20)(s+4)}$$

2. The transfer function of the forward path of a unity-feedback system is given by

$$G(s) = \frac{Ke^{-0.2s}}{s(s+2)(s+8)}$$

Determine the value of K so that the system may be stable with (*a*) gain margin equal to 6 dB, and (*b*) phase margin equal to 45°.

3. The polar plots of the open-loop frequency response of two unity-feedback systems for positive values of frequency are shown in Fig. P9.3. It is known that the open-loop transfer function has no pole in the right half of the *s*-plane. Draw the complete Nyquist plot for each case and determine whether the closed-loop system will be stable.

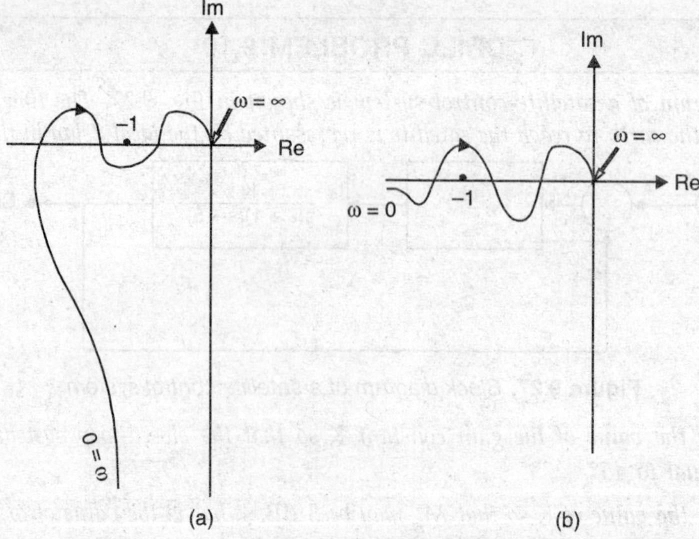

Figure P9.3. *Polar plots*

4. The frequency response of the forward path of a unity-feedback control system was obtained experimentally, and is given in Table P9.4.
   (*a*) Draw the Bode plots of the frequency response.
   (*b*) Increase the loop gain to make the gain margin equal to 5 dB. Determine the values of M_m and ω_m for this value of the gain using the Nichols chart.

TABLE P9.4: Frequency Response of the Transfer Function given by Equation (9.17)

ω	*Gain (dB)*	*Phase shift*	ω	*Gain (dB)*	*Phase shift*
0.1	20.0	–90.30°	10	–13.98	–180.0°
0.2	13.98	–90.57°	15	–26.8	–239.0°
0.4	7.97	–91.15°	20	–36.0	–251.6°
0.6	4.46	–91.73°	30	–47.75	–259.4°
1.0	0.08	–92.9°	40	–55.64	–262.4°
2	–5.71	–95.9°	50	–61.3	–264.1°
4	–10.77	–103.4°	60	–66.48	–265.1°
6	–12.55	–115.1°	80	–74.04	–266.4°
8	–12.68	–138.0°	100	–79.92	–267.1°

5. The forward transfer function of a unity-feedback system is given by

$$G(s) = \frac{Ke^{-0.2s}}{s(s+2)}$$

Draw the polar plot of the frequency response for $K = 10$ and obtain the value of K that will make $M_m = 6$ dB.

6. The polar plot of the frequency response of the forward path of a unity-feedback system is shown in Fig. P9.6. Determine (*a*) the phase margin, (*b*) the gain margin, (*c*) M_m and (*d*) ω_m.

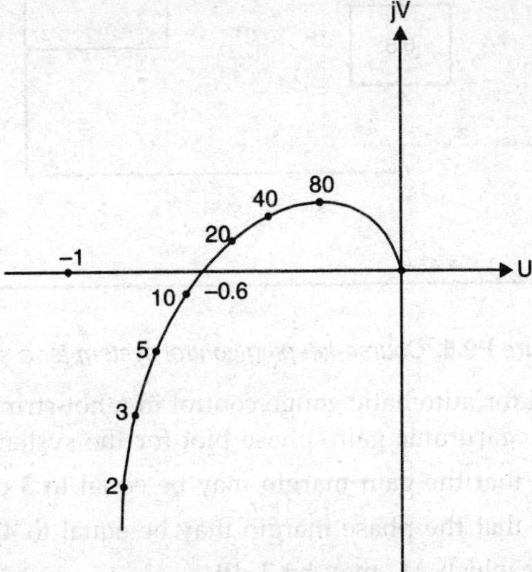

Figure P9.6. *Polar plot*

7. A feedback system with two delays is shown in Fig. P9.7. It represents a typical chemical reactor.

(*a*) Investigate the stability of this system for $K = 1$.

(*b*) What value of K will make the phase margin equal to 45°?

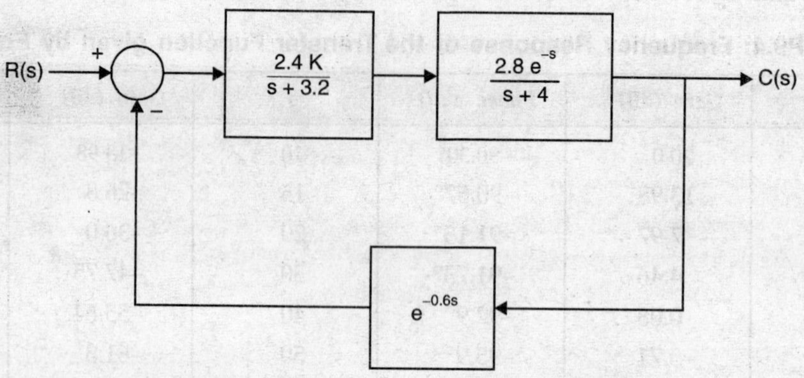

Figure P9.7. *A feedback system with two delays*

8. The block diagram of a course-keeping control system for a ship is shown in Fig. P9.8. Determine the value of K so that the maximum magnitude of the frequency response of the closed-loop system is 3 dB. What is the phase margin for this value of K?

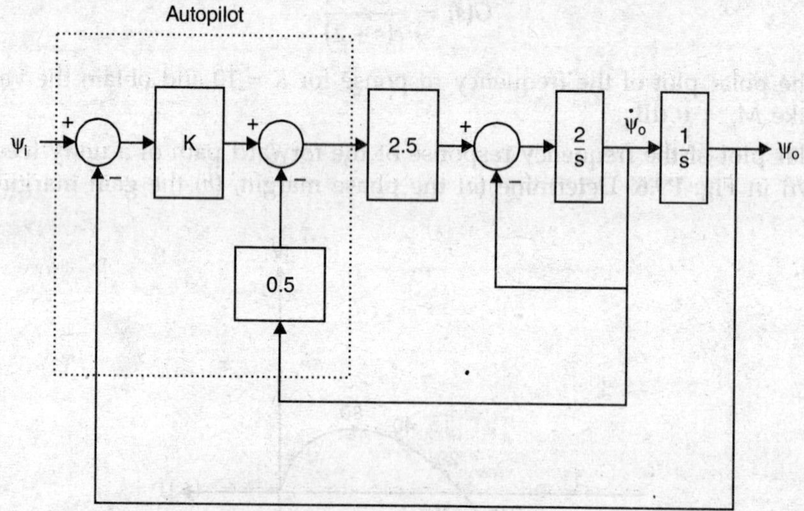

Figure P9.8. *Course-keeping control system for a ship*

9. The linearized model for automatic gauge control in a hot-strip finishing mill is shown in Fig. P9.9. Draw the logarithmic gain/phase plot for the system and determine

(*a*) the value of K so that the gain margin may be equal to 3 dB.

(*b*) the value of K so that the phase margin may be equal to 45°.

(*c*) the value of K for which M_m may be 3 dB.

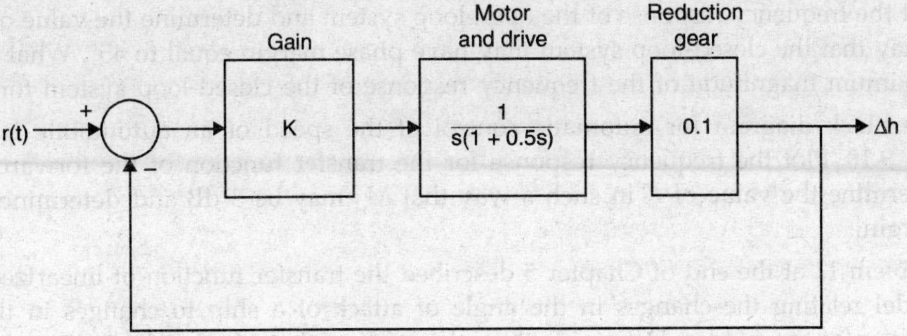

Figure P9.9. *Gauge control system for a hot-strip finishing mill*

10. A linearized model of a driver steering-control system is shown in Fig. P9.10, where the approximate transfer function of the human driver includes a gain K, as well as a time delay due to the reaction time of the neuro-muscular system.
    (*a*) Investigate the stability of the closed-loop system if $K = 5$.
    (*b*) How should K be changed to obtain phase margin equal to 45°?

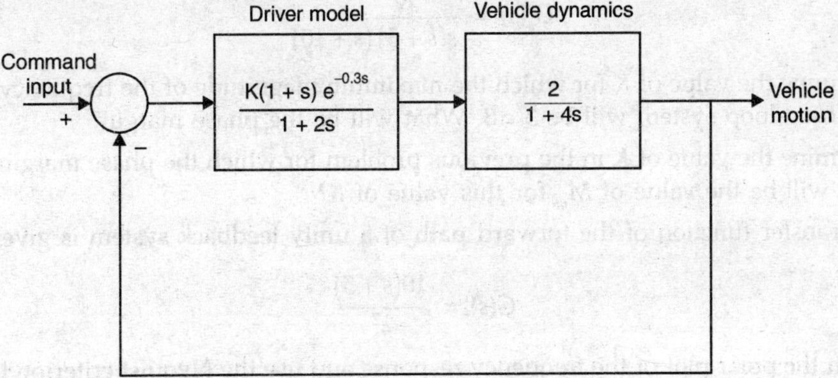

Figure P9.10. *Driver steering-control system*

11. The block diagram for speed control of a motor using a phase-locked loop is shown in Fig. P9.11. It has been claimed that such a system has a higher accuracy than a conventional speed control system, while requiring fewer components.

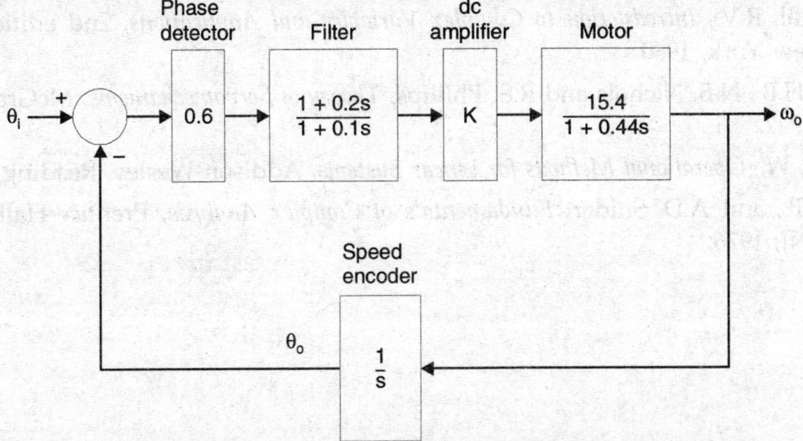

Figure P9.11. *Speed control using a phase-locked loop*

Plot the frequency response of the open-loop system and determine the value of K in such a way that the closed-loop system may have phase margin equal to 45°. What will be the maximum magnitude of the frequency response of the closed-loop system for this K?

12. The block diagram for automatic control of the speed of an automobile is shown in Fig. 9.10. Plot the frequency response for the transfer function of the forward path and determine the value of K in such a way that M_m may be 3 dB and determine the phase margin.
13. Problem 12 at the end of Chapter 5 described the transfer function of linearized dynamic model relating the changes in the angle of attack of a ship to changes in the angular position of the rudder. Determine the value of K so that M_m may be 3 dB and determine the phase margin for this value of K.
14. Repeat for the type 2 system shown in Fig. P6.4 (see Chapter 6, Problem 4), when the value of α is fixed at 0.1, but the gain K is adjustable.
15. The transfer function of the forward path of a unity-feedback system representing a chemical process is given by

$$G_p(s) = \frac{Ke^{-0.05s}}{s(s+5)(s+10)}$$

Determine the value of K for which the maximum magnitude of the frequency response of the closed-loop system will be 3 dB. What will be the phase margin?
16. Determine the value of K in the previous problem for which the phase margin will be 45°. What will be the value of M_m for this value of K?
17. The transfer function of the forward path of a unity feedback system is given by

$$G(s) = \frac{10(s+5)}{s^2}$$

Sketch the polar plot of the frequency response and use the Nyquist criterion to determine the stability of the closed-loop system.

References

1. Churchill, R.V., *Introduction to Complex Variables and Applications*, 2nd edition, McGraw-Hill, New York, 1960.
2. James, H.B., N.B. Nichols and R.S. Phillips, *Theory of Servomechanisms*, McGram-Hill, New York, 1947.
3. Kaplan, W., *Operational Methods for Linear Systems*, Addison-Wesley, Reading, Mass., 1962.
4. Saff, E.B., and A.D. Snider, *Fundamentals of Complex Analysis*, Prentice-Hall, Englewood Cliffs, NJ, 1976.

CHAPTER 10

Design and Compensation of Control Systems

10.1 INTRODUCTION

Control systems are always designed for a specific purpose. A good control system should have the following properties:

1. It should operate with as little error as possible.
2. It should exhibit suitable damping, *i.e.*, the controlled output should follow the changes in the reference input without unduly large oscillations or overshoots.
3. Its performance should not be affected by small changes in certain parameters.
4. It should be able to mitigate the effect of undesirable disturbances.

It has already been explained earlier that requirements 1 and 2 are contradictory, since increase in the loop gain reduces the steady-state error but causes a deterioration in the transient performance. For example, it has been shown in Chapters 3 and 4 that by increasing the loop gain, we can increase the speed of response and reduce the steady-state error to a step input, but we also tend to increase the overshoot in the response and reduce the damping ratio of the dominant poles. This was further clarified by the root locus method in Chapter 7, which showed that in general, the damping ratio of the dominant poles is reduced as the gain is increased. Requirement 3 makes it possible to produce the actual system economically.

A control system usually requires some adjustment so that the various conflicting and demanding specifications may be met. This adjustment is called *compensation,* and can be done in several ways. For example, an additional component may be inserted in the forward path, as shown in Fig. 10.1. This is called *series* or *cascade compensation.* The transfer function of the compensator is denoted by $G_c(s)$, whereas that of the original process or plant is denoted as $G_p(s)$. Alternatively, the compensator may be inserted in the feedback path, as shown in Fig. 10.2. This is called *feedback compensation.* A combination of these two schemes is shown in Fig. 10.3.

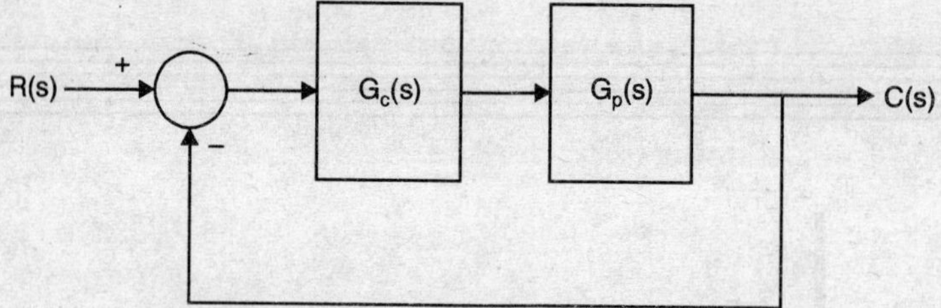

Figure 10.1. *Cascade compensation*

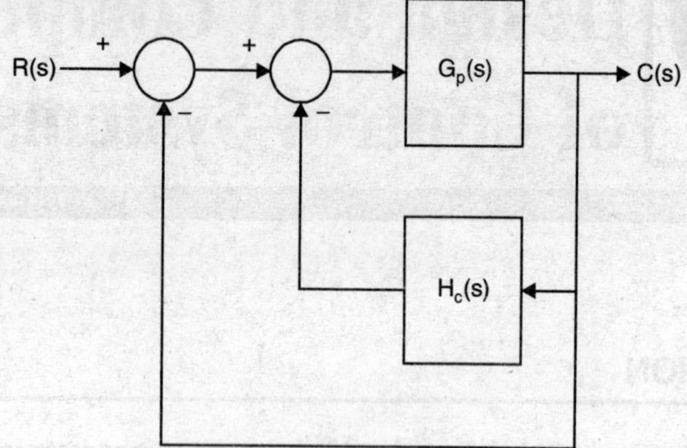

Figure 10.2. *Feedback compensation*

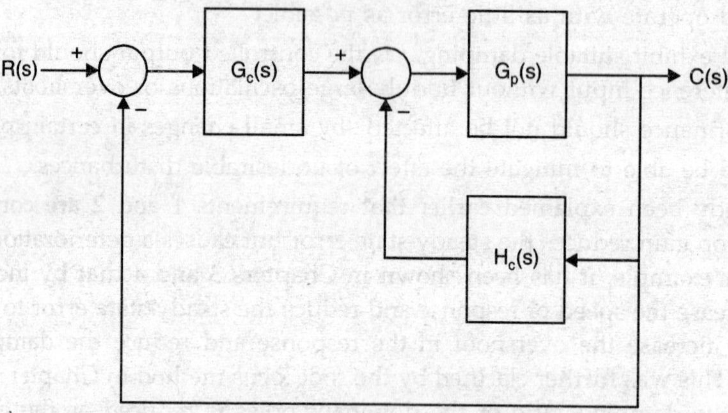

Figure 10.3. *Combined cascade and feedback compensation*

The compensator, like the system components, may be an electrical, mechanical, hydraulic, pneumatic, or other type of device. Electrical networks are often used as compensators in many control systems. The simplest among these are the lead, lag, and lag-lead networks, which will be described in the next section.

10.2 TYPICAL COMPENSATORS

Although many different types of compensators can be used, the most common are the three basic compensators discussed in the section. Each of these can be realized using an *RC* network (with an amplifier in one case). We shall discuss their characteristics and design in detail.

10.2.1 Lead Compensators

One of the simplest compensators has the first-order transfer function

$$G(s) = \frac{1+s/a}{1+s/\alpha a} = \frac{\alpha(s+a)}{s+\alpha a} \qquad ...(10.1)$$

with pole-zero plot shown in Fig. 10.4.

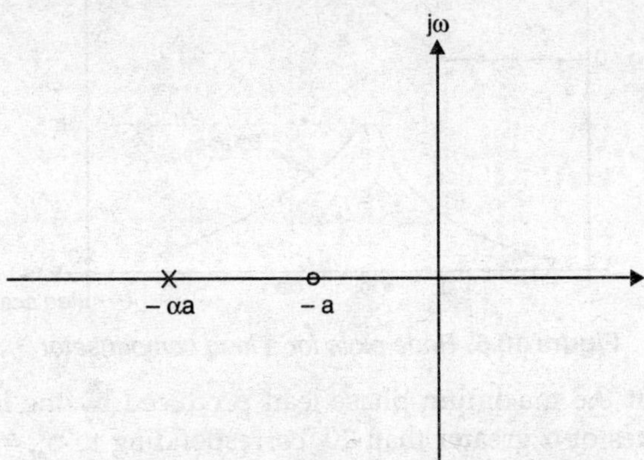

Figure 10.4. *Pole-zero plot of lead compensator*

It is evident that the compensator provides a phase lead between the output and the input, given at any frequency, ω, by

$$\phi(\omega) = \tan^{-1}\frac{\omega}{a} - \tan^{-1}\frac{\omega}{\alpha a} \qquad ...(10.2)$$

Hence, it is called a lead compensator. The Bode plots of the gain and the phase shift of this transfer function are shown in Fig. 10.5. It is seen that at frequencies below a, the gain is nearly zero decibels and at frequencies above αa, the gain is nearly 20 log αa decibels. Similarly, the phase shift approaches zero at low frequencies as well as high frequencies, and has a maximum value of less than 90° at some frequency between a and αa.

Let us now determine the frequency at which the phase lead is maximum. From Equation (10.2), we have

$$\phi = \tan^{-1}\frac{\frac{\omega}{a} - \frac{\omega}{\alpha a}}{1+\frac{\omega}{a}\cdot\frac{\omega}{\alpha a}} = \tan^{-1}\frac{\omega a(\alpha-1)}{\omega^2+\alpha a^2} = \tan^{-1}\frac{\alpha-1}{\frac{\omega}{a}+\frac{\alpha a}{\omega}} \qquad ...(10.3)$$

It is easily shown that ϕ is maximum when

$$\frac{\omega}{a} = \frac{\alpha a}{\omega}, \qquad \text{or } \omega_m = a\sqrt{\alpha} \qquad ...(10.4)$$

Also, the maximum value of ϕ is given by

$$\phi_m = \tan^{-1}\frac{\alpha-1}{2\sqrt{\alpha}} = \sin^{-1}\frac{\alpha-1}{\alpha+1} \qquad ...(10.5)$$

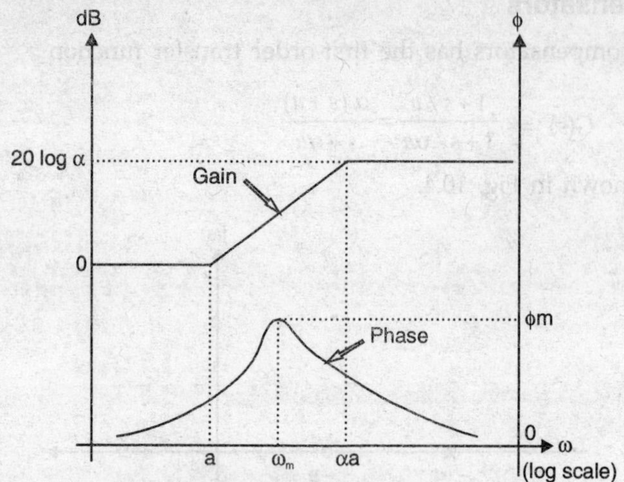

Figure 10.5. *Bode plots for a lead compensator*

It may be noted that the maximum phase lead produced by the lead compensator is less than 90°. In practice, values of α greater than 20 (corresponding to $\phi_m = 64.8°$) are seldom used.

It will now be shown that given the values of a specified phase lead ϕ_c, and gain M_c at a particular frequency ω_c, one may always determine uniquely the transfer function of the corresponding lead compensator if some conditions are satisfied. Let

$$q = \tan \phi_c = \frac{\Omega(\alpha - 1)}{\Omega^2 + \alpha} \qquad \text{...(10.6)}$$

where

$$\Omega = \frac{\omega_c}{a} \qquad \text{...(10.7)}$$

and let

$$c = 10^{M_c/10} = \frac{\alpha^2\left(\omega_c^2 + a^2\right)}{\omega_c^2 + \alpha^2 a^2} = \frac{\alpha^2\left(1 + \Omega^2\right)}{\alpha^2 + \Omega^2} \qquad \text{...(10.8)}$$

Solving Equation (10.8) for Ω, we get

$$\Omega^2 = \frac{\alpha^2 (c - 1)}{\alpha^2 - c} \qquad \text{...(10.9)}$$

From Equation (10.6), we obtain, by squaring both sides and rearranging,

$$q^2 (\Omega^2 + \alpha)^2 = \Omega^2 (\alpha - 1)^2 \qquad \text{...(10.10)}$$

Substituting for Ω^2 from Equation (10.9) into (10.10) and simplifying, we obtain

$$q^2 [\alpha^2 (c - 1) + \alpha (\alpha^2 - c)]^2 = \alpha^2 (c - 1) (\alpha^2 - c) (\alpha - 1)^2 \qquad \text{...(10.11)}$$

It is easily seen that the left-hand side of Equation (10.11) may be written as

$$q^2 [a^2 (c - 1) + \alpha (\alpha^2 - c)]^2 = q^2 [\alpha^2 c - \alpha^2 + \alpha^3 - \alpha c]^2 \qquad \text{...(10.12)}$$

$$= q^2\alpha^2 (\alpha - 1)^2 (\alpha + c)^2$$

so that Equation (10.11) may be simplified as

$$q^2\alpha^2 (\alpha - 1)^2 (\alpha + c)^2 = \alpha^2 (c - 1) (\alpha^2 - c) (\alpha - 1)^2 \qquad \text{...(10.13)}$$

Since $\alpha > 1$, we can cancel the common factors in equation (10.13) to get

$$q^2 (\alpha + c)^2 = (c - 1) (\alpha^2 - c) \quad ...(10.14)$$

Since q and c are known, we can rearrange equation (10.14) to yield a quadratic Equation for the unknown quantity α. This is shown below.

$$(q^2 - c + 1) \alpha^2 + 2q^2 c\alpha + (q^2 c + c - 1) = 0 \quad (10.15)$$

This quadratic will have one positive real solution for α if

$$(i)\ q > 0, \qquad (ii)\ c > 1, \text{ and} \qquad (iii)\ c > q^2 + 1 \quad ...(10.16)$$

The first two conditions are implied for the lead compensator. Consequently, the third condition is necessary and sufficient for the existence of the lead compensator that will satisfy the given requirements.

Finally, after α has been evaluated as the solution of (10.15), we can determine the value of a by using Equation (10.9), so that

$$a = \frac{\omega_c}{\alpha} \sqrt{\frac{\alpha^2 - c}{c - 1}} \quad ...(10.17)$$

EXAMPLE 10.1

Determine the transfer function of a lead compensator that will provide phase of 45° and gain equal to 10 dB at the angular frequency $\omega = 8$ rad/s.

SOLUTION:

Here, $\phi_c = 45°$, $M_c = 10$ and $\omega_c = 8$. Consequently,

$$q = \tan 45° = 1,\ c = 10^{M_c/10} = 10,\ q^2 - c + 1 = -8$$

$$2q^2 c = 20,\ (q^2 c + c - 1)c = 190$$

Thus, we get the quadratic

$$-8\alpha^2 + 20\alpha + 190 = 0$$

which has a positive root $\alpha = 6.2812$.

Hence, we now evaluate a as

$$a = \frac{\omega_c}{\alpha} \sqrt{\frac{\alpha^2 - c}{c - 1}} = 2.304$$

and the transfer function is given by

$$G_c(s) = \frac{\alpha (s + a)}{s + \alpha a} = \frac{6.2812 (s + 2.304)}{s + 14.472}$$

DRILL PROBLEM 10.1

Determine the transfer function of a lead compensator that will provide a phase lead of 50° and gain of 8 dB at $\omega = 5$ rad/s.

Ans. $\dfrac{7.6389 (s + 2.0492)}{s + 15.654}$

DRILL PROBLEM 10.2

While designing a suitable control system for a missile, it was found necessary to introduce a lead of 35° and gain of 6.5 dB at 2.8 rad/s. What will be the transfer function of the lead compensator that will satisfy the above requirements?

Ans. $\dfrac{3.7408\,(s+1.2408)}{s+4.6416}$

An electrical network realization of a lead compensator is shown in Figure 10.6.

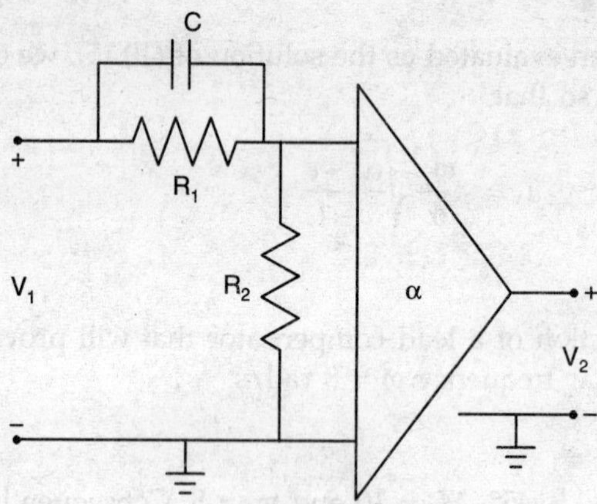

Figure 10.6. *Lead network*

The transfer function is obtained as

$$\frac{V_2(s)}{V_1(s)} = \frac{R_2}{R_2 + \dfrac{R_1 \cdot \dfrac{1}{sC}}{R_1 + \dfrac{1}{sC}}} \cdot \alpha = \alpha \frac{R_2\left(R_1 + \dfrac{1}{sC}\right)}{R_2\left(R_1 + \dfrac{1}{sC}\right) + R_1 \cdot \dfrac{1}{sC}}$$

$$= \alpha \frac{R_2 + R_1 R_2 sC}{R_1 + R_2 + R_1 sC} = \alpha \frac{s + \dfrac{1}{R_1 C}}{s + \dfrac{R_1 + R_2}{C(R_1 R_2)}} = \frac{\alpha(s+a)}{s+\alpha a} \qquad \text{...(10.18)}$$

where $a = \dfrac{1}{R_1 C}$ and $\alpha = \dfrac{R_1 + R_2}{R_2}$.

It should be noted that an amplifier with gain α is required in cascade with an *RC* netwrok to obtain the transfer function in Equation (10.1). This can be realized either by including an operational amplifier as shown, or by increasing the gain of the main amplifier by α.

10.2.2 Lag Compensator

The transfer function of a lag compensator is given by

$$G_c(s) = \frac{1}{\alpha} \cdot \frac{s+\alpha a}{s+a} \qquad \alpha > 1 \qquad \text{...(10.19)}$$

$$= \frac{\beta(s+b)}{s+\beta b} \qquad \beta < 1$$

where $\beta = 1/\alpha$, $b = \alpha a$, and $\beta b = a = b/\alpha$.

The pole-zero plot of this transfer function is shown in Fig. 10.7.

It may be noted that the transfer function in Equation (10.19) is identical to that for the lead compensator, if α is replaced with β and a is replaced with b. From the pole-zero plot, it is evident that there will be a phase lag for all values of ω. The Bode plots will be the mirror image of those for the lead compensator, and are shown in Fig. 10.8.

As will be seen, the gain (in decibels) will also be negative for all ω. The maximum phase lag will occur at

$$\omega_m = b\sqrt{\beta} \qquad \text{...(10.20)}$$

and

$$\phi_m = -\sin^{-1}\frac{\alpha-1}{\alpha+1} = -\sin^{-1}\frac{1-\beta}{1+\beta} \qquad \text{...(10.21)}$$

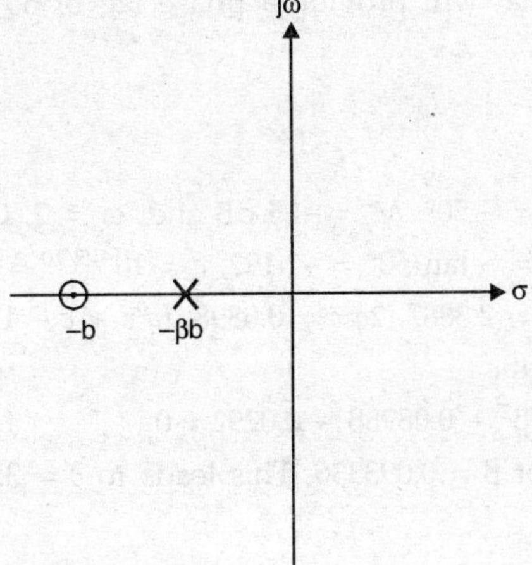

Figure 10.7. *Pole-zero plot for lag compensator*

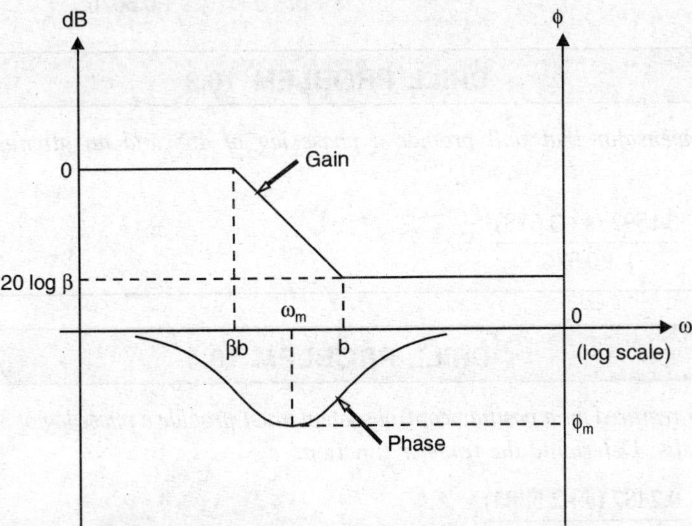

Figure 10.8. *Bode plots for lag compensator*

Therefore, the design of a compensator to provide a given phase lag, ϕ_c, and attenuation, M_c, at a given angular frequency ω_c can be carried out in a similar manner. In fact, we get the same quadratic as in Equation (10.15) but in terms of β, that is,

$$(q^2 - c + 1)\beta^2 + 2q^2c\beta + (q^2c + c - 1)c = 0 \quad ...(10.22)$$

Note that for this case, $0 < c < 1$, and $q < 0$. Hence, the coefficients of β^2 and β will both be positive. Consequently, a positive and real solution for β will exist if and only if

$$q^2c + c - 1 < 0 \quad ...(10.23)$$

Also, the value of b can be obtained from the equation

$$b = \frac{\omega_c}{\beta}\sqrt{\frac{\beta^2 - c}{c-1}}.$$

EXAMPLE 10.2

Design a lag compensator that will provide a phase lag of 50° and attenuation of 15 dB at 2 rad/s.

SOLUTION:

Here, we have

$$\phi_c = -50°,\ M_c = -15 \text{ dB and, } \omega_c = 2. \text{ Consequently,}$$

$$q = -\tan 50° = -1.192,\ c = 10^{-15/10} = 0.0316,$$

$$q^2 - c + 1 = 2.3887,\ 2q^2c = 0.0898,\ (q^2c + c - 1)c = -0.0292$$

Thus, we get the quadratic

$$2.3887\beta^2 + 0.0898\beta - 0.0292 = 0$$

which has a positive real root β = 0.093336. This leads to b = 3.295 and the resulting transfer function is given by

$$G_c(s) = \frac{\beta(s+b)}{s+\beta b} = \frac{0.09336\,(s+3.295)}{s+0.3076}$$

DRILL PROBLEM 10.3

Design a compensator that will provide a phase lag of 45° and an attenuation of 10 dB at 2 rad/s.

Ans. $\dfrac{0.1592\,(s+3.618)}{s+0.576}$

DRILL PROBLEM 10.4

A compensator required for a position control system must provide a phase lag of 35° and attenuation of 8 dB at 1.9 rad/s. Determine the transfer function.

Ans. $\dfrac{0.2487\,(s+2.5883)}{s+0.6438}$

The electrical network shown in Fig. 10.9 can be ultilized to obtain the phase-lag characteristic. It should be noted that we do not need an amplifier in this case, since the dc gain of this network is one, which corresponds to that obtained from the transfer function given in Equation (10.19).

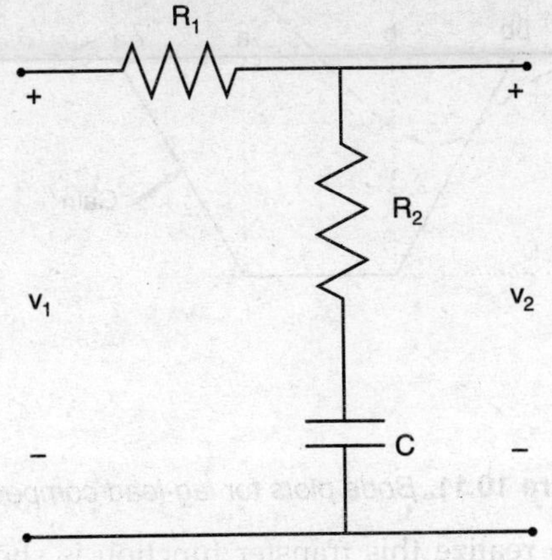

Figure 10.9. *Lag network*

10.2.3 Lag-lead Compensator

The transfer function of the lag-lead compensator is given by

$$G_c(s) = \frac{\alpha(s+a)}{s+\alpha a}\cdot\frac{\beta(s+b)}{s+\beta b}, \qquad \alpha b = 1,\ \alpha > 1,\ a > b$$

$$= \frac{(s+a)(s+b)}{(s+\alpha a)\left(s+\dfrac{b}{\alpha}\right)} \qquad \text{...(10.24)}$$

It can be considered as a combination of lag and lead compensators. The pole-zero plot of the transfer function is shown in Fig. 10.10.

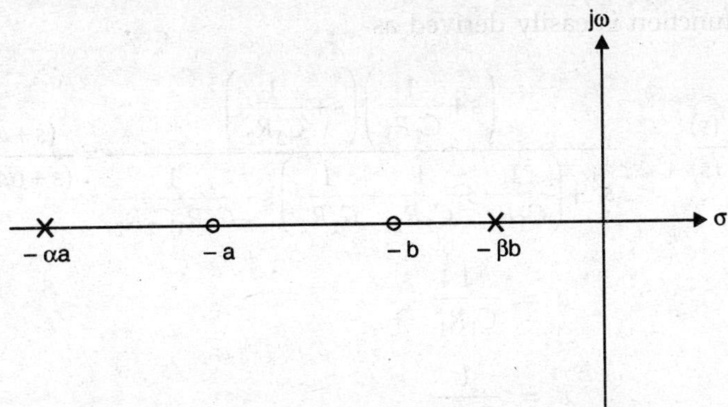

Figure 10.10. *Pole-zero plot for lag-lead compensator*

It is seen to consist of one pole-zero pair for the lag part much closer to the $j\omega$-axis, and another pole-zero pair for the lead part. The Bode plots of the frequency response are shown in Fig. 10.11.

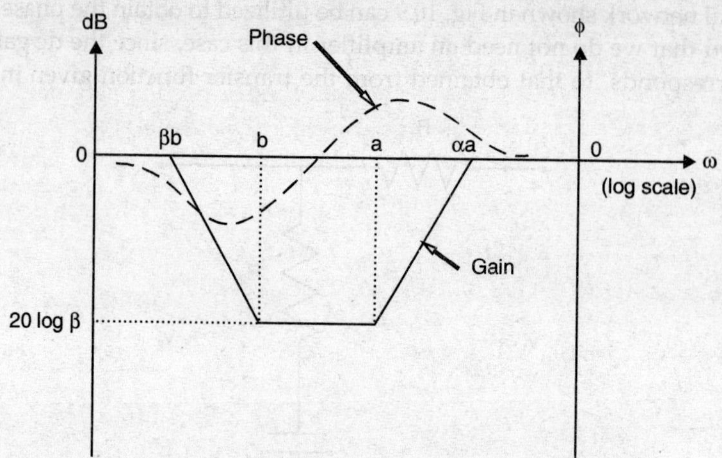

Figure 10.11. *Bode plots for lag-lead compensator*

An electrical network to realize this transfer function is shown in Fig. 10.12.

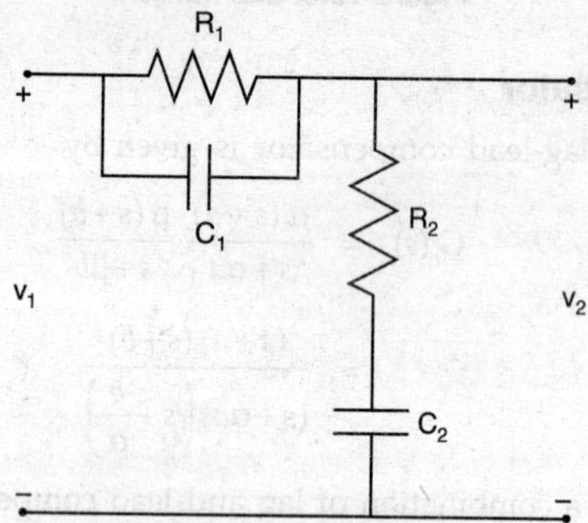

Figure 10.12. *Lag-lead network*

The transfer function is easily derived as

$$\frac{V_2(s)}{V_1(s)} = \frac{\left(s+\frac{1}{C_1R_1}\right)\left(s+\frac{1}{C_2R_2}\right)}{s^2+\left(\frac{1}{C_1R_1}+\frac{1}{C_2R_2}+\frac{1}{C_1R_2}\right)s+\frac{1}{C_1R_1C_2R_2}} = \frac{(s+a)(s+b)}{(s+\alpha a)(s+\beta b)} \quad \text{...(10.25)}$$

where

$$a = \frac{1}{C_1R_1}$$

$$b = \frac{1}{C_2R_2}$$

and

$$\frac{1}{C_2R_1} = (\alpha-1)a-(1-\beta)b$$

We shall now derive the equations for determining a, b and α so that the system may provide a specified lead ϕ_c and gain M_c dB at a given frequency ω_c. Note that M_c must be negative and ω_c must be in the lead portion of the Bode plots. Consequently, ω_c must be much larger than b.

We can simplify the design by considering that the lag portion produces an attenuation of 20 log β decibels at ω_c and a small phase lag, ϕ_2 (say, 1° to 5°). A unique design will be obtained for a given value of ϕ_2. The lead part will then be designed to produce a gain taking these into account, that is,

$$M_c = -20\log\alpha + 20\log\frac{\alpha\sqrt{\omega_c^2+a^2}}{\sqrt{\omega_c^2+\alpha^2 a^2}}, \qquad \text{since } \alpha = \frac{1}{\beta} \qquad \text{...(10.26)}$$

so that

$$\Omega^2 = \frac{\alpha^2 c - 1}{1-c} \qquad \text{...(10.27)}$$

where

$$\Omega = \frac{\omega_c}{a} \quad \text{and} \quad c = 10^{\frac{M_c}{10}} < 1 \qquad \text{...(10.28)}$$

Also,

$$q = \tan(\phi_c + \phi_2) = \frac{\Omega(\alpha-1)}{\alpha+\Omega^2} \qquad \text{...(10.29)}$$

Eliminating Ω from the last two equations, and cancelling out the common factor $(\alpha - 1)^2$, since $\alpha \neq 1$, we get the following quadratic:

$$(q^2c + c - 1)c\alpha^2 + 2q^2c\alpha + (q^2 - c + 1) = 0 \qquad \text{...(10.30)}$$

Since the constant term as well as the coefficient of α in the above equation are positive, a positive root for α will exist if and only if

$$(q^2c + c - 1) < 0, \text{ or } c < \frac{1}{q^2+1} \qquad \text{...(10.31)}$$

After α is known, we obtain a from the following equation:

$$a = \omega_c\sqrt{\frac{1-c}{\alpha^2c-1}} \qquad \text{...(10.32)}$$

The final step is to determine the value of b, that satisfies the stipulation that the phase lag at ω_c is, in fact, equal to ϕ_2. This is done by noting that

$$\tan(-\phi_2) = \frac{\dfrac{\omega_c}{b} - \dfrac{\omega_c}{\beta b}}{1+\dfrac{\omega_c}{b}\cdot\dfrac{\omega_c}{\beta b}} = \frac{\dfrac{\omega_c}{b}(1-\alpha)}{1+\left(\dfrac{\omega_c}{b}\right)^2\alpha} \qquad \text{...(10.33)}$$

This can be simplified to obtain the following quadratic equation:

$$\left(\frac{b}{\omega_c}\right)^2 + \frac{\alpha-1}{\tan(-\phi_2)}\left(\frac{b}{\omega_c}\right) + \alpha = 0 \qquad \text{...(10.34)}$$

This equation has two real and positive roots. We take the smaller of the two to obtain b.

Note that this design is unique for the chosen value of ϕ_2, so that different transfer functions will be obtained if ϕ_2 is changed. Also, there is a minor approximation involved in assuming that the lag portion contributes 20 log β decibles at ω_c. If desired, it is possible to go through another iteration with the value of the actual gain of the lag portion thus obtained. The computer programme included with the book obtains the exact solution in this manner.

EXAMPLE 10.3

Determine the transfer function of a lag-lead compensator that will provide a lead of 50° and an attenuation of 15 dB at $\omega_c = 6$ rad/s.

SOLUTION:

Let the contribution of the lag part at $\omega_c = 6$ be given by $\phi_2 = 2°$. Then

$$q = \tan(50 + 2)° = 1.28$$

$$c = 10^{-1.5} = 0.0316$$

$$(q^2c + c - 1) = -0.02898$$

$$2q^2c = 0.1036$$

$$q^2 - c + 1 = 2.6066$$

Consequently, we get the following quadratic for α.

$$-0.02898\,\alpha^2 + 0.1036\,\alpha + 2.6066 = 0$$

The positive root of the above gives α = 11.4376. We now determine

$$a = \omega_c\sqrt{\frac{1-c}{\alpha^2 c - 1}} = 3.3337$$

and

$$\frac{\alpha - 1}{\tan(-\phi_2)} = -298.893$$

Thus, we obtain the following quadratic:

$$\left(\frac{b}{\omega_c}\right)^2 - 298.893\left(\frac{b}{\omega_c}\right) + 11.4376 = 0$$

Solving and using the smaller root, we get

$$\frac{b}{\omega_c} = 0.0383 \quad \text{and} \quad b = 0.2296$$

Finally, the transfer function of the lag-lead compensator is obtained as

$$G_c(s) = \frac{(s+a)(s+b)}{(s+\alpha a)(s+\beta b)} = \frac{(s+3.3337)(s+0.2296)}{(s+38.1295)(s+0.0201)}.$$

DRILL PROBLEM 10.5

Determine the transfer function of a lag-lead compensator that will provide a phase lead of 55° and an attenuation of 20 dB at 4 rad/s.

Ans. $\dfrac{(s+2.1210)(s+0.1466)}{(s+45.1288)(s+0.00689)}$

10.3 APPROACHES TO COMPENSATION

A number of different approaches for compensating a control system have been proposed in the literature. Some of these will be described here. The earliest approaches were based on modifying the frequency response of the forward path of a unity-feedback system in such a manner that the performance of the closed-loop system is satisfactory. For this method of design, the steady-state accuracy is specified in terms of the error coefficients K_p, K_v, etc., and the transient performance is specified in terms of the gain margin, the phase margin, and the maximum magnitude of the closed-loop system frequency response, M_m. The speed of response is specified either in terms of the crossover frequency ω_c or the frequency ω_m at which the maximum closed-loop frequency response is obtained. It may be noted that although these quantities are related to the transient performance in an indirect manner, they are conveniently determined from the open-loop frequency response.

Another approach to compensation is through the root locus method, where the root locus is reshaped through the use of compensators so that the dominant poles of the closed-loop system are at suitable locations. In this case, we get a more direct correlation between the specifications and the desired transient response.

A more direct approach to the design of a control system is to determine the transfer function of the closed-loop system that will give a satisfactory performance. The compensator is then designed so that the system is forced to have this transfer function. The selection of the desired closed-loop transfer function is the most important step in this procedure. One must also make the design physically realizable and economical. Other desirable properties are insensitivity to small variations in parameters (or *robustness*), as well disturbance rejection.

It should be evident to the reader that in all these methods there is a certain degree of arbitrariness involved in deciding what is the best control system for a particular application. Once 'suitable' specifications have been obtained, there are many possible solutions that will satisfy these and one may use other criteria for selection. A unique solution is obtained when a quantitative performance index, called the 'cost function', is minimized. This has led to the development of *optimal control theory*.

In the following sections, we shall study some of the methods of compensation described earlier. These methods, and typical corresponding performance specifications, are summarized in Table 10.1.

TABLE 10.1: Typical Performance Specifications for Different Methods of Compensation

Method of compensation	*Performance criteria*	*Type of compensation*
Frequency response (Bode plots)	Steady-state error coefficients, phase margin, gain crossover frequency	Cascade lead, lag, or lag-lead compensation
Frequency response (Nichols charts)	Steady-state error coefficients, closed-loop frequency response M_m and frequency ω_m	Cascade lead, lag, or lag-lead compensation
Root locus	Steady-state error coefficients, location of dominant poles of the closed-loop, root sensitivity	Cascade lead or lag compensation, feedback compensation

Pole-placement compensation	Desired closed-loop system transfer function	Two compensators in special configuration
Time-domain methods	Integral performance criteria	Optimal control theory is ultilized to determine the input to the plant.

10.4 COMPENSATION USING BODE PLOTS

Historically, this is the earliest approach to the design of control system. Its main attraction is that one does not have to know precisely the transfer function of the plant to be controlled. The frequency response of the open-loop transfer function is measured experimentally and then reshaped, if necessary, to satisfy the performance specifications. These consist of the steady-state coefficients, the phase margin, and the gain crossover frequency.

The first step in the design is to adjust the loop gain so that steady-state accuracy requirement is satisfied. The frequency response of the uncompensated open-loop transfer function is then calculated. If it does not satisfy the phase margin specification, a lag, lead, or lag-lead compensator must be included in cascade in the forward path. The choice of the type of compensation depends on the gain-phase characteristics at the desired gain crossover frequency. For a lag compensator, the gain (in decibels) should be positive at the desired frequency and the phase lag less than $180°–\phi_m$, where ϕ_m is the specified phase margin. For a lead compensator the gain at the desired crossover frequency must be negative and the phase lag must be more than $180° – \phi_m$. For each case, the magnitude of the phase shift to be provided by the compensator must be less than 90° and, preferably be less than 70°. For a lag-lead compensator, gain at the desired crossover frequency should be positive and the phase lag should be more than $180° – \phi_m$. This is illustrated in the polar plot shown in Fig. 10.13.

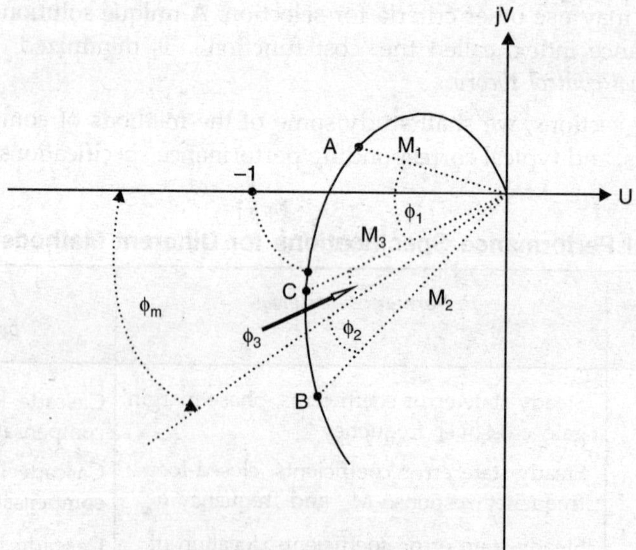

Figure 10.13. *Polar plot of a typical transfer function requiring compensation to satisfy the phase margin specifications*

If we want to use a lead compensator to satisfy the design specifications, we should select ω_1, corresponding to the frequency at the point A, as the gain crossover frequency; so that the compennsator will be designed to provide a lead equal to ϕ_1 and a gain equal to $1/M_1$ at the frequency ω_1 (note that $M_1 < 1$). Similarly, for a lag compensator to satisfy the specifications, we should select ω_2, corresponding to the point B as the gain crossover frequency. In this case, the compensator should provide a phase lag ϕ_2 and gain $1/M_2$ (note that $M_2 > 1$) at the frequency ω_2. Finally, for a lag-lead compensator, the crossover frequency may be selected as ω_3, corresponding to the point C. In this case, the compensator should provide a phase lead of ϕ_3 and gain $1/M_3$ (note that $M_3 > 1$) at the frequency ω_3.

It will be seen from the above that with a lead compensator, we get a higher gain crossover frequency and thus a larger bandwidth than possible with a lag compensator. The larger bandwidth results in a faster response, but it also makes the system susceptible to more noise. The bandwidth of the lag-lead compensator is intermediate between these two. The actual choice will depend upon what is more important for a particular application.

The procedure can be easily understood through the following examples:

EXAMPLE 10.4

The transfer function of a plant is given by

$$G_p(s) = \frac{K}{s(s+5)(s+10)} \qquad ...(10.35)$$

Design a suitable compensator to meet the following specifications:

(*a*) $K_v = 10$

(*b*) Phase margin = 45°.

Solution: For $K_v = 10$, we must have $K = 500$. The frequency response of

$$G_p(s) = \frac{500}{s(s+5)(s+10)} \qquad ...(10.36)$$

is shown in Table 10.2. It will be seen that the phase margin specification is not met.

TABLE 10.2: Frequency Response of the Transfer Function given by Equation (10.36)

ω	ϕ	*M (dB)*	ω	ϕ	*M (dB)*
1	– 107.02°	19.79	4	– 150.46°	5.166
1.5	– 115.21°	16.01	6	– 171.16°	– 0.772
2	– 123.11°	13.164	8	– 186.65°	– 5.725
2.5	– 130.60°	10.81	9	– 192.93°	– 7.935
3	– 136.66°	8.75	10	– 198.43°	– 10.00

If we want to design a lead compensator, the gain crossover frequency (that is the frequency where the open-loop transfer function crosses the zero-decibel line) should be more than 6 and less than 10. These limits are set by the fact that we must have a negative gain at that frequency so that when the positive gain of the lead compensator is added to it, the total gain becomes equal to zero decibels. It is also desirable that the phase lead required of the compensator does not exceed 65°, otherwise the required value of α, the gain of the compensator at high frequencies, would be very large. If we select ω_c = 8, the compensator must provide a phase lead of 186.65° – 135° = 51.65° and gain equal to 5.725 dB at this frequency. Following the procedure developed in Section 10.21, the transfer function of the compensator is obtained as

$$G_c(s) = \frac{12.725\,(s+4.78)}{s+60.82} \quad ...(10.37)$$

The frequency response of the open-loop transfer function of the compensated system is shown in Fig. 10.14, along with that of the uncompensated system. It will be seen that in addition to satisfying the phase margin requirement, the compensated system has also a larger bandwidth.

It may be added that one could have been chosen ω_c = 7, or ω_c = 9 to design the lead compensator. These give different values of α, the additional gain required for the compensator. One may choose the value of the crossover frequency that minimizes α. Another practical criterion is the bandwidth of the system, which is related to the gain crossover frequency.

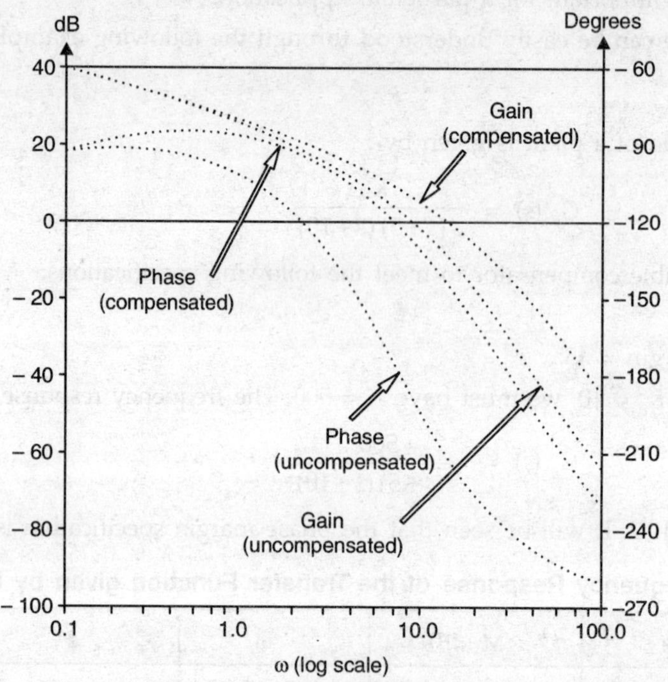

Figure 10.14. *Frequency response of open-loop system with and without lead compensator*

We may also satisfy the design requirements using a lag compensator, but we must choose ω_c equal to or less than 2.5. For ω_c = 2, the lag compensator must produce a phase lag equal to 11.89° and gain equal to –13.164 dB at that frequency. Following the procedure that was described in Section 10.2.2, we obtain

$$G_c(s) = \frac{0.2124\,(s+0.543)}{s+0.1153} \quad ...(10.38)$$

The frequency response of the open-loop transfer function of the system is shown in Fig. 10.15, with and without the lag compensator. It will be seen that the bandwidth of the compensated system is much smaller.

Although both the lead and the lag compensator satisfy the specifications in this example, it is seen that with lead compensation we get a larger crossover frequency. This implies a lager bandwidth, which results in a faster response, as expected.

We can also obtain a lag-lead compensator by selecting the crossover frequency equal to 4. In this case, the compensator must provide a phase lead of 15.46° and gain equal to – 5.166 dB at $\omega_c = 4$. The transfer function of the compensator is calculated using the procedure described in Section 10.2.3, with $\phi_2 = 2°$, and is given by

$$G_c(s) = \frac{(s+5.36236)(s+0.26331)}{(s+11.44735)(s+0.1233)} \quad ...(10.39)$$

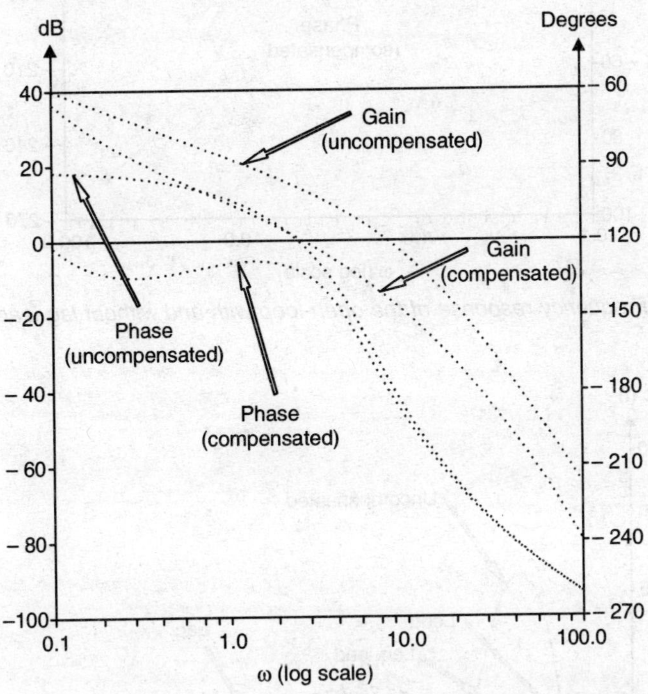

Figure 10.15. *Open-loop frequency response of system with and without lag compensation*

The frequency response of the open-loop transfer function of the system is shown in Fig. 10.16, with and without the lag-lead compensator. This time the bandwidth of the compensated system is slightly smaller than that of the system before compensation.

Since the design of the compensators using this approach is based on the idea that a good phase margin will lead to a good transient response, it would be instructive to compare the step responses of the uncompensated closed-loop system and the various compensated closed-loop systems. These are shown Fig. 10.17.

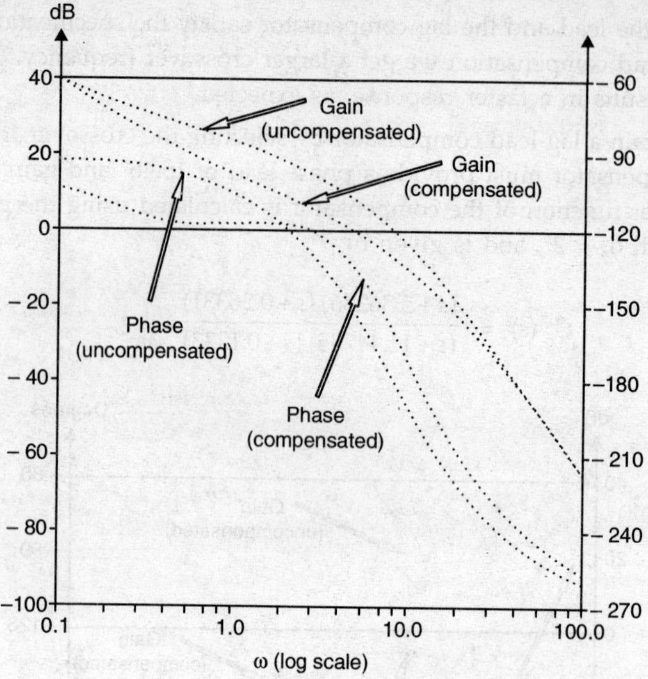

Figure 10.16. *Frequency response of the open-loop with and without lag-lead compensator*

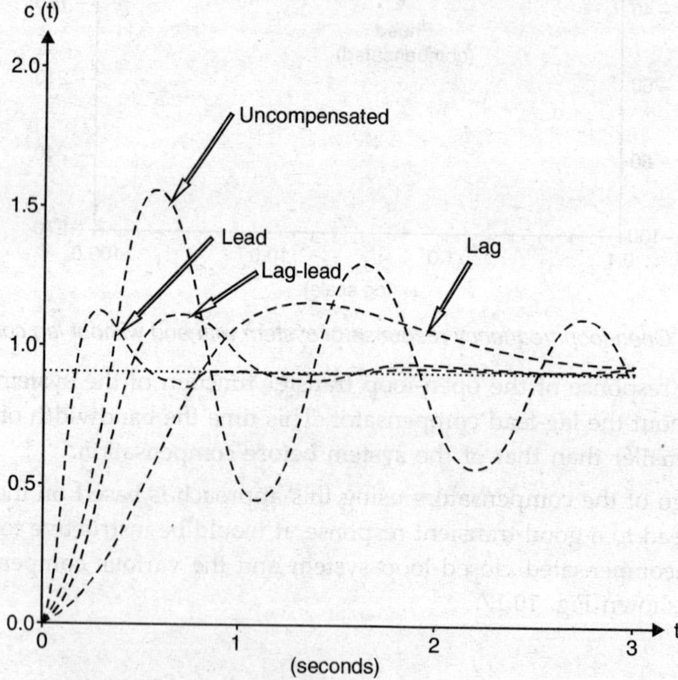

Figure 10.17. *Comparison of responses to a unit step*

It will be seen that the maximum overshoot is reduced from over 70% in the uncompensated to about 30% for each of the compensated system. As expected, the system with the lead compensator has the fastest response, whereas that with the lag compensator has the slowest response.

In the example considered here, we have been able to find compensators of all three types. This may not always be possible, as will be seen from the following example:

EXAMPLE 10.5

Design a suitable compensator for the plant in Example 10.4, but with the following specifications:

(*a*) $K_v = 100$

(*b*) Phase margin = 45°.

SOLUTION:

To meet the K_v requirement, we must make $K = 5000$, and the corresponding frequency response of the open-loop transfer function is given in Table 10.3.

TABLE 10.3: Frequency Response of the Transfer Function Given by Equation (10.36)

ω	ϕ	M (dB)	ω	ϕ	M (dB)
2	– 123.11°	33.16	10	– 198.43°	10.00
2.5	– 130.66°	30.81	12	– 207.57°	6.242
4	– 150.46°	25.166	15	– 217.87°	1.359
6	– 171.16°	19.228	16	– 220.64°	– 0.105
8	– 186.65°	14.275	20	– 229.40°	– 5.315

Clearly, the phase margin specification is not satisfied and a compensator is required. However, a lead compensator cannot be used, since the phase lag is very large for negative gain. A lag compensator is possible, but it requires a pole very near the origin, with a pole-zero ratio of more than 30. In practice, a pole very close to the origin is not desirable, since the corresponding compensator would require *RC* network with a large time constant. Furthermore, as would follow from the root locus plot, since we also have a pole at the origin, this would cause a pole of the closed-loop system to be very close to the origin as well. Although this would not affect the transient response very much (due to the presence of the zero of the compensator very near this pole), the performance can be degraded considerably, due to a small variation in the parameters.

A lag-lead compensator can, however, be obtained if we select the crossover frequency between 4 and 10. For $\omega_c = 8$, the following compensator is obtained, with $\alpha = 13.09$:

$$G_c(s) = \frac{(s+3.422)(s+0.7587)}{(s+44.499)(s+0.05835)}.$$

10.5 COMPENSATION WITH NICHOLS CHARTS

The procedure discussed in the previous section is based on the assumption that a good phase margin will produce a desirable transient response for the closed-loop system. As would be evident from Example 10.4, although the lead, lag, and lag-lead compensators were designed to give the same phase margin in each case, the step responses of the resulting closed-loop systems are quite different, as shown in Fig. 10.17. Another approach is based on the fact that if the closed-loop system transfer function has poles with a small damping ratio, the resulting frequency response curves will have a large peak in the magnitude. Therefore, it appears logical that in order to obtain a desirable transient response, one should require that M_m, the maximum magnitude ratio of the closed-loop frequency response, does not exceed a specified value, say 1.5 to 4 dB. The specifications for a compensator for this purpose can be obtained using the Nichols chart. The procedure will be evident from the following example:

EXAMPLE 10.6

Design a suitable compensator for the same plant, as in Example 10.4, but with the following specifications:

(*a*) $K_v = 10$, (*b*) $M_m = 3$ dB, and (*c*) $\omega_m = 4$ rad/s.

SOLUTION:

For K_v to be 10, we must make $K = 500$. The frequency response of the open-loop transfer function was calculated in Problem 10.4, and is given in Table 10.2. A plot of this response on the Nichols chart is shown in Fig. 10.18. It will be seen that in order that this plot be tangential to the 3 dB *M*-curve at $\omega = 4$, it must be shifted down by 5 dB and to the right by 10°. Thus, we need a compensator that provides a gain of – 5 dB and a phase lead of 10° at this frequency. The transfer function of this lag-lead compensator is readily calculated using the procedure described in Section 10.2.3, and is given by

$$G_c(s) = \frac{(s+7.99967)(s+0.29161)}{(s+15.39369)(s+0.1515)} \quad ...(10.40)$$

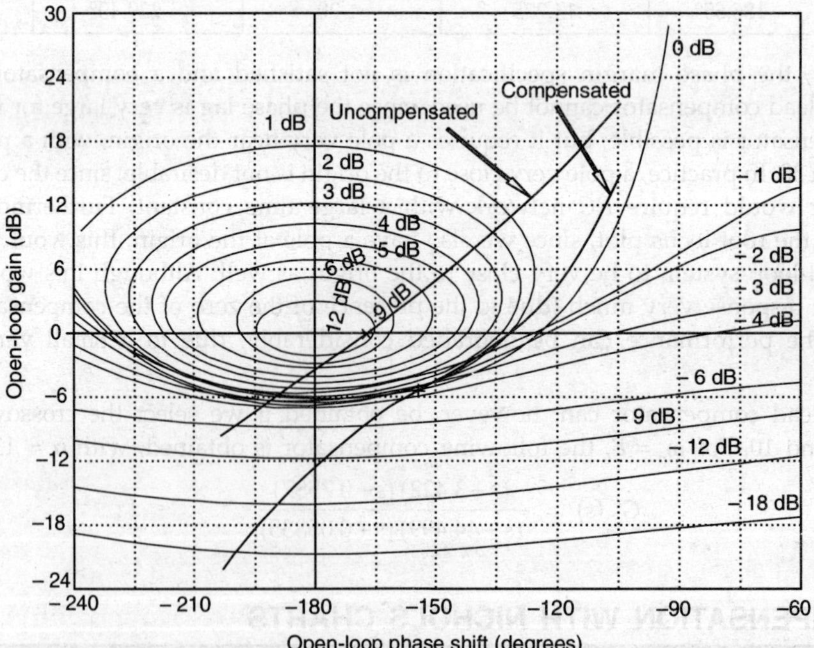

Figure 10.18. *Nichols chart design*

The frequency response of the forward path of the compensated system is also shown in Fig. 10.18. It will be seen that it touches the 3 dB *M*-curve at ω = 4 (marked by a × on the plot) and thus satisfies the M_m specification.

The responses of the uncompensated as well as the compensated system to a unit step input are shown in Fig. 10.19. It will be seen that the maximum overshoot for the compensated system is almost the same as was obtained in Fig. 10.17, for the case when the phase margin of 45° was used for designing the compensator.

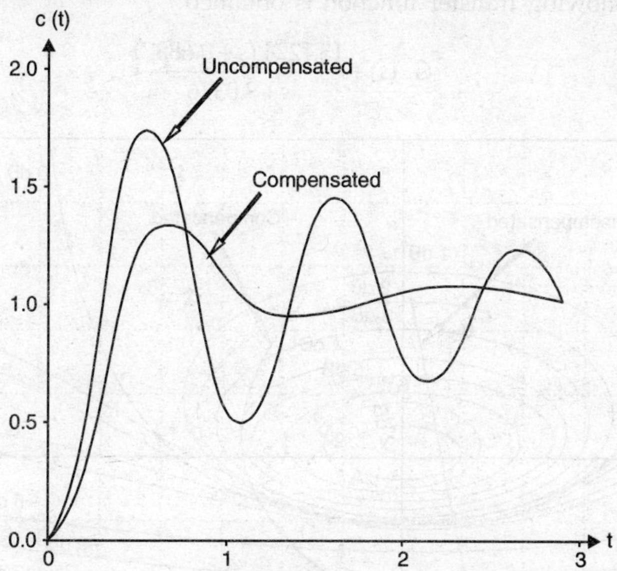

Figure 10.19. *Step responses of the compensated and uncompensated systems*

EXAMPLE 10.7

The forward transfer function of a unity-feedback system is given by

$$G_p\ (s) \ = \frac{10\ e^{-0.05\ s}}{s^2\ (s+10)} \qquad \text{...(10.41)}$$

Design a suitable compensator so that M_m does not exceed 3.5 dB and ω_m is nearly 1.4 rad/s.

SOLUTION:

The open-loop frequency response of the forward path is shown in Table 10.4.

TABLE 10.4: Frequency Response of the Transfer Function Given by Equation (10.41)

ω	ϕ	*M (dB)*	ω	ϕ	*M (dB)*
0.5	– 186.87°	3.849	1.8	– 195.36°	– 10.35
1.0	– 188.57°	– 0.043	2.0	– 197.04°	– 12.21
1.2	– 190.28°	– 3.229	2.5	– 201.20°	– 16.18
1.4	– 191.98°	– 5.929	3.0	– 205.29°	– 19.46
1.6	– 193.67°	– 8.275			

These values are plotted in the Nichols chart in Fig. 10.20. For M_m to be 3.5 dB at ω_m = 1.4, the plot should be shifted to the right by 55° and raised by 7 dB. This was determined by first locating the point corresponding to ω = 1.4 in the plot of the frequency response of the uncompensated system, and then deciding on the amount of lead and gain required to make this point tangential to the 3.5–dB closed-loop gain curve. (Since the Nichols chart has only 3-dB and 4-dB closed-loop gain curves, one must do a little interpolation.) Hence, the lead compensator that will provide a lead of 55° and gain of 7 dB at ω = 1.4 is calculated using the procedure in Section 10.2.1. The following transfer function is obtained:

$$G_c(s) + \frac{13.1224\,(s+0.6887)}{s+9.0376} \qquad ...(10.42)$$

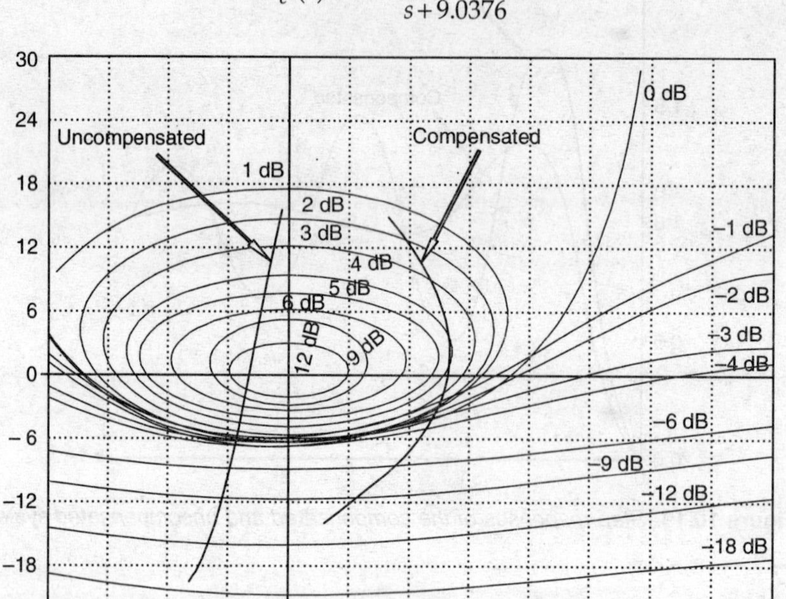

Figure 10.20. *Nichols chart design*

The frequency response of the forward path of the compensated system is shown in Table 10.5, and has been plotted along with that of the uncompensated system on the Nichols chart in Fig. 10.20. It is seen that the M_m specification is satisfied, with ω_m nearly equal to 1.4.

TABLE 10.5: Frequency Response of Forward Path of the Compensated System

ω	φ	*M (dB)*	ω	φ	*M (dB)*
0.5	– 151.5°	13.86	1.8	– 137.6°	– 1.58
0.7	– 145°	9.2	2.0	– 138.5°	– 2.67
1.0	– 139.4°	4.83	2.5	– 142.06°	– 5.0
1.2	– 137.7°	2.75	3.0	– 146.6°	– 6.9
1.4	– 137°	1.07	5.0	– 167.7°	– 12.8

10.6 COMPENSATION USING ROOT LOCUS

The compensation methods discussed in the two preceding sections specify transient performance in an indirect manner. A more direct approach is to specify the location of the dominant poles of the transfer function of the closed-loop system. This is usually done in terms of the damping ratio of these poles requiring that they be to the left of a certain vertical line in the s-plane in order that the settling time will be less than a specified value. Other poles of the transfer function do not have much effect on the transient response if they are much farther to the left. We shall now discuss the use of the root locus method for selecting a suitable compensator.

10.6.1 Cascade Compensation Using Root Locus

We start by plotting the root locus for the uncompensated system and determine the value of the loop gain K to obtain the dominant poles with the desired damping ratio. If these poles are suitably located, but the steady-state specification is not met, one may add a lag compensator with pole and zero sufficiently close to the origin of the s-plane so that the root locus is not altered appreciably, but the low-frequency gain is suitably increased. On the other hand, if the dominant poles must be moved to the left, a lead compensator must be designed. The procedure for the two cases will be evident from the following examples:

EXAMPLE 10.8

The forward transfer function of a unity-feedback system is given by

$$G_p(s) = \frac{K}{s(s+3)(s+6)} \qquad ...(10.43)$$

It is desired that the dominant poles of the closed-loop system transfer function have a damping ratio of 0.5 and the magnitude of the real part of the pole be not less than 1. Also, K_v must be at least 10. Design a suitable compensator.

SOLUTION:

The root locus plot for the open-loop transfer function is shown in Figure 10.21. From this plot, it is determined that for $K = 28$, the characteristic polynomial of the closed-loop system has roots with damping ratio 0.5, and is given by $(s + 7)(s^2 + 2s + 4)$. The real part of the complex poles is -1, which just meets the specification. However, for $K = 28$, we get

$$K_v = \frac{28}{18} = \frac{14}{9}$$

which is less than the desired value of 10.

To raise the value of K_v to 10, we must include a lag compensator, which does not appreciably alter the root locus near the dominant poles, and has a dc gain equal to 45/7 in order to provide the desired K_v. Selecting the pole of the compensator at $s = 0.01$ leads to

It may be noted that the distances of the compensator pole and zero from the origin of the s-plane are chosen to be small compared with the distance of the dominant poles from the origin so that the compensator will not affect the locus near the dominant poles. A plot of the root locus for the system with the lag compensator is shown in Fig. 10.22.

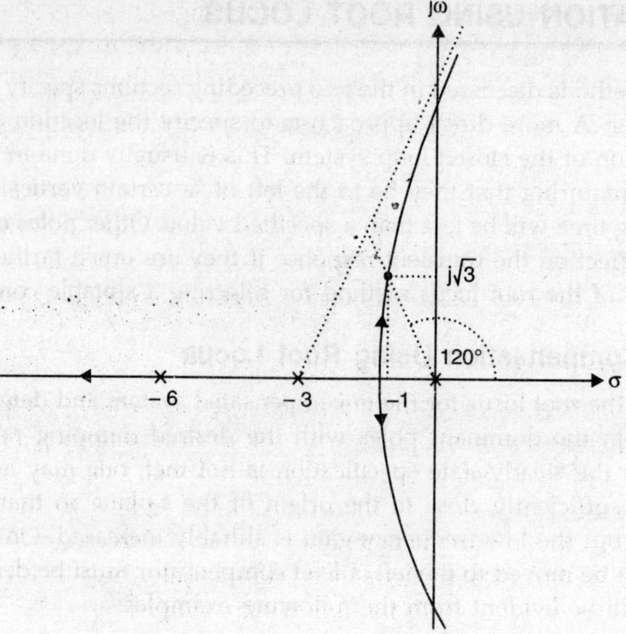

Figure 10.21. *Roots locus plot for G_p (s) in Equation (10.43)*

$$G_c(s) = \frac{s + (0.01) \times \frac{45}{7}}{s + 0.01} = \frac{s + 0.0643}{s + 0.01} \qquad \text{...(10.44)}$$

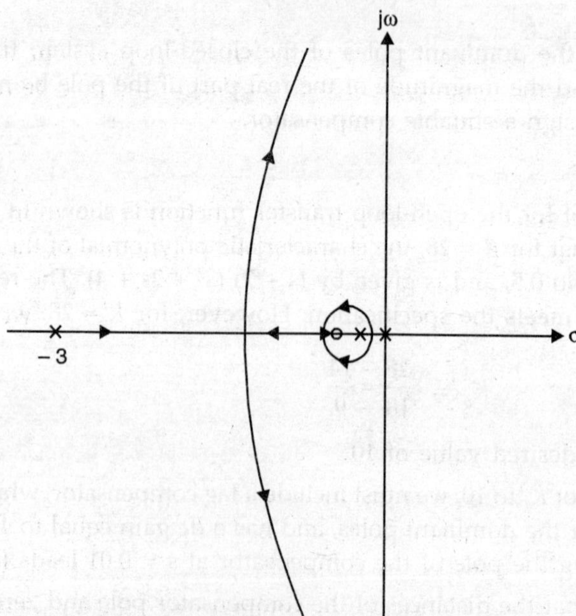

Figure 10.22. *Root locus plot for G_p (s) G_c (s)*

It will be seen that the root locus plot is almost unchanged near the dominant poles. Although we are adding a pole and a zero to the transfer function of the forward path, their contribution to the argument near the dominant pole is negligible, so that the angle condition is still satisfied at that point.

EXAMPLE 10.9

Consider the same system, as in Example 10.8, but with the specification that the magnitude of the real part of the dominant poles must not be less than 4.

SOLUTION:

In this case, we must use a lead compensator to shift the root locus to the left. The first step in this procedure is to determine the total angle subtended by the poles and zeros of $G_p(s)$ at the desired location, which, in this case, will be taken as $-4 + j4\sqrt{3}$, in order to satisfy the specifications for the dominant poles. This is shown in Fig. 10.23.

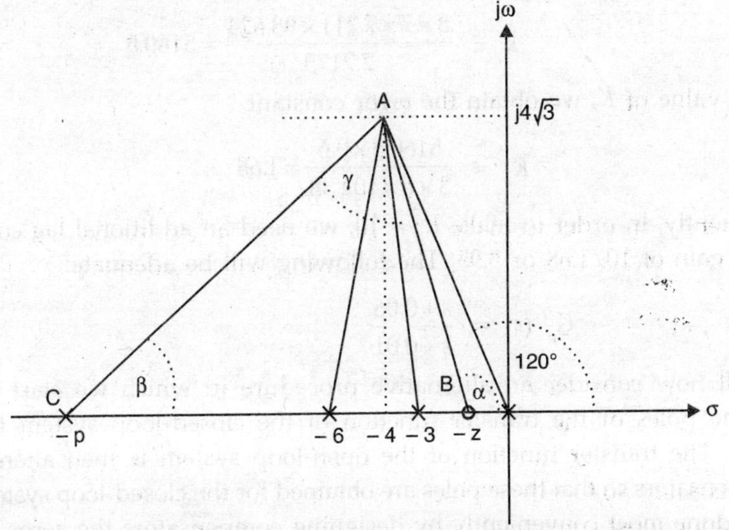

Figure 10.23. *Determination of the pole and zero of the compensator*

This angle is given by

$$\phi = 120°+\tan^{-1}\frac{4\sqrt{3}}{2}+\left(180°-\tan^{-1}\frac{4\sqrt{3}}{1}\right)=292.11°$$

Now, if the zero of the compensator is located at $-z$ and the pole at $-p$, these must be selected in such a manner that the angle condition is satisfied at the point A, that is,

$$\phi + \beta - \alpha = 180°$$

Hence,

$$\gamma = \alpha - \beta = \phi - 180° = 112.111°$$

This suggests a simple construction for obtaining p for a given choice of z. For example, we can arbitrarily select the location of the zero at the point B, and then determine the location of the pole by drawing the line AC making an angle $\gamma = 112.111°$, as shown in Fig. 10.23. A solution is possible as long as the point C lies on the negative real axis, the limiting case will occur when AC is parallel to the real axis. Consequently, a solution will not exist if z is greater than $4 - 4\sqrt{3}$ tan 22.111°, that is, 1.18. Some possible solutions are given under.

z	p	p/z	z	p	p/z
0.2	63.579	317.9	0.8	152.0	190.09
0.4	78.04	195.1	0.9	202.9	225.44
0.6	102.38	170.6	1.0	308.78	308.78
0.7	122.08	174.4	1.15	1195.6	1387.5

If we select $z = 0.6$, which provides the smallest p/z ratio in this table, we get the following compensated open-loop transfer function:

$$G_p(s)\, G_c(s) = \frac{K(s+0.6)}{s(s+3)(s+6)(s+102.38)} \qquad \text{...(10.45)}$$

and the value of K to obtain the complex poles at $s = -4 + j4\sqrt{3}$ is determined from Fig. 10.23 as

$$K = \frac{8 \times 7 \times 7.211 \times 98.624}{7.7175} = 5160.5$$

For this value of K, we obtain the error constant

$$K_v = \frac{5160.5 \times 0.6}{3 \times 6 \times 102.38} = 1.68$$

Consequently, in order to make $K_v = 10$, we need an additional lag compensator that will provide a dc gain of 10/1.68 or 5.95. The following will be adequate:

$$G_c'(s) = \frac{s+0.06}{s+0.01} \qquad \text{...(10.46)}$$

We shall now consider an alternative procedure in which we start by determining the location of the poles of the transfer function of the closed-loop system that will satisfy the specifications. The transfer function of the open-loop system is then altered by the choice of suitable compensators so that these poles are obtained for the closed-loop system transfer function. This may be done most conveniently by designing compensators the zeros of which cancel the poles of open-loop transfer function. Thus, in effect, the poles of the open-loop transfer function are moved so that the transfer function of the closed-loop system is forced to have the specified poles. For example, the specification for this problem can be met if the dominant poles of the closed-loop system transfer function are located at $s = -8 \pm j8\sqrt{3}$.

This can be done by using a compensator of the from

$$G_c(s) = \frac{s+3}{s+p_1} \cdot \frac{s+6}{s+p_2} \qquad \text{...(10.47)}$$

where the values of p_1 and p_2 can be selected by using the geometric construction discussed earlier. As will be seen from Fig. 10.24, we must select the locations of these poles in such a way that the angles α and β add up to 60° in order that the angle condition is satisfied at the point A. Many solutions are possible. Perhaps, the most convenient solution is to make p_1 and p_2 equal so that $\alpha = \beta = 30°$ and both the poles of the compensator are located at $s = -32$. From Fig. 10.24, by using the rules of the root locus, it is easily that for this case $K = 12{,}288$, and the resulting $K_v = 12$.

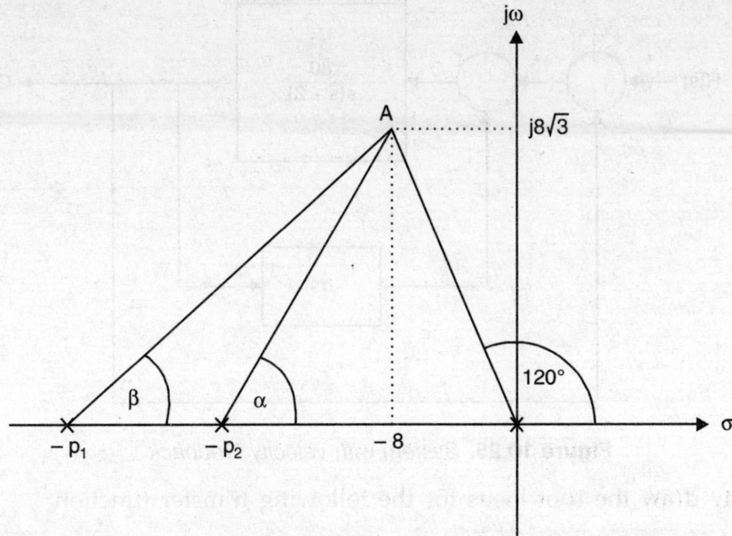

Figure 10.24. *Construction for locating the poles of the compensator*

Although the pole-zero cancellation approach discussed above appears quite simple as well as elegant, a word of caution is necessary. In most practical situations, we must allow some tolerance in the values of components. As a result, exact pole-zero cancellation may not always be obtained. Hence, to be realistic, one must take this fact into consideration and draw the root locus for the case when the poles of the transfer function of the plant are not exactly cancelled by the zeroes of the compensator. In such cases, we may get some poles of the transfer function of the closed-loop system that are much closer to the $j\omega$-axis and will lead to larger time constant. However, since these poles are very close to the zeros of the compensator, their contribution to the transient response may still be quite small. This would require careful investigation for a practical design.

10.6.2 Feedback Compensation Using Root Locus

The root locus method can also be used for the design of feedback compensation. The idea, again, is to shape the locus so that the dominant poles have desired locations. It will be illustrated through the following example:

EXAMPLE 10.10

The forward transfer function of unity-feedback position-control system is

$$G_p(s) = \frac{60}{s(s+2)} \qquad \text{...(10.48)}$$

It is desired to add velocity feedback as shown in Fig. 10.25, so that the damping ratio of the poles of the closed-loop system is 0.5. Determine the value of α for this case.

SOLUTION:

The characteristic polynomial of the system with velocity feedback is obtained as

$$P(s) = s^2 + (60\alpha + 2)s + 60 = (s^2 + 2s + 60) + 60\alpha s \qquad \text{...(10.49)}$$

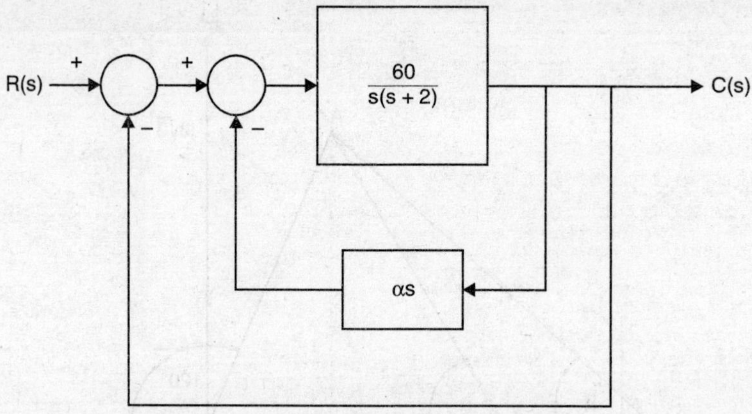

Figure 10.25. *System with velocity feedback*

Thus, we may draw the root locus for the following transfer function:

$$\frac{60\,\alpha s}{s^2 + 2s + 60}$$

and determine the value of α for which the damping ratio will be 0.5. The root locus plot is shown in Fig. 10.26. The value of α for the desired damping ratio is found to be 0.0958.

In practice, there may be some difficulty with velocity feedback if the system is subject to noise. Since velocity feedback requires differentiation of the output, the result will be to accentuate the noise that may be present in the output. Furthermore, although velocity feedback improves the damping in the system, it does not increase the steady-state accuracy. Nevertheless, velocity feedback is often used for improving the transient response of systems.

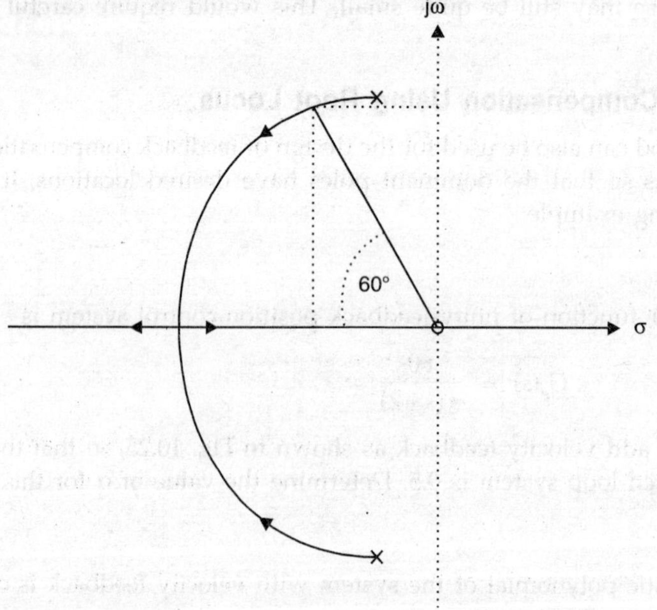

Figure 10.26. *Root locus plot for system with velocity feedback*

10.7 POLE-PLACEMENT COMPENSATION

The philosophy of design in this approach is to select all the poles of the transfer function of the closed-loop system in such a manner that the specifications for steady-state accuracy as well as good transient response are satisfied. Suitable compensation is then designed to force the closed-loop system to have this transfer function. The general scheme for compensation is shown in Fig. 10.27. As will be seen, two compensators are required.

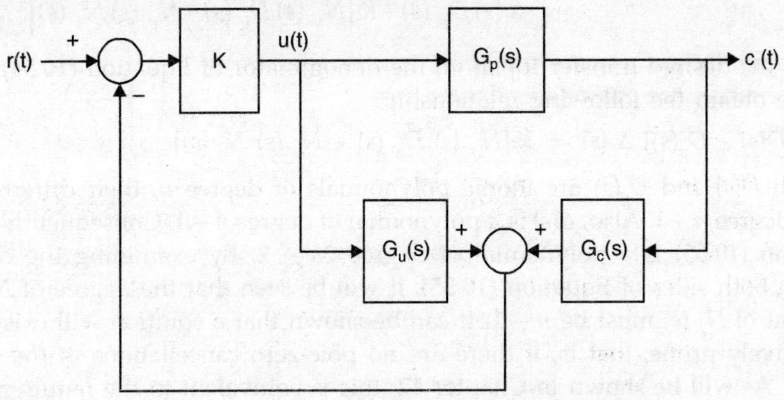

Figure 10.27. *Compensators for pole-placement*

The transfer functions of the two compensators have the same denominator. Let

$$G_p\ (s) = \frac{N_p\ (s)}{D_p\ (s)} \qquad ...(10.50)$$

$$G_u\ (s) = \frac{N_u\ (s)}{\Delta\ (s)} \qquad ...(10.51)$$

and

$$G_c\ (s) = \frac{N_c\ (s)}{\Delta\ (s)} \qquad ...(10.52)$$

where $D_p\ (s)$ is a monic polynomial of degree n, and $\Delta\ (s)$ is an arbitrarily selected monic polynomial of degree $n - 1$. Usually, $\Delta\ (s)$ is selected so that all its roots are negative real numbers. This ensures stable compensators, which may be realized using resistors and capacitors.

Applying Mason's rule, the overall transfer function is obtained as

$$T\ (s) = \frac{C\ (s)}{R\ (s)} = \frac{KG_p\ (s)}{1+KG_u\ (s)+KG_p\ (s)\ G_c\ (s)} \qquad ...(10.53)$$

Let the desired closed-loop system transfer function have the same numerator as in Equation (10.53), but its denominator is an arbitrarily specified monic polynomial $D\ (s)$, of degree n. Then, from Equation (10.53), we have

$$T\ (s) = \frac{N_p\ (s)}{D\ (s)} = \frac{KG_p\ (s)}{1+KG_u\ (s)+KG_p\ (s)\ G_c\ (s)}$$

$$= \frac{\dfrac{KN_p(s)}{D_p(s)}}{1+\dfrac{KN_u(s)}{\Delta(s)}+\dfrac{KN_p(s)}{D_p(s)}\cdot\dfrac{N_c(s)}{D(s)}}$$

$$= \frac{KN_p(s)\,\Delta(s)}{\Delta(s)\,D_p(s)+K\left[N_u(s)\,D_p(s)+N_p(s)\,N_c(s)\right]} \qquad ...(10.54)$$

To obtain the desired transfer function, the denominator of Equation (10.54) must be $D(s)\,\Delta(s)$. Hence, we obtain the following relationship:

$$[D(s) - D_p(s)]\,\Delta(s) = K[N_u(s)\,D_p(s) + N_p(s)\,N_c(s)] \qquad ...(10.55)$$

Since both $D(s)$ and $D_p(s)$ are monic polynomials of degree n, their difference will be a polynomial of degree $n - 1$. Also, $\Delta(s)$ is a polynomial of degree $n - 1$. Consequently, the left-hand side of Equation (10.55) is a polynomial of degree $2n - 2$. By examining the degrees of the polynomials on both sides of Equation (10.55), it will be seen that the degree of $N_u(s)$ must be $n - 2$, while that of $N_c(s)$ must be $n - 1$. It can be shown that a solution will exist if $N_p(s)$ and $D_p(s)$ are relatively prime, that is, if there are no pole-zero cancellations in the plant transfer function $G_p(s)$. As will be shown in Chapter 12, this is equivalent to the requirements that the system be both controllable and observable. A formal proof of the result is given in Kailath [1]. We shall illustrate the procedure through two simple examples.

EXAMPLE 10.11

The transfer function of a system is given by

$$G_p(s) = \frac{s+2}{s^2\,(s+10)} \qquad ...(10.56)$$

It is desired to have the following transfer function for the closed-loop system:

$$T(s) = \frac{1000\,(s+2)}{\left(s^2+10s+50\right)(s+40)} \qquad ...(10.57)$$

The transfer function of the closed-loop system is designed to make the damping ratio of the dominant poles equal to 0.707, and the setttling time is less than 1 second. We have also made $K = 1000$ to make the steady-state error a unit step input is zero, since $T(0) = 1$. Here, we have

$$N_p(s) = s + 2,\ D_p(s) = s^3 + 10s^2$$

$$D(s) = s^3 + 50s^2 + 450s + 2000$$

$$D(s) - D_p(s) = 40s^2 + 450s + 2000$$

Let us make $\Delta(s) = (s + 10)(s + 20) = s^2 + 30s + 200$

Then from Equation (10.55), we get the following relationship:

$$\frac{1}{K}\left[D(s)-D_p(s)\,\Delta(s)\right] = 0.04s^4 + 1.65s^3 + 23.5s^2 + 150s + 400$$

$$= N_c(s)\,N_p(s) + N_u(s)\,D_p(s) \qquad ...(10.58)$$

Since $n = 3$, the degree of $N_c(s)$ must be two and that of $N_u(s)$ must be one. This is seen to be necessary so that the degrees of the polynomials on the two sides of Equation (10.55) match and also in order that we have as many equations as unknowns. Let

$$N_c(s) = b_0s^2 + b_1s + b_2,\ N_u\ (s) = a_0s + a_1$$

Thus, we get the following relationships:

$$N_c(s)\ N_p(s) = b_0s^3 + (b_1 + b_0)s^2 + (b_2 + 2b_1)s + 2b_2$$

$$N_u(s)\ D_p(s) = a_0s^4 + (10a_0 + a_1)s^3 + 10a_0s^2$$

The coefficients b_0, b_1, b_2, a_0 and a_1 must be selected in such a manner that they match the coefficients in Equation (10.58). Equating the corresponding coefficients, we get a set of $2n - 1$ linear equations for the unknown coefficients. These equations can be arranged in the form of a matrix, as shown below:

$$\begin{bmatrix} 1 & 0 & 0 & 0 & 0 \\ 10 & 1 & 1 & 0 & 0 \\ 0 & 10 & 2 & 1 & 0 \\ 0 & 0 & 0 & 2 & 1 \\ 0 & 0 & 0 & 0 & 2 \end{bmatrix} \begin{bmatrix} a_0 \\ a_1 \\ b_0 \\ b_1 \\ b_2 \end{bmatrix} = \begin{bmatrix} 0.04 \\ 1.65 \\ 23.5 \\ 150 \\ 400 \end{bmatrix} \qquad ...(10.59)$$

A solution to these equations will exist if the matrix of the left-hand side of Equation (10.59) is nonsingular. It should be noted that the matrix is made up from the coefficients of the numerator and the denominator of the plant transfer function, $KG_p(s)$. The first two columns of the matrix are from the coefficients of $D_p(s)$, and the remaining three are formed from the coefficients of N_p (s). It turns out that this is the well-known Sylvester matrix and will be nonsingular if N_p (s) and D_p (s) are relatively prime. For details, the interested reader may see Section 4.5 of Kailath's book [1].

The solutoin of Equation (10.59) gives $a_0 = 0.04$, $a_1 = 5.75$, $b_0 = -4.5$, $b_1 = -25$, and $b_2 = 200$. The transfer functions of the two compensators are now obtained from these coefficients. It will be seen that both of them are proper rational functions with poles on the negative real axis [due to out choice of Δ (s)]. Hence, they can be realized as *RC* networks, with amplifiers in cascade, if necessary.

Example 10.12

Consider the transfer function

$$G_p\ (s) = \frac{1}{s\,(s+2)} \qquad ...(10.60)$$

It is desired that the transfer function of the closed-loop system is given by

$$T\ (s) = \frac{K\,(s+\delta)}{(s+10)\left(s^2+4s+8\right)} \qquad ...(10.61)$$

where K and δ are to be selected so that the steady-state error to a ramp input is zero.

It may be noted that the order of the transfer function T (s) is three while that of the plant is only two. This can be accomplished by not cancelling the common factor Δ (s), as implied in Equation (10.54) and (10.55). The price we pay is that we must make $\Delta(s)$ equal to $(s + \delta)$.

We shall first determine the values of K and δ so that $K_v = \infty$. This requires that the dc gain of $T(s)$ be one, and that the sum of the reciprocals of the poles of T (s) be equal to the reciprocal of its zero (see Section 5.7). The equations are obtained.

$$\frac{1}{\delta} = \frac{1}{10} + \frac{4}{8}$$

$$T(0) = \frac{K\delta}{80} = 1$$

These are solved to obtain $K = 48$ and $\delta = 5/3$, so that we have $\Delta(s) = s + 5/3$. We may now design the compensators as before. Substituting the values of $N_p(s)$ and $\Delta(s)$ into Equation (10.55), we get

$$N_c(s) + (s^2 + 2s)\, N_u(s) = \frac{1}{48}\left[(s+10)\left(s^2+4s+8\right) - \left(s^2+2s\right)\left(s+\frac{5}{3}\right)\right]$$

$$= \frac{31}{144}s^2 + \frac{67}{72}s + \frac{5}{3} \qquad \text{...(10.62)}$$

Let $N_c(s) = b_0 s + b_1$, and $N_u(s) = a_0$. Using these values in Equation (10.62), we can equate the corresponding coefficients of both sides and obtain three linear simultaneous equations. These are arranged in the matrix form shown below:

$$\begin{bmatrix} 1 & 0 & 0 \\ 2 & 1 & 0 \\ 0 & 0 & 1 \end{bmatrix} \begin{bmatrix} a_0 \\ b_0 \\ b_1 \end{bmatrix} = \begin{bmatrix} \frac{31}{144} \\ \frac{67}{72} \\ \frac{5}{3} \end{bmatrix} \qquad \text{...(10.63)}$$

Solving Equation (10.63), we obtain $a_0 = 31/144$, $b_0 = 0.5$, and $b_1 = 5/3$.

Some important properties of this design should be noted. Here we have been able to increase the degrees of the numerator and denominator polynomials of the transfer function of the closed-loop system and avoid the cancellation of $\Delta(s)$, required in the original design.

In some cases, it is possible to make $N_u(s) = 0$, and simplify the design to the standard case of a system with a feedback compensator. This requires making $\Delta(s) = N_p(s)$, and is possible only if the degree of $N_p(s)$ is $n - 1$, and all its roots have negative real parts. We do not have the freedom of selecting $\Delta(s)$ in this case.

In general, the approach presented in this section is better than the pole-zero cancellation method described in Section 10.6.1 since we do not attempt to cancel the poles of the plant transfer function. However, it should be noted that the overall transfer function has hidden modes, the roots of $\Delta(s)$. Although these are cancelled, since $\Delta(s)$ is present in both the numerator and the denominator of Equation (10.54), there may be some difficulty due to imperfect cancellation caused by tolerance in the values of the components. Hence, $\Delta(s)$ must be selected carefully. Nevertheless, it is much easier to design the compensator with greater precision than to attempt to cancel the poles of the transfer function of the plant (which may actually be non-linear) by the zeros of the transfer function of the compensator.

As stated earlier, the requirement that the numerator and the denominator of the plant transfer function be relatively prime, it is equivalent to the system being both controllable and observable. These terms will be introduced in Chapter 12, where we shall return to the design of pole-placement compensators using the powerful concepts of state-variable feedback and asymptotic state observers.

10.8 PID CONTROLLERS

In our deliberations so far we have considered that the transfer function of the plant to be controlled is completely known to us. In practice, this is not always the case. It may still be possible to obtain good performance of the closed-loop system by introducing a PID (proportional, integral and derivative) controller as shown in the block diagram of Fig. 10.28.

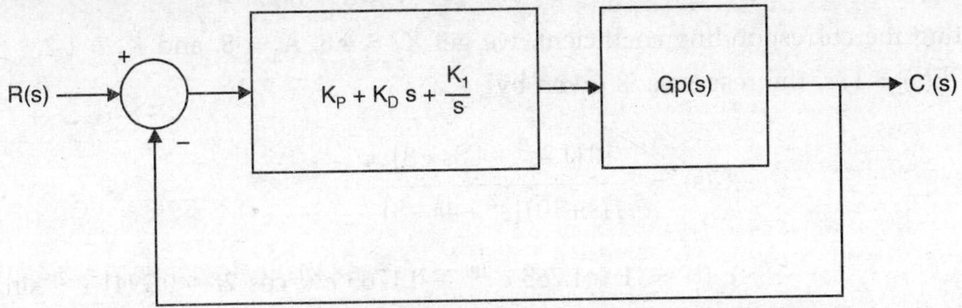

Figure 10.28. *A PID Controller*

The input to the plant consists of three components: (1) K_pE, which is proportional to the error, (2) K_IE/s, which is proportional to the integral of the error, and (3) K_DsE, which is proportional to the derivative of the error. The first components increases the loop gain of the system thereby reducing its sensitivity to plant parameter variations. The second component increases the order of the system and reduces the steady-state error of the system by adding a pole at the origin of the *s*-plane. The last component tends to stabilize the system by introducing the derivative of the error.

The main attraction of this approach is that the values of the three gain constant, K_p, K_I and K_D can often be determined by trial and error if G_p (*s*) is not known exactly. If the parameters of the plant are subject to large variations, the gain constants can be adjusted to improve the performance. This scheme is suitable for adaptive control of systems that the incompletely known or are subject to large variations in parameters.

In practical implementation, sometimes a problem may be caused by noise, which is invariably present in all physical systems. Differentiation of noise is seldom desirable, since it has the effect of accentuating it. As a result, the derivative component of the PID controller is often difficult to implement, except in the case of discrete-time systems. This will be further discussed in Chapter 11.

EXAMPLE 10.13

Consider the case when

$$G_p\,(s) = \frac{10}{s\,(s+2)} \qquad \text{...(10.64)}$$

With a PID controller, the order of the transfer function of the closed-loop system is increased from two or three. It is desired to locate the poles of this transfer function at – 10 and –2 ± j2. We shall determine the values of K_1, K_2 and K_3 to accomplish this, and calculate the response of the compensated system to a unit step. From Fig. 10.28,

$$T\,(s) = \frac{\left(K_1 + \frac{K_2}{s} + K_3 s\right) G_p\,(s)}{1 + \left(K_1 + \frac{K_2}{s} + K_3 s\right) G_p\,(s)}$$

$$= \frac{10\left(K_3 s^2 + K_1 s + K_2\right)}{s^3 + 2s^2 + 10K_3 s^2 + 10K_1 s + 10K_2} \quad ...(10.65)$$

Hence, in order to get the desired poles, we must have

$$s^3 + (10K_3 + 2)s^2 + 10K_1 s + 10K_2 = (s + 10)(s^2 + 4s + 8)$$
$$= s^3 + 14s^2 + 48s + 80$$

Equating the corresponding coefficient, we get $K_1 = 4.8$, $K_2 = 8$, and $K_3 = 1.2$.

Since $R(s) = 1/s$, the response is given by

$$C(s) = \frac{10\left(1.2s^2 + 4.8s + 8\right)}{s\,(s+10)\left(s^2 + 4s + 8\right)}$$

$$c\,(t) = 1 - 1.765\,e^{-10t} + 0.1765\,e^{-2t}\cos 2t + 0.2941\,e^{-2t}\sin 2t$$

It may also be noted that this system will have zero steady-state error to a step as well a ramp input. This is caused by the addition of a pole at the origin by the compensator, making it a type two system.

Practical Implementation of PID Controllers

Practical implementation of differentiator in the PID controller requires the inclusion of a pole on the negative real axis. Consequently, the transfer function of the controller will be of the form

$$G_c(s) = K_P + \frac{K_I}{s} + \frac{K_D s}{1 + sT} \quad ...(10.66)$$

It may also be noted that transfer function representing the differentiator can be realized by a simple RC network of the form shown in Fig. 10.29. The transfer function is readily evaluated as

$$\frac{V_2\,(s)}{V_1\,(s)} = \frac{Ts}{1 + Ts}$$

where $T = RC$.

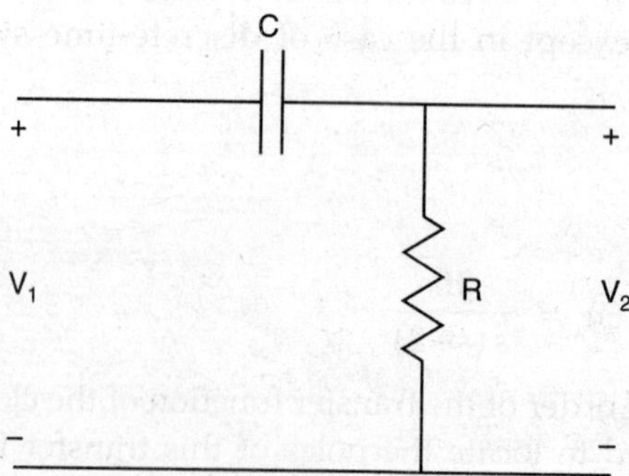

Figure 10.29. *Realization of transfer function in Equation (10.66)*

By selecting the value of T suitably, the frequency response will approach that of the ideal differentiator in the frequency range of interest. This condition will be satisfied if ωT is about 10 times the bandwidth of the system. In practice, it may be necessary to include an amplifier to adjust the value of K_I.

Ziegler-Nichols Tuning

If the system model is not known, one can determine the values of K_p, K_I and K_D by trial and error. This is called 'tuning' the controller. Ziegler and Nichols have proposed an approach for estimating the parameters of the controller based on simple tests that can be performed with the process to be controlled. These will be described here for the continuous-time case, but are also applicable to the discrete-time case provided that the sampling frequency is fairly high, say more than 10 times the bandwidth. For this purpose, the transfer function of the compensator will be written as

$$G_c(s) = K_p\left[1+\frac{1}{T_1 s}+T_D s\right] \quad ...(10.67)$$

(a) Step Response Method

This ultilizes the transient response of the system to a unit step input, with the loop open. The typical response of the system, that may also include a transport lag, is illustrated in Fig. 10.30, from which the 'reaction time' R and the time lag L, are estimated. The values of the various parameters in Equation (10.67) are given in the following Table for three possible types of controllers: P (proportional only), PI (proportional and integrator) and PID (proportional, integral and derivative).

	K_p	T_I	T_D
P	$1/(RL)$		
PI	$0.9/(RL)$	$3L$	
PID	$1.2/(RL)$	$2L$	$0.5\ L$

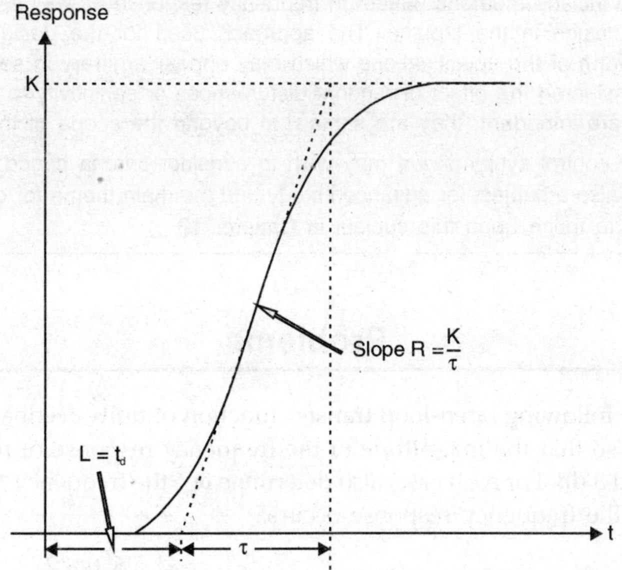

Figure 10.30. *Typical open-loop step response*

The parameters of the controller listed above normally lead to a system that has dominant poles with damping ratio of about 0.2.

(b) Stability Limit Method

In this case, the system is first tested with proportional control only and the gain is increased until the system starts oscillating. Let this value of the gain be denoted by K_u and the period of oscillations by P_u. Then the various parameters in Equation (10.67) are selected as in the following table:

	K_p	T_I	T_D
P	$0.5\ K_u$		
PI	$0.45\ K_u$	$P_u/1.2$	
PID	$0.6\ K_u$	$P_u/2$	$P_u/8$

In practice, the Ziegler-Nichols tuning method provides a good set of initial values for the various terms in the compensators, which can often be improved by on-line tuning. The main advantage is that by using this method one can completely bypass the need to obtain an exact mathematical model for the process to be controlled.

As will be seen in Chapter 11, the Ziegler-Nichols approach can also be applied to the design of PID controllers for discrete-time systems with very little modification.

SUMMARY

- We have discussed several method for compensating a control system so that its response is desirable. These include methods based on frequency response as well as those which attempt to perform the design in the *s*-plane. The approach used for the design depends, to some degree, on the form of the specifications which may appear arbitrary in some cases. Moreover, we have not considered the effect of random disturbances or sensitivity to parameter variations. Although these are important, they are somewhat beyond the scope of this book.
- In the design of control systems, we may wish to consider criteria based on optimizing a cost function. This is also a subject for advanced study and the main theme for optimal control theory. We shall briefly to touch upon this subject in Chapter 12.

Problems

1. For each of the following open-loop transfer function of unity-feedback systems, determine the value of K so that the magnitude of the frequency response of the closed-loop system does not exceed 3 dB. For each case, also determine ω_m, the frequency at which the maximum magnitude of the frequency response occurs.

(*a*) $\dfrac{K}{s(s+2)(s+8)}$ (*b*) $\dfrac{K(s+2)}{s\left(s^2+16s+400\right)(s+4)}$.

2. For Problem 1, design a suitable compensator for each case so that the steady-state error of the closed-loop system to a unit ramp input does not exceed 0.1, and the maximum magnitude of the frequency response of the closed-loop system does not exceed 3 dB.

3. The transfer function of the forward path of a unity feedback system representing a typical chemical process is given by

$$G_p(s) = \frac{Ke^{-0.05s}}{s(s+5)(s+10)}$$

It is desired that the system should have $K_v = 5$ and phase margin equal to 45°.

(*a*) Design a compensator that will achieve gain crossover at 5 rad/s.

(*b*) Design a compensator that will achieve gain crossover at 2 rad/s.

4. The forward transfer function of a unity-feedback system is given by

$$G_p(s) = \frac{1}{s^2(1+0.1s)}$$

Design a suitable compensator so that the system is stable with $M_m \le 1.5$, and ω_m is approximately equal to 1.4 rad/s.

5. The forward transfer function of a unity-feedback system is given by $KG_p(s)\,G_c(s)$. It is known that

$$G_p(s) = \frac{1}{s(s+1)(s+12)} \quad \text{and } G_c(s) = \frac{\alpha(s+1)}{s+\alpha}$$

(*a*) Determine K and α so that the transfer function of the closed-loop system has dominant poles with real part equal to – 4, and damping ratio 0.5.

(*b*) What will be the response of the system to a unit step input?

6. Use the Nichols chart to determine the gain margin, the phase margin, M_m and ω_m for the system in Problem 5 with and without compensation. [For the latter case take $G_c(s)$ equal to one and $K = 224/3$.]

7. Design a suitable compensator for Problem 3 so that it has $K_v = 5$ and $M_m = 3$ dB, with (*a*) $\omega_m = 5$/rad/s, and (*b*) $\omega_m = 2$ rad/s, approximately.

8. The transfer function of the forward path of a unity feedback system is given by

$$G(s) = \frac{K}{s(s+3)(s+5)}$$

It is desired that the closed-loop system should satisfy the following specifications:

(*a*) The steady-state error for a unit step input should be zero.

(*b*) The steady-state error for a unit ramp input should not exceed 0.1.

(*c*) The damping ratio of the complex poles of the transfer function should be 0.5.

Design a suitable compensator and determine the sensitivity of the complex poles of the transfer function of the closed-loop system to changes in K.

9. For Problem 3, determine the range of gain crossover frequencies over which it is possible to design (*a*) a lead compensator, (*b*) a lag compensator, and (*c*) a lag-lead compensator to satisfy the specification given in that problem.

10. The linearized transfer function of a helicopter near hover can be approximated as

$$G_p(s) = \frac{s+0.3}{(s+0.65)\left(s^2-0.2s+0.1\right)}$$

It is desired to stabilize the system by adding compensation so that the poles of the transfer function of the closed-loop system are located at $s = -2$, $-1 + j1$ and $-1 - j1$. Also, the system is required to give zero steady-state error to a step input. Use the method of Section 10.7 and locate the poles of the compensator at $s = -4$ and -5.

11. The control system for a tracking antenna used in radar can be represented by a unity-feedback system with the forward transfer function

$$G_p(s) = \frac{1}{s(s+4)(s+8)}$$

The closed-loop system is required to satisfy the following specifications:

(*a*) The steady-state error to a unit ramp should be 0.05.

(*b*) The damping ratio of the complex poles should be 0.707.

Design a suitable compensator using the procedure of Section 10.7. (*Hint:* First determine the transfer function of the closed-loop system that will satisfy the specifications.)

12. Consider the feedback system with two delays shown in Fig. 9.30 (see Chapter 9, Problem 7), which represents a chemical reactor. Design a suitable compensator to meet the following specifications:

(*a*) The steady-state error to a unit step input should be 0.05.

(*b*) The phase margin should be 45°.

13. Consider the block diagram for an automatic gauge control system for a hot-strip finishing mill shown in Fig. 9.32 (see Chapter 9, Problem 9). It is desired to design a compensator to satisfy the following specifications:

(*a*) The steady-state error to a unit ramp input should not exceed 0.05.

(*b*) The phase margin should be at least 45°.

Discuss what type of compensator will be most suitable if the bandwidth of the system is to be kept is as small as possible. Design the compensator and determine the response of the compensated system to a unit step.

14. Repeat Problem 13 if the following specifications have to be met:

(*a*) The steady-state error to a unit ramp input should not exceed 0.05.

(*b*) The value of M_m must not exceed 3 dB.

15. Repeat Problem 13 if the following specifications have to be met:

(*a*) The steady-state error to a unit ramp input should not exceed 0.05.

(*b*) The damping ratio of the complex poles should be 0.5 and the undamped natural frequency should not be less than 1.5 rad/s.

16. The block diagram of a driver steering control system is shown in Fig. P9.10 (Chapter 9, Problem 10). Design a compensator so that the following specifications are satisfied:

(*a*) The steady-state error to a unit step input should be 0.05.

(*b*) The phase margin should be 45°.

17. The block diagram of a speed-control system using a phase-locked loop is shown in Fig. P9.11 (see Chapter 9, Problem 11). It is desired to replace the filter by a lead compensator and adjust the value of K so that the following specifications are met.
   (*a*) The steady-state error to a unit step input does not exceed 0.01.
   (*b*) The phase margin is not less than 45°.
   If it is possible, design such a compensator. [*Hint:* The error in this problem is defined as $e(t) = \theta_i(t) - \theta_o(t)$.]
18. Figure P7.8 (see Chapter 7, Problem 8) shows the simplified block diagram for the pitch control system of a supersonic aircraft. Design a compensator so that the steady-state error to a unit step is less than 0.02, and the damping ratio of the dominant poles is 0.6.
19. For Problem 18 determine the transfer function of the closed-loop system so that dominant poles are located at $-5 \pm j5$ and the steady-state error to a ramp input is zero. Use the method of Section 10.7 to design suitable compensators with poles at $s = -10$.
20. The approximate transfer function of a helicopter near hover is given in Problem 10. It is desired to design a PID controller to stabilize the system as well as to make the steady-state error to a step input equal to zero. This controller will be connected in cascade with $G(s)$ in a unity-feedback system, and its transfer function has the following form:
$$G_c(s) = \frac{K(s+0.5)(s+1)}{s}$$
   (*a*) What value of K will cause the complex poles of the transfer function of the closed-loop system to have damping ratio equal to 0.707?
   (*b*) Calculate the response of the compensated system to a unit step.
   (*c*) What will be steady-state error of this system to a unit ramp input?
21. Reconsider Problem 20 with the PID controller of the form
$$G_c(s) = \frac{K(s+\alpha)(s+\beta)}{s}$$
   The three adjustable parameters can be selected in many ways. For example, one may make $\beta = 0.65$ to cancel the pole at $s = -0.65$. In this case, what must be the values of K and α so that the complex poles of the closed-loop system are located at $-1 \pm j1$? Calculate the response of the compensated system to a unit step input and determine the steady-state error to a unit ramp input.
22. The transfer function of a servomotor (including the amplifier and the load) is given by
$$G_p(s) = \frac{1}{s(s+4)}$$
   Use the method of Section 10.7 to design compensators so that the transfer function of the closed-loop system is given by
$$T(s) = \frac{K(s+\delta)}{(s+10)(s^2+4s+13)}$$
   What must be the values of K and δ so that the system has zero steady-state error to a ramp-input? Also, determine the response of the compensated system to a unit step.
23. A video tracking system (see Fig. P10.23) is used for monitoring the automatic docking of a satellite in a space rendezvous. A video camera is geared to a motor. The position of the camera with respect to an inertial reference is θ_1. The position of the point to be tracked

with respect to the same reference is θ_c. It has been determined that a satisfactory performance can be obtained if the following specifications are met:

(*i*) The settling time for response to a step input is less than 0.5 second.

(*ii*) The maximum overshoot in the step response is less than 16 per cent.

(*iii*) The steady-state error to a unit ramp input does not exceed 0.01.

(*a*) Show that K and α can be chosen to meet the first two specifications.

(*b*) Design a suitable compensator so that the last specification can also be satisfied.

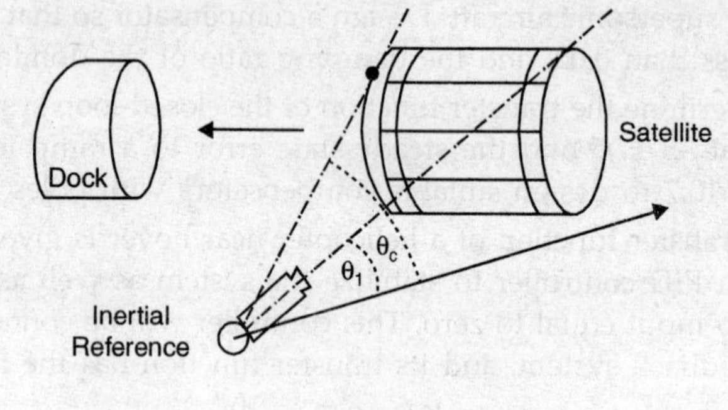

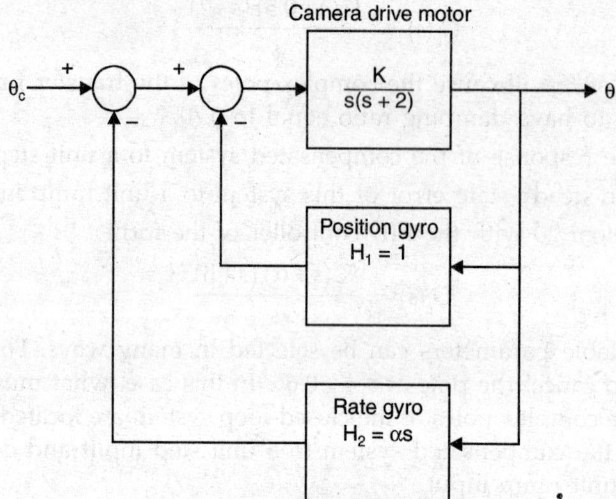

Figure P10.23. *A video tracking system*

References

1. Dazzo, J.J., and C.H. Houpis, *Feedback Control Systems Analysis and Synthesis*, 2nd ed., McGraw-Hill, New York, 1966.

2. Kailath, T., *Linear System*, Prentice-Hall, Englewood Cliffs, N.J., 1980 (section 4.5).

3. Truxal, J.G., *Automatic Feedback Control System Synthesis*, McGraw-Hill, New York, 1955.

4. Ziegler, J.G., and N.B. Nichols, "Optimum settings for Automatic Controllers", Trans. ASME, vol. 66, 1942, pp. 759-768.

❑❑❑

CHAPTER 11

Digital Control

11.1 INTRODUCTION

The use of a digital computer as a part of a control system has many advantages. Some of these are listed below:

1. The need for amplification of analog signals is eliminated, and this results in the reduction of noise in the system.
2. Higher resolution and accuracy can be obtained with digital transducers. These are less affected by noise.
3. Digitally coded data can be stored for any length of time and transmitted or used, as many times as necessary, without any loss in accuracy.
4. Overall costs are reduced due to the possibility of time-sharing.
5. Implementation of optimal and adaptive control is much more convenient with a digital computer as part of the control loop.

With the rapid advances in the area of microprocessors in recent years, digital control has become even more attractive and economical in a large number of applications. One effect of the microelectronics revolution has been a significant reduction in the size and weight of microcomputers to the extent that now it is quite practical to incorporate them as components of space vehicles, automobiles, ships, cameras, and other devices without appreciable increase in cost or physical dimensions. This has provided a great impetus to the design of control systems with one or more microprocessors as integral components.

The block diagram of a typical computer control system is shown in Fig. 11.1. In this diagram, the analog error signal, $e(t)$, is transformed into a sequence of numbers, $m(kT)$, by an analog-to-digital (A/D) converter. This requires sampling the analog signal at the instants $t = kT$, where k is an integer and replaces $e(kT)$ by a sequence of number, $m(kT)$. The sequenece $m(kT)$ is then processed by the computer to obtain another sequence of numbers, $u(kT)$. This is then transformed into an analog signal $u(t)$ by a digital-to-analog (D/A) converter before it is applied to the process.

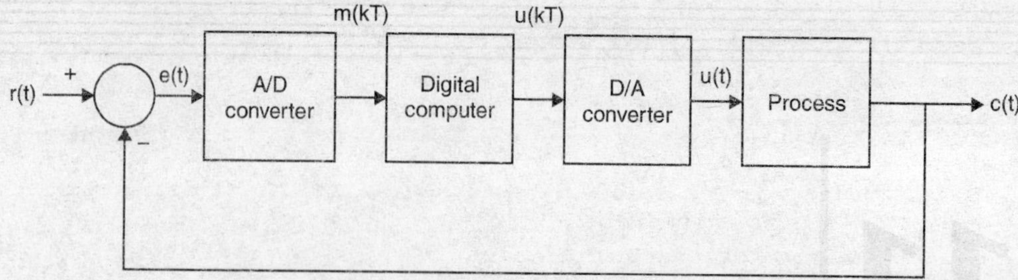

Figure 11.1. *A typical computer control system*

Although, in practice, the sampling period is not always maintained constant, we shall make the simplifying assumption that it is. Typically, as a part of the computer hardware, there is a clock that sends a pulse every T seconds and then the A/D converter sends a number to the computer. Thus, the input sequence, $m(kT)$ supplied to the computer can change only at discrete instants of time, that is, at $t = kT$, $k = 0, 1, 2, ...$ In addition to the process of discretizing, the A/D converter also quantizes the discrete-time signal to obtain a digital output, or a sequence of numbers suitable for use by the computer. It is common practice to call the resulting system a discrete-time or sampled-data system.

The actual input to the process, denoted as $u(t)$, is usually a continuous-time function, reconstructed from the output $u(kT)$ of the digital computer. Several methods are used for reconstructing the continuous-time function $u(t)$ from the sequence $u(kT)$. All of them require a D/A converter, followed by a data extrapolator. The most commonly used data extrapolator is the zero-order hold which will be described in the next solution.

The mathematical analysis of discrete-time systems can be carried out conveniently by many methods. Two of the most outstanding methods are: (1) those based on using z-transforms and (2) the state variable method. The former is a modification of the Laplace-transform method used for continuous-time systems, whereas the latter follows directly from the state equations for continuous-time system, if the input is held constant between the sampling instants. The state transition equations of a discrete-time system were derived in Section 3.8 in Chapter 3. In this chapter, we shall study the z-transform method of analysis and design of discrete-time systems. The state-variable approach will be discussed in Chapter 12. It is expected that the reader has some familiarity with the theory of z-transforms. For the sake of completeness, a brief review is presented in Appendix C.

11.2 DATA EXTRAPOLATORS

A data extrapolator is a device that reconstructs a sampled function into a continuous-time signal based on a knowledge of the past samples. This device, which is also called a data hold, follows the digital computer in the system, and is usually a part of the D/A converter.

Data extrapolators are classified according to the number of prior samples utilized for predicting the sampled function during waiting intervals. The simplest case is the 'zero-order hold,' in which the value of the reconstructed function during any waiting period is simply equal to the value of the sampled function at the beginning of the interval. This type of hold, therefore, leads to the 'staircase' approximation of the continuous-time function, as shown in Fig. 11.2. Its operation is similar to an electronic clamping circuit, which keeps its output level equal to the magnitude of an input pulse and then resets itself when a new pulse arrives.

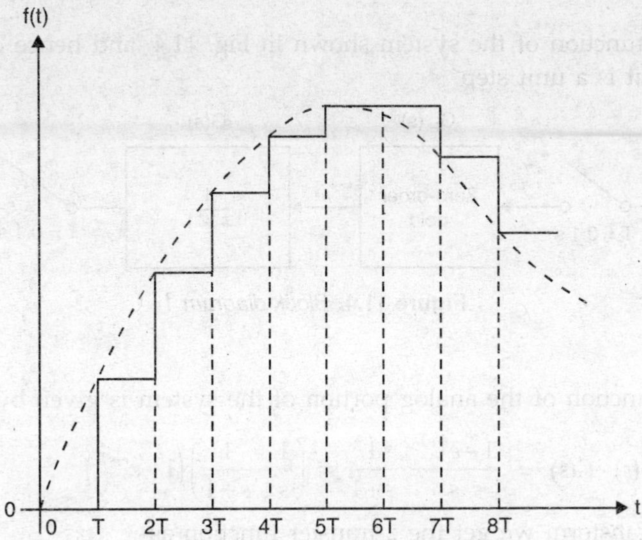

Figure 11.2. *Staircase approximation*

The transfer function of a zero-order hold can be obtained easily from its impulse response, which is a rectangular pulse of unit height and duration T, as shown in Fig. 11.3. It can be regarded as the result of subtracting from the unit step function occurring at $t = 0$, another unit step function delayed by the time T. Hence, taking the Laplace transform, we get

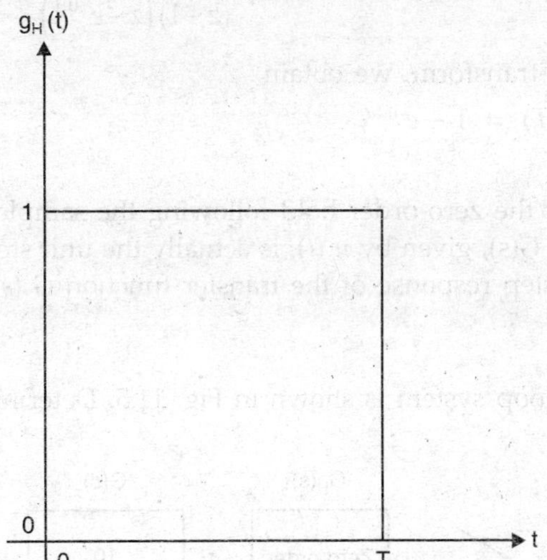

Figure 11.3. *Impulse response of a zero-order hold*

$$G_H(s) = \frac{1}{s}\left(1 - e^{-sT}\right) \qquad \text{...(11.1)}$$

An extrapolator that depends upon the present and one prior sample is known as the 'first-order hold'. In this case, during the waiting interval, the function is extrapolated as a straight-line determined by these two values. In the same way, one may envisage higher-order holds. These are seldom used in feedback control systems.

EXAMPLE 11.1

Find the z transfer function of the system shown in Fig. 11.4, and hence determine the output sequence if the input is a unit step.

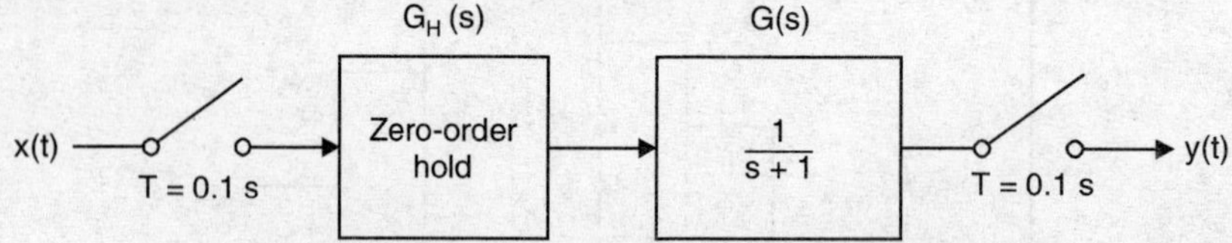

Figure 11.4. *Block diagram 1*

SOLUTION:

The transfer function of the analog portion of the system is given by

$$G_H(s)\cdot G(s) = \frac{1-e^{-sT}}{s}\cdot\frac{1}{s+1} = \left(\frac{1}{s}-\frac{1}{s+1}\right)\left(1-e^{-sT}\right) \qquad \text{...(11.2)}$$

Taking the z-transform we get the z transfer function as

$$G_HG(z) = \left(\frac{z}{z-1}-\frac{z}{z-e^{-0.1}}\right)\left(1-z^{-1}\right) = \frac{1-e^{-0.1}}{z-e^{-0.1}} \qquad \text{...(11.3)}$$

The z-transform of the output can be written as

$$Y(z) = G_H\,G(z)\cdot X(z) = \frac{\left(1-e^{-0.1}\right)z}{(z-1)\left(z-e^{-0.1}\right)} = \frac{z}{z-1}-\frac{z}{z-e^{0.1}} \qquad \text{...(11.4)}$$

Taking the inverse z-transform, we obtain

$$y(nT) = 1 - e^{-0.1n},$$

as expected.

It may be noted that the zero-order hold following the sampler reconstructs the unit step accurately, so the input to $G(s)$, given by $u(t)$, is actually the unit step. Hence, $y(nT)$ is identical to the nth sample of the step response of the transfer function $G(s) = 1/(s+1)$.

EXAMPLE 11.2

An error-sampled closed-loop system is shown in Fig. 11.5. Determine the output sequence for a unit step input.

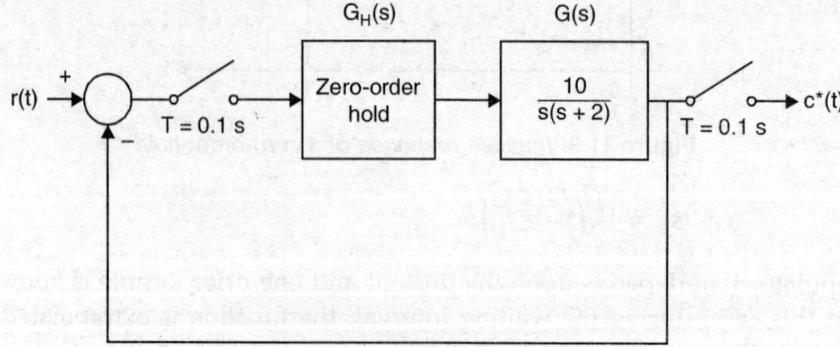

Figure 11.5. *Block diagram of a closed-loop sampled-data system*

SOLUTION:

The transfer function of the forward path is obtained as

$$G_H(s) \cdot G(s) = \frac{1-e^{-sT}}{s} \cdot \frac{10}{s(s+2)} = \left(\frac{5}{s^2} - \frac{2.5}{s} + \frac{2.5}{s+2}\right)\left(1-e^{-sT}\right) \quad \text{...(11.5)}$$

so that

$$G_H G(z) = \left(\frac{0.5z}{(z-1)^2} - \frac{2.5z}{z-1} + \frac{2.5z}{z-e^{-0.2}}\right)\left(1-z^{-1}\right) = \frac{\left(2.5e^{-0.2}-2\right)z + 2.5 - 3e^{-0.2}}{(z-1)\left(z-e^{-0.2}\right)} \quad \text{...(11.6)}$$

The transfer function of the closed-loop system is given by

$$\frac{C(z)}{R(z)} = \frac{G_H\,G(z)}{1+G_h\,G(z)} = \frac{\left(2.5e^{-0.2}-2\right)z+2.5-3e^{-0.2}}{(z-1)\left(z-e^{-0.2}\right)+\left(2.5e^{-0.2}-2\right)z+2.5-3e^{-0.2}}$$

$$= \frac{0.046827\,z + 0.043808}{z^2 - 1.771904\,z + 0.862538} \quad \text{...(11.7)}$$

Since the *z*-transform of the input is

$$R(z) = \frac{z}{z-1} \quad \text{...(11.8)}$$

the *z*-transform of the output is given by

$$C(z) = \frac{0.046827z^2 + 0.043808\,z}{(z-1)\left(z^2 - 1.771904\,z + 0.8625385\right)}$$

$$= \frac{z}{z-1} - \frac{z\,(z-0.81873)}{z^2 - 2\,(0.92873)\times(0.95394)\,z + (0.92873^2)} \quad \text{...(11.9)}$$

and $$c(nT) = 1 - 0.92873^n\,(\cos 0.30469\,n + 0.24127 \sin 0.30469\,n) \quad \text{...(11.10)}$$

DRILL PROBLEM 11.1

An error-sampled control system is shown in Fig. 11.6. For K = 2, find the output if the input is a unit step. Determine the values of the output for t equal to 0.5, 1, 2, 3, and ∞ second.

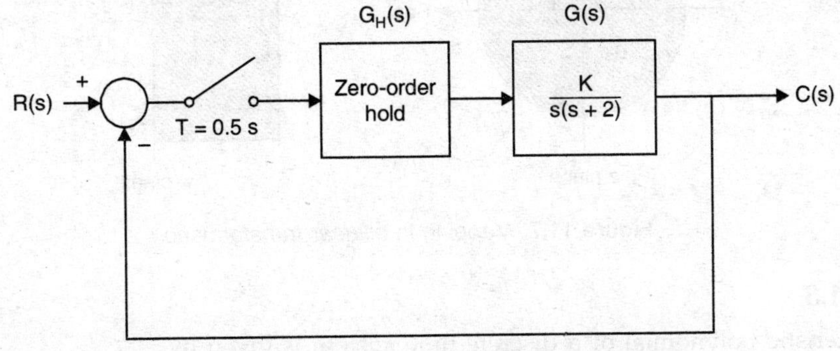

Figure 11.6. *An error-sampled closed-loop system*

Ans. $c(nT) = 1 - 0.7072^n (\cos 0.5787n + 0.5794 \sin 0.5787n)$

$c(0.5) = 0.1838,\ c(1) = 0.5337,\ c(2) = 1.0628,\ c(3) = 1.1419,\ c(\infty) = 1$

11.3 STABILITY OF CLOSED-LOOP DISCRETE-TIME SYSTEMS

The necessary and sufficient condition for the stability of a discrete-time system is that all the poles of its z transfer function lie inside the unit circle of the z-plane. This fact follows from the transformation $z = e^{sT}$, which maps the left half of the s-plane inside the unit circle.

Keeping this mind, the three basic techniques studied for continuous-time systems are: (*a*) the Routh-Hurwitz criterion, (*b*) the Nyquist criterion, and (*c*) the root locus method, can be used with slight modifications for discrete-time systems.

11.3.1 The Routh-Hurwitz Criterion

The Routh-Hurwitz criterion, discussed in Chapter 6, is an attractive method for investigating system stability without requiring evaluation of the roots of the characteristic polynomial. Since it enables us to determine the number of roots with positive real parts, it cannot be used directly for discrete-time systems, where we need to find the number of roots outside the unit circle. It is possible to use the Routh-Hurwitz criterion to determine if a polynomial $Q(z)$ has roots outside the unit circle by using the *bilinear transformation*

$$w = u + jv = \frac{z+1}{z-1}, \quad \text{or} \quad z = \frac{w+1}{w-1} = \frac{u+1+jv}{u-1+jv} \qquad ...(11.11)$$

(also called the Mobius transformation) which maps the unit circle of the z-plane into the imaginary axis of the w-plane and the interior of the unit cirlce into the left of the w-plane. This is shown in Fig. 11.7. We can now apply the Routh criterion to $Q(w)$ as in the s-plane.

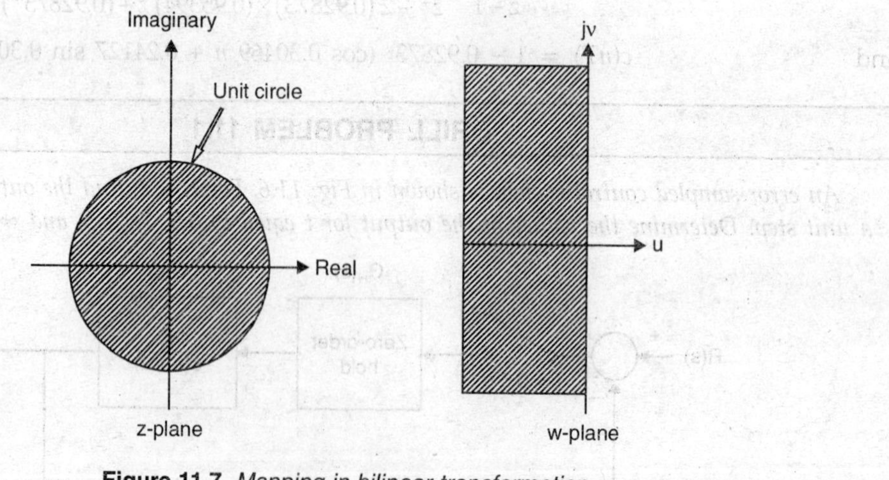

Figure 11.7. *Mapping in bilinear transformation*

EXAMPLE 11.3

The characteristic polynomial of a discrete-time system is given by

$$Q(z) = z^3 - 2z^2 + 1.5z - 0.4 = 0 \qquad ...(11.12)$$

Then

$$Q(w) = \left(\frac{w+1}{w-1}\right)^3 - 2\left(\frac{w+1}{w-1}\right)^2 + 1.5\left(\frac{w+1}{w-1}\right) - 0.4 = 0$$

or $(w + 1)^3 - 2(w + 1)^2 (w - 1) + 1.5(w + 1)(w - 1)^2 - 0.4(w - 1)^3 = 0$

This can be further simplified to obtain the polynomial

$$0.1 w^3 + 0.7 w^2 + 2.3 w + 4.9 = 0$$

The Routh table is shown below. ...(11.13)

w^3	0.1	2.3
w^2	0.7	4.9
w^1	1.6	
w^0	4.9	

No change in sign in the first column of the Routh table indicates that all roots of $Q(w)$ are in the left half of the w-plane, and correspondingly, all roots of $Q(z)$ are inside the unit circle. Hence, $Q(z)$ is the characteristic polynomial of a stable system.

It will be seen that this method requires a lot of algebra. The computer programme 'Stability. EXE' on the disk available with this book can perform all the necessary computations and display the Routh table in the w-plane for a characteristic polynomial $Q(z)$. In the next subsection we shall discuss another test for stability that can be performed directly in the z-plane, in a manner reminiscent of the conventional Routh test in the s-plane, although requiring more terms and more computation.

EXAMPLE 11.4

Determine the maximum value of K for which a unity feedback system with the following forward path transfer function will be stable.

$$G(z) = \frac{K(0.368z + 0.264)}{(z-1)(z-0.368)} \quad ...(11.14)$$

SOLUTION:

The characteristic equation for the closed-loop system is given by

$$Q(z) = z^2 + (0.368K - 1.368)z + (0.264K + 0.368)$$

Transformation of this equation to the w-plane leads to the following polynomial after simplification

$$0.632Kw^2 + 2(0.632 - 0.264K)w - (2.736 - 0.104K) = 0$$

Every term in the corresponding Routh table will be positive if and only if the gain K is positive and less than 2.394.

DRILL PROBLEM 11.2

Use the Routh-Hurwitz criterion to determine the maximum value of K for which the system shown in Fig. 11.6 will be stable.

Ans. $K = 9.5686$

11.3.2 The Jury Stability Test

The is similar to the Routh test, and can be used directly in the z-plane. Let the characteristic polynomial be given by

$$Q(z) = a_0 z^n + a_1 z^{n-1} + \dots + a_{n-1} z + a_n, \; a_0 > 0 \qquad \dots(11.15)$$

Then we form the array shown in the following table:

a_0	a_1	a_2	...	a_k	...	a_{n-1}	a_n	
a_n	a_{n-1}	a_{n-2}	...	a_{n-k}	...	a_1	a_0	$\alpha_n = a_n/a_0$
b_0	b_1	b_2	...	b_{n-k}	...	b_{n-1}		
b_{n-1}	b_{n-2}	b_{n-3}	...	b_{k-1}	...	b_0		$\alpha_{n-1} = b_{n-1}/b_0$
c_0	c_1	c_2	...	c_{n-k}	...			
c_{n-2}	c_{n-3}	c_{n-4}	...	c_{k-2}	...			$\alpha_{n-2} = c_{n-2}/c_0$
.	.	.	...	.	...			
.	.	.	...	.	...			

The elements of the first and second rows are the coefficients of $Q(z)$, in the forward and the reverse order respectively. The third row is obtained by multiplying the second row by $\alpha_n = a_n/a_0$ and subtracting this from the first row. The fourth row is the third row written in the reverse order. The fifth row is obtained in a similar manner, that is by multiplying the fourth row by b_{n-1}/b_0 and subtracting this from the third row. The sixth row is the fifth row written in the reverse order. The process is continued until there are $2n + 1$ rows. The last row has only one element.

Jury's stability test states that if $a_0 > 0$, then all roots of $Q(z)$ are inside the unit circle if and only if b_0, c_0, d_0, etc. (that is the first element in each odd-numbered row) are all positive. Furthermore, the number of negative elements in this set is equal to the number of roots of $Q(z)$ outside the unit circle and a zero element implies a root on the circle.

EXAMPLE 11.5

Consider the characteristics polynomial, $Q(z)$, given in Example 11.3. The Jury table is shown below:

1	– 2	1.5	– 0.4	
– 0.4	1.5	–2	1	$\alpha_3 = -0.4$
0.84	– 1.4	0.7		
0.7	– 1.4	0.84		$\alpha_2 = 5/6$
0.2567	– 0.2333			
– 0.2333	0.2567			$\alpha_1 = -0.909$
0.0445				

Since the coefficients a_0, b_0, c_0 and d_0 are all positive, it follows that all roots of $Q_t(z)$ are inside the unit circle and it is the characteristic polynomial of a stable system.

EXAMPLE 11.6

We shall now use Jury's test to determine the conditions for stability of a second-order system. Consider the polynomial

$$Q(z) = z^2 + a_1 z + a_2 = 0 \qquad ...(11.16)$$

The Jury table is shown below:

1	a_1	a_2	
a_2	a_1	1	$\alpha_2 = a_2$
$1 - a_2^2$	$a_1(1 - a_2)$		
$a_1(1 - a_2)$	$1 - a_2^2$		$\alpha_2 = a_1(1 + a_2)$
$\dfrac{1 - a_2(1 + a_2 - a_1)(1 + a_2 + a_1)}{1 + a_2}$			

Thus, the following conditions must be satisfied in order that all roots of $Q(z)$ may be within the unit circle.

(*a*) $|a_2| < 1$

(*b*) $a_2 > a_1 - 1$

(*c*) $a_2 > -(a_1 + 1)$

These equations define the stability region shown shaded in Fig. 11.8.

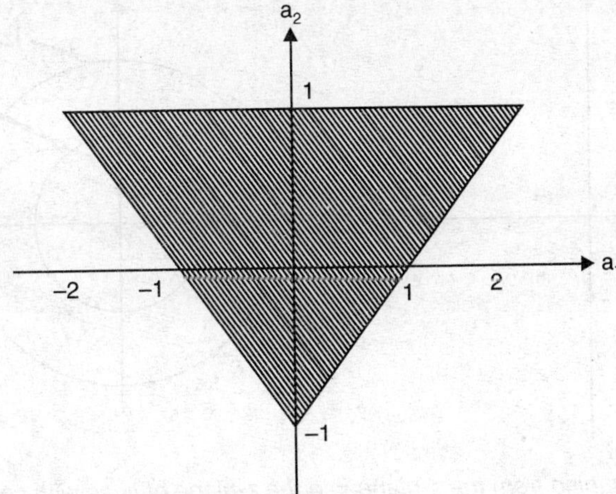

Figure 11.8. *The stability region for second-order Equation (11.16)*

It is interesting to compare this result with the fact that the characteristic polynomial, $Q(s) = s_2 + a_1 s + a_2$, for a continuous-time system is stable for all positive a_1 and a_2.

It is possible to reduce the number of rows required in the Jury table from $2n + 1$ to $2n - 3$ by including the following conditions:

$$Q(1) > 0 \qquad ...(11.17)$$

$$(-1)^n Q(-1) > 0$$

This is especially useful if the degree of $Q(z)$ is two. In that case, we need only to apply the tests in Equation (11.17), along with the requirement that $|a_2| < a_0$. For instance, the results of Example 11.6 can be obtained with much less efforts using these tests.

11.4 THE ROOT LOCUS METHOD

The root locus method for continuous-time systems was discussed in Chapter 7. It can be extended to discrete-time systems with the same rules for construction. The only difference arises in the manner in which the resulting root locations are related to stability and time response.

The closed-loop system will be unstable when the root locus enters the region outside the unit circle of the *z*-plane, whereas for the continuous-time case, instability occurs when the locus enters crosses the $j\omega$-axis to enter the right half of the *s*-plane. Furthermore, lines of constant damping in the *s*-plane are parallel to the $j\omega$-axis in the left half of the *s*-plane. These transform into circles of radii smaller than one. For example, the line $\sigma = -a$ in the *s*-plane will map into a circle with centre at the origin and of radius e^{-aT}. On the other hand, a line of constant damping ratio, which is a radial line from the origin in the *s*-plane, maps into a logarithmic spiral in the *z*-plane, given by the following equation:

$$z = e^{-\frac{2\pi\omega}{\omega_s \tan\phi}} e^{j2\pi\frac{\omega}{\omega_s}} \quad \text{where } \phi = \cos^{-1}\zeta \qquad \text{...(11.18)}$$

These mappings are shown in Fig. 11.9.

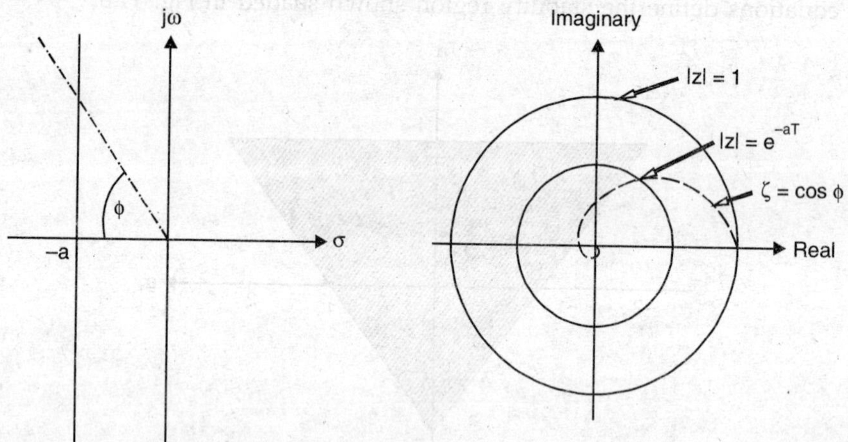

Figure 11.9. *Mapping from the s-plane into the z-plane of lines with constant damping and constant damping ratio*

EXAMPLE 11.7

The forward transfer function of a unity feedback sampled-data system is

$$G(z) = \frac{K(z+0.2)}{(z-1)(z-0.2)} \qquad \text{...(11.19)}$$

Determine the value of K so that all the poles of the closed-loop system are inside a circle of radius 0.7071.

SOLUTION

A sketch of the root locus is shown in Fig. 11.10. The largest value of K that will satisfy the above condition is 1.5, with poles at $-0.15 \pm j0.691$.

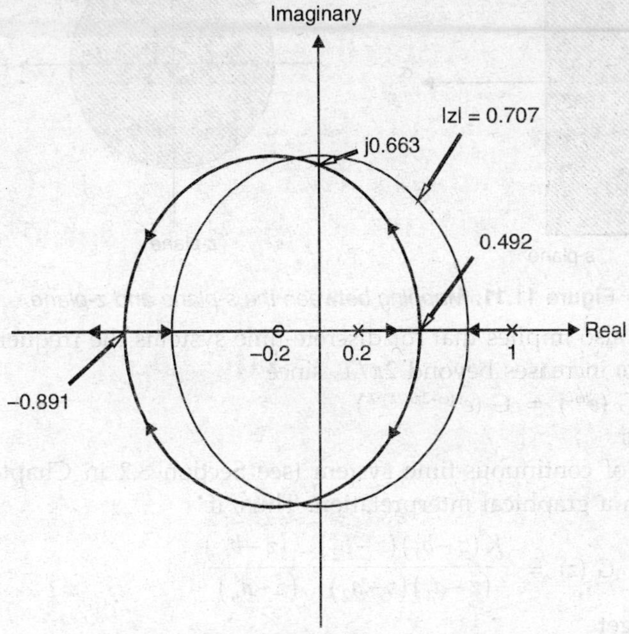

Figure 11.10. *Root locus in the z-plane*

DRILL PROBLEM 11.3

Repeat the problem in Example 11.7 if

$$G(z) = \frac{K(z+0.3)}{(z-0.3)(z-1)}$$

Ans. $K = 0.66$ gives poles at $0.32 \pm j0.629$.

11.5 FREQUENCY RESPONSE OF DISCRETE-TIME SYSTEMS

The frequency response of a discrete-time system is obtained by replacing z by $e^{j\omega T}$ in the transfer function and evaluating by resulting expression for different values of ω. An important feature of the frequency response of discrete-time system is evident. Here, we evaluate $G(z)$ along the unit circle of the z-plane, whereas for continuous-time systems, the function $G(s)$ is evaluated along the $j\omega$-axis of the s-plane. This follows due to the transformation between the two planes according to the relationship

$$z = e^{sT} \qquad \text{...(11.20)}$$

and the corresponding mapping shown in Fig. 11.11.

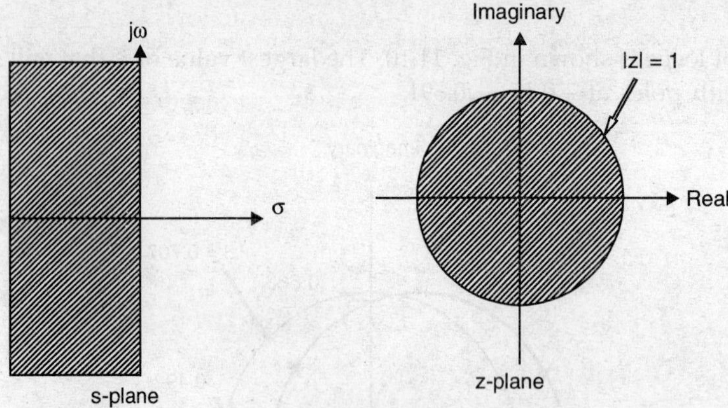

Figure 11.11. *Mapping between the s-plane and z-plane*

Equation (11.20) also implies that for discrete-time systems the frequency response repeats itself as the value of ω increases beyond $2\pi/T$, since

$$G\,(e^{j\omega T}) = G\,(e^{j\,(\omega+2\pi n/T)\,T}) \qquad ...(11.21)$$

where n is any integer.

As for the case of continuous-time system (see Section 8.2 in Chapter 8), the frequency response can be given a graphical interpretation. Thus, if

$$G\,(z) = \frac{K(z-b_1)(z-b_2)\ldots(z-b_m)}{(z-a_1)(z-a_2)\ldots(z-a_n)} \qquad ...(11.22)$$

then for $z = e^{j\omega_1 T}$, we get

$$G\,(e^{j\omega_1 T}) = \frac{K\left(e^{j\omega_1 T}-b_1\right)\left(e^{j\omega_1 T}-b_2\right)\ldots\left(e^{j\omega_1 T}-b_m\right)}{\left(e^{j\omega_1 T}-a_1\right)\left(e^{j\omega_1 T}-a_2\right)\ldots\left(e^{j\omega_1 T}-a_n\right)}$$

$$= K\cdot\frac{\text{product of directed distances each zero to } z = e^{j\omega_1 T}}{\text{product of directed distances each pole to } z = e^{j\omega_1 T}} \qquad ...(11.23)$$

The graphical interpretation of Equation (11.22) is shown in Fig. 11.12 for a transfer function with two real poles and a real zero.

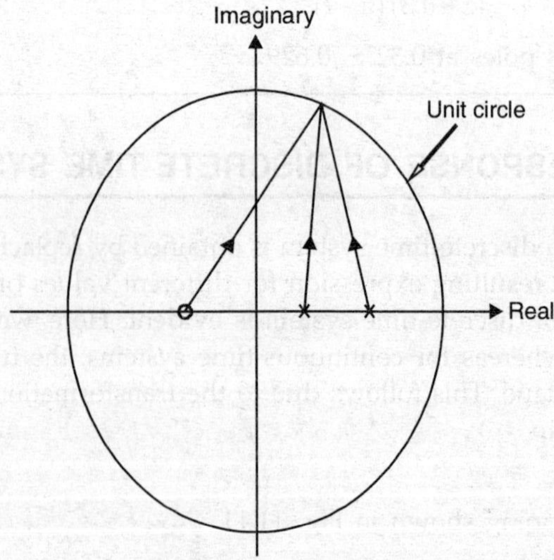

Figure 11.12. *Graphical interpretation of frequency response*

From the symmetry of Fig. 11.12 about the real axis, it follows that

$$G(e^{j\omega T}) = G^*(e^{-j\omega T}) = G^*(e^{j(2\pi - \omega T)}) \quad ...(11.24)$$

where G^* is the complex conjugate of G. Consequently, we need to calculate the frequency response only within the range ω varying from 0 to π/T, instead of from 0 to ∞, as required for continuous-time systems. The following example will illustrate the procedure:

EXAMPLE 11.8

Consider the second-order transfer function given by

$$G(z) = \frac{2(z+0.2)}{(z-0.5)(z-0.8)} = \frac{2z+0.4}{z^2-1.3z+0.4} \quad ...(11.25)$$

Replacing z by $e^{j\omega T}$ we obtain

$$G(e^{j\omega T}) = \frac{2e^{j\omega T}+0.4}{e^{j2\omega T}-1.3e^{j\omega T}+0.4}$$

$$= \frac{(2\cos\omega T+0.4)+j2\sin\omega T}{(\cos 2\omega T-1.3\cos\omega T+0.4)+j(\sin 2\omega T-1.3\sin\omega T)} \quad ...(11.26)$$

Equation (11.26) can now be evaluated for different values of ωT ranging between 0 and π to obtain the frequency response, as shown in Table 11.1. These values can be utilized to get different types of frequency response plots, as discussed in Chapter 8. The computational aspects are discussed in Appendix E and a computer programme for plotting the frequency response is contained on the disk available with this book.

TABLE 11.1: Frequency Response of the Transfer Function Given by Equation (11.25)

ωT	M	ϕ	ωT	M	ϕ
0	24.0	0°	1.0	3.02	−145.75°
0.1	21.68	− 33.74°	1.5	1.54	−164.64°
0.2	17.18	− 60.74°	2.0	0.95	−175.06°
0.3	13.14	− 80.51°	2.5	0.69	−179.67°
0.4	10.15	− 96.7°	3.0	0.597	−179.81°
0.5	7.96	− 108.2°	π	0.593	−180.00°

The polar plot of the frequency response is shown in Figure 11.13.

DRILL PROBLEM 11.4

Sketch the gain-phase plot (that is the log-magnitude/phase plot) for the frequency response of the transfer function given in Equation (11.25), for ωT varying between 0 to π.

*(**Hint:** You may use the values given in Table 11.1. Note that the gain must be converted into decibels for the plot.)*

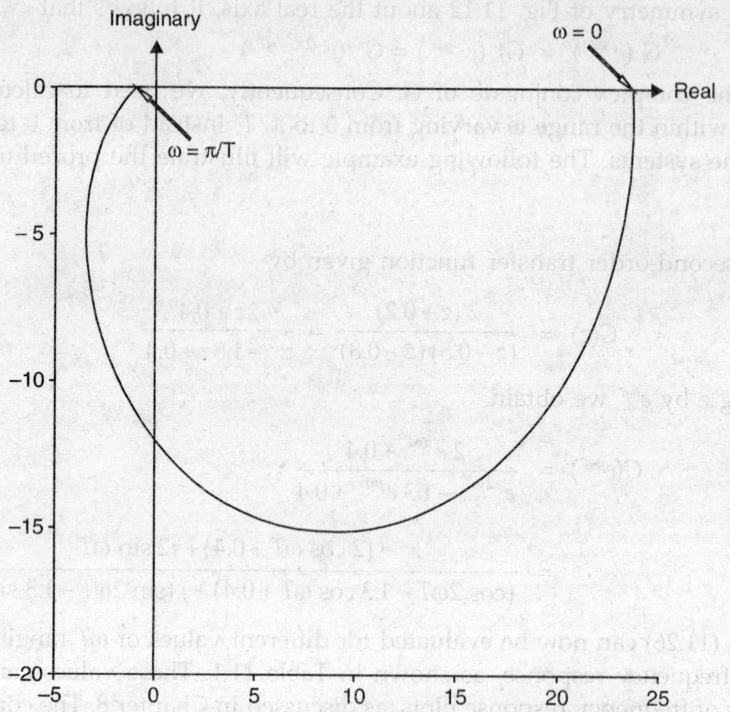

Figure 11.13. *Polar plot of frequency response for transfer function in Equation (11.25)*

11.6 BILINEAR TRANSFORMATION WITH TUSTIN'S APPROXIMATION

In Section 11.3, we had discussed a bilinear transformation which maps the interior of the unit circle of the z-plane into the left half of the w-plane. Another bilinear transformation, proposed by Tustin is given by

$$w = \frac{2}{T}\frac{z-1}{z+1}$$

$$z = \frac{\frac{2}{T}+w}{\frac{2}{T}-w} \qquad ...(11.27)$$

The main attraction of this transformation is that it can be used for analysis and design treating $G(w)$ in the same way as $G(s)$ for a continuous-time system. Consequently, one can use all the frequency response methods for compensator design discussed in Chapter 10, without any modification. Tustin had originally used this transformation for approximate design of digital filters.

To understand the implications of this transformation, let us consider the stability boundary in the w-plane by writing

$$w = \frac{2}{T}\frac{e^{sT}-1}{e^{sT}+1} = \frac{2}{T}\tan h\frac{sT}{2} \qquad ...(11.28)$$

For $s = j\omega$, we get $z = e^{j\omega T}$, which is the unit circle in the z-plane, and

$$w \triangleq jv = j\frac{2}{T}\tan\frac{\omega T}{2} \qquad ...(11.29)$$

Thus, the w-plane frequency, v, is related to the s-plane frequency ω through the relationship

$$v = \frac{2}{T}\tan\left(\frac{\omega T}{2}\right) \approx \omega \qquad ...(11.30)$$

Note that v is approximately equal to ω for small values of ωT. The approximation is valid as long as $\tan(\omega T/2) \approx \omega T/2$. The error in this approximation is less than 4 per cent for

$$\frac{\omega T}{2} \le \frac{\pi}{10}, \quad \text{or} \quad \omega \le \frac{2\pi}{10T} = \frac{\omega_s}{10} \qquad ...(11.31)$$

Thus, the sampling frequency should be chosen so that Equation (11.31) is satisfied over the bandwidth of the system. In such cases, it is possible to use the w-plane frequency response of a discrete-time system exactly like the s-plane frequency of a continuous-time system, both for the study of system stability as well as for the design of compensators based on frequency response.

Furthermore, we see that while we go around the unit circle in the z-plane, the frequency v in the w-plane moves from 0 to ∞, making it convenient to apply the w-plane frequency response for the Nyquist criterion of stability. The choice of the scale factor $2/T$ ensures that the error constants remain unchanged from the z-plane to the s-plane to the w-plane, since

$$\lim_{z\to 1} G(z) = \lim_{s\to 0} G(s) = \lim_{w\to 0} G(w) \qquad ...(11.32)$$

This will be studied in detail in the next section.

11.7 STEADY-STATE ACCURACY

In Chapter 5, we had studied the steady-state error of continuous-time systems in response to standard test inputs and defined the error consants, K_p, K_v, and K_a. The expressions for these constants were derived in Sections 5.6 and 5.7. We shall now study these for discrete-time systems. Consider the block diagram shown in Fig. 11.14.

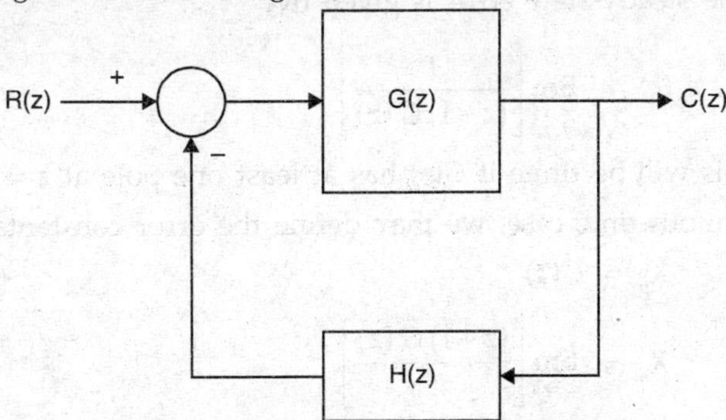

Figure 11.14. *A general closed-loop system*

Define the error of the closed-loop system as

$$e(kT) = r(kT) - c(kT) \qquad ...(11.33)$$

Taking z-transforms of both sides of Equation (11.33), we get

$$E(z) = R(z) - C(z) = R(z)\left[1 - \frac{G(z)}{1+G(z)H(z)}\right] \qquad ...(11.34)$$

$$= \frac{1+G(z)H(z)-G(z)}{1+G(z)H(z)} R(z)$$

Applying the final-value theorem (assuming the system is stable), we get

$$e_{ss}(k) = \lim_{n\to\infty} e(n) = \lim_{z\to 1}\left[(z-1)E(z)\right]$$

$$= \lim_{n\to 1}\left[\frac{1+G(z)H(z)-G(z)}{1+G(z)H(z)}.(z-1)R(z)\right] \quad ...(11.35)$$

The above equation can be utilized for evaluating the steady-state error for the general case. We shall now consider the special case of a *unity-feedback* system. In this case, we have

$$e_{ss} = \lim_{z\to 1}\left[\frac{(z-1)R(z)}{1+G(z)}\right] \quad ...(11.36)$$

For a *unit step* input,

$$R(z) = \frac{z}{z-1} \quad ...(11.37)$$

Hence, the steady-state error is given by

$$e_{ss}(k) = \lim_{z\to 1}\left[\frac{z}{1+G(z)}\right] = \frac{1}{1+G(1)} \quad ...(11.38)$$

Thus, the steady-state error to a unit step input will be zero if $G(z)$ has at least one pole at $z = 1$. Note that this corresponds to a continuous-time system having at least one pole at the origin of the s-plane.

For a *unit ramp* input

$$R(z) = \frac{Tz}{(z-1)^2} \quad ...(11.39)$$

Consequently, the steady-state error is given by

$$e_{ss}(k) = \lim_{z\to 1}\left[\frac{T}{(z-1)G(z)}\right] \quad ...(11.40)$$

It is seen that this will be finite if $G(z)$ has at least one pole at $z = 1$.

As for the continuous-time case, we may define the error constants. These are

$$K_p = G(1) \quad ...(11.41)$$

and

$$K_v = \lim_{z\to 1}\left[\frac{(z-1)G(z)}{T}\right] \quad ...(11.42)$$

EXAMPLE 11.9

Consider a unity feedback discrete-time system, with sampling interval $T = 0.1$, and forward path transfer function given by

$$G(z) = \frac{0.6(z+0.4)}{(z-1)(z-0.4)} \quad ...(11.43)$$

Since $G(z)$ has a pole at $z = 1$, the system will have $K_p = \infty$, and the steady-state error to a step input will be zero. Also, we have

$$K_v = \lim_{z\to 1}\left[\frac{(z-1)\,G(z)}{T}\right] = 14 \quad ...(11.44)$$

so that the steady-state error to a unit ramp input will be 1/14.

Let us now consider transformation to the w-plane. Since $T = 0.1$, we get

$$z = \frac{\frac{2}{T}+w}{\frac{2}{T}-w} = \frac{20+w}{20-w} \quad ...(11.45)$$

Substituting this in the expression for $G(z)$, given in Equation (11.43), we obtain, after some simplification,

$$G(w) = \frac{0.6\,(28+0.6\,w)\,(20-w)}{2w\,(12+1.4\,w)} \quad ...(11.46)$$

As expected, $G(w)$ has a pole at the origin of the w-plane, corresponding to the pole at one in the z-plane. We can calculate K_v from $G(w)$ as

$$K_v = \lim_{w\to 0}\left[wG(w)\right] = 14 \quad ...(11.47)$$

which is identical to the value in Equation (11.44) obtained from the z-plane transfer function.

EXAMPLE 11.10

For the system shown in Fig. 11.15, determine the value of K so that the closed-loop system will be stable with gain margin equal to 6 dB. What will be the steady-state error of the system to a unit-ramp input?

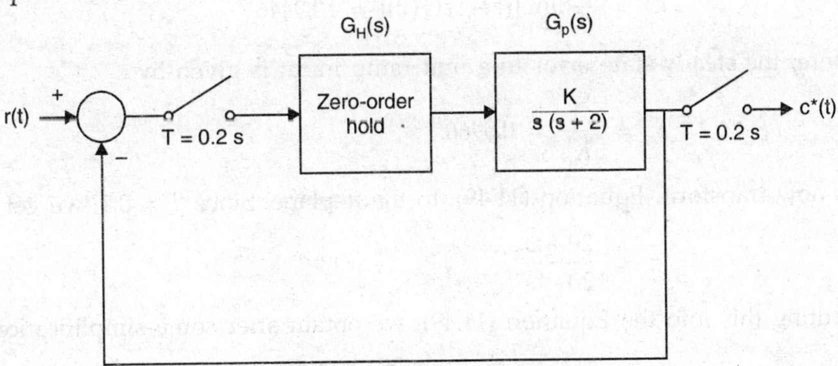

Figure 11.15. *Block diagram*

SOLUTION:

We shall first solve this problem in the z-plane. The result will then be verified by solving it in w-plane.

The transfer function of the forward path can be calculated as in Section 11.2 (see Example 11.2). This is shown below.

$$G_H(s)\,G_p(s) = \frac{1-e^{-sT}}{s}\cdot\frac{K}{s(s+2)} = \left(1-e^{-sT}\right)\frac{K}{2}\left[\frac{1}{s^2}-\frac{0.5}{s}+\frac{0.5}{s+2}\right] \quad ...(11.48)$$

and

$$G_HG_p(z) = \frac{K}{2}\left(1-z^{-1}\right)\left[\frac{zT}{(z-1)^2}-\frac{0.5z}{z-1}+\frac{0.5z}{z-e^{-2T}}\right]$$

$$= \frac{K}{2}\left[\frac{0.1}{z-1}-0.5+\frac{0.5(z-1)}{z-e^{-0.2}}\right]$$

$$= \frac{K(0.0046827\,z+0.0043808)}{(z-1)(z-0.8187308)} \quad ...(11.49)$$

The characteristic polynomial of the closed-loop system is obtained as

$$\Delta(z) = z^2 + (0.0046827\,K - 1.8187308)z + (0.8187308 + 0.0043808\,K)$$

Applying Jury's test to the case of this quadratic we obtain the following:

$$\Delta(1) = 0.0090633\,K > 0$$

$$\Delta(-1) = 3.6374616 - 0.0003019\,K > 0$$

$$0.8187308 + 0.00438081\,K < 1$$

These are all satisfied if $0 < K < 41.3781$. Hence, for gain margin of 6 dB,

$$K = 20.689.$$

For this value of K, we have

$$G(z) = G_HG_p(z) = \frac{0.098807\,z+0.0906346}{(z-1)(z-0.8187308)}$$

and the steady-state error constant

$$K_v = \frac{1}{T}\lim_{z\to 1}\left[(z-1)\,G(z)\right] = 10.3446$$

Therefore, the steady-state error to a unit ramp input is given by

$$e_{ss} = \frac{1}{K_v} = 0.09667$$

Let us now transform Equation (11.49), to the w-plane. Since $T = 0.1$, we get

$$z = \frac{20+w}{20-w}$$

Substituting this into the Equation (11.49), we obtain after some simplification,

$$G(w) = \frac{K\left(-0.000083\,w^2-0.0488041\,w+0.996684\right)}{w(w+1.9933598)} \quad ...(11.50)$$

The characteristic polynomial is now obtained as

$$\Delta(w) = (1 - 0.000083\,K)w^2 + (1.9933598 - 0.0488014\,K)w + 0.9966841\,K \quad ...(11.51)$$

In order to ensure that all the roots of this polynomial have negative real parts, it is necessary and sufficient that all the coefficients of $\Delta(z)$ be positive. This will be possible if $0 < K < 41.378$. Hence, for the system to be stable with gain margin equal to 6 dB, we must make $K = 20.689$. Note that the same answer was obtained from Δ (z).

Finally, the steady-state error constant

$$K_v = \lim_{w \to 0} [wG(w)] = 10.3446.$$

is found to be identical with the value calculated in the z-plane. Hence, the steady-state error to a unit ramp imput will be

$$e_{ss} = \frac{1}{K_v} = 0.09667.$$

DRILL PROBLEM 11.5

For the system shown in Fig. 11.16, determine the value of K, so that the gain margin may be 6 dB. What will be the steady-state error to a unit ramp input for this K? (Solve this problem both in the z-plane and the w-plane.)

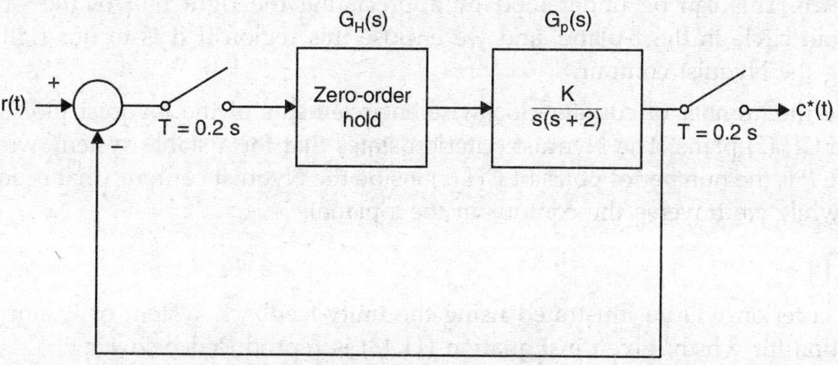

Figure 11.16. *Block diagram*

Ans. $K = 10.712$, $e_{ss} = 0.1867$

11.8 THE NYQUIST CRITERION

It was seen in Chapter 9 that one can use the frequency response of the open-loop to determine stability of a closed-loop continuous-time system using the Nyquist criterion. The main objective was to determine the number of zeros of $1 + GH(s)$ in the right half of the s-plane from the number of encirclements of the point $-1 + j0$ in the GH-plane.

To apply the Nyquist criterion to determine stability of discrete-time systems, we need to determine the number of roots of $1 + GH(z)$ in the region outside the unit circle of the z-plane. The Nyquist contour in this case is the unit circle of the z-plane, with detours around poles on the path, traversed in the counter-clockwise direction, as shown in Fig. 11.17.

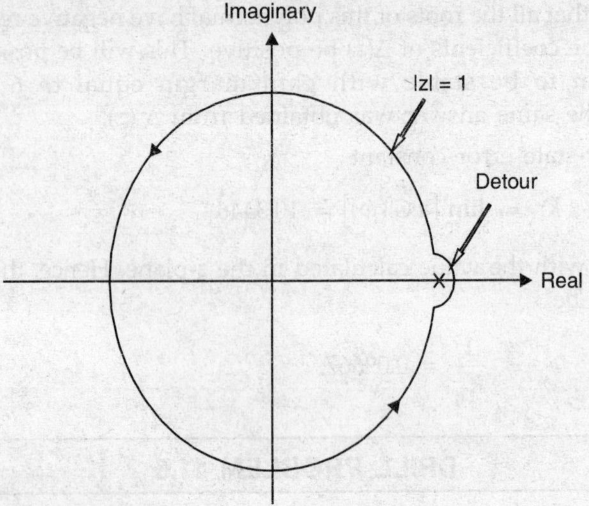

Figure 11.17. *The Nyquist contour in the z-plane*

Note that in the *s*-plane the Nyquist contour was taken in the clockwise direction (see Section 9.3). This can be understood by appreciating the right half of the *s*-plane maps outside the unit circle in the *z*-plane, and we enclose this region if it is to our right while we traverse along the Nyquist contour.

Let N be the number of counter-clockwise encirclements of the Nyquist plot of the point $-1 + j0$ in the $GH(z)$-plane. The Nyquist criterion states that for a stable system, we must have $N = -P$, where P is the number of poles of $GH(z)$ inside the Nyquist contour (that is, in the region to our right while we traverse the contour in the *z*-plane).

Example 11.11

The Nyquist criterion will be illustrated using the unity-feedback system of Example 11.9, the transfer function for which, given in Equation (11.43) is reproduced below.

$$G(z) = \frac{0.6\,(z+0.4)}{(z-1)\,(z-0.4)}$$

Since the transfer function has a pole at $z = 1$, the Nyquist contour will be as shown in Fig. 11.17, with a detour around that point. On this detour

$$z = 1 + \in e^{j\theta},\ \in \ll 1 \qquad ...(11.52)$$

and

$$G(z)\,|_{z=1+\in e^{j\theta}} \approx \frac{(0.6)\,(1.4)}{\in e^{j\theta}\,(0.6)} = \frac{1.4}{\in} e^{-j\theta} \qquad ...(11.53)$$

This leads to an arc of infinite radius on the Nyquist plot. For z on the unit circle, we have

$$G(z)\,|_{z=e^{j\omega T}} = \frac{0.6\left(e^{j\omega t}+0.4\right)}{\left(e^{j\omega T}-1\right)\left(e^{j\omega T}-0.4\right)} \qquad ...(11.54)$$

In Equation (11.54), $0 < \omega T < 2\pi$. Since $G(e^{j\omega T})$ is the complex conjugate of $G(e^{-j\omega T})$, it is necessary to calculate the frequency response only for $0 < \omega T < \pi$. Some values of the frequency response are shown in Table 11.2.

TABLE 11.2: Frequency Response of the Transfer Function Given by Equation (11.43)

ωT	M	ϕ	ωT	M	ϕ
0.1	13.91	– 98.3°	1.2	0.69	– 166.0°
0.2	6.83	– 106.4°	1.4	0.52	– 173.3°
0.3	4.43	– 114.3°	1.6	0.41	– 179.3°
0.4	3.20	– 121.8°	1.8	0.33	– 184.4°
0.5	2.45	– 128.9°	2.0	0.26	– 188.2°
0.6	1.94	– 135.5°	2.2	0.22	– 190.6°
0.7	1.57	– 141.6°	2.4	0.18	– 191.5°
0.8	1.30	– 147.3°	2.6	0.16	– 190.6°
0.9	1.09	– 152.5°	2.8	0.14	– 187.9°
1.0	0.93	– 157.4	π	0.13	– 180.0°

The polar plot of the frequency response is the map of the upper half of the unit circle of the z-plane. The lower half maps mirror image of this response about the real axis. The plot is completed from the map of the detour, which is a semicircle of infinite radius, obtained from Equation (11.53). Figure 11.18 shows the Nyquist plot for this system, with the Nyquist contour as in Fig. 11.17. From the Nyquist plot, it is evident that the system is stable. It can, however, be forced into instability if the gain is increased by a factor of 2.5 so that the point $-1 + j0$ is enclosed within the Nyquist plot.

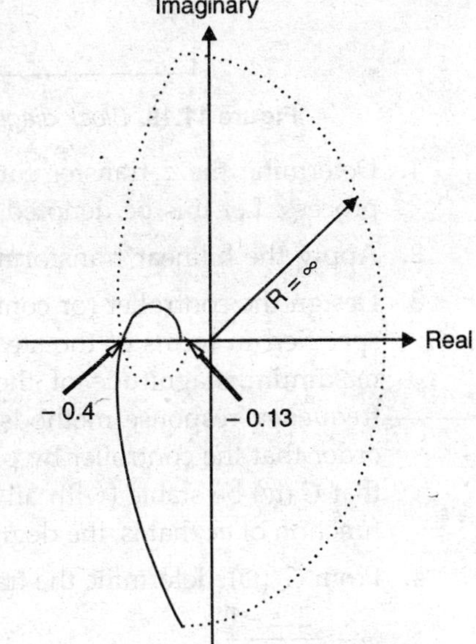

Figure 11.18. *Nyquist plot*

The concepts of gain margin and phase margin, described in Chapter 9, can be utilized with discrete-time systems without any change. Moreover, they may be used either in the z-plane, or in the w-plane (using the bilinear transformation due to Tustin). For instance, in the problem of Example 11.11, the gain margin is 2.5 and the phase margin is about 25°.

Alternatively, one may transform $G(s)$ to the w-plane using the bilinear transformation due to Tustin. It was seen in Example 11.9 that this is obtained as

$$G(w) = \frac{0.6\,(28+0.6\,w)\,(20-w)}{2\,w\,(12+1.4\,w)} = \frac{-0.36\,w^2 - 9.6\,w + 336}{2.8\,w^2 + 24\,w}$$

The Nyquist plot is obtained by calculating the frequency response $G(jv)$ for v varying between 0 to ∞ and completing the plot by taking a detour around the plot at the origin of the w-plane, as was done is Chapter 9 in similar cases with continuous-time systems. This Nyquist plot is identical to that in Fig. 11.18, and leads to the same conclusion about stability as well as the gain margin and the phase margin.

11.9 DESIGN OF DIGITAL CONTROLLERS

In a large number of practical situations the sampled and quantized error must be digitally processed before it is reconstructed and applied to the process. The block diagram of such a system is shown in Fig. 11.19.

The transfer function $G_c(z)$ of the digital controller is selected in such a way that the performance of the system is improved and satisfaction is specified. As in the case of continuous-time systems, this can be done in several ways.

For example, we apply the frequency response method, using Bode plots or Nichols charts. This is done most conveniently by first carrying out the bilinear transformation from the z-plane to the w-plane, denoted by Equation (11.27), although it is possible to design directly in the z-plane. The procedure is outlined below.

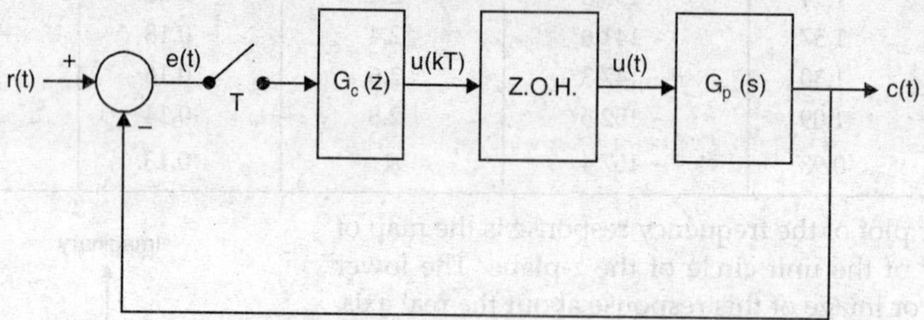

Figure 11.19. *Block diagram of system with digitally processed error signal*

1. Determine the z transfer function of the combination of the zero-order hold and the process. Let this be denoted by $G_HG_p\,(z)$.
2. Apply the bilinear transformation $z = (2 + wT)/(2 - wT)$ to obtain $G_HG_p\,(w)$.
3. Design the controller (or compensator) $G_c\,(w)$ that will satisfy the desired characteristics specified in terms of the steady-state error coefficients, phase margin, bandwidth, and maximum magnitude of the closed-loop frequency response M_m, ω_m, etc., using the frequency response methods discussed in Chapter 10 for continuous-time systems. In order that the controller by physically realizable and stable, it is necessary and sufficient that $G_c(w)$ be stable (with all poles in the left half of the w-plane) and a proper rational function of w; that is, the degree of its numerator must not exceed that of its denominator.
4. From $G_c\,(w)$, determine the transfer function of the digital controller $G_c\,(z)$ by substituting $w = \dfrac{2}{T}\dfrac{z-1}{z+1}$.

It is also possible to carry out the entire design in the z-plane, using the frequency response method. This can be done in the same manner, but it is necessary to develop design equations for determining the z-transfer function of lead, lag or lag-lead compensators that will provide a given gain and phase shift at a specified frequency. Although this can be done, it is usually better to do the computations in the w-plane instead of the z-plane. Since the entire left half of the s-plane maps inside the unit circle of the z-plane, we may run into problems of numerical instability due to the crowding of poles and zeros in a small region. This difficulty is avoided by the transformation to the w-plane.

Instead of using frequency-response methods, it is also possible to design the compensator using the root locus or pole-placement methods discussed in Chapter 10. These can be designed directly in the z-plane, but it is necessary to include many significant digits in the computation to avoid numerical problems caused by the crowding of the poles and zeros inside the small region of the unit circle. A PID controller can also be designed. Some examples to illustrate the various methods will be discussed in the sequel.

We shall now discuss the implementation of the digital controller. Considering the block diagram of Fig. 11.19, suppose that we find that a suitable compensator has the following transfer function:

$$G_c(z) = \frac{U(z)}{E(z)} = \frac{b_0 z^2 + b_1 z + b_2}{z^2 + a_1 z + a_2} = \frac{b_0 + b_1 z^{-1} + b_2 z^{-2}}{1 + a_1 z^{-1} + a_2 z^{-2}} \quad \text{...(11.55)}$$

We may write Equation (11.55) in the form of the following difference equation:

$$u_k = b_0 e_k + b_1 e_{k-1} + b_1 e_{k-2} - a_1 u_{k-1} - a_2 u_{k-2} \quad \text{...(11.56)}$$

where $u_i \triangleq u(iT)$ and $e_i \triangleq e(iT)$ for any integer i.

Equation (11.56) tells us that the input $u(kT)$ at the kth sampling instant is obtained as a linear combination of the present and two past values of the error, $e(kT)$, $e(kT - T)$ and $e(kT - 2T)$, as well as two previous value of the input, $u(kT - T)$ and $u(kT - 2T)$. This simple numerical alogrithm can be carried out on any microprocessor.

An important feature that should be noted is the ease with which the parameters of the controller can be changed as it merely implies altering the values of the coefficients a_i and b_i in the program. On the other hand, the controller in a continuous-time system cannot be altered so easily, because the values of physical components must now be changed. In practice, however, we must take into consideration the limitations of a microprocessor caused by finite word length and quantization errors.

EXAMPLE 11.12

As an example, consider the block diagram shown in Fig. 11.20. It is required to design a compensator so that the steady-state error to a unit ramp input does not exceed 0.05, and the phase margin is 45° with the gain cross-over frequency $\omega_c = 2$ rad/s.

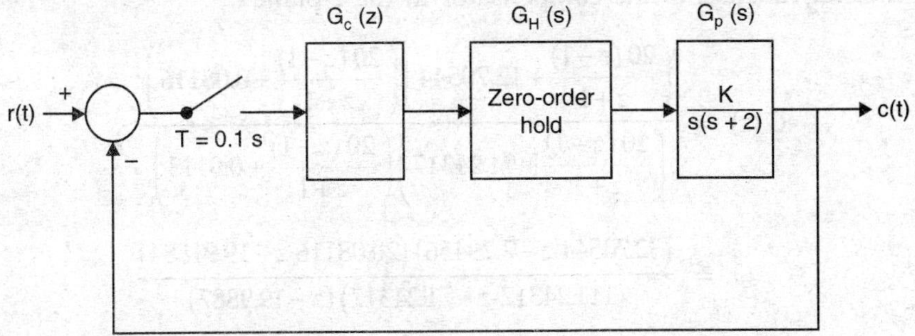

Figure 11.20. *Block diagram for compensator design*

It will be seen that the transfer function of the forward path is similar to that in the block diagram in Example 11.9. Consequently, we obtain

$$G_H G_p(z) = \frac{K\,(0.0046872\,z + 0.0043808)}{(z-1)\,(z-0.8187308)} \qquad \text{...(11.57)}$$

Using the bilinear transformation

$$z = \frac{2/T + w}{2/T - w} = \frac{20+w}{20-w} \qquad \text{...(11.58)}$$

the w-plane transfer function is found to be (see Example 11.9)

$$G_h G_p(w) = \frac{K\left(-0.000083\,w^2 - 0.0484041\,w + 0.996684\right)}{w\,(w+1.9933598)} \qquad \text{...(11.59)}$$

The specification for steady-state accuracy will be satisfied if we make $K_v = 20$. Since we have a pole at the origin of the w-plane (corresponding to a pole at -1 in the z-plane), we must make

$$K \geq \frac{20 \times 1.9933598}{0.9966799} = 39.9998235 \qquad \text{...(11.60)}$$

Thus, making $K = 40$ will be satisfactory. Consequently, we have

$$G_h G_p(w) = \frac{-0.00332\,w^2 - 1.9521656\,w + 39.867192}{w\,(w+1.9933598)} \qquad \text{...(11.61)}$$

We now calculate the frequency response of this transfer function for $w = jv = j2$. It is found that at this frequency, the gain is 17.02 dB and the phase shift is $-140.69°$. Therefore, we need a lag-lead compensator that will provide a lead of 5.69° and gain of -17.02 dB for this value of v. Following the procedure discussed in Chapter 10 (Section 10.2), we get

$$G_c(w) = \frac{(w+12.70544)\,(w+0.08116)}{(w+91.24317)\,(w+0.0013)} \qquad \text{...(11.62)}$$

Finally, we transform back to the z-plane using the transformation

$$w = \frac{2}{T}\,\frac{z-1}{z+1} = \frac{20\,(z-1)}{z+1} \qquad \text{...(11.63)}$$

to obtain the transfer function of the compensator in the z-plane

$$G_c(z) = \frac{\left(\dfrac{20\,(z-1)}{z+1} + 12.70544\right)\left(\dfrac{20\,(z-1)}{z+1} + 0.08116\right)}{\left(\dfrac{20\,(z-1)}{z+1} + 91.24317\right)\left(\dfrac{20\,(z-1)}{z+1} + 0.0113\right)}$$

$$= \frac{(32.70544\,z - 7.29456)\,(20.08116\,z - 19.91884)}{(111.24317\,z + 71.24317)\,(z - 19.9887)}$$

$$= \frac{0.295039\,(z-0.2230381)\,(z-0.9919168)}{(z+0.6404274)\,(z-0.99988706)}$$

It should be noted that although the poles and zeros of the transfer function of the compensator appear well separated in the w-plane, they are rather close together for the transfer function in the z-plane. In particular, one pole and one zero for the latter are both very near the point $1 + j0$, since they are both close to the origin in the w-plane. This is the reason for using seven significant digits in the transfer function in the z-plane.

DRILL PROBLEM 11.6

Design a suitable compensator for the problem in Example 11.12 if the gain cross-over frequency is required to be 10 rad/s, while the other specifications are unchanged.

Ans. $$G_c(z) = \frac{4.4268914\,(z - 0.6093859)}{z + 0.7292049}$$

EXAMPLE 11.13

As another example, consider the block diagram of an attitude control system of a satellite shown in Fig. 11.21. The transfer function $G_p(z)$ includes a zero-order hold. It is desired to design a digital compensator for the forward path so that the maximum magnitude of the frequency response of the closed-loop system may be 3 dB, with $\omega_m \simeq 0.4$.

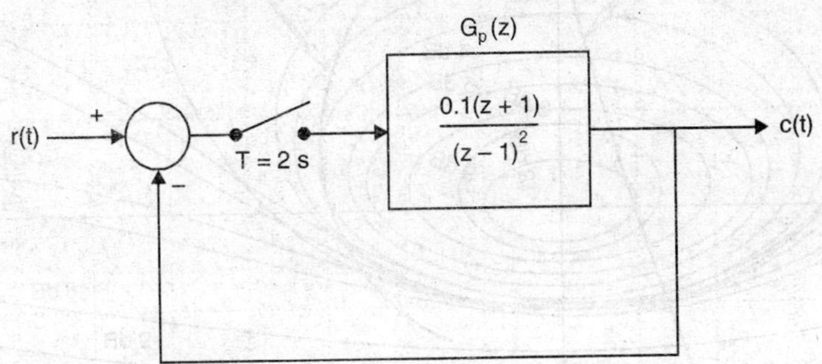

Figure 11.21. *Block diagram of an uncompensated attitude control system for a satellite*

Using the bilinear transformation

$$z = \frac{2 + wT}{2 - wT}$$

we get

$$G_p(w) = \frac{-0.05\,w + 0.05}{w^2} \qquad ...(11.64)$$

The frequency response of this transfer function is easily calculated by replacing w by jv and is given in Table 11.3. The frequency response is plotted on the Nichols chart in Fig. 11.22. Clearly, the system is unstable.

From the plot it is estimated that providing a phase lead of 66.8° and gain equal to 9.46 dB at $v = 0.4$, will shape the frequency response curve so that it will be tangential to the 3 dB M-curve. The transfer function of the resulting lead compensator is calculated using the method described in Chapter 10, and is given by

$$G_c(w) = \frac{44.884361\,(w + 0.1426267)}{w + +6.401741} \qquad ...(11.65)$$

TABLE 11.3: Frequency Response of the Transfer Function Given by Equation (11.64)

v	*Gain* (dB)	*Phase*	v	*Gain* (dB)	*Phase*
0.1	14.02	−185.7°	0.5	−13.01	−206.6°
0.2	2.11	−191.3°	0.7	−18.09	−215.0°
0.3	− 4.73	−196.7°	1.0	−23.01	−225.0°
0.4	− 9.46	−201.8°	2.0	−31.07	−243.5°

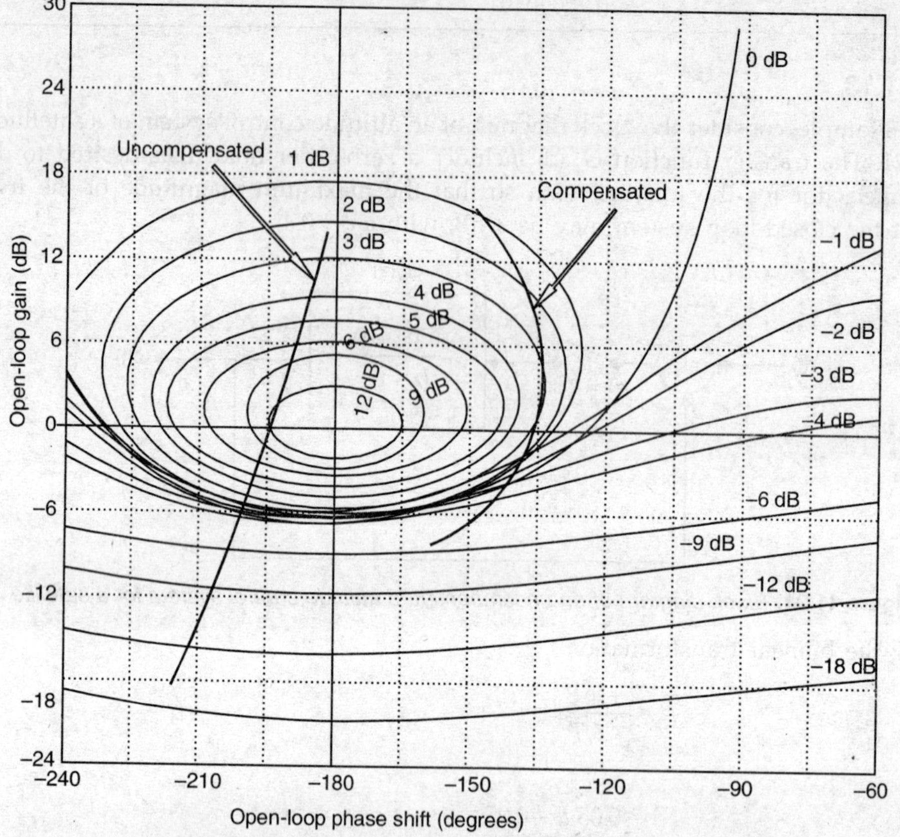

Figure 11.22. *Frequency response plots on Nichols chart*

The frequency response for the compensated system is shown in Table 11.4, and is also plotted in Fig. 11.22. This plot shows that the compensated system will be stable with $M_m = 3$ dB, and $v_m \simeq 0.4$ rad/s.

We shall now reverse the transformation to obtain the transfer function of the digital compensator. Recalling that

$$w = \frac{2}{T}\,\frac{z-1}{z+1}$$

we get

$$G_c(z) = \frac{6.92892\,(z-0.75033)}{z+0.729793} \qquad ...(11.66)$$

TABLE 11.4: Frequency Response of the Transfer Function of the Compensated System

v	Gain (dB)	Phase	v	Gain (dB)	Phase
0.1	15.78	−151.6°	0.6	− 3.13	−139.7°
0.2	6.83	−138.6°	0.7	− 4.15	−142.7°
0.3	2.60	−134.7°	0.8	− 4.95	−145.9°
0.4	0.0	−135.0°	1.0	− 6.11	−152.0°
0.5	−1.8	−137.0°	2.0	− 8.52	−174.9°

The transfer function of the compensated closed-loop system is evaluated as

$$\frac{C(z)}{R(z)} = \frac{0.6928292\,(z+1)\,(z-0.75035)}{(z+0.5704)\,(z-0.5738+j0.1967)\,(z-0.5738-j0.1967)} \qquad ...(11.67)$$

As expected, the poles of this transfer function are well within the unit circle of the z-plane. Consequently, the system will be stable.

DRILL PROBLEM 11.7

Repeat the problem of Example 11.12 if the transfer function of the forward path before compensation (including the zero-order hold) is given by

$$G_p(z) = \frac{0.5\,(z+0.7)}{(z-1)\,(z-0.8)}$$

Ans. $G_c(z) = \dfrac{0.1655076\,(z-0.8510634)}{z-0.9742557}$

11.10 DIGITAL PID CONTROLLERS

PID controllers for continuous-time systems were discussed in Chapter 10 (Section 10.8). As stated there, they are particularly useful if the transfer function of the system is not known. It was noted that the proportional part of the compensator improves the sensitivity to parameter variations, the integral part improves the steady-state accuracy and the derivative part improves the stability of the system by increasing the damping. The same basic of concepts is also applicable to digital PID controllers. It was pointed out in Section 10.8 that, in practice, the ideal differentiator had to be replaced by a transfer function that has a zero at the origin and pole on the negative real axis placed so that its frequency response approximates that of the differentiator in the frequency range of interest. For a discrete-time system, the differentiator may by approximated by the $\dfrac{1-z^{-1}}{T}$, or $\dfrac{z-1}{Tz}$. Since integration is the inverse of differentiation, the integrator may be represented by $\frac{Tz}{z-1}$. Consequently, the digital PID controller will have the following transfer function

$$G_c(z) = K_p + K_D \frac{z-1}{Tz} + K_I \frac{Tz}{z-1} \qquad ...(11.68)$$

Note that the differentiator adds a pole at the origin of the *z*-plane and a zero at $z = 1$. The pole at the origin may sometimes cause a problem due to the phase shift introduced by it. As in the case of continuous-time systems, it is possible to replace this by a pole on the positive real axis, chosen in such a way that the frequency response is unchanged within the bandwidth of interest.

It is possible to use the Ziegler-Nichols tuning alogrithm (described in Section 10.8) for digital PID controllers as well. For this purpose, the transfer function of the compensator is written as

$$G_c(z) = K_P \left[1 + \frac{Tz}{T_I(z-1)} + \frac{T_D(z-1)}{Tz}\right] \qquad ...(11.69)$$

One may again use either the step response method or the stability limit method to estimate the values of the various constants in Equation (11.69), according to the two tables given in Section 10.8. The values given there lead to a system with damping ratio of about 0.2, if the sampling frequency is sufficiently high.

In practice, the Ziegler-Nichols tuning method provides a good set of initial values for the various terms in the compensators, which can often be improved by on-line tuning. The main advantage is that by using this method one can completely bypass the need to obtain an exact mathematical model for the process to be controlled.

Alternatively, the digital PID controller can be designed in the *w*-plane with Tustin's transformation.

11.11 DESIGN OF A DIGITAL CONTROLLER FOR DEAD-BEAT RESPONSE

It is also possible to design a digital controller so that the response to a unit step is of the dead-beat type; that is, the output reaches the value of 1 in the minimum number of sampling intervals and then stays there. This is achieved by selecting $G_c(z)$ so that

$$G_c(z)\, G_H G_p(z) = \frac{1}{z^m - 1} \qquad ...(11.70)$$

where *m* is the difference between the degrees of the numerator and the denominator of $G_H G_p(z)$. With unity feedback the transfer function of the closed-loop system is then obtained as z^{-m}, which implies that the response will be the unit step delayed by *m* sampling intervals. This will be evident from Example 11.14.

It should be emphasized that the response of the system reaches the steady-state value exactly after *m* sampling intervals. This is an important property of discrete-time systems and should be contrasted with the fact that theoretically it requires infinite amount of time for a stable continuous-time system to reach the steady-state in response to a step input.

EXAMPLE 11.14

Consider the system shown in Fig. 11.23, where $G_H(z)$ is a zero-order hold and

$$G_p(s) = \frac{10}{s(s+2)} \quad \text{...(11.71)}$$

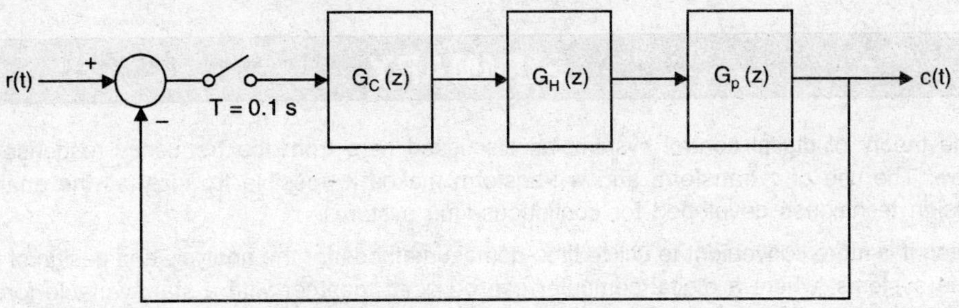

Figure 11.23. *A closed-loop system*

As in Example 11.2, the transfer function of the forward path is

$$G_H G_p(z) = \frac{0.046827\,z + 0.043808}{(z-1)\left(z-e^{-0.2}\right)} \quad \text{...(11.72)}$$

$$= \frac{0.46827\,(z+0.93553)}{(z-1)\left(z-e^{-0.2}\right)}$$

Since here $m = 1$, we shall make

$$G_c(z) = \frac{z-e^{-0.2}}{0.046827\,z+0.043808} \quad \text{...(11.73)}$$

so that $$G_c(z)\, G_H G_p(z) = \frac{1}{z-1} \quad \text{...(11.74)}$$

and $$\frac{C(z)}{R(z)} = \frac{1}{z} \quad \text{...(11.75)}$$

Consequently, the output of the system is given by

$$C(z) = \frac{1}{z-1} = z^{-1} + z^{-2} + z^{-3} + \ldots \quad \text{...(11.76)}$$

It should be noted that although the response is exactly one at the sampling instants, there is no guarantee that $c(t)$ has no oscillation between the sampling instant. In this example, the sampling period of 0.1 second is much smaller than the system time constant, and hence we may expect that $c(nT)$ gives a reasonable accurate description of $c(t)$. One may use the method of modified z-transform to obtain the output between sampling instant. The interested reader can find more details in references [1] and [2].

Another difficulty with this design is caused by the requirement that the compensator $G_c(z)$ cancels the poles and zeros of $G_H G_p(z)$ in order that the transfer function of the forward path be given by Equation (11.74). This may not always be possible. In particular, if $G_H G_p(z)$ is non-minimum phase (with one or more zeros outside the unit circle of the z-plane), this cancellation would require an unstable compensator, with a serious problem in case of imperfect cancellation due to to tolerance.

A more practical approach to the design of a dead-beat control system will be discussed in Chapter 12.

SUMMARY

- The theory of digital control system, as discussed here from the frequency response point of view. The use of z transform and w transform makes it possible to apply all the analysis and design techniques developed for continuous-time systems.
- Often it is more convenient to utilize time-domain methods for the analysis and design of discrete-time systems, where a digital computer can be used together with a state-variable formulation. This approach will be described in Chapter 12.
- In this chapter, we have only been able to discuss some of the basic concepts of digital control. Many details, especially those of interfacing between the analog and digital components, and the effects of finite word length in digital devices, have been left out. The interested reader is referred to the books listed in the next section.

Problems

1. An error-sampled control system can be represented by the block diagram shown in Fig. 11.5, with T = 0.2 second and

$$G(s) = \frac{K}{s(s+1)}$$

   (*a*) Determine the maximum value of K for which the system will be stable.

   (*b*) Determine the response of this system to a unit step with gain margin set to 2.

2. Design a digital compensator for the system in the previous problem that will give a deadbeat response to a unit step (with K set for gain margin of 2).
3. The block diagram of the control system for a tracking antenna is shown in Fig. P11.3.

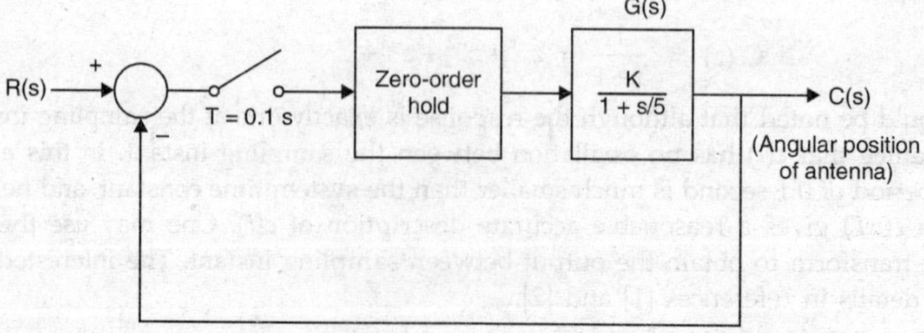

Figure P11.3. *Block diagram of the control system for a tracking antenna*

(*a*) Design a suitable digital compensator so that for $K = 10$, the phase margin is not less than 45°.

(*b*) Determine the response of the compensated system of a unit step input.

4. The block diagram for a digital cruise control system for an automobile is shown in Fig. P11.4.

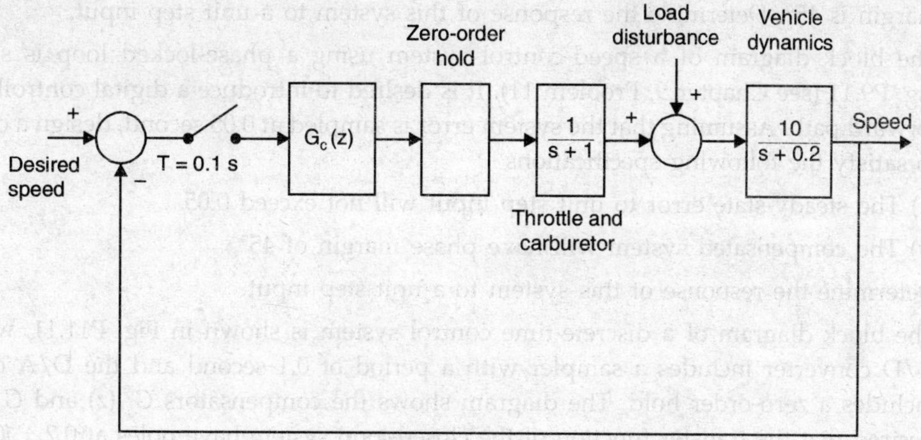

Figure P11.4. *A digital cruise control system*

(*a*) For $G_c(z) = K$, determine the range of values of K for which the system will be stable.

(*b*) Design a suitable digital compensator so that the steady-state error to a unit step input does not exceed 0.02 and the phase margin is 45°.

5. For Problem 4, design a digital compensator to obtain dead-beat response to a step load disturbance.

6. Skecth the root locus for the system described in Problem 4, where $G_c(z) = K$ varies linearly from zero to infinity. Use both the z-plane and the w-plane to suggest a suitable value for K.

7. The procedure for designing compensators that will place all the poles of the transfer function of the closed-loop system at specified locations, provided that there are no common factors between the numerator and the denominator of the forward path, as discussed in Section 10.7, can also be applied to discrete-time systems. The only important difference is that the z transfer function is used and the poles should lie inside the unit circle for a stable system.

Use this procedure to design compensator with poles at 0.1 in the z-plane if the forward path transfer function is

$$G_p(z) = \frac{12z(z-0.6)}{(z-1)(z-0.95)(z-0.9)}$$

and it is desired that the poles of the transfer function of the closed-loop system be placed at 0.3, and $0.2 \pm j0.2$.

8. The block diagram of the gauge control system for a hot-strip finishing mill is shown in Fig. P9.9 (see Chapter 9, Problem 9). It is desired to introduce a digital controller in the forward path with the system error sampled at 0.1 second. Design a controller that will provide deadbeat response to a step input.
9. For the gauge control system discussed in the previous problem design a digital controller so that the steady-state error to a unit ramp input does not exceed 0.1 and the phase margin is 45°. Determine the response of this system to a unit step input.
10. The block diagram of a speed-control system using a phase-locked loop is shown in Fig. P9.11 (see Chapter 9, Problem 11). It is desired to introduce a digital controller in the forward path. Assuming that the system error is sampled at 0.05 second, design a controller to satisfy the following specifications

    (*a*) The steady-state error to unit step input will not exceed 0.05.

    (*b*) The compensated system will have phase margin of 45°.

    Determine the response of this system to a unit step input.
11. The block diagram of a discrete-time control system is shown in Fig. P11.11, where the A/D converter includes a sampler with a period of 0.1 second and the D/A converter includes a zero-order hold. The diagram shows the compensators $G_u(z)$ and $G_c(z)$. It is desired that the transfer function of the closed-loop system have poles at $0.2 \pm j0.2$ in the z-plane and the steady-state error to a step input be zero. Design suitable compensators, with poles at 0.15, if the transfer function $G_p(s)$, which represents a dc servomotor, is given by

$$G_p(s) = \frac{4}{s(s+1)}$$

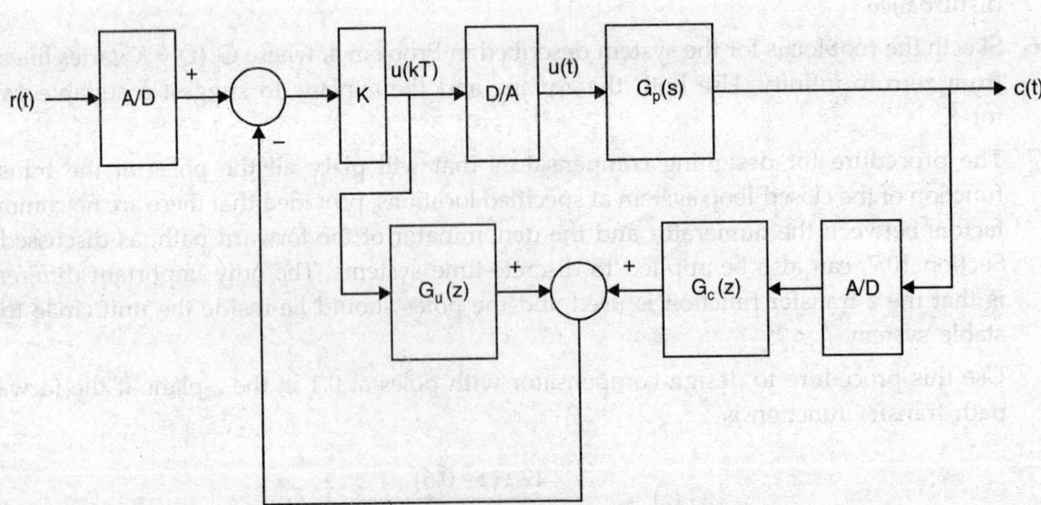

Figure P11.11. *Discrete-time system with compensators for pole placement*

12. Consider the block diagram of a digital cruise control system shown in Fig. P11.4. It is desired that the poles of the transfer function of the closed-loop system be placed at $0.2 \pm j0.2$ and the steady-state error to a step input be zero. Design suitable compensators using the configuration shown in Fig. P11.11, placing the poles of the compensators at 0.15. Determine the response of the compensated system to a unit step input.

13. The transfer function of the forward path of a discrete-time system is given by

$$G(z) = \frac{Kz(z-0.6)}{(z-1)(z-0.95)(z-0.9)}$$

It is to be used with unity feedback. What value of K will make the system stable with phase margin equal to 45° if the sampling period is 0.1 second?

14. For the system described in Problem 13, design a digital compensator to satisfy the following specifications.

(*a*) The steady-state error to a unit ramp input should not exceed 0.1.

(*b*) The maximum magnitude of the frequency response of the closed-loop system must not exceed 3 dB.

Determine the response of the compensated system to a unit step input.

15. The transfer function of the forward path of a unity-feedback control system is given by

$$G(z) = \frac{K(z+0.5)}{(z-1)(z-0.4)}$$

(*a*) Skecth the root locus of the system as K varies from zero to infinity.

(*b*) Design suitable compensators so that the poles of the transfer function of the compensated system are located at $0.3 \pm j0.3$. The poles of the compensators are to be located at – 0.1.

(*c*) Determine the steady-state error of the compensated system to a unit ramp input if the sampling interval is 0.2 second.

16. The transfer function of the forward path of a unity-feedback discrete-time system, with sampling interval of 0.5 s, is given by

$$G(z) = \frac{K(z+0.1)}{(z-1)(z-0.8)}$$

(*a*) Determine the value of K so that the closed-loop system will be stable with gain margin of 6 dB.

(*b*) What is the phase margin of the system for this value of K?

(*c*) Design a suitable compensator so that the phase-margin of the system is 55°.

17. Show that the steady-state error of a closed-loop discrete-time system to a step input will be zero if the overall transfer function $T(z)$ has the value of 1 for $z = 1$. Also, if the above condition is met, the steady-state error to a unit ramp input is

$$e_{ss} = T\left[\sum_{i=1}^{m}\frac{1}{z_i - 1} - \sum_{i=1}^{n}\frac{1}{p_i - 1}\right]$$

where T is the sampling interval, z_i, $i = 1, ..., m$, are the zeros of $T(z)$ and p_i, $i = 1, ..., n$, are the poles of $T(z)$. (**Hint:** See Section 5.7.)

References

1. Aström, K.J., and B. Wittenmark, *Computer Controlled System*, Prentice-Hall, Englewood Cliffs, New Jersey, 1984.
2. Gene, F., Franklin, J., David Powell and Michael L. Workman, *Digital Control of Dynamic Systems*, Addison-Wesley Publishing Company, Reading, Mass., Second Edition, 1990.
3. Isermann, R., *Digital Control Systems*, Springer-Verlag, Berlin, 1981.
4. Kuo, Benjamin C., *Digital Control Systems*, Holt, Rinehart and Winston, Inc., New York, 1990.
5. Phillips, Charles L., and H. Troy Nagel, *Digital Control System Analysis and Design*, Prentice-Hall, Englewood Cliffs, New Jersey, Second Edition, 1989.

CHAPTER 12

Design with State-Space Representation

12.1 INTRODUCTION

The transfer function representation of physical systems has led to the development of the root locus and other s-plane methods for analysis and design. Moreover, it relates directly to the frequency response approach for the analysis and design of closed-loop systems. Both these approaches, however, are limited to linear time-invariant systems.

On the other hand, time-domain methods, based on the state-space representation, are more powerful. In addition to being applicable to nonlinear and time-varying systems, they are readily extended to multivariable systems, *i.e.*, systems with several inputs and outputs. Their main attraction is that they are natural and convenient for computer solutions. In particular, design and analysis of discrete-time systems can be carried out more easily with digital computers when the state-space representation is employed.

In this chapter, we shall study the application of state-space methods to the design of control systems. These are based on the key concepts of controllability and observability. To explain the idea of controllability we shall first compute the input sequence that will transfer the state of a discrete-time system from a given initial value of a specified final value.

12.2 DETERMINATION OF THE INPUT SEQUENCE FOR TRANSFERRING THE STATE OF A DISCRETE-TIME SYSTEM

If a linear time-invariant single-input discrete-time system of order n is controllable, it is then possible to determine uniquely the input sequence that will transfer the state of the system from any given initial state to any specified final state in no more than n sampling intervals. The concept of controllability will be discussed in the next section. Here, we shall see how to obtain the required sequence of inputs.

Consider a single-input system with the state transition equation

$$x(kT + T) = Fx(kT) + Gu(kT) \qquad ...(12.1)$$

where x is an n-dimensional state vector and T is the sampling interval. As discussed in Section 3.8, the matrices F and G are obtained from the state-space description of the corresponding continuous-time system, *i.e.*, if

$$\dot{x} = Ax + Bu \quad \text{...(12.2)}$$

then
$$F = e^{AT} \quad \text{...(12.3)}$$

and
$$G = \int_0^T e^{At}\, dt\, B \quad \text{...(12.4)}$$

If we assume that the initial state of the system is given as $x(0)$, then

$$x(T) = Fx(0) + Gu(0) \quad \text{...(12.5)}$$

$$\begin{aligned} x(2T) &= Fx(T) + Gu(T) \\ &= F^2x(0) + FGu(0) + Gu(T) \end{aligned} \quad \text{...(12.6)}$$

Continuing in this manner, it is easily shown that

$$\begin{aligned} x(nT) &= F^n x(0) + F^{n-1}Gu(0) + F^{n-2}Gu(T) \\ &\quad + ... + FGu(nT - 2T) + Gu(nT - T) \end{aligned} \quad \text{...(12.7)}$$

Since the final state $x(nT)$ and the initial state $x(0)$ are given, we may rearrange Equation (12.7) to obtain the following matrix equation.

$$\begin{bmatrix} F^{n-1}G & F^{n-2}G & ... & FG & G \end{bmatrix} \begin{bmatrix} u(0) \\ u(T) \\ \vdots \\ u(nT-T) \end{bmatrix} = x(nT) - F^n x(0) \quad \text{...(12.8)}$$

A solution to this equation will exist if the matrix on the left-hand side is nonsingular, and in that case we shall get a unique solution to the desired input sequence. It will be seen in the next section that this condition implies complete controllability of the system.

Example 12.1

A control system with a sampler and zero-order hold is shown in Fig. 12.1. The elements of the state vector of the continuous-time system are the output y and its derivative, that is, $x_1 = y$ and $x_2 = \dot{y}$. Determine the input sequence that will transfer the state of the system from the origin of the state space to the point $\begin{bmatrix} 1 \\ 0 \end{bmatrix}$ in two steps.

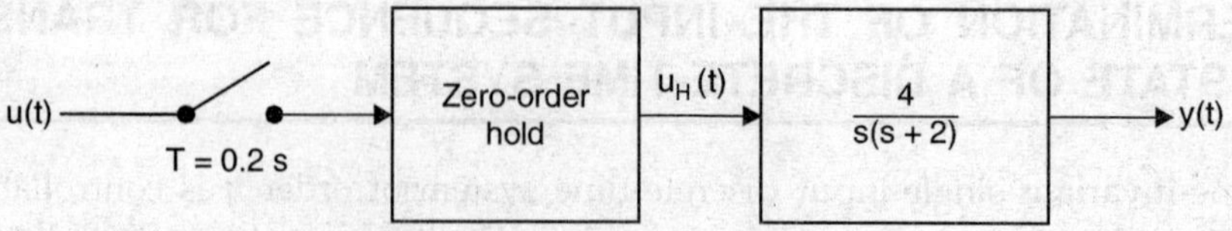

Figure 12.1. *A discrete-time system*

Solution:

The state-space representation of the continuous-time portion of the system is

$$\begin{bmatrix} \dot{x}_1 \\ \dot{x}_2 \end{bmatrix} = \begin{bmatrix} 0 & 1 \\ 0 & -2 \end{bmatrix} \begin{bmatrix} x_1 \\ x_2 \end{bmatrix} + \begin{bmatrix} 0 \\ 4 \end{bmatrix} u_H(t) \quad \text{...(12.9)}$$

$$y(t) = \begin{bmatrix} 1 & 0 \end{bmatrix} \begin{bmatrix} x_1 \\ x_2 \end{bmatrix} \qquad \text{...(12.10)}$$

where $u_H(t)$ is the output of the zero-order hold, and is a staircase approximation to $u(t)$, as discussed in Chapter 11, Section 11.2.

The state transition matrices for the system are easily calculated as

$$F = e^{AT} = \begin{bmatrix} 1 & 0.16484 \\ 0 & 0.67032 \end{bmatrix} \qquad \text{...(12.11)}$$

$$G = \int_0^T e^{AT}\, dt\, B = \begin{bmatrix} 0.07032 \\ 0.65936 \end{bmatrix} \qquad \text{...(12.12)}$$

Here we need two sampling intervals for the transfer. Hence,

$$x(2T) - F^2 x(0) = \begin{bmatrix} 1 \\ 0 \end{bmatrix} \qquad \text{...(12.13)}$$

Also $$\begin{bmatrix} FG & G \end{bmatrix} = \begin{bmatrix} 0.179 & 0.07032 \\ 0.442 & 0.6593 \end{bmatrix} \qquad \text{...(12.14)}$$

and $$\begin{bmatrix} u(0) \\ u(T) \end{bmatrix} = \begin{bmatrix} FG & G \end{bmatrix}^{-1} \begin{bmatrix} 1 \\ 0 \end{bmatrix} = \begin{bmatrix} 7.5831 \\ -5.0831 \end{bmatrix} \qquad \text{...(12.15)}$$

It can be easily verified that this input sequence will transfer the state as desired. The output of the system will therefore correspond to a *dead-beat and ripple-free* response to a unit step. This follows from the fact that the output increases monotonically from 0 to and 1, and after it has reached that value, its rate of change becomes zero. Therefore, the output will remain at this value for all time in the future until a new input is applied. Note that in the case of a continuous-time system, the steady-state is reached only after an infinite amount of time has elapsed (at least theoretically, although in the case of a stable system, 'for all practical purposes' the steady-state is reached after an elapsed time equal to five times the largest time constant of the system).

Instead of generating the input sequence calculated above in an 'open-loop' mode, it is also possible to close the loop with a digital compensator, as shown in Fig. 12.2.

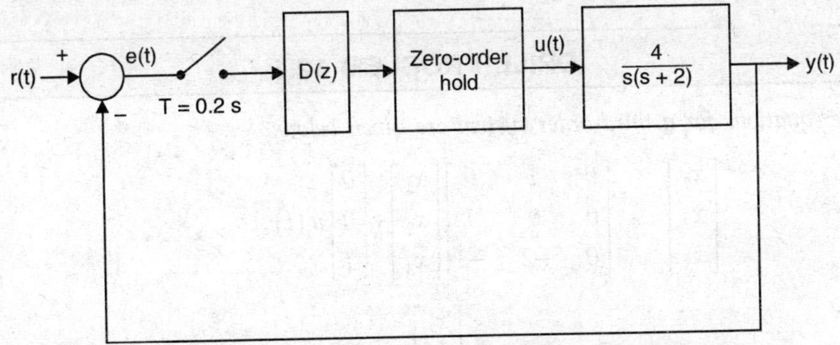

Figure 12.2. *Closed-loop realization of dead-beat response*

The transfer function, $D(z)$ is calculated in such a way that with a unit step input, the error sequence $e(kT)$ is transformed into the desired input sequence $u(kT)$ calculated in Equation (12.15).

We shall first calculate the states at T and $2T$. We obtain

$$x(T) = Fx(0) + Gu(0) = \begin{bmatrix} 0.5332 \\ 5.0 \end{bmatrix} \quad ...(12.16)$$

and

$$x(2T) = Fx(T) + Gu(T) = \begin{bmatrix} 1 \\ 0 \end{bmatrix} \quad ...(12.17)$$

Consequently, $y(0) = 0$, $y(T) = 0.5332$, $y(2T) = 1$ and the error sequence is

$$\begin{aligned} e(0) &= 1 - y(0) = 1.0 \\ e(T) &= 1 - y(T) = 0.4668 \end{aligned} \quad ...(12.18)$$

The z transfer function of the digital compensator is obtained by dividing the z transform of the input sequence by that of the error sequence. Since both of them are finite sequences (zero for $t \geq 2T$), we get

$$D(z) = \frac{u(0)+u(T)z^{-1}}{e(0)+e(T)z^{-1}} = \frac{7.5831-5.0831\,z^{-1}}{1+0.4668\,z^{-1}} \quad ...(12.19)$$

An advantage of this closed-loop implementation is that $u(kT) = 0$ for $k > 1$, so that the system is kept at the state $\begin{bmatrix} 1 \\ 0 \end{bmatrix}$ after being brought there. Thus, the output will be ripple-free. The approach discussed in Chapter 11 did not guarantee that there will be no ripples between the sampling instants. Also, note that with this controller, we get dead-beat and ripple-free response for all step inputs.

It may appear to the reader that using this approach, it is possible to transfer the state of the system from any given state to the desired final state in as small a time as desired by simply selecting the sampling interval as $1/n$ times the desired transition time. In practice, however, there will be some difficulties. The reduction in transition time will cause an increase in the magnitude of the input signal applied to the system. Hence, there will be a physical limitation imposed by the allowable magnitude of the input to the system. For instance, there is always a limit to the magnitude of the voltage that we can safety apply to a servomotor.

DRILL PROBLEM 12.1

The state equations for a third-order system are given below:

$$\begin{bmatrix} \dot{x}_1 \\ \dot{x}_2 \\ \dot{x}_3 \end{bmatrix} = \begin{bmatrix} 0 & 1 & 0 \\ 0 & 0 & 1 \\ 0 & -2 & -3 \end{bmatrix} \begin{bmatrix} x_1 \\ x_2 \\ x_3 \end{bmatrix} + \begin{bmatrix} 0 \\ 0 \\ 1 \end{bmatrix} u(t)$$

$$y(t) = \begin{bmatrix} 1 & 0 & 0 \end{bmatrix} \begin{bmatrix} x_1 \\ x_2 \\ x_3 \end{bmatrix}$$

(a) *Determine the input sequence that will transfer the state from the origin to* $\begin{bmatrix} 1 \\ 0 \\ 0 \end{bmatrix}$ *in three sampling intervals if the sampling interval is 0.5 second.*

(b) *If the system is connected in the closed-loop form, as shown in Fig. 12.2, determine the transfer function of the digital controller that will provide a dead-beat ripple-free response to a step input.*

Ans. (a) $u(0) = 16.082354,\ u(T) = -15.670808,\ u(2T) = 3.588458$

(b) $$D(z) = \frac{16.082354 - 15.670808\,z^{-1} + 3.588458\,z^{-2}}{1 + 0.7658\,z^{-1} + 0.0407\,z^{-2}}$$

DRILL PROBLEM 12.2

Repeat Drill Problem 12.1 if the sampling interval is reduced to 0.1 second.

Ans. (a) $u(0) = 1159.3961,\ u(T) = -1998.2979,\ u(2T) = 858.9016$

(b) $$D(z) = \frac{1159.3961 - 1998.2979\,z^{-1} + 858.9016\,z^{-2}}{1 + 0.8203\,z^{-1} + 0.1546\,z^{-2}}$$

12.3 CONTROLLABILITY AND OBSERVABILITY

We are now ready to study the very important concepts of controllability and observability. They tell us whether it is at all possible to control all the states of the system completely by a suitable choice of an input and whether it is possible to reconstruct the states of a system from its input and output. A precise knowledge of the state equations of the system is assumed.

12.3.1 Controllability

A system is said to be completely controllable if it is possible to find an input $u(t)$ that will transfer the system from any given initial state $x(t_0)$ to any given final state $x(t_f)$ over a specified interval of time $t_f - t_0$. (Note that for the sake of brevity we shall use the term 'uncontrollable' for a system that is not completely controllable.)

THEOREM 12.1

A linear time-invariant continuous-time system described by the state equations

$$\dot{x} = Ax + Bu \qquad x \in \Re^n,\ u \in \Re^m$$
$$y = Cx \qquad y \in \Re^p$$

is completely controllable if and only if the rank of the controllability matrix, defined as

$$U \triangleq [B \quad AB \quad A^2B \quad \ldots \quad A^{n-1}B] \qquad \ldots(12.20)$$

is equal to n.

To appreciate this theorem, consider the discrete-time system described by

$$x(kT + T) = F(T)x(kT) + G(T)u\,(kT) \qquad \ldots(12.21)$$

It was seen in Section 12.2 that, for a single-input single-output system, an input sequence $u(0)$, $u(T)$, ..., $u(nT - T)$ that will transfer the state of the system from any given initial state to any given final state can be determined uniquely by inverting the following matrix

$$U_d = [F^{n-1}G \quad F^{n-2}G \quad \dots \quad FG \quad G] \qquad \text{...(12.22)}$$

It can be shown (by recognizing that $F = e^{AT}$, and by using the Cayley-Hamilton theorem) that the inverse of U_d will exist if and only if the rank of U is n (that is, if U is nonsingular for the case when B is a column vector.)

If a system is not completely controllable, it implies that it has one or more natural modes that cannot be affected by the input, directly, or indirectly. This is illustrated below.

EXAMPLE 12.2

Consider the system described by the state equations

$$\begin{bmatrix} \dot{x}_1 \\ \dot{x}_2 \end{bmatrix} = \begin{bmatrix} -0.5 & 0 \\ 0 & -2 \end{bmatrix} \begin{bmatrix} x_1 \\ x_2 \end{bmatrix} + \begin{bmatrix} 0 \\ 1 \end{bmatrix} u(t) \qquad \text{...(12.23)}$$

The controllability matrix for this system is

$$U = [B \quad AB] = \begin{bmatrix} 0 & 0 \\ 1 & -2 \end{bmatrix} \qquad \text{...(12.24)}$$

which is seen to be singular. Hence, the system is uncontrollable. This is more obvious if we write the two differential equations separately; that is,

$$\dot{x}_1 = -0.5\, x_1 \qquad \text{...(12.25)}$$

$$\dot{x}_2 = -2x_2 + u(t)$$

It is evident that whereas x_2 can be changed by $u(t)$, the state x_1 is unaffected by our choice of the input, since it is not coupled either directly to the input or to the state x_2. Hence, this state (or the mode $x_1(0)\, e^{-0.5t}$) is uncontrollable. On the other hand, if we had

$$\dot{x}_1 = -0.5x_1 + x_2$$

we would have had $A = \begin{bmatrix} -0.5 & 1 \\ 0 & -2 \end{bmatrix}$ $B = \begin{bmatrix} 0 \\ 1 \end{bmatrix}$ $U = \begin{bmatrix} 0 & 1 \\ 1 & -2 \end{bmatrix}$ ...(12.26)

giving us a controllable system since we can now control x_1 indirectly through x_2.

12.3.2 Observability

A system is said to be completely observable if its state can be determined from a knowledge of the input $u(t)$ and the output $y(t)$ over a finite interval of time.

Since, by definition, if we know the initial state of the system we can determine all future outputs and states for a specified input, it follows that a system is observable if the initial state can be determined from the input/output records for a finite interval.

(*Note:* For the sake of brevity we shall use the term "unobservable" for a system that is not completely observable.)

THEOREM 12.2

A linear system is completely observable if the rank of the matrix V is equal to n, where V, called the observability matrix, is defined as

$$V \triangleq \begin{bmatrix} C \\ CA \\ CA^2 \\ \vdots \\ CA^{n-1} \end{bmatrix} \qquad ...(12.27)$$

To appreciate this, let us again consider the discrete-time single-input single-output system described by the equations

$$\left.\begin{aligned} x(kT + T) &= Fx(kT) + Gu(kT) \; x \in \Re^n \\ y(kT) &= Cx(kT) \end{aligned}\right\} \qquad ...(12.28)$$

We shall assume that the initial state $x(0)$ is not known, and we want to determine it from observations of the output $y(kT)$, $k = 0, 1, ..., n - 1$. To simplify matters, let us make $u(kT) = 0$ for all k. We shall then get the following equations for the output at different sampling instants.

$$\begin{aligned} y(0) &= Cx(0) \\ y(T) &= Cx(T) = CFx(0) \\ y(2T) &= Cx(2T) = CF^2x(0) \\ y(nT - T) &= CF^{n-1}x(0) \end{aligned} \qquad ...(12.29)$$

which may also be written as

$$\begin{bmatrix} C \\ CF \\ CF^2 \\ \vdots \\ CF^{n-1} \end{bmatrix} x(0) = \begin{bmatrix} y(0) \\ y(T) \\ \vdots \\ y(nT-T) \end{bmatrix} \qquad ...(12.30)$$

It follows from the above equation that the initial state $x(0)$ can be uniquely determined if and only if the matrix on the left is nonsingular. Invoking the Cayley-Hamilton theorem we can show that this will be possible only if V is nonsingular.

EXAMPLE 12.3

Consider the system described by the state equations

$$\begin{bmatrix} \dot{x}_1 \\ \dot{x}_2 \end{bmatrix} = \begin{bmatrix} -0.5 & 0 \\ 0 & -2 \end{bmatrix} \begin{bmatrix} x_1 \\ x_2 \end{bmatrix} + \begin{bmatrix} 0 \\ 1 \end{bmatrix} u(t) \qquad ...(12.31)$$

$$y(t) = \begin{bmatrix} 0 & 1 \end{bmatrix} \begin{bmatrix} x_1 \\ x_2 \end{bmatrix} \qquad ...(12.32)$$

Note that the Equation (12.31) is identical to Equation (12.23), considered in Example 12.1. It was seen that this system is uncontrollable, since the state x_1 is not affected either by $u(t)$ or $x_2(t)$. We now form the observability matrix

$$V = \begin{bmatrix} C \\ CA \end{bmatrix} = \begin{bmatrix} 0 & 1 \\ 0 & -2 \end{bmatrix} \qquad \text{...(12.33)}$$

which is seen to be singular. The reason this system is unobservable is that the state x_1 does not affect the output, nor does it affect the state x_2 (which is coupled to the output).

EXAMPLE 12.4

Consider the system described by the state equations

$$\begin{bmatrix} \dot{x}_1 \\ \dot{x}_2 \end{bmatrix} = \begin{bmatrix} 0 & 1 \\ -2 & -3 \end{bmatrix} \begin{bmatrix} x_1 \\ x_2 \end{bmatrix} + \begin{bmatrix} 0 \\ 1 \end{bmatrix} u(t) \qquad \text{...(12.34)}$$

$$y(t) = \begin{bmatrix} 1 & 1 \end{bmatrix} \begin{bmatrix} x_1 \\ x_2 \end{bmatrix} \qquad \text{...(12.35)}$$

The controllability matrix for this system is given by

$$U = \begin{bmatrix} 0 & 1 \\ 1 & -3 \end{bmatrix} \qquad \text{...(12.36)}$$

It is seen that U is nonsingular, its determinant is –1. Therefore, the system is controllable. Also, the observability matrix is given by

$$V = \begin{bmatrix} 1 & 1 \\ -2 & -2 \end{bmatrix} \qquad \text{...(12.37)}$$

which is singular. Hence, this system is unobservable.

This is explained easily by examining its transfer function

$$\frac{Y(s)}{U(s)} = \frac{s+1}{s^2+3s+2} = \frac{s+1}{(s+1)(s+2)} = \frac{1}{s+2} \qquad \text{...(12.38)}$$

Since the factor $(s + 1)$ is cancelled from the numerator and the denominator, the mode at $s = -1$, giving rise to a term of the form Ke^{-t} does not appear in the output, although it is present as one of the states within the system. This is similar to the situation in many electronic circuits which have 'parasitic' oscillations that do not appear at the output.

For the particular case when the state equations are in the controller form (see Section 3.4 in Chapter 3), we have

$$A = \begin{bmatrix} -a_1 & -a_2 & -a_3 & \cdots & -a_{n-1} & -a_n \\ 1 & 0 & 0 & \cdots & 0 & 0 \\ 0 & 1 & 0 & \cdots & 0 & 0 \\ \vdots & \vdots & \vdots & \cdots & \vdots & \vdots \\ 0 & 0 & 0 & \cdots & 1 & 0 \end{bmatrix} \quad B = \begin{bmatrix} 1 \\ 0 \\ 0 \\ \vdots \\ 0 \\ 0 \end{bmatrix} \qquad \text{...(12.39)}$$

The controllability matrix is

$$U = \begin{bmatrix} 1 & -a_1 & * & \cdots & 0 \\ 0 & 1 & * & \cdots & 0 \\ \vdots & \vdots & \vdots & \cdots & \vdots \\ 0 & 0 & 0 & \cdots & 1 \end{bmatrix} \qquad \text{...(12.40)}$$

where the asterisks (*s) indicate elements that are, in general, different from 0 to 1.

Because of the triangular nature of U, it is nonsingular with determinant equal to one. Thus, a system is always controllable if it can be transformed to this form.

As was shown in Chapter 3, for any given transfer function

$$\frac{Y(s)}{U(s)} = \frac{b_1 s^{n-1} + b_2 s^{n-2} + \ldots + b_{n-1} s + b_n}{s^n + a_1 s^{n-1} + \ldots + a_{n-1} s + a_n} \qquad \text{...(12.41)}$$

it is possible to write the state equations in the controller canonical form directly. The matrices A and B are as in Equation (12.39), and

$$C = \begin{bmatrix} b_1 & b_2 & \cdots & b_{n-1} & b_n \end{bmatrix} \qquad \text{...(12.42)}$$

Furthermore, the inverse of U for this case can also be written by inspection. It is given by

$$U^{-1} = \begin{bmatrix} 1 & a_1 & a_2 & \cdots & a_{n-1} \\ 0 & 1 & a_1 & \cdots & a_{n-2} \\ \vdots & \vdots & \vdots & \cdots & \vdots \\ 0 & 0 & 0 & \cdots & 1 \end{bmatrix} \qquad \text{...(12.43)}$$

which is an upper triangular Toeplitz matrix. This can be verified for $n = 2$ and 3 and then proved by induction.

In the same way, one can write the state equations for any given transfer function in the observer canonical form by inspection. In this case, the observability matrix and its inverse will be the transpose of the corresponding matrices in Equations (12.40) and (12.43).

Corollary 1

A linear time-invariant single-input single-output system is completely controllable and observable if the numerator and denominator polynomials of its transfer function are coprime that is, they do not have a common factor, with the exception of a constant.

It can be proved by writing the state equations for the given transfer function in the diagonal form (or the Jordan form in the case of multiple poles) by performing a partial fraction expansion, as shown in Example 3.5. If there is a common factor between the numerator and the denominator, the residue at that particular pole (which is cancelled by a zero) will be zero. The resulting state equations will have the form shown in Example 12.3, with a zero in the corresponding column of C. This will make that state unobservable.

In general, if the numerator and the denominator of the transfer function have common factors, we can write the state equations either in the controller or the observer form by inspection. However, the state equations in the controller form will not be completely observable. Similarly, the state equations in the observer form will not be completely controllable. The state equations will be both controllable and observable if there are no pole-zero cancellations. In this case, the state equations are said to be a minimal realization of the given transfer function.

Corollary 2

A linear time-invariant single-input single-output system is completely controllable and observable if the diagonal form of its state equations has no zero entries in both B and C.

DRILL PROBLEM 12.3

The state equations of a system are given below. Determine if the system is completely controllable and observable.

$$\dot{x} = \begin{bmatrix} -1 & 0 & 3 \\ 2 & -1 & -1 \\ -3 & 1 & -2 \end{bmatrix} x + \begin{bmatrix} 1 \\ 0 \\ 0 \end{bmatrix} u$$

$$y = \begin{bmatrix} 1 & 2 & 2 \end{bmatrix} x$$

Ans. The system is completely controllable and completely observable.

DRILL PROBLEM 12.4

Repeat for the state equations given below:

$$\dot{x} = \begin{bmatrix} -6 & 2 & -4 \\ -18 & 3 & -8 \\ -6 & 1 & -3 \end{bmatrix} x + \begin{bmatrix} 1 \\ 3 \\ 1 \end{bmatrix} u$$

$$y = \begin{bmatrix} 1 & -1 & 2 \end{bmatrix} x$$

Ans. The system is completely controllable but not completely observable.

DRILL PROBLEM 12.5

Repeat for the state equations given below:

$$\dot{x} = \begin{bmatrix} -6 & -18 & -6 \\ 2 & 3 & 1 \\ -4 & -8 & -3 \end{bmatrix} x + \begin{bmatrix} 2 \\ -3 \\ 7 \end{bmatrix} u$$

$$y = \begin{bmatrix} 1 & 3 & 1 \end{bmatrix} x$$

Ans. The system is completely observable but not completely controllable.

12.4 STATE-VARIABLE FEEDBACK

Consider the block diagram shown in Fig. 12.3. We have a linear system described by the state equations

$$\dot{x} = Ax + Bu \quad \text{...(12.44)}$$

$$y = Cx \quad \text{...(12.45)}$$

The system is assumed to be of the single-input single-output type.

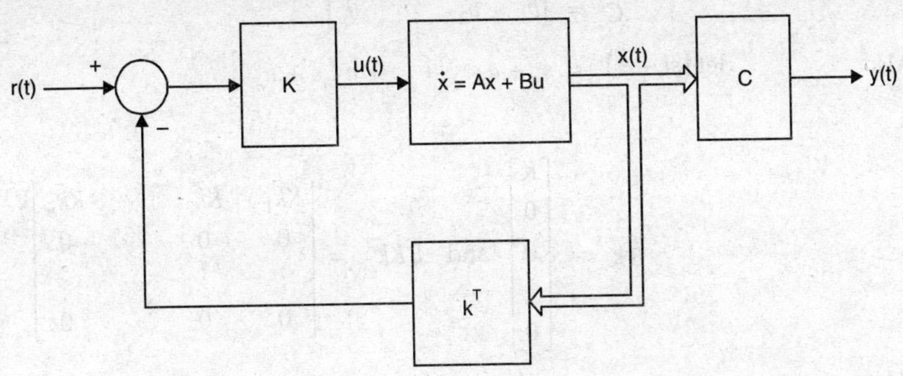

Figure 12.3. *System with state-variable feedback*

The state vector is fed back and thus

$$u = K\,[r - k^T x] \quad \text{...(12.46)}$$

where $$k^T = \begin{bmatrix} k_1 & k_2 & \cdots & k_n \end{bmatrix} \quad \text{...(12.47)}$$

a row vector of the same dimension as x.

Substitution of Equation (12.46) into Equation (12.44) leads to

$$\dot{x} = Ax + BK\,[r - k^T x] \quad \text{...(12.48)}$$

which can be written as $$\dot{x} = A_f x + BKr \quad \text{...(12.49)}$$

where $$A_f = A - BKk^T \quad \text{...(12.50)}$$

Note that the output equation remains unchanged as Equation (12.45).

THEOREM 12.3

If a system is completely controllable, one can always determine the product Kk^T so that the eigenvalues of A_f (corresponding to the poles of the closed-loop system) will be placed at arbitrarily specified locations.

Proof: If the system is completely controllable, one can always find a nonsingular transformation matrix P such that $w = P^{-1}x$ and we get the equations

$$\dot{w} = \overline{A}w + \overline{B}u \quad \text{...(12.51)}$$

and

$$y = \overline{C}w \quad \text{...(12.52)}$$

where $\overline{A} = P^{-1}AP$, $\overline{B} = P^{-1}B$ and $\overline{C} = CP$ are in the controller canonical form. The method for determining P will be described in Section 12.4.1. Consequenlty,

$$\overline{A} = \begin{bmatrix} -a_1 & -a_2 & -a_3 & \cdots & -a_{n-1} & -a_n \\ 1 & 0 & 0 & \cdots & 0 & 0 \\ 0 & 1 & 0 & \cdots & 0 & 0 \\ \vdots & \vdots & \vdots & \cdots & \vdots & \vdots \\ 0 & 0 & 0 & \cdots & 1 & 0 \end{bmatrix} \quad \overline{B} = \begin{bmatrix} 1 \\ 0 \\ 0 \\ \vdots \\ 0 \\ 0 \end{bmatrix} \quad ...(12.53)$$

and $$\overline{C} = \begin{bmatrix} b_1 & b_2 & \cdots & b_n \end{bmatrix}$$

Also $$\det\left(sI - \overline{A}\right) = s^n + a_1 s^{n-1} + \dots + a_{n-1} s + a_n \quad ...(12.54)$$

$$\overline{B}K = \begin{bmatrix} K \\ 0 \\ 0 \\ \vdots \\ 0 \end{bmatrix} \text{ and } \overline{B}K\overline{k}^T = \begin{bmatrix} K\overline{k}_1 & K\overline{k}_2 & \cdots & K\overline{k}_n \\ 0 & 0 & \cdots & 0 \\ \vdots & \vdots & \cdots & \vdots \\ 0 & 0 & \cdots & 0 \end{bmatrix} \quad ...(12.55)$$

Hence,

$$\overline{A}_f = \overline{A} - \overline{B}K\overline{k}^T = \begin{bmatrix} -a_1 - K\overline{k}_1 & -a_2 - K\overline{k}_2 & \cdots & -a_{n-1} - K\overline{k}_{n-1} & -a_n - K\overline{k}_n \\ 1 & 0 & \cdots & 0 & 0 \\ 0 & 1 & \cdots & 0 & 0 \\ \vdots & \vdots & \cdots & \vdots & \vdots \\ 0 & 0 & \cdots & 1 & 0 \end{bmatrix} \quad ...(12.56)$$

From the desired locations of the poles of the closed-loop system, we can obtain the characteristic polynomial of $\overline{A}_f$. Let this be given by

$$\det\left(sI - \overline{A}_f\right) = s^n + \alpha_1 s^{n-1} + \dots + \alpha_{n-1} s + \alpha_n \quad ...(12.57)$$

Comparison between Equation (12.57) and the first row of Equation (12.56) yields the following equations

$$K\overline{k}_1 = \alpha_1 - a_1,\ K\overline{k}_2 = \alpha_2 - a_2, \dots, K\overline{k}_n = \alpha_n - a_n \quad ...(12.58)$$

If we select the gain constant K so that the dc gain of the closed-loop system is unity (corresponding to zero steady-state error to a step input), we get a unique solution for K and $\overline{k}$. Also, since

$$\overline{k}^T w = \overline{k}^T P^{-1} x = k^T x \quad ...(12.59)$$

we get $kT = \overline{k}^T p^{-1}$ as the feedback vector for the system before transformation.

EXAMPLE 12.5

Consider a system described by the state equations

$$\dot{x} = \begin{bmatrix} 0 & -6 & -20 \\ 1 & 0 & 0 \\ 0 & 1 & 0 \end{bmatrix} x + \begin{bmatrix} 1 \\ 0 \\ 0 \end{bmatrix} u \quad \text{...(12.60)}$$

$$y = \begin{bmatrix} 0 & 1 & 2 \end{bmatrix} x \quad \text{...(12.61)}$$

Since these are in the controller canonical form, the transfer function of the open-loop system is obtained by inspection as

$$\frac{Y(s)}{U(s)} = \frac{s+2}{s^3 + 6s + 20} \quad \text{...(12.62)}$$

Evidently, this is unstable. We want to stabilize the system so that the poles of the transfer function of the closed-loop system are located at $s = -4$, and $-3 \pm j4$. We also want the closed-loop system to have zero steady-state error to step inputs. Consequently, the transfer function of the closed-loop system is obtained as

$$\frac{Y(s)}{R(s)} = \frac{K(s+2)}{(s+4)\left(s^2 + 6s + 25\right)} = \frac{K(s+2)}{s^3 + 10s^2 + 49s + 100} \quad \text{...(12.63)}$$

From the denominator of Equation (12.63), we get

$$A_f = \begin{bmatrix} -10 & -49 & -100 \\ 1 & 0 & 0 \\ 0 & 1 & 0 \end{bmatrix} \quad \text{...(12.64)}$$

Subtracting A_f from A, we obtain

$$BKk^T = A - A_f = \begin{bmatrix} 10 & 43 & 80 \\ 0 & 0 & 0 \\ 0 & 0 & 0 \end{bmatrix} \quad \text{...(12.65)}$$

Since we must have $K = 50$ for unity dc gain, the feedback gain vector is

$$k^T = \begin{bmatrix} 0.2 & 0.86 & 1.6 \end{bmatrix} \quad \text{...(12.66)}$$

EXAMPLE 12.6

Consider the state equations

$$\dot{x} = \begin{bmatrix} -1 & 1 & 0 \\ 0 & -4 & 2 \\ 0 & 0 & -10 \end{bmatrix} x + \begin{bmatrix} 1 \\ 0 \\ -1 \end{bmatrix} u \quad \text{...(12.67)}$$

$$y = \begin{bmatrix} 1 & 0 & 1 \end{bmatrix} x \quad \text{...(12.68)}$$

These can be transformed to the controller from using $w = P^{-1}x$, where

$$P = \begin{bmatrix} 1 & 14 & 38 \\ 0 & -2 & -2 \\ -1 & -5 & -4 \end{bmatrix} \quad \text{...(12.69)}$$

The procedure for determining P is shown in the next section (see Example 12.7). Hence,

$$\bar{A} = P^{-1}AP = \begin{bmatrix} -15 & -54 & -40 \\ 1 & 0 & 0 \\ 0 & 1 & 0 \end{bmatrix}, \quad \bar{B} = P^{-1}B = \begin{bmatrix} 1 \\ 0 \\ 0 \end{bmatrix} \qquad \text{...(12.70)}$$

and

$$\bar{C} = CP = \begin{bmatrix} 0 & 9 & 34 \end{bmatrix} \qquad \text{...(12.71)}$$

Also,

$$\frac{Y(s)}{U(s)} = \frac{9s+34}{s^3+15s^2+54s+40} \qquad \text{...(12.72)}$$

It is desired that the poles of the transfer function of the closed-loop system be located at –8 and $-6 \pm j10$, and the steady-state error to a step input be zero. Hence, the denominator of the desired transfer function is given by

and

$$\alpha(s) = (s^2 + 8)\,(s^2 + 12s + 136) = s^3 + 20s^2 + 232s + 1088 \qquad \text{...(12.73)}$$

$$\frac{Y(s)}{R(s)} = \frac{K(9s+34)}{s^3+20s^2+232s+1088} \qquad \text{...(12.74)}$$

Therefore,

$$\bar{A}_f = \begin{bmatrix} -20 & -232 & -1088 \\ 1 & 0 & 0 \\ 0 & 1 & 0 \end{bmatrix}$$

$$\bar{B}K\bar{k}^T = \begin{bmatrix} 5 & 178 & 1048 \\ 0 & 0 & 0 \\ 0 & 0 & 0 \end{bmatrix} \qquad \text{...(12.75)}$$

Since we must have the dc gain equal to one,

$$K = 32 \qquad \text{...(12.76)}$$

Hence, from Equation (12.75),

$$\bar{k}^T = \begin{bmatrix} 0.15625 & 5.5625 & 32.75 \end{bmatrix} \qquad \text{...(12.77)}$$

and

$$k^T = \bar{k}^T P^{-1} = \begin{bmatrix} 1.09375 & 2.53125 & 0.9375 \end{bmatrix} \qquad \text{...(12.78)}$$

Note that the numerator of the transfer function of the closed-loop system is K times that of the transfer function of the open-loop. The latter is usually chosen to make the dc gain of the closed-loop system equal to one to provide zero steady-state error to a step input.

12.4.1 Transformation to the Controller Form

We shall now consider the problem of transformation of the state equations to the controller form. First, we note that the controllability matrix of the transformed equations is

$$\begin{aligned} \bar{U} &= \begin{bmatrix} \bar{B} & \bar{A}\bar{B} & \bar{A}^2\bar{B} & \cdots & \bar{A}^{n-1}\bar{B} \end{bmatrix} \\ &= P^{-1}\begin{bmatrix} B & AB & A^2B & \cdots & A^{n-1}B \end{bmatrix} \\ &= P^{-1}U \end{aligned} \qquad \text{...(12.79)}$$

where the second step follows by substituting $P^{-1}AP$ for $\bar{A}$ and $P^{-1}B$ for $\bar{B}$, and simplifying.

Hence, $$P = U\overline{U}^{-1} \quad ...(12.80)$$

Given A and B, we can easily calculate U. It was shown in Section 12.3 [see Equation (12.43)], that $\overline{U}^{-1}$ is also determined directly from the determinant of $(sI - A)$. Let

$$\det(sI - A) = s^n + a_1 s^{n-1} + a_2 s^{n-2} + ... + a_{n-1} s + a_n \quad ...(12.81)$$

Then, we know that

$$\overline{A} = \begin{bmatrix} -a_1 & -a_2 & -a_3 & \cdots & -a_{n-1} & -a_n \\ 1 & 0 & 0 & \cdots & 0 & 0 \\ 0 & 1 & 0 & \cdots & 0 & 0 \\ \vdots & \vdots & \vdots & \cdots & \vdots & \vdots \\ 0 & 0 & 0 & \cdots & 1 & 0 \end{bmatrix} \quad \overline{B} = \begin{bmatrix} 1 \\ 0 \\ 0 \\ \vdots \\ 0 \\ 0 \end{bmatrix} \quad ...(12.82)$$

and $$\overline{U}^{-1} = \begin{bmatrix} 1 & a_1 & a_2 & \cdots & a_{n-1} \\ 0 & 1 & a_1 & \cdots & a_{n-2} \\ \vdots & \vdots & \vdots & \cdots & \vdots \\ 0 & 0 & 0 & \cdots & 1 \end{bmatrix} \quad ...(12.83)$$

This enables us to determine P from Equation (12.80).

EXAMPLE 12.7

Consider the system in Example 12.6. Here

$$A = \begin{bmatrix} -1 & 1 & 0 \\ 0 & -4 & 2 \\ 0 & 0 & -10 \end{bmatrix} \quad B = \begin{bmatrix} 1 \\ 0 \\ -1 \end{bmatrix} \quad C = \begin{bmatrix} 1 & 0 & 1 \end{bmatrix} \quad ...(12.84)$$

Hence $$U = \begin{bmatrix} B & AB & A^2B \end{bmatrix} = \begin{bmatrix} 1 & -1 & -1 \\ 0 & -2 & 28 \\ -1 & 10 & -100 \end{bmatrix} \quad ...(12.85)$$

and $$\det(sI - A) = \begin{bmatrix} s+1 & -1 & 0 \\ 0 & s+4 & -2 \\ 0 & 0 & s+10 \end{bmatrix} \quad ...(12.86)$$

$$= (s + 1)(s + 4)(s + 10) = s^3 + 15s^2 + 54s + 40$$

Consequently,

$$\overline{U}^{-1} = \begin{bmatrix} 1 & 15 & 54 \\ 0 & 1 & 15 \\ 0 & 0 & 1 \end{bmatrix}, \quad P = U\overline{U}^{-1} = \begin{bmatrix} 1 & 14 & 38 \\ 0 & -2 & -2 \\ -1 & -5 & -4 \end{bmatrix} \quad ...(12.87)$$

Finally, we obtain

$$\bar{A} = \begin{bmatrix} -15 & -54 & -40 \\ 1 & 0 & 0 \\ 0 & 1 & 0 \end{bmatrix}, \; \bar{B} = \begin{bmatrix} 1 \\ 0 \\ 0 \end{bmatrix} \; \bar{C} = \begin{bmatrix} 0 & 9 & 34 \end{bmatrix} \quad ...(12.88)$$

DRILL PROBLEM 12.6

The system described by the state equations in Drill Problem 12.3 is to be used as a part of a closed-loop system with state-variable feedback, as shown in the block diagram of Fig. 12.3. Determine the gain K and the state feedback vector so that the transfer function of the closed-loop system is given by

$$\frac{5\left(s^2+4s+16\right)}{(s+5)\left(s^2+6s+16\right)}$$

Ans. $K = 5,\ k_1 = 1.4,\ k_2 = 149/95,\ k_3 = 36/95.$

12.5 STATE-VARIABLE FEEDBACK: A TRANSFER FUNCTION APPROACH

The method for calculating the state-feedback vector described in the previous section is very convenient if the state equations of the system are given in the controller canonical form. If this is not the case, then the equations must first be transformed to the canonical form. We shall now study an alternative approach based on the use of transfer functions. In addition to being computationally simpler, it also gives better insight into the problem.

Consider the block diagram shown in Fig. 12.4, where the effect of state-variable feedback is to return.

$$f(t) = K^T x(t) \quad ...(12.89)$$

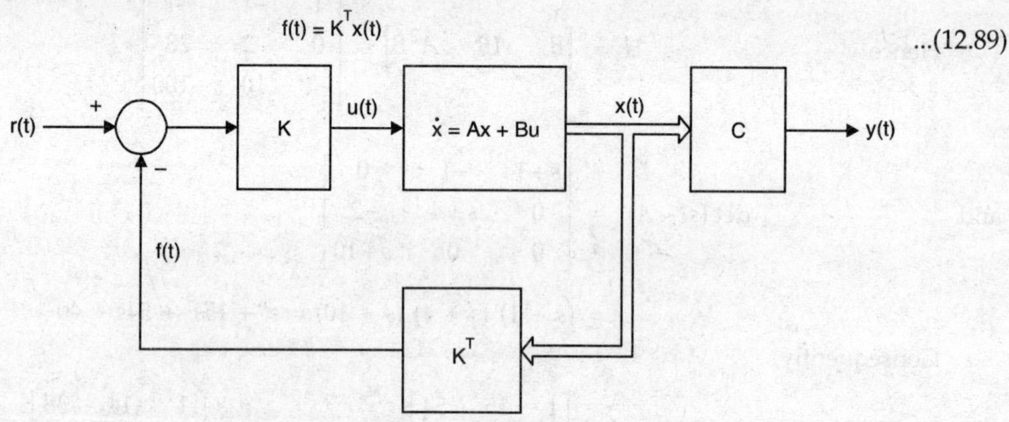

Figure 12.4. *System with state-variable feedback*

It also follows that the input to the process is

$$u(t) = K[r(t) - f(t)] \quad ...(12.90)$$

Taking the Laplace transform of both sides of Equation (12.89), we have

$$F(s) = k^T X(s) = k^T (sI - A)^{-1} BU(s) \qquad ...(12.91)$$

so that

$$\frac{F(s)}{U(s)} = k^T (sI - A)^{-1} B \triangleq \frac{f(s)}{a(s)} \qquad ...(12.92)$$

where

$$a(s) = \det (sI - A) \qquad ...(12.93)$$

$$f(s) = k^T \operatorname{adj} (sI - A)B \qquad ...(12.94)$$

Similarly, for the process, we have

$$\frac{Y(s)}{U(s)} = C (sI - A)^{-1} B \triangleq \frac{b(s)}{a(s)} \qquad ...(12.95)$$

where

$$b(s) = C \operatorname{adj} (sI - A)B \qquad ...(12.96)$$

From the block diagram in Fig. 12.4, we also have

$$U(s) = K[R(s) - F(s)] \qquad ...(12.97)$$

Thus,

$$\frac{F(s)}{U(s)} = \frac{F(s)}{K[R(s) - F(s)]} = \frac{f(s)}{a(s)} \qquad ...(12.98)$$

which can be solved to obtain

$$\frac{F(s)}{R(s)} = \frac{Kf(s)}{a(s) + Kf(s)} \qquad ...(12.99)$$

Hence

$$\frac{Y(s)}{R(s)} = \frac{Y(s)}{U(s)} \cdot \frac{U(s)}{F(s)} \cdot \frac{F(s)}{R(s)} = \frac{Kb(s)}{a(s) + Kf(s)} \qquad ...(12.100)$$

Equation (12.100) shows that the numerator of the transfer function of the closed-loop system is just K times the numerator of the transfer function of the open-loop *i.e.*, the zeros of the transfer function are not affected by state feedback. It also shows that the denominator of the transfer function of the closed-loop system can be altered arbitrarily by the choice of $Kf(s)$. Let the desired transfer function of the closed-loop system be given by

$$\frac{Y(s)}{R(s)} = \frac{Kb(s)}{\alpha(s)} \qquad ...(12.101)$$

Comparing Equation (12.101) with Equation (12.100), we get

$$f(s) = \frac{\alpha(s) - a(s)}{K} \qquad ...(12.102)$$

Hence, $f(s)$ can be determined from the denominator of the transfer function of the open-loop system and the desired transfer function of the closed-loop system.

Returning to Equation (12.92) and utilizing Equation (12.102), we obtain

$$\frac{f(s)}{a(s)} = \frac{\alpha(s) - a(s)}{Ka(s)} = k^T (sI - A)^{-1} B \qquad ...(12.103)$$

$$= k^T Bs\text{–}1 + k^T ABs^{-2} + k^T A^2 Bs^{-3} + \ldots$$

The last equation is obtained from the power series expansion

$$(sI - A)^{-1} = s^{-1}(I - s^{-1}A)^{-1} = s^{-1}(I + s^{-1}A + s^{-2}A^2 + ...) \quad ...(12.104)$$

Thus we have a simple procedure for obtaining the elements of k, through the following steps:

1. From the specified locations of the poles of the transfer function of the closed-loop system, determine $\alpha(s)$. Also, determine K from the numerator of the transfer function of the open-loop and $\alpha(s)$ so that the dc gain is equal to one.
2. Determine $f(s)$ by subtracting $a(s)$ from $\alpha(s)$ and dividing by K.
3. Divide $f(s)$ by $a(s)$ to obtain a power series in nagative powers of s. Only n terms are needed, where n is the order of the system. Let this be given by

$$\frac{f(s)}{a(s)} = f_1 s^{-1} + f_2 s^{-2} + ... + f_n s^{-n} + ... \quad ...(12.105)$$

4. In view of Equation (12.104), we get the following equation:

$$k^T \begin{bmatrix} B & AB & A^2B & \cdots & A^{n-1}B \end{bmatrix} = \begin{bmatrix} f_1 & f_2 & f_3 & \cdots & f_n \end{bmatrix} \quad ...(12.106)$$

which may also be written as

$$U^T \begin{bmatrix} k_1 \\ k_2 \\ \vdots \\ k_n \end{bmatrix} = \begin{bmatrix} f_1 \\ f_2 \\ \vdots \\ f_n \end{bmatrix} \quad ...(12.107)$$

where U is the controllability matrix for the system.

We may now solve Equation (12.107) for the feedback vector k. It follows that a solution to Equation (12.107) will exist if U is nonsingular

EXAMPLE 12.8

For Example 12.6 the open-loop transfer function is given by

$$\frac{Y(s)}{U(s)} = \frac{9s+34}{s^3+15s^2+54s+40} \quad ...(12.108)$$

as in Equation (12.72). Also, from Equation (12.73), the desired transfer function of the closed-loop system is

$$\frac{Y(s)}{R(s)} = \frac{K(9s+34)}{s^3+20s^2+232s+1088} \quad ...(12.109)$$

We must make $K = 32$ in order that the dc gain may be done. Thus, we have

$$f(s) = \frac{\alpha(s) - a(s)}{K} = \frac{5}{32}s^2 + \frac{89}{16}s + \frac{131}{4} \quad (12.110)$$

By long division of $f(s)$ by $a(s)$, we obtain

$$\frac{f(s)}{a(s)} = \frac{5}{32}s^{-1} + \frac{103}{32}s^{-2} - \frac{767}{32}s^{-3} + ... \quad ...(12.111)$$

The controllability matrix, U of this system is given in Equation (12.85), and is repeated below.

$$U = \begin{bmatrix} 1 & -1 & -1 \\ 0 & -2 & 28 \\ -1 & 10 & -100 \end{bmatrix}$$

We may now use Equation (12.107) to obtain

$$\begin{bmatrix} 1 & 0 & -1 \\ -1 & -2 & 10 \\ -1 & 28 & -100 \end{bmatrix} \cdot \begin{bmatrix} k_1 \\ k_2 \\ k_3 \end{bmatrix} = \begin{bmatrix} \dfrac{5}{32} \\ \dfrac{103}{32} \\ -\dfrac{767}{32} \end{bmatrix} \qquad ...(12.112)$$

Solution of this set of equations leads to

$$k_1 = 1.09375,\ k_2 = 2.53125,\ k_3 = 0.9375 \qquad ...(12.113)$$

which agrees with the answer obtained in Equation (12.78).

DRILL PROBLEM 12.7

Repeat Drill Problem 12.6 using the transfer function approach discussed in this section.

12.6 ASYMPTOTIC STATE OBSERVERS

The concept of state feedback discussed in Section 12.4 is very powerful since it allows us to place the poles of the transfer function of the closed-loop system at desired locations provided that the system is completely controllable. In practice, however, all the states of a system are seldom available for feedback. Hence, in order to implement the state feedback, we must somchow obtain (or reconstruct) the states from the input and the output of the system. This will be possible if the system is completely observable.

One simple (but impractical) solution to the problem of state reconstruction is to differentiate the output of the system $n - 1$ times, where n is the order of the system, to obtain a state vector consisting of the output and these derivatives. This is seldom desirable due to the inevitable presence of noise, which gets accentuated by differentiation. A better alternative is to use an analog or digital computer model of the system, and obtain the states by applying the same input to the model and the system. Thus, if the system is described by Equations (12.44) and (12.45), we may set up a model given by the equations

$$\frac{d\hat{x}}{dt} = A\hat{x} + Bu \qquad ...(12.114)$$

$$\hat{y} = C\hat{x} \qquad ...(12.115)$$

where $\hat{x}(t)$ is the state vector for this model.

Note that in general, $x(t)$ and $\hat{x}(t)$ will not be equal unless the initial conditions $x(0)$ and

$\hat{x}(0)$ are also equal. Since $x(0)$ is not known, this is not possible. We may overcome this difficulty by utilizing a correction term based on our knowledge of the output $y(t)$. Consider changing the differential Equation (12.114) to

$$\frac{d\hat{x}}{dt} = A\hat{x}+Bu+L\left(y-\hat{y}\right) \qquad ...(12.116)$$

where L is a suitably chosen gain vector of the same dimension as x. The introduction of this feedback will, as we shall see, reduce the error between $x(t)$ and $\hat{x}(t)$ asymptotically. A block diagram for the resulting asymptotic state observer is shown in Fig. 12.5.

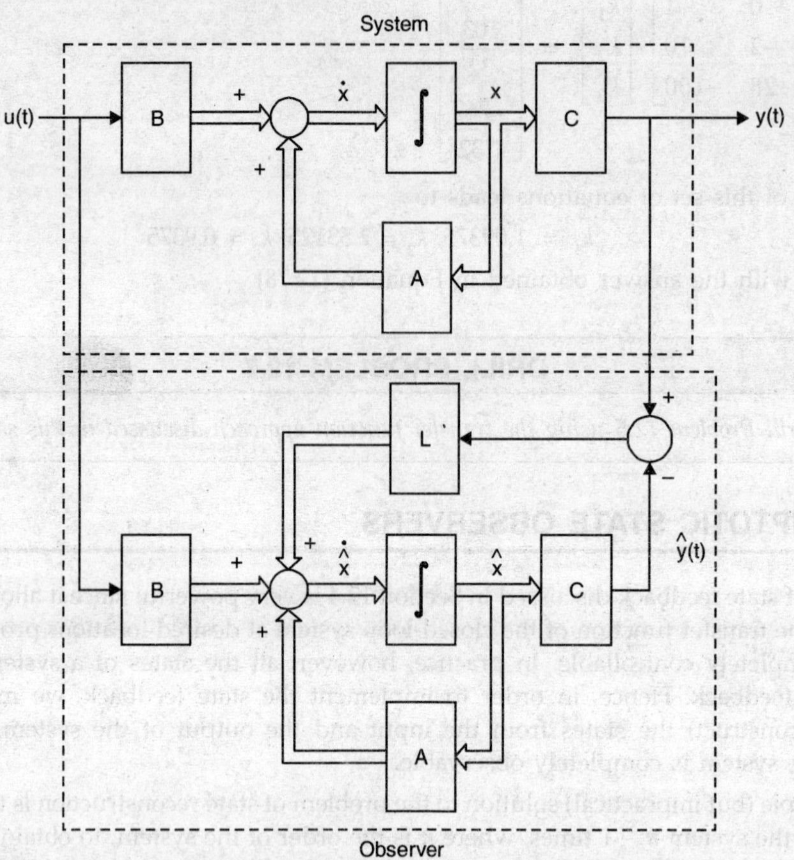

Figure 12.5. *Asymptotic state observer*

Define the error in state reconstruction as

$$\tilde{x}(t) = x(t)-\hat{x}(t) \qquad ...(12.117)$$

Differentiating both sides of Equation (12.117) and substituting for the derivatives of x and $\hat{x}$ from Equations (12.44) and (12.116), respectively, we have

$$\frac{d\tilde{x}}{dt} = x\left(x-\hat{x}\right)-L\left(y-\hat{y}\right) \qquad ...(12.118)$$

Expressing y and $\hat{y}$ in terms of x and $\hat{x}$ through Equations (12.45) and (12.115), we obtain

$$\frac{d\tilde{x}}{dt} = (A - LC)\,\tilde{x} \qquad \text{...(12.119)}$$

Equation (12.119) can be solved to give the state reconstruction error as

$$\tilde{x}(t) = e^{(A-LC)t}\,\tilde{x}(0) \qquad \text{...(12.120)}$$

where

$$\tilde{x}(0) = x(0) - \hat{x}(0) \qquad \text{...(12.121)}$$

It follows from Equation (12.120) that although initially the state reconstruction error may not be zero, we can make it decrease asymptotically by proper choice of the vector L so that all the eigenvalues of the matrix $(A - LC)$ have negative real parts. We now have the following theorem, which is similar to Theorem 12.3 stated earlier in connection with the design of the state feedback controller.

THEOREM 12.4

If a single-input single-output system is completely observable, it is always possible to determine a gain vector L so that the eigenvalues of $(A - LC)$ are placed at arbitrarily specified locations.

Proof: If the system is completely observable, one can always find a transformation matrix P so that $\bar{x} = P^{-1}x$, and the resulting matrices $\bar{A} = P^{-1}AP$, $\bar{B} = P^{-1}B$, and $\bar{C} = CP$ are in the observer canonical form. The method for determining P will be described in Section 12.6.1.

Thus, we get

$$\frac{d\bar{x}}{dt} = \bar{A}\bar{x} + \bar{B}\dot{u} \qquad \text{...(12.122)}$$

and

$$y = \bar{C}\bar{x} \qquad \text{...(12.123)}$$

where

$$\bar{A} = \begin{bmatrix} -a_1 & 1 & 0 & 0 & \cdots & 0 \\ -a_2 & 0 & 1 & 0 & \cdots & 0 \\ \vdots & \vdots & \vdots & \vdots & \cdots & \vdots \\ -a_{n-1} & 0 & 0 & 0 & \cdots & 1 \\ -a_n & 0 & 0 & 0 & \cdots & 0 \end{bmatrix} \qquad \text{...(12.124)}$$

$$\bar{C} = [1 \quad 0 \quad 0 \quad 0 \quad \cdots \quad 0] \qquad \text{...(12.125)}$$

and

$$\det\left(sI - \bar{A}\right) = \det(sI - A) = s^n + a_1 s^{n-1} + a_2 s^{n-2} + \dots + a_{n-1}s + a_n \qquad \text{...(12.126)}$$

It follows that if

$$\bar{L} = \begin{bmatrix} \ell_1 \\ \ell_2 \\ \vdots \\ \ell_n \end{bmatrix} \qquad \text{...(12.127)}$$

then

$$\bar{A} - \bar{L}\bar{C} = \begin{bmatrix} -a_1 - \ell_1 & 1 & 0 & 0 & \cdots & 0 \\ -a_2 - \ell_2 & 0 & 1 & 0 & \cdots & 0 \\ \vdots & \vdots & \vdots & \vdots & \cdots & \vdots \\ -a_{n-1} - \ell_{n-1} & 0 & 0 & 0 & \cdots & 1 \\ -a_n - \ell_n & 0 & 0 & 0 & \cdots & 0 \end{bmatrix} \qquad \text{...(12.128)}$$

Hence

$$\det\left[sI-\left(\bar{A}-\bar{L}\bar{C}\right)\right] = s^n + (a_1 + \ell_1)s^{n-1} + (a_2 + \ell_2)s^{n-2} + \ldots + (a_{n-1} + \ell_{n-2})s + (a_n + \ell_n) \quad \ldots(12.129)$$

It is evident that by a proper choice of the elements of $\bar{L}$, we can place the eigenvalues of $\left(\bar{A}-\bar{L}\bar{C}\right)$ at desired locations. Also, if we make

$$L = P\bar{L} \quad \ldots(12.130)$$

or

$$\bar{L} = P^{-1}L \quad \ldots(12.131)$$

then

$$\bar{A}-\bar{L}\bar{C} = P^{-1}AP - (P^{-1}L)\,(CP)$$
$$= P^{-1}\,(A - LC)\,P \quad \ldots(12.132)$$

which shows that $\left(\bar{A}-\bar{L}\bar{C}\right)$ and $(A - LC)$ have the same eigenvalues.

Note that the proof for Theorem 12.4 is very similar to that for Theorem 12.3. This is due to the duality between controllability and observability, and is the result of the fact that $\bar{A}$ and $\bar{B}$ for the controller form can be obtained by transposing $\bar{A}$ and $\bar{C}$ of the observer form respectively.

EXAMPLE 12.9

Consider a system described by the state equations

$$\dot{x} = \begin{bmatrix} -2 & 1 & 0 \\ -2 & 0 & 1 \\ -1 & 0 & 0 \end{bmatrix} x + \begin{bmatrix} 0 \\ 2 \\ 1 \end{bmatrix} \quad \ldots(12.133)$$

$$y = \begin{bmatrix} 1 & 0 & 0 \end{bmatrix} x \quad \ldots(12.134)$$

These equations are already in the observer form. We shall design an observer that has eigenvalues at $s = -4, -5, -6$. Consequently, the characteristic polynomial for the observer is

$$\Delta_o(s) = (s + 4)\,(s + 5)\,(s + 6) = s^3 + 15s^2 + 74s + 120 \quad \ldots(12.135)$$

Thus we must have

$$A - LC = \begin{bmatrix} -15 & 1 & 0 \\ -74 & 0 & 1 \\ -120 & 0 & 0 \end{bmatrix} \quad \ldots(12.136)$$

which requires that the observer gain vector be

$$L = \begin{bmatrix} 13 \\ 72 \\ 119 \end{bmatrix} \quad \ldots(12.137)$$

12.6.1 Transformation to the Observer Form

We shall now consider the problem of transformation of the state equations to the observer canonical form. As in the case of transformation to the controller form, we note that

$$\overline{V} = \begin{bmatrix} \overline{C} \\ \overline{C}\,\overline{A} \\ \vdots \\ \overline{C}\,\overline{A}^{n-1} \end{bmatrix} = VP \qquad ...(12.138)$$

where P is the matrix required for the transformation; that is $\overline{A} = P^{-1}AP$, $\overline{B} = P^{-1}B$ and $\overline{C} = CP$. Hence, we get the following relationships:

$$P = V^{-1}\overline{V} \qquad ...(12.139)$$

and

$$P^{-1} = \overline{V}^{-1}V \qquad ...(12.140)$$

The latter is more suitable since $\overline{V}^{-1}$ can be obtained directly from the characteristic polynomial for A. Let

$$\det(sI - A) = s^n + a_1 s^{n-1} + a_2 s^{n-2} + \ldots + a_{n-1}s + a_n \qquad ...(12.141)$$

It can then be shown that

$$\overline{V}^{-1} = \begin{bmatrix} 1 & 0 & 0 & \cdots & 0 \\ a_1 & 1 & 0 & \cdots & 0 \\ a_2 & a_1 & 1 & \cdots & 0 \\ \vdots & \vdots & \vdots & \cdots & \vdots \\ a_{n-1} & a_{n-2} & a_{n-3} & \cdots & 1 \end{bmatrix} \qquad ...(12.142)$$

which can be written by inspection if the characteristic polynomial is known.

EXAMPLE 12.10

Consider the system described in Example 12.6, for which

$$A = \begin{bmatrix} -1 & 1 & 0 \\ 0 & -4 & 2 \\ 0 & 0 & -10 \end{bmatrix} \quad C = \begin{bmatrix} 1 & 0 & 1 \end{bmatrix} \qquad ...(12.143)$$

The observability matrix for the system is given by

$$V = \begin{bmatrix} 1 & 0 & 1 \\ -1 & 1 & -10 \\ 1 & -5 & 102 \end{bmatrix} \qquad ...(12.144)$$

Also, as shown in Example 12.7,

$$\Delta(s) = \det(sI - A) = s^3 + 15s^2 + 54s + 40 \qquad ...(12.145)$$

This gives

$$\overline{V}^{-1} = \begin{bmatrix} 1 & 0 & 0 \\ 15 & 1 & 0 \\ 54 & 15 & 1 \end{bmatrix} \qquad ...(12.146)$$

so that

$$P^{-1} = \overline{V}^{-1}V = \begin{bmatrix} 1 & 0 & 1 \\ 14 & 1 & 5 \\ 40 & 10 & 6 \end{bmatrix} \qquad ...(12.147)$$

It is easily verified that

$$P^{-1}A = \bar{A}P^{-1}, \quad \text{and} \quad C = \bar{C}P^{-1} \qquad \text{...(12.148)}$$

DRILL PROBLEM 12.8

Design an observer for the system described in Drill Problem 12.3 so that the eigenvalues of $(A - LC)$ are located at -4 and $-4 \pm j2$.

Ans. $L = \begin{bmatrix} -3 \\ 16 \\ -21 \end{bmatrix}$

12.7 COMBINED OBSERVER-CONTROLLER COMPENSATOR

Our main objective in designing an observer was to reconstruct the states so that they may be utilized for state-variable feedback. The block diagram of a compensator combining the observer with state feedback is shown in Fig. 12.6.

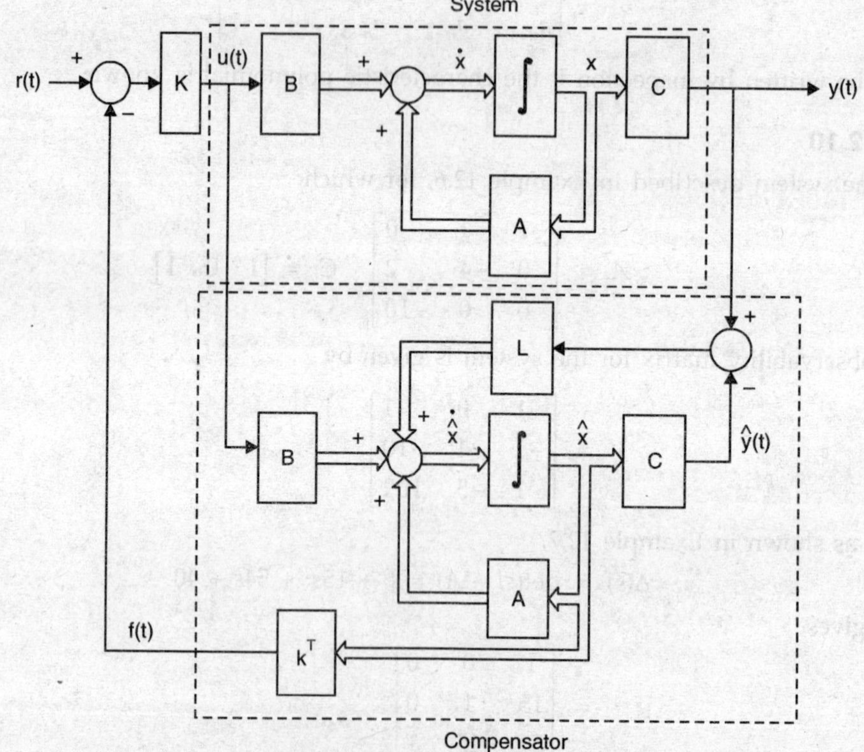

Figure 12.6. *Compensator combining observer with state feedback*

It will be seen that the observer, driven by the system input $u(t)$ and the output $y(t)$, reconstructs the state $\hat{x}(t)$, which is utilized for state feedback so that

$$u(t) = K\,[r(t) - k^T \hat{x}(t)] \qquad \text{...(12.149)}$$

A basic question that arises here is whether the use of $\hat{x}(t)$ for state variable feedback instead of the actual state $x(t)$ will affect the stability and overall performance of the system. We shall now attempt to answer this question.

In our design of the observer, we had seen that $\hat{x}(t)$ approaches $x(t)$ asymptotically if all the eigenvalues of $(A - LC)$ have negative real parts. Hence, in the steady-state, for all practical purposes $\hat{x}(t)$ and $x(t)$ will be identical. It can be shown that the transfer function relating $Y(s)$ to $R(s)$ of the system shown in Fig. 12.6 will be identical to the transfer function of the system shown in Fig. 12.4 (for example, see Section 4.2 of Kailath [1]). This follows intuitively from the fact that the transfer functions are based on zero initial conditions, with the result that $\hat{x}(t)$ and $x(t)$ are identical for this case. The transfer function of the overall system is given by

$$\frac{Y(s)}{R(s)} = \frac{Kb(s)\,\Delta_o(s)}{\alpha(s)\,\Delta_o(s)} \qquad \text{...(12.150)}$$

where $b(s)$ is the numerator of the transfer function of the original system, $\alpha(s)$ is the desired characteristic polynomial, and $\Delta_o(s)$ is the characteristic polynomial of the observer, that is,

$$\Delta_o(s) = \det\,[sI - (A - LC)] \qquad \text{...(12.151)}$$

This shows that, as long as the roots of $\Delta_o(s)$ have negative real parts, we can design the observer and the state feedback controller separately, and combine them as shown in Fig. 12.6.

The actual implementation of the combined observer-controller compensator requires the use of a model for the system. This is not a problem, since one can easily construct an electronic analog from the given state equations (as described in Chapter 3) and generate the states using integrators. Alternatively, one may implement the scheme through a microprocessor designed specially to integrate the state equations for the observer and hence calculate $k^T \hat{x}$.

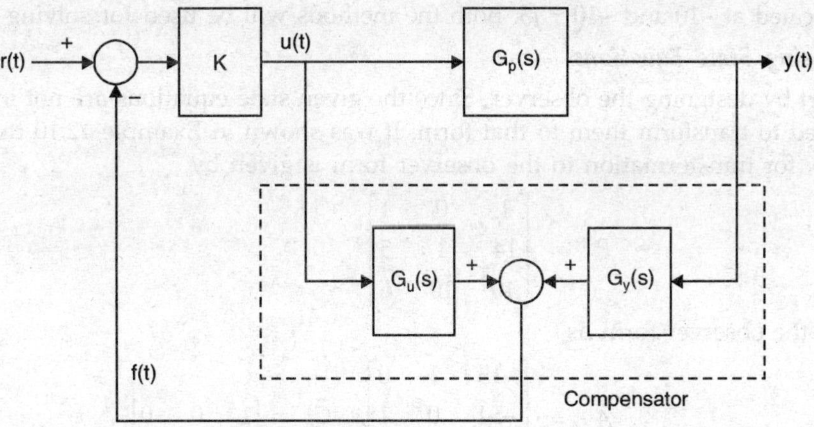

Figure 12.7. *Compensator implementation of observer-controller*

It is very revealing to redraw the block diagram of Fig. 12.6 as in Fig. 12.7. This follows from the fact that the input to the compensator consists of $u(t)$ and $y(t)$ and its output is $f(t) = k^T \hat{x}(t)$. Hence, the block marked compensator in Fig. 12.6 can be replaced by the two transfer functions shown in Fig. 12.7 so that

$$F(s) = G_u(s)\,U(s) + G_y(s)\,Y(s) \qquad \text{...(12.152)}$$

Therefore, one may also solve the problem of compensator design by selecting the two compensators.

$$G_u(s) = \frac{n_u(s)}{\Delta_o(s)} \quad ...(12.153)$$

and
$$G_y(s) = \frac{n_y(s)}{\Delta_o(s)} \quad ...(12.154)$$

so that
$$\frac{Y(s)}{R(s)} = \frac{Kb(s)\,\Delta_o(s)}{\alpha(s)\,\Delta_o(s)} \quad ...(12.155)$$

This is precisely the problem that was discussed in Chapter 10 (Section 10.7). Note that although there is no essential difference between the two schemes for compensator design, the transfer function approach requires less computation in addition to the fact that it is intuitively more appealing. It should be appreciated, however, that the block diagram shown in Fig. 12.7 was motivated by the state variable approach shown in Fig. 12.6.

EXAMPLE 12.11

Consider a system described by the following state equations.

$$\dot{x} = \begin{bmatrix} -1 & 1 & 0 \\ 0 & -4 & 2 \\ 0 & 0 & -10 \end{bmatrix} x + \begin{bmatrix} 1 \\ 0 \\ -1 \end{bmatrix} u \quad ...(12.156)$$

$$y = \begin{bmatrix} 1 & 0 & 1 \end{bmatrix} x \quad ...(12.157)$$

We shall design a compensator so that the poles of the closed-loop system are placed at – 8 and – 6 ± *j*10 while providing a dc gain of one. It is also specified that the poles of the observer should be located at –10 and –10 ± *j*5. Both the methods will be used for solving this problem.

Method I: Using State Equations

We shall start by designing the observer. Since the given state equations are not in the observer form, we need to transform them to that form. It was shown in Example 12.10 that the inverse of the matrix for transformation to the observer form is given by

$$P^{-1} = \begin{bmatrix} 1 & 0 & 1 \\ 14 & 1 & 5 \\ 40 & 10 & 6 \end{bmatrix} \quad ...(12.158)$$

which gives the observer form as

$$\overline{A}_o = \begin{bmatrix} -15 & 1 & 0 \\ -54 & 0 & 1 \\ -40 & 0 & 0 \end{bmatrix}, \quad \overline{C}_o = \begin{bmatrix} 1 & 0 & 0 \end{bmatrix} \quad ...(12.159)$$

From the specified locations of the poles of the observer, we have

$$\Delta_o(s) = (s + 10)(s^2 + 20s + 125) = s^3 + 30s^2 + 325s + 1250 \quad ...(12.160)$$

In order to get this characteristic polynomial, we must have

$$\overline{A}_o - \overline{L}\overline{C}_o = \begin{bmatrix} -30 & 1 & 0 \\ -325 & 0 & 1 \\ -1250 & 0 & 0 \end{bmatrix} \quad ...(12.161)$$

Subtracting the first column of $\bar{A}_o - \bar{L}\bar{C}_o$ from the first column of $\bar{A}_o$, we get

$$\bar{L} = \begin{bmatrix} 15 \\ 271 \\ 1210 \end{bmatrix} \qquad ...(12.162)$$

Finally, the observer gain matrix for the original state equations is found as

$$L = P\bar{L} = \begin{bmatrix} 15 \\ 61 \\ 0 \end{bmatrix} \qquad ...(12.163)$$

Transformation of the state equations for this example to the controller form was done in Example 12.7. Also, the state feedback vector for placing the poles at the specified locations was calculated in Examples 12.6 as well as 12.8. The final results are given below

$$K = 32 \qquad ...(12.164)$$

$$k^T = \begin{bmatrix} 1.09375 & 2.53125 & 0.9375 \end{bmatrix} \qquad ...(12.165)$$

With the values of L, K and k^T thus obtained, we may set up the scheme for combined observer-controller compensation shown in Fig. 12.6.

Method II: Using Transfer Functions

Following the development in Section 10.7, and using the transfer function evaluated in Example 12.6, we note that

$$b(s) = N_p(s) = 9s + 34 \qquad ...(12.166)$$

and

$$a(s) = D_p(s) = s^3 + 15s^2 + 54s + 40 \qquad ...(12.167)$$

Also,

$$\alpha(s) = D(s) = s^3 + 20s^2 + 232s + 1088 \qquad ...(12.168)$$

and

$$\Delta(s) = s^3 + 30s^2 + 325s + 1250$$

Hence,

$$[D(s) - D_p(s)]\,\Delta(s)/K = (5s^2 + 178s + 1048)(s^3 + 30s^2 + 325s + 1250)/32 \qquad ...(12.169)$$

$$= 0.15625\, s^5 + 10.25\, s^4 + 250.40625\, s^3 + 2985.625\, s^2 + 17596.875\, s + 40937.5$$

Let

$$N_y(s) = q_0 s^2 + q_1 s + q_2 \qquad ...(12.170)$$

and

$$N_u(s) = r_0 s^2 + r_1 s + r_2 \qquad ...(12.171)$$

The coefficients of $N_u(s)$ and $N_y(s)$ must be determined so that

$$N_y(s)\, N_p(s) + N_u(s)\, D_p = [D(s) - D_p(s)]\,\Delta(s)/K \qquad ...(12.172)$$

These equations can be arranged in the matrix form (see Section 10.7) so that

$$\begin{bmatrix} 1 & 0 & 0 & 0 & 0 & 0 \\ 15 & 1 & 0 & 0 & 0 & 0 \\ 54 & 15 & 1 & 9 & 0 & 0 \\ 40 & 54 & 15 & 32 & 9 & 0 \\ 0 & 40 & 54 & 0 & 34 & 9 \\ 0 & 0 & 40 & 0 & 0 & 34 \end{bmatrix} \begin{bmatrix} r_0 \\ r_1 \\ r_2 \\ q_0 \\ q_1 \\ q_2 \end{bmatrix} = \begin{bmatrix} 0.15625 \\ 10.25 \\ 250.40625 \\ 2985.625 \\ 17596.875 \\ 40937.5 \end{bmatrix} \qquad ...(12.173)$$

Solution of these equations leads to

$$N_u(s) = 0.15625s^2 + 7.90625s - 1413.9375 \quad ...(12.174)$$

and

$$N_y(s) = 170.8125s^2 + 1994.875s + 2867.5 \quad ...(12.175)$$

It may be noted that it is possible to reduce the order of the observer from n to $n - 1$. This follows from the fact that the output can be treated as one of the states, as will be seen from the observer form, where the first state is the output. Consequently, we only need to reconstruct the remaining $n - 1$ states. As the design of the reduced-order observer, using the state variable approach is a little involved, we shall not discuss it here. The interested reader can find more details in the reference [1], listed at the end of this chapter.

The transfer function approach allows the design of the reduced-order observer-compensator without any additional complication. As a matter of fact, the discussion in Section 10.7 of Chapter 10 assumes that the degree of $\Delta(s)$ is $n - 1$, and then calculates the compensator transfer function. For instance, we shall use this approach to obtain minimal order compensators for the system that was considered in Example 12.11.

EXAMPLE 12.12

For the system described in Example 12.11, design compensators with poles at $-10 \pm j5$, placing the poles of the closed-loop system at -8 and $-6 \pm j10$.

We shall proceed, as in the previous example, but now we have

$$\Delta(s) = s^2 + 20s + 125 \quad ...(12.176)$$

Consequently, as discussed in Section 10.7 in Chapter 10, we now make

$$N_y(s) = q_0 s^2 + q_1 s + q_2 \quad ...(12.177)$$

and

$$N_u(s) = r_0 s + r_1 \quad ...(12.178)$$

Also, we have

$$[D(s) - D_p(s)]\,\Delta(s) = (5s^2 + 178s + 1048)\,(s^2 + 20s + 125) \quad ...(12.179)$$

$$= 5s^4 + 278s^3 + 5233s^2 + 43210s + 131000$$

The coefficients of $N_u(s)$ and $N_y(s)$ are determined by solving the following matrix equation:

$$\begin{bmatrix} 1 & 0 & 0 & 0 & 0 \\ 15 & 1 & 288 & 0 & 0 \\ 54 & 15 & 1088 & 288 & 0 \\ 40 & 54 & 0 & 1088 & 288 \\ 0 & 40 & 0 & 0 & 108 \end{bmatrix} \begin{bmatrix} r_0 \\ r_1 \\ q_0 \\ q_1 \\ q_2 \end{bmatrix} = \begin{bmatrix} 5 \\ 278 \\ 5233 \\ 43210 \\ 131000 \end{bmatrix} \quad ...(12.180)$$

Solution of these equations leads ot the following:

$$N_u(s) = 0.15625s - 231.09152 \quad ...(12.181)$$

and

$$N_y(s) = 26.3817s^2 + 302.72098s + 392.27679 \quad ...(12.182)$$

DRILL PROBLEM 12.9

Consider a system described by the following state equations:

$$\dot{x} = \begin{bmatrix} 0 & 0 & 0 \\ 1 & 0 & 0 \\ 0 & 1 & -1 \end{bmatrix} x + \begin{bmatrix} 1 \\ 0 \\ 0 \end{bmatrix} u$$

$$y = \begin{bmatrix} 0 & 0 & 1 \end{bmatrix} x$$

A compensator is to be designed so that the poles of the closed-loop system are placed at –2 and –2 ± j2 while providing a dc gain of one. It is also specified that the poles of the observer should be located at –5 and –4 ± j2. Design the compensator using both the methods discussed in Section 12.7.

Ans. $K = 16,\ k^T = \begin{bmatrix} \frac{5}{16} & \frac{11}{16} & \frac{5}{16} \end{bmatrix},\ L = \begin{bmatrix} 100 \\ 60 \\ 12 \end{bmatrix}$

$$G_y(s) = \frac{76.25\,s^2 + 160\,s + 100}{s^3 + 13s^2 + 60s + 100}$$

$$G_u(s) = \frac{0.3125\,s^2 + 4.75\,s + 28}{s^3 + 13\,s^2 + 60\,s + 100}$$

DRILL PROBLEM 12.10

Design a reduced-order compensator for the system described in Drill Problem 12.9. The poles of the observer are to be located at – 4 ± j2. [Hint: Use the transfer function approach.]

Ans. $K = 16,\ G_u(s) = \dfrac{0.3125\,s + 3.1875}{s^2 + 8s + 20}$

$$G_y(s) = \frac{12.0625\,s^2 + 28\,s + 20}{s^2 + 8s + 20}$$

12.8 INTRODUCTION TO OPTIMAL CONTROL

The design of compensators discussed in the previous sections was based on placing the poles of the closed-loop system at specified locations. In optimal control theory, the approach is to determine the control input which minimizes an integral cost function. Here, we shall study the case of the optimal state regulator, where we seek the control input to minimize the function

$$J = \frac{1}{2}\int_0^\infty \left(x^T Q x + u^T R u\right) dt \qquad ...(12.183)$$

where Q and R are positive definite weighting matrices. The terms x^TQx and u^TRu in the integrand are quadratic forms, which provide a measure of the performance and the cost of control, respectively. In the simplest case Q and R are diagonal matrices with positive elements and for a single-input single-output system, R is a positive scalar. The choice of Q and R often requires careful consideration.

We shall investigate the special case of a linear system described by the standard state equations

$$\dot{x} = Ax + Bu \qquad y = Cx \qquad \text{...(12.184)}$$

THEOREM 12.5

The optimal control for the system described by Equation (12.184), which minimizes the cost function in Equation (12.183), is obtained by state feedback, that is

$$u = -Kx \qquad \text{...(12.185)}$$

where the constant feedback matrix K is given by

$$K = R^{-1}B^TP \qquad \text{...(12.186)}$$

and P is a symmetric positive definite matrix obtained by solving the algebraic matrix Riccati equation

$$PA + A^TP + Q - PBR^{-1}B^TP = 0 \qquad \text{...(12.187)}$$

(**Note:** It is assumed that the system is completely controllable)

We shall not prove this theorem in this book. The interested reader may see one of the books listed in Section 12.10.

The Riccati equation can be solved analytically only for simple cases and in, general, a computer solution is required. Computer programmes for solving this equation are available in most computer libraries (see reference [4] in Section 12.10). We shall illustrate the main idea by a simple example of a second-order system.

EXAMPLE 12.13

Consider a system described by the state equations

$$\dot{x} = \begin{bmatrix} -2 & 0 \\ 1 & 0 \end{bmatrix} x + \begin{bmatrix} 1 \\ 0 \end{bmatrix} u \qquad \text{...(12.188)}$$

It is required to find the feedback vector that will minimize the cost function given by Equation (12.183) if

$$Q = \begin{bmatrix} 1 & 0 \\ 0 & 4 \end{bmatrix} \quad R = 1 \qquad \text{...(12.189)}$$

SOLUTION:

Let

$$P = \begin{bmatrix} p_1 & p_2 \\ p_2 & p_3 \end{bmatrix} \qquad \text{...(12.190)}$$

Then

$$PA = \begin{bmatrix} p_1 & p_2 \\ p_2 & p_3 \end{bmatrix} \begin{bmatrix} -2 & 0 \\ 1 & 0 \end{bmatrix} = \begin{bmatrix} p_2 - 2p_1 & 0 \\ p_3 - 2p_2 & 0 \end{bmatrix} \qquad \text{...(12.191)}$$

$$A^TP = \begin{bmatrix} -2 & 1 \\ 0 & 0 \end{bmatrix} \begin{bmatrix} p_1 & p_2 \\ p_2 & p_3 \end{bmatrix} = \begin{bmatrix} p_2 - 2p_1 & p_3 - 2p_1 \\ 0 & 0 \end{bmatrix} \qquad \text{...(12.192)}$$

$$PBR^{-1}B^TP = \begin{bmatrix} p_1 & p_2 \\ p_2 & p_3 \end{bmatrix}\begin{bmatrix} 1 \\ 0 \end{bmatrix}\begin{bmatrix} 1 & 0 \end{bmatrix}\begin{bmatrix} p_1 & p_2 \\ p_2 & p_3 \end{bmatrix} \quad ...(12.193)$$

$$= \begin{bmatrix} p_1^2 & p_1p_2 \\ p_1p_2 & p_2^2 \end{bmatrix}$$

Substituting these into the matrix Riccati equation, we obtain

$$\begin{bmatrix} 2p_2 - 4p_1 + 1 - p_1^2 & p_3 - 2p_2 - p_1p_2 \\ p_3 - 2p_2 - p_1p_2 & 4 - p_2^2 \end{bmatrix} = 0 \quad ...(12.194)$$

Equation (12.194) leads to three nonlinear scalar equations. Several solutions are possible, but only one of these gives a positive definite P. This is given below:

$$P = \begin{bmatrix} 1 & 2 \\ 2 & 6 \end{bmatrix} \quad ...(12.195)$$

The resulting feedback matrix is obtained as

$$K = R^{-1}B^TP = \begin{bmatrix} 1 & 0 \end{bmatrix}\begin{bmatrix} 1 & 2 \\ 2 & 6 \end{bmatrix} = \begin{bmatrix} 1 & 2 \end{bmatrix} \quad ...(12.196)$$

It is interesting to note that while the eigenvalues of A are located a –2 and 0, those of $(A - BK)$ are located at –1 and –2. Thus, the optimal control has minimized the cost function by placing the poles of the closed-loop system at suitable locations. One may say that this theory has determined the location of these poles in such a way that the given cost function is minimized. In general, the location of the eigenvalues of $(A - BK)$ will depend on the choice of Q and R.

Optimal control theory can be applied to general nonlinear systems with quite complicated integral performance criteria. It has been applied to minimum-time and minimum-energy problems, and has many applications. These are beyond the scope of this book. Our main aim was to give the reader a flavour of this interesting but highly mathematical subject.

SUMMARY

- State-space representation provides a powerful approach to the design of control systems. The important concepts of controllability and observability lead to the application of state-variable feedback and reconstruction of system states from the system input and output. If a system is completely controllable, state variable feedback allows us to place the poles of a closed-loop system at suitable locations in order to improve system performance.
- For example, consider the simple position control system with two poles in the open-loop transfer function, one at the origin and the other at $-1/\tau_m$, where τ_m = motor time constant. Then by adjusting the forward gain only, we can locate the poles of the closed-loop system transfer function at any point on the corresponding root locus plot. On the other hand, if we can feedback the velocity as well as the position, this will enable us to place the poles of the transfer function of the closed-loop system anywhere in the s-plane.
- Note, however, that we have to differentiate the position to get the velocity. Although it is not much of a problem in this case, for a higher-order system many differentiations may be required. This is not desirable, due to the fact that noise, which is invariably present, will cause problems, Moreover, feeding back all the state variables requires measuring them by using sensors. Not

only will this increase the cost of the control system, but will also have the effect of introducing into the system noise produced by non-ideal sensors.

- There is another way to generate the state of the system without requiring differentiation or measurement of all the states of the system. If the system is completely observable, this is possible by including the so-called 'asymptotic observer.' Its use, however, requires that we know the state equations of the system exactly. The same assumption is also made in designing state-variable feedback. One important point to consider, then, is the effect on this control scheme of small changes in system parameters. This is the parameter sensitivity problem. In particular, it would be desirable to know how the eigenvalues move in the *s*-plane with small variations in parameters. This is a topic for further study, and is usually referred to as the problem of robust control.
- A transfer function approach to the design of state-variable feedback provides more insight into the subject and also gives a better idea of pole-placement. It is seen how by using low compensators, one driven by the system input and the other by the system output, we can generate the feedback function that will place the poles at specified locations provided that the system is completely controllable and observable.
- The chapter concluded by an introduction to the powerful optimal control theory, which is based mainly on the use of state-space representation. Instead of assuming any special type of compensator, the idea here is to determine the control input to the system which will minimize a given cost function. We considered, briefly, the case of the optimal linear regulator.

Problems

1. The state equations of a linear system are given below:

$$\dot{x} = \begin{bmatrix} 0 & 0 & -8 \\ 1 & 0 & -14 \\ 0 & 1 & -7 \end{bmatrix} x + \begin{bmatrix} 2 \\ 1 \\ 0 \end{bmatrix} u$$

$$y = \begin{bmatrix} 0 & 0 & 1 \end{bmatrix} x$$

Determine whether the system is (*a*) controllable, (*b*) observable, and (*c*) stable.

2. For the error-sampled control system shown in Fig. P12.2, the response to a step input is required to be deadbeat and ripple-free, in addition to being as fast as possible. Determine the transfer function *D*(*z*) of a suitable compensator.

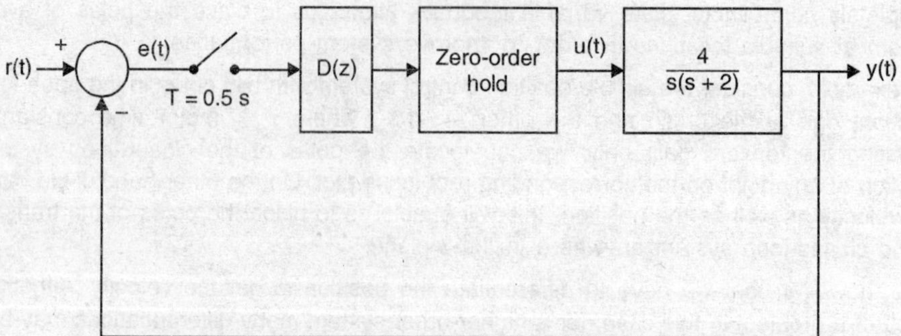

Figure P12.2. *Block diagram of a closed-loop discrete-time system*

3. The state equations of a linear system are given by

$$\dot{x} = \begin{bmatrix} -2 & 0 & 1 \\ 1 & -3 & 0 \\ 1 & 1 & -1 \end{bmatrix} x + \begin{bmatrix} 1 \\ 0 \\ 1 \end{bmatrix} u$$

$$y = [1 \quad 1 \quad -2]\, x$$

Determine the state feedback law $u = K\,(r - k^T x)$ so that the closed-loop system has the following transfer function.

$$\frac{Y(s)}{R(s)} = \frac{25\left(s^2 + 6s + 12s\right)}{(s+5)\left(s^2 + 10s + 60\right)}$$

4. If only the output is available for feedback in the previous problem, what compensator in the feedback path will give the same transfer function for the closed-loop system?

5. The state equations for a linear system are given below:

$$\dot{x} = \begin{bmatrix} 0 & 1 & -6 \\ 1 & 0 & 5 \\ 0 & -1 & 2 \end{bmatrix} x + \begin{bmatrix} 1 \\ 1 \\ 0 \end{bmatrix} u$$

$$y = [0 \quad 1 \quad 2]\, x$$

(*a*) Transform them to the controller canonical form, and hence, determine the transfer function of the system.

(*b*) Determine the state feedback vector that will move the poles to -2 and $-1 \pm j1$.

6. The state equations for a helicopter near hover are as follows:

$$\dot{x} = \begin{bmatrix} -0.002 & -1.4 & 9.8 \\ -0.01 & -0.04 & 0 \\ 0 & 1 & 2 \end{bmatrix} x + \begin{bmatrix} 9.8 \\ 6.3 \\ 1 \end{bmatrix} u$$

$$y = \begin{bmatrix} 0 & 0 & 1 \end{bmatrix} x$$

Determine the state feedback vector that will move the poles of the transfer function of the closed-loop system to -3 and $-2 \pm j2$.

7. Design an asymptotic state observer for the previous problem so that the poles of the observer are located at -4 and $-4 \pm j4$.

8. Consider the system described in Problem 3. It is assumed that only the system output is available. Design an observer with poles at -10 and $-10 \pm j5$.

9. Design a compensator for Problem 3 using the transfer function approach. The poles of the observer are to be located at -10 and $-10 \pm j5$.

10. Repeat Problem 9 with a second-order observer, having poles at $-10 \pm j5$.

11. Consider the state equations of a helicopter at hover, as given in Problem 6. We shall now investigate the case when the input to the system is sampled and held constant between the sampling instants. The sampling interval is 0.1 second.

(*a*) Calculate the input sequence that will transfer the state of the system from $\begin{bmatrix} 0 \\ 0 \\ 0 \end{bmatrix}$ to $\begin{bmatrix} 1 \\ 0 \\ 0 \end{bmatrix}$ in three sampling intervals.

(*b*) Design a digital controller so that the response of this system to a step input is deadbeat and ripple-free, using the state-variable approach presented in Section 12.2.

12. For Problem 11 determine the state feedback vector that will place the eigenvalues at 0.25 and $0.2 \pm j0.2$ in the *z*-plane. What will be the response of the system to a unit step input?

13. Design an observer for the discrete-time system in Problem 11 that will reconstruct the states from the input and the output. The eigenvalues of the observer are to be located at 0.1, 0.12 and 0.15 in the *z*-plane.

14. Design an observer-compensator of the minimum order for the discrete-time system of Problem 11 using the method of Section 10.7. The poles of the system are to be located as in Problem 12 and those of the observer are to be at 0.1 and 0.2 in the *z*-plane.

15. The inverted pendulum shown in Fig. P12.15 has been the subject of considerable study. Although the system is inherently unstable, it is possible to keep the stick from falling by applying a suitable input *u* (*t*) to the motorized cart. The linearized state equations are of the form

$$\dot{x} = Ax + Bu$$

where

$$A = \begin{bmatrix} 0 & 1 & 0 & 0 \\ 0 & 0 & -1 & 0 \\ 0 & 0 & 0 & 1 \\ 0 & 0 & 10 & 0 \end{bmatrix} \text{ and } B = \begin{bmatrix} 0 \\ 1 \\ 0 \\ -1 \end{bmatrix}$$

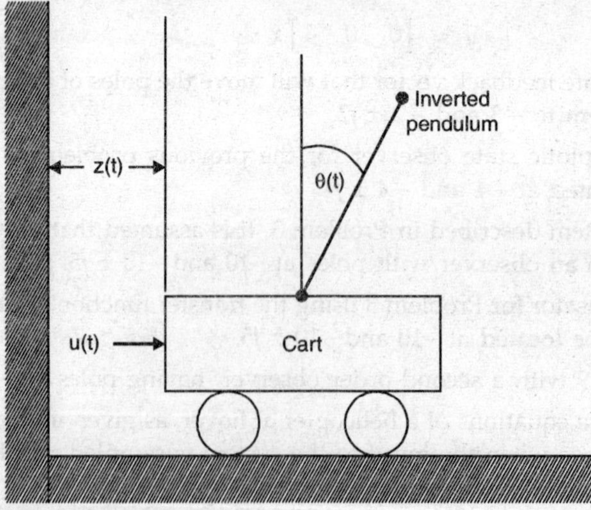

Figure P12.15. *Cart with inverted pendulum*

The state variables are as follows:

x_1 = The horizontal displacement, $z(t)$

x_2 = The horizontal velocity, $\dot{z}(t)$

x_3 = The angular rotation, $\theta(t)$

x_4 = The angular velocity, $\dot{\theta}(t)$

Determine the state feedback vector that will stabilize the system and place its poles at –1, –2, and $-1 \pm j1$.

16. Design an observer to reconstruct the states of the system described in Problem 15, assuming that only the output $z(t)$ is available. The eigenvalues of the observer are to be located at –3, –3, –4 and –4.

17. The linearized equations for a satellite in a circular equatorial orbit are given below:

$$\dot{x} = \begin{bmatrix} 0 & 1 & 0 & 0 \\ 3 & 0 & 0 & 2 \\ 0 & 0 & 0 & 1 \\ 0 & -2 & -3 & 0 \end{bmatrix} x + \begin{bmatrix} 0 \\ 1 \\ 0 \\ 0 \end{bmatrix} u$$

where the state variables are

x_1 = The distance from the centre of the Earth

x_2 = The rate of change of x_1

x_3 = Angular displacement in the equatorial plane

x_4 = The rate of change of x_3

The input $u(t)$ is the thrust produced by a rocket engine.

(*a*) Verify that this is a controllable system.

(*b*) Determine the state feedback vector that will place the poles of the transfer function of the closed-loop system at –1, –2 and $-1 \pm j1$.

18. Determine the state transition equation for the system described in the previous problem assuming a sampling interval of 0.1 second and that the input is held constant between the sampling instants. Calculate the state variable feedback that will cause the poles of the resulting discrete-time system to be placed at 0.15, 0.2 and $0.1 \pm j0.12$ in the *z*-plane.

19. Design a minimal-order observer-compensator for Problem 18 if only the state x_3 is available. Use the transfer function approach in the *z*-plane, placing the poles of the observer at 0.1 and $0.1 \pm j0.1$.

20. The discrete-time model of the mixing system of a rotary cement kiln is given below.

$$x(k+1) = \begin{bmatrix} 0 & 1 & 0 & 0 \\ 0.24 & 0 & 0 & 0.76 \\ 0 & 0.07 & 0.93 & 0 \\ 0 & 0 & 0 & 1 \end{bmatrix} x(k) + \begin{bmatrix} 1 \\ 0.76 \\ 0 \\ 1 \end{bmatrix} u(k)$$

(*a*) Is this system controllable?

(*b*) If the answer to part (*a*) is yes, then determine the state feedback vector that will place the eigenvalues at 0.2, 0.3 and $0.2 \pm j0.2$.

21. The state equations for the control of the quality of water in a river by effluent treatment are given below,

$$\dot{x} = \begin{bmatrix} -12 & -0.32 & 0 \\ 0 & -1.32 & 0 \\ -1 & 0 & 0 \end{bmatrix} x + \begin{bmatrix} 0 \\ 0.1 \\ 0 \end{bmatrix} u$$

where the states are

x_1 = Dissolved hydrogen

x_2 = Biochemical oxygen demand

x_3 = Integral of error between the specified reference level and dissolved oxygen level.

The input $u(t)$ is to be obtained from state feedback; that is

$$u(t) = -k^T x$$

where k is the state feedback vector.

Determine the value of k so that the characteristic polynomial of the resulting system is given by

$$P(s) = s^3 + 7.2s^2 + 21.6s + 27$$

22. Repeat Problem 21 if all the poles of the transfer function of the closed-loop system are to be located at $s = -3$.

23. The state equations for a linear system are given below:

$$\dot{x} = \begin{bmatrix} -1 & -2 & 0 & 0 & 0 & 0 \\ 2 & -1 & 0 & 0 & 0 & 0 \\ 0 & 0 & -5 & 0 & 0 & 0 \\ 0 & 0 & 0 & -2 & 1 & 0 \\ 0 & 0 & 0 & 0 & -2 & 1 \\ 0 & 0 & 0 & 0 & 0 & -2 \end{bmatrix} x + \begin{bmatrix} 1 \\ 0 \\ 0 \\ 0 \\ 0 \\ 1 \end{bmatrix} u$$

$$y = \begin{bmatrix} 0 & 11 & -1 & 2 & 0 & 7 \end{bmatrix} x$$

(*a*) Show, how you will simulate this system on an analog computer?

(*b*) Which of the states of the system are uncontrollable?

(*c*) Which of the states of the system are unobservable?

24. A discrete-time system is described by the transfer function

$$G(z) = \frac{z^2 + \alpha z}{z^3 - 0.8z^2 - 0.21z + 0.01}$$

(*a*) Derive the state equations for the system in the controller form and sketch an analog computer realization diagram.

(*b*) For what values of the parameter α will this realization be unobservable? What happens to the transfer function in that case?

(*c*) Derive a set of state equations in the observer canonical form and sketch an analog computer realization diagram.

(*d*) For what values of the parameter α will this realization be uncontrollable? What happens to the transfer function for these values of α?

25. Sometimes it is desirable to 'decouple' multiple-input multiple-output systems with equal numbers of inputs and outputs so that the resulting transfer function matrix is of the diagonal form. This is of considerable practical importance in systems where we want to adjust the value of one output without affecting the others. It can be accomplished by state feedback in systems that satisfy certain conditions. The control input for these systems is obtained as

$$u = Fx + Gu$$

The state equations for a particular two-input two-output system are given below:

$$\dot{x} = \begin{bmatrix} 0 & 1 & 0 & 0 & 0 \\ 0 & 0 & 1 & 0 & 0 \\ -6 & 5 & 2 & -5 & -6 \\ 0 & 0 & 0 & 0 & 1 \\ 0 & 0 & 0 & -2 & 3 \end{bmatrix} x + \begin{bmatrix} 0 & 0 \\ 0 & 0 \\ 1 & 0 \\ 0 & 0 \\ 0 & 1 \end{bmatrix} u$$

$$y = \begin{bmatrix} 3 & 1 & 0 & 0 & 0 \\ 0 & 1 & 2 & 0 & 1 \end{bmatrix} x$$

(*a*) Determine the transfer function matrix relating $Y(s)$ to $U(s)$.

(*b*) Determine the transfer function matrix relating $Y(s)$ to $V(s)$ if the matrices F and G are as given below:

$$F = \begin{bmatrix} 4 & -10 & 6 & 5 & 6 \\ 3 & 7 & 5 & 1 & 4 \end{bmatrix} \qquad G = \begin{bmatrix} 1 & 0 \\ -2 & 1 \end{bmatrix}$$

26. The regulation of glucose level in the blood can be expressed by the following linearized differential equations:

$$\dot{x}_1 = -ax_1 - bx_2$$
$$\dot{x}_2 = -cx_1 - dx_2 + u$$

where x_1 is the deviation of the extracellular insulin level from the mean, x_2 is the deviation of the extracellular glucose level from the mean, and u is the rate of glucose intravenous injection. It is desired to stabilize the glocose level by state feedback, that is by making u a linear combination of x_1 and x_2. Determine the state feedback vector that will locate the eigenvalues of the closed-loop system at $-1 + j1$ if the parameters are $a = 0.78$, $b = 0.208$, $c = 4.34$ and $d = 2.92$.

27. A linear system is described by the state equations

$$\begin{bmatrix} \dot{x}_1 \\ \dot{x}_2 \end{bmatrix} = \begin{bmatrix} 0 & 1 \\ 0 & 0 \end{bmatrix} \begin{bmatrix} x_1 \\ x_2 \end{bmatrix} + \begin{bmatrix} 0 \\ 1 \end{bmatrix} u$$

where $$u = -k_1x_1 - k_2x_2$$

(*a*) Determine the value of k_1 and k_2 that will minimize the cost function

$$J = \int_0^\infty \left(x_1^2 + x_2^2 + u^2\right) dt$$

(*b*) Determine the minimum value of J if $x_1(0) = 1$ and $x_2(0) = 0$.

References

1. Chen, C.T., *Linear System Theory and Design*, Holt, Rinehart and Winston, New York, 1984.
2. DeCarlo, R.A., *Linear Systems: A State Variable Approach with Numerical Implementation*, Prentice-Hall, Englewood Cliffs, New Jersey, 1989.
3. Fallside, F., (ed.), *Control System Design by Pole-Zero Assignment*, Academic Press, London, 1977.
4. Kailath, T., *Linear Systems*, Prentice-Hall, Englewood-Cliffs, New Jersey, 1980.
5. O' Reilly, J., *Observers for Linear Systems*, Academic Press, London, 1983.
6. Ogata, K., *Modern Control Engineering*, Prentice-Hall, Englewood Cliffs, New Jersey, 1990 (second edition).
7. Petkov, P. Hr., N.D. Christov, and M.M. Konstantinov, *Computational Methods for Linear Control Systems*, Prentice-Hall International, London, 1991.

CHAPTER 13

Nonlinear Systems

13.1 INTRODUCTION

In our discussions so far we have considered only linear system. In practice, however, all physical systems have some kind of nonlinearities. Sometimes it may even be desirable to introduce a nonlinearity deliberately in order to improve the performance of a system and make its operation safer. This may also result in making the system more economical than is possible with linear components alone.

One of the simplest examples of a system with an intentionally introduced non-linearity is a relay-controlled or on/off system. For instance, in a typical home-heating system, a furnace is turned on when the temperature falls below a certain specified value and it is off when the temperature exceeds another given value. Another example is a nonlinear controller designed to realize a damping ratio that varies with the magnitude of the actuating signal.

Nonlinear systems differ from the usual linear systems in several ways. Perhaps the most significant of these is the fact that the principle of superposition is not applicable to nonlinear systems. For example, the output of a nonlinear system, with zero initial conditions, to step function inputs of different magnitudes is shown in Fig. 13.1. Altering the size of the input does not change the shape of the response of the linear system, whereas for the nonlinear system there is considerable change in both the percentage overshoot and the frequency of oscillation.

Similar observations may also be made about stability. In linear systems, stability is a characteristic of the system, independent of the magnitude of the input or the initial conditions. In the case of nonlinear systems, stability may depend on the magnitude of the input as well as the initial conditions. Furthermore, application of a sinusoidal input to a stable linear system causes the steady-state output to be a sinusoid of the same frequency, which will, in general, differ from the input in phase and magnitude. In nonlinear systems, on the other hand, the steady-state output may contain harmonics of the input, and in some cases even subharmonics may arise.

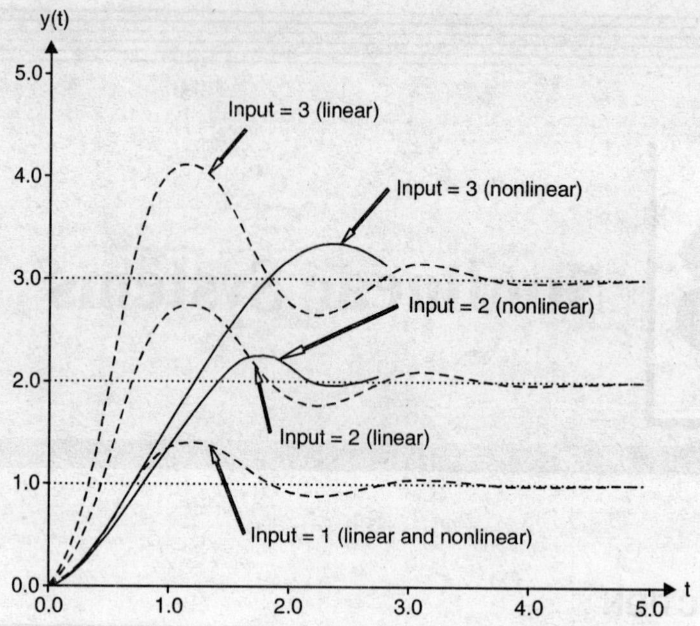

Figure 13.1. *Step responses of linear and nonlinear systems*

Other unusual features of nonlinear systems include such phenomena as limit cycles and jump resonance. The former means that independently of the magnitude of the input or initial conditions, the system may produce oscillations of a certain period and amplitude which may not be sinusoidal. The latter imply jumps in magnitude and phase as the frequency is changed resonance. This is illustrated in Fig. 13.2. If we apply a sinusoid of constant amplitude to a linear system, and change the frequency of the input, we get a frequency response curve as shown, in Fig. 13.2(*a*). It will be seen that the plot of the magnitude of the frequency response system is continuous. On the other hand, for some nonlinear systems, as the frequency is increased, a discontinuity or jump occurs as shown in Fig. 13.2(*b*). Another jump in the amplitude ratio occurs when the frequency is decreased.

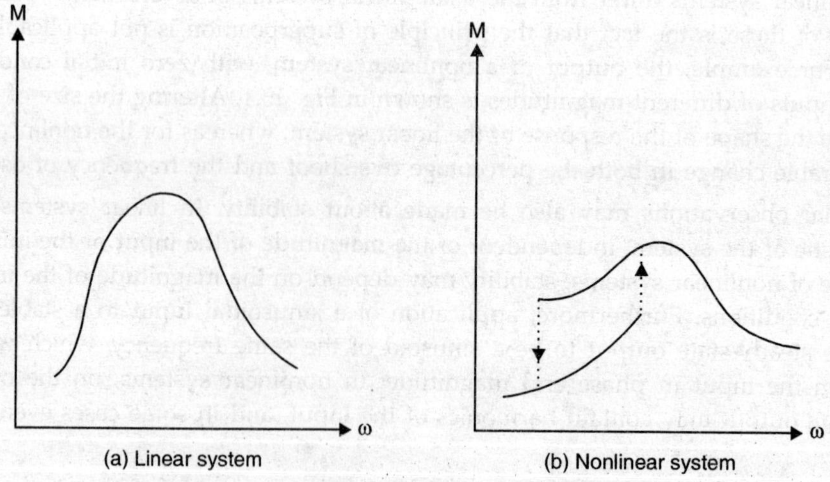

Figure 13.2. *Jump resonance in frequency response*

Although our list of special features of nonlinear systems is far from complete, it shows that the world of nonlinear systems is much richer than that of linear systems. Since, in practice, all physical systems have some nonlinearities we need an understanding of the basic methods for the analysis of nonlinear systems. The chapter is an attempt to do so.

13.2 COMMON NONLINEARITIES

In most control systems, we cannot avoid the presence of certain types of nonlinearities. These can be classified as *static or dynamic.* There is a nonlinear relationship between the input, $x(t)$, and the output, $y(t)$, that does not involve a differential equation and it is called a static nonlinearity. On the other hand, the input and the output may be related through a non-linear differential equation. Such a device is called a dynamic nonlinearity. In this section, we shall discuss briefly the basic features of some common nonlinearities.

13.2.1 Saturation

This is one of the most common static nonlinearities. A simple example is an amplifier for which the output is proportional to the input only for a limited range of values of the input. As the magnitude of the input exceeds the range, the output approaches a constant, as shown in Fig. 13.3. Although the change from one range to the other is usually gradual, it is often sufficiently accurate in most cases to approximate the curve by a set of straight-lines, as shown.

Besides amplifiers, many other physical devices exhibit saturation. Another well-known example of saturation is the relationship between magnetic flux and current in an iron-cored coil.

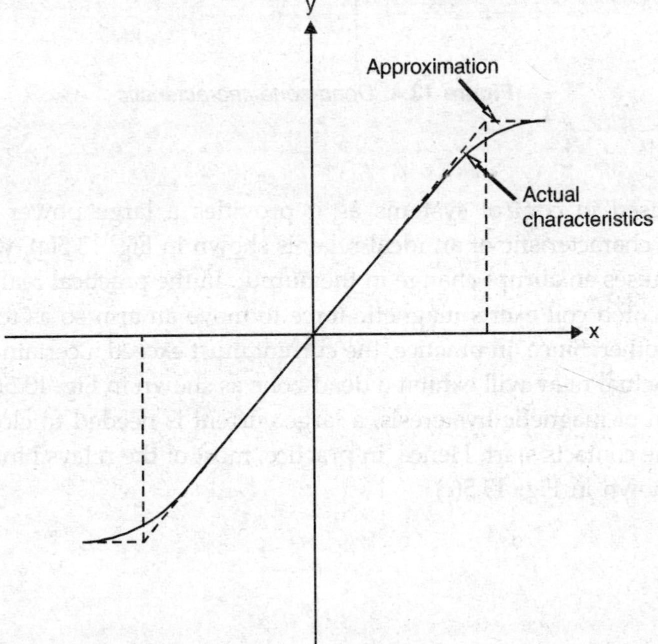

Figure 13.3. *Saturation characteristic and piece-wise linear approximation*

13.2.2 Dead Zone

In many physical devices the output is zero until the magnitude of the input exceeds a certain value. For example, while developing the mathematical model for a dc servomotor, we had assumed that any voltage applied to the armature windings will cause the armature to rotate, if the field current is maintained constant. In practice, rotation will result only if the torque produced by the motor is sufficient to overcome the static friction. As a result, if we plot the relationship between the steady-state angular velocity and the applied voltage, we get the characteristic shown in Fig. 13.4, which exhibits the dead-zone phenomenon. Many other devices have similar characteristics.

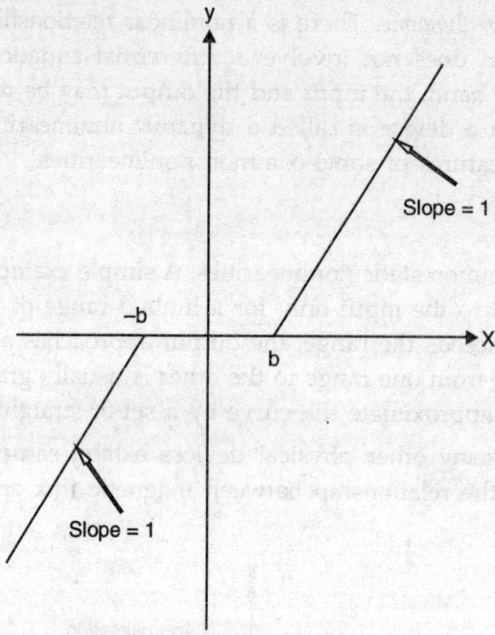

Figure 13.4. *Dead-zone characteristic*

13.2.3 Relays

A relay is often used in control systems as it provides a large power amplification rather inexpensively. The characteristic of an ideal relay is shown in Fig. 13.5(*a*), where a change in the sign of the input causes an abrupt change in the output. In the practical realization of a relay, the current in an iron-cored coil exerts magnetic force to move an arm so as to make the contact in one direction or another. Since, in practice, the current must exceed a certain value before the arm can be moved, an actual relay will exhibit a dead-zone as shown in Fig. 13.5(*b*). Furthermore, due to the phenomenon of magnetic hysteresis, a large current is needed to close the relay than the current at which the contacts start. Hence, in practice, most of the relays have dead-zones as well as hysteresis, as shown in Fig. 13.5(*c*).

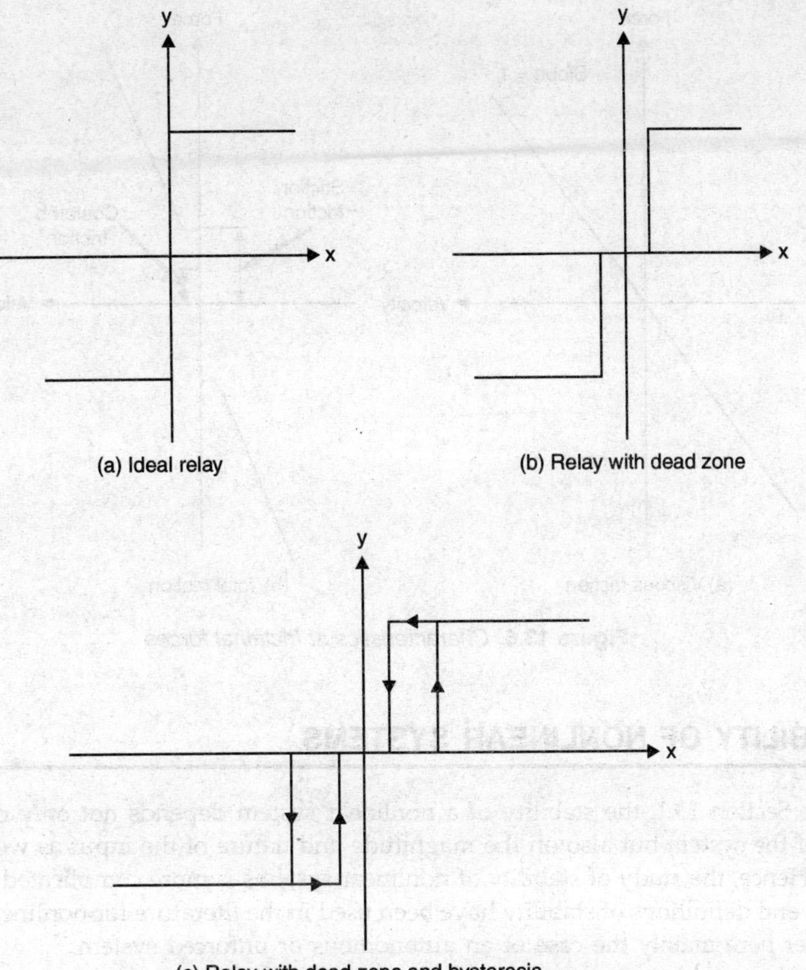

Figure 13.5. *Characteristics of a relay*

13.2.4 Friction

Frictional forces oppose motion whenever there is sliding contact between mechanical surfaces. The predominant part of the frictional force is called viscous friction, which is proportional to the relative velocity between the moving surfaces. This is linear in nature, as shown in Fig. 13.6(*a*). In addition to this viscous friction, there are two components of the total frictional forces that are nonlinear. One of them is coulomb friction, which produces a constant force opposing motion. The other is called stiction, which is the force required to initiate motion, and is always greater than the force of coulomb friction. The characteristic relating the total frictional force to velocity is shown in Fig. 13.6(*b*).

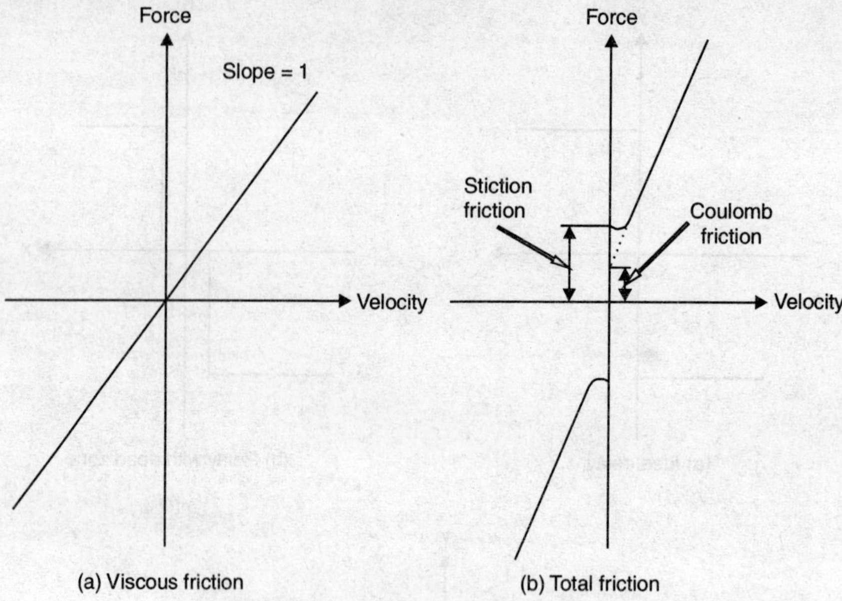

Figure 13.6. *Characteristics of frictional forces*

13.3 STABILITY OF NONLINEAR SYSTEMS

As stated in Section 13.1, the stability of a nonlinear system depends not only on the physical properties of the system but also on the magnitude and nature of the input as well as the initial conditions. Hence, the study of stability of nonlinear systems is more complicated than for linear systems. Several definitions of stability have been used in the literature for nonlinear systems. We shall consider here mainly the case of an autonomous or unforced system.

13.3.1 Autonomous Systems

In general, a nonlinear continuous-time system can be represented by the state equations

$$\dot{x} = f(x, u) \quad \text{...(13.1)}$$

$$y = g(x, u) \quad \text{...(13.2)}$$

where $x(t)$ is the n-dimensional state vector, $u(t)$ is the m-dimensional input vector, $y(t)$ is the p-dimensional output vector, and the vectors f and g are nonlinear functions of x and u. The system is said to be autonomous if the input $u(t)$ is identically zero. For this case, Equation (13.1) is reduced to

$$\dot{x} = f(x) \quad \text{...(13.3)}$$

A point of equilibrium is obtained for any value of the vector x that makes $\dot{x} = 0$. In general a nonlinear system may have many points of equilibrium. Some of these may be points of stable equilibrium, while others may be points of unstable equilibrium. A good example is the bistable multivibrator, an electronic circuit with three states of equilibrium, two of which are stable and one unstable. Consequently, it is necessary to examine stability at each point of equilibrium. It is common practice to transform coordinates in the state space so that the origin becomes the point of equilibrium. This is convenient for examining *local stability* and can be done for each point of equilibrium.

Let us now consider a hypersphere of finite radius surrounding the orign of the state space (the point of equilibrium), that is the set of points described by the equation

$$x_1^2 + x_2^2 + \ldots + x_n^2 = R^2 \qquad \ldots(13.4)$$

in the n-dimensional state space. Let this region be denoted by $S\ (R)$.

The system is said to be stable in the sense of Lyapunov[1] if there exists a region $S\ (\varepsilon)$ so that a trajectory starting from any point $x\ (0)$ in this region does not go outside the region $S\ (R)$. This is illustrated in Fig. 13.7 for the two-dimensional case, $n = 2$.

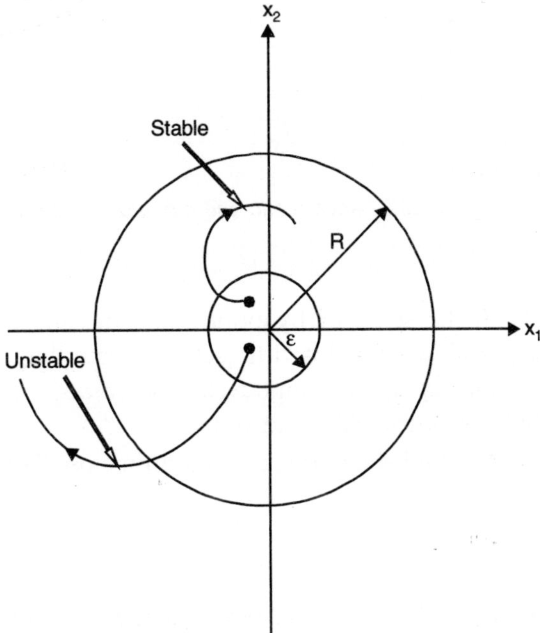

Figure 13.7. *Stability in the sense of Lyapunov*

The system is said to be *asymptotically stable* if there exists a $\delta > 0$ so that the trajectory starting from any point $x\ (0)$ within $S\ (\delta)$ does not leave $S\ (R)$ at any time and finally returns to the origin. The trajectory in Fig. 13.8(a) shows asymptotic stability.

The system is said to be *monotonically stable* if it is asymptotically stable and the distance of the state from the origin decreases monotonically with time. The trajectory in Fig. 13.8(b) shows monotonic stability.

A system is said to be *globally stable* if the regions $S\ (\delta)$ and $S\ (R)$ extend to infinity. A system is said to be *locally stable* if the region $S\ (\delta)$ is small and when subjected to small perturbations the state remains within the small specified region $S\ (R)$.

1. After A.M. Lyapunov, A Russian mathematician who did pioneering work in this area during late nineteenth century.

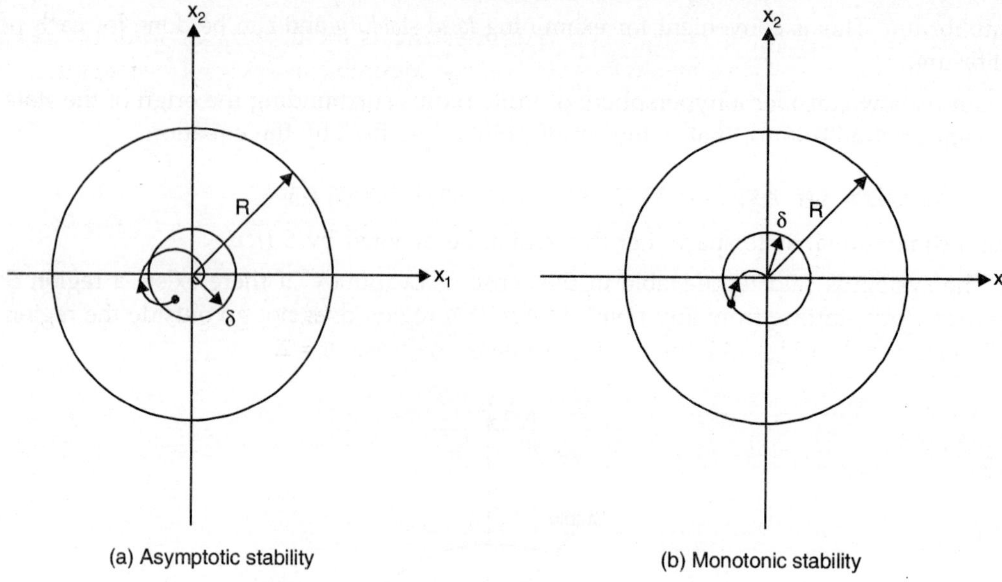

Figure 13.8. *Asymptotic and monotonic stability*

13.3.2 Limit Cycles

The definition of stability in the sense of Lyapunov includes the possibility of the state of a perturbed nonlinear system following a closed trajectory within the tolerance limits specified by the region $S(R)$. This behaviour is called a limit cycle, and corresponds to an oscillation of fixed amplitude and period, but not necessarily sinusoidal.

As an example, consider the behaviour of an electronic oscillator, which can be described by Van der Pol's differential equation

$$\frac{d^2x}{dt^2} - \mu\left(1 - x^2\right)\frac{dx}{dt} + x = 0 \quad \text{...(13.5)}$$

This differential equation describes the electrical circuit shown in Fig. 13.9, where R_n represents a nonlinear resistor and x is the voltage across a capacitor. For small x, the resistance is negative, which implies that the amplitude of oscillations will increase with an exponential envelope. As x increases, the resistance also increases, and is positive for $x > 1$. For an intermediate value of x, the oscillations are stable. The waveform of x is not sinusoidal.

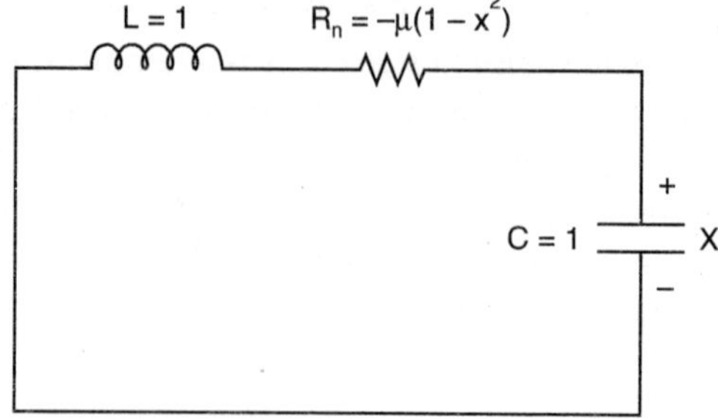

Figure 13.9. *Equivalent circuit for Equation (13.5)*

An important aim in the study of nonlinear systems is the determination of the existence and location of limit cycles. In general, a limit cycle is undesirable in a control system but in some cases it can be tolerated if the amplitude of oscillations is small.

13.4 METHODS OF STUDYING NONLINEAR SYSTEMS

Nonlinear systems can be studied by (1) hardware, (2) simulation and (3) analysis. Whereas in the final analysis one must test the complete hardware, it is usually more economical to build a system only after it has been studied by simulation or analysis. It may be noted that both simulation and analysis consider the performance of the system model.

With modern analog, digital, and hybrid computers, one can undertake exhaustive simulation of complex systems. However, simulation without preliminary analysis can often be wasteful and ineffective, since each computer run merely provides a set of outputs for given parameters and inputs. Theoretical analysis, on the other hand, provides at least in a qualitative manner, considerable insight into the behaviour of the system for different values of the parameters as well as various types of disturbances. The study can then be completed by simulation on a computer in an intelligent manner, since the choice of the type of computer and the simulation program will be based on the preliminary analysis.

Although a number of methods have been employed by control engineers for the study of nonlinear systems, we shall consider here three methods that are most well-known. These are (1) linearization, (2) describing function method, and (3) the phase-plane method. The linear incremental model used in the analysis of electronic circuits is a common example of linearization about an operating point. The describing function method is an attempt to extend frequency response methods to closed-loop systems containing only one nonlinearity. The phase-plane method provides a graphical technique for obtaining the solution of a second-order nonlinear differential equation.

A more general method for the study of the stability of nonlinear systems, based on the classical work of A.M. Lyapunov, will be presented in a later section in this chapter.

13.5 LINEARIZATION

Linearization is based on the Taylor series expansion of a nonlinear function about an operating point. For example, consider a nonlinear function, $f(x)$. It can be written as

$$f(x) = f(x_0) + \left.\frac{df}{dx}\right|_{x=x_0}(x-x_0) + \left.\frac{d^2 f}{dx^2}\right|_{x=x_0}\frac{(x-x_0)^2}{2!} + \ldots \qquad \text{...(13.6)}$$

We get a linear approximation of Equation (13.6) if we ignore all terms except the first-two. Clearly, this will be a good approximation if either $(x - x_0)$ is very small, or the higher order derivatives of f are very small. This is the main idea behind the incremental linear models used for the analysis of electronic circuits.

We shall now generalize this to the case of the state equations for nonlinear systems. Assuming that the dimension of x is n, we shall re-write Equation (13.1) as

$$\dot{x} = f(x, u) \begin{vmatrix} f_1(x, u) \\ f_2(x, u) \\ \vdots \\ f_n(x, u) \end{vmatrix} \qquad ...(13.7)$$

Ignoring the higher-order terms in the Taylor series expansion of this vector differential equation leads to the linearized model (assuming $x_0 = 0$, that is, the coordinates have been transformed to make the origin the point of equilibrium, and $u_0 = 0$)

$$\dot{x} = Ax + Bu \qquad ...(13.8)$$

where

$$A = \begin{vmatrix} \frac{\partial f_1}{\partial x_1} & \frac{\partial f_1}{\partial x_2} & \cdots & \frac{\partial f_1}{\partial x_n} \\ \frac{\partial f_2}{\partial x_1} & \frac{\partial f_2}{\partial x_2} & \cdots & \frac{\partial f_2}{\partial x_n} \\ \vdots & \vdots & \cdots & \vdots \\ \frac{\partial f_n}{\partial x_1} & \frac{\partial f_n}{\partial x_2} & \cdots & \frac{\partial f_n}{\partial x_n} \end{vmatrix} \qquad ...(13.9)$$

and

$$B = \begin{vmatrix} \frac{\partial f_1}{\partial u_1} & \frac{\partial f_1}{\partial u_2} & \cdots & \frac{\partial f_1}{\partial u_m} \\ \frac{\partial f_2}{\partial u_1} & \frac{\partial f_2}{\partial u_2} & \cdots & \frac{\partial f_2}{\partial u_m} \\ \vdots & \vdots & \cdots & \vdots \\ \frac{\partial f_n}{\partial u_1} & \frac{\partial f_n}{\partial u_2} & \cdots & \frac{\partial f_n}{\partial u_m} \end{vmatrix} \qquad ...(13.10)$$

A and B are said to be Jacobian matrices. Again, this linear model will be valid only for small deviations around the point of equilibrium. Nevertheless, it can be used for investigating local stability around the point of equilibrium (for the autonomous) case by simply applying the Routh criterion to the characteristic polynomial of A. The following example will illustrate the main idea behind this approach.

Note that in Equation (13.8), we have used x instead of the deviation $x - x_0$. This is a common practice, even if x_0 is not the origin of the state space. It is to be understood that x represents the variation of the state from the point of equilibrium, or 'set-point' in the terminology of process control.

EXAMPLE 13.1

Consider the following second-order nonlinear differential equation:

$$\frac{d^2x}{dt^2}+x^2\left(\frac{dx}{dt}-1\right)+x = 0 \quad ...(13.11)$$

(*a*) Determine the points of equilibrium.

(*b*) Investigate the stability of the system near each point of equilibrium.

SOLUTION

We shall first derive state equations for the given differential equation. Let $x_1 = x$ and $x_2 = \dot{x}$. Then, we obtain

$$f_1 = \dot{x}_1 = x_2 \quad ...(13.12)$$

$$f_2 = \dot{x}_2 = x_1^2 (1 - x_2) - x_1 \quad ...(13.13)$$

The points of equilibrium are obtained by setting the two derivatives in Equation (13.13) to zero, and are readily found as (0, 0) and (1, 0).

The Jacobian matrix is obtained as

$$\begin{bmatrix} \frac{\partial f_1}{\partial x_1} & \frac{\partial f_1}{\partial x_2} \\ \frac{\partial f_2}{\partial x_1} & \frac{\partial f_2}{\partial x_2} \end{bmatrix} = \begin{bmatrix} 0 & 1 \\ 2\dot{x}_1 (1-x_2)-1 & -x_1^2 \end{bmatrix} \quad ...(13.14)$$

Hence, for the equilibrium point given by $x_1 = 0$, $x_2 = 0$, we have

$$A = \begin{bmatrix} 0 & 1 \\ -1 & 0 \end{bmatrix} \quad ...(13.15)$$

The characteristic polynomial for this case is

$$\det (sI - A) = \begin{bmatrix} s & -1 \\ 1 & s \end{bmatrix} = s^2 + 1 \quad ...(13.16)$$

indicating an oscillatory system, although stable in the sense of Lypunov. Hence, we can expect a limit cycle in the neighbourhood of this point.

For the equilibrium point given by $x_1 = 1$, $x_2 = 0$, we have

$$A = \begin{bmatrix} 0 & 1 \\ 1 & -1 \end{bmatrix} \quad ...(13.17)$$

The characteristic polynomial for this case is

$$\det (sI - A) = \begin{vmatrix} s & -1 \\ -1 & s+1 \end{vmatrix} = s^2 + s - 1 \quad ...(13.18)$$

which shows that the system is unstable in the neighbourhood of this point.

DRILL PROBLEM 13.1

The state equations of a nonlinear system are given below. Determine all points of equilibrium and investigate the nature of stability in the neighbourhood of these points.

$$\dot{x}_1 = x_2$$

$$x_2 = -x_1^2 - x_2^2 + 2x_1 - x_2$$

Ans. The points of equilibrium are (0, 0) and (2, 0). The former is a point of unstable equilibrium, wheras the latter is a point of stable equilibrium.

13.6 DESCRIBING FUNCTION ANALYSIS

If we apply a sine wave of period T to a nonlinear device, the steady-state output will have the same period, but will not necessarily be sinusoidal. Fourier analysis of the nonsinusoidal waveform will give us a fundamental component and several harmonics. Note that it is assumed that subharmonics do not exist. Thus, if the input is given by

$$x(t) = X \sin \omega t, \quad \text{where } \omega = \frac{2\pi}{T} \quad \text{...(13.19)}$$

the steady-state output may be expressed in the form of the following Fourier series:

$$y(t) = A_0 + \sum_{n=1}^{\infty} A_n \cos n\omega t + \sum_{n=1}^{\infty} B_n \sin n\omega t + \quad \text{...(13.20)}$$

where (for details, see reference [5], or any other book on 'Signals and Systems') the Fourier coefficients are obtained as

$$A_0 = \frac{1}{T}\int_0^T y(t)\, dt \quad \text{...(13.21)}$$

$$A_n = \frac{2}{T}\int_0^T y(t) \cos n\omega t\, dt \quad \text{...(13.22)}$$

and

$$B_n = \frac{2}{T}\int_0^T y(t) \sin n\omega t\, dt \quad \text{...(13.23)}$$

From Equation (13.20), the fundamental component of the output is obtained as

$$y_1(t) = A_1 \cos \omega t + B_1 \sin \omega t = Y_1 \sin(\omega t + \phi) \quad \text{...(13.24)}$$

where

$$Y_1 = \sqrt{A_1^2 + B_1^2} \quad \text{...(13.25)}$$

and

$$\phi = \tan^{-1} \frac{A_1}{B_1} \quad \text{...(13.26)}$$

The describing function of a nonlinear device is defined as the ratio of the complex number (or phasor) representing the fundamental component of its output to the complex number representing the sinusoidal input; that is,

$$N(X, \omega) = \frac{Y_1}{X} e^{j\phi} \qquad ...(13.27)$$

Note that, in general, the describing function depends on both the magnitude as well as the frequency of the input.

If we assume that the linear part of the feedback system, as shown in Fig. 13.10, is of the low-pass nature, then the harmonics will be attenuated and may be neglected to obtain a reasonable approximation to the performance of the system.

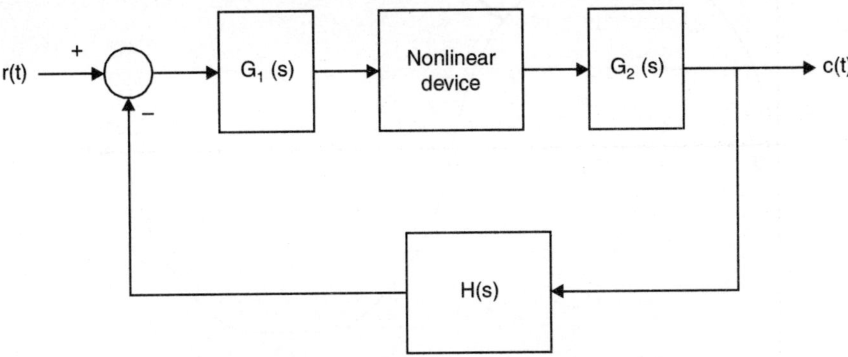

Figure 13.10. *A typical nonlinear feedback system*

In such cases, one may treat the describing function of a nonlinear device in a manner similar to the transfer function of a linear device, except that it must be kept in mind that now both the magnitude and the frequency of the input have to be taken into account. Thus, it is possible to use the describing function for stability analysis. This is based on the idea that if at some frequency, the loop gain is one, with a phase shift of 180°, possibilities of oscillation exist. We shall discuss stability analysis in the sequel, after deriving the describing functions of some common nonlinearities.

13.6.1 Saturation

The saturation characteristic is shown in Fig. 13.11, simplified so that it can be represented by straight-line segments.

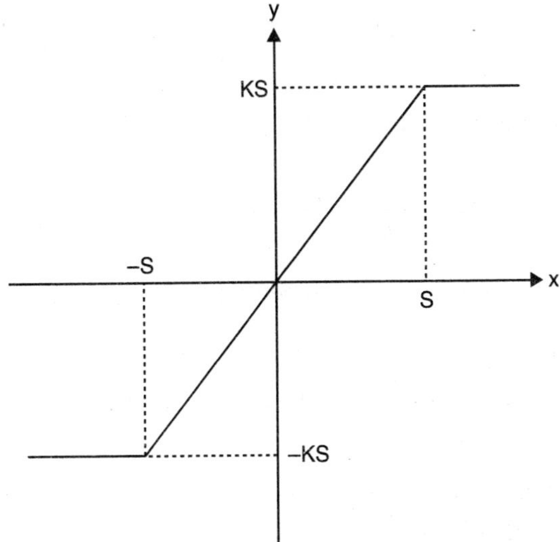

Figure 13.11. *Piecewise linear approximation of saturation*

Let the input be denoted by $x = X \sin \omega t$, where $X > S$. The waveform of the output is shown in Fig. 13.12, where $\theta = \omega t$.

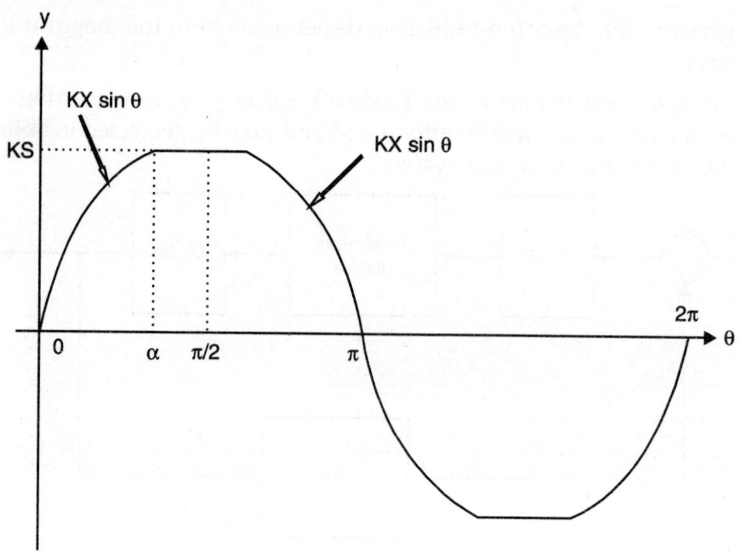

Figure 13.12. *The output for a sine wave input*

Since the output in Fig. 13.12 has odd symmetry, its Fourier series will have only sine terms. Taking advantage of the half-wave symmetry, we can obtain the fundamental component by integrating over only quarter cycle. Hence,

$$Y_1 = \frac{4}{\pi}\int_0^{\pi/2} y \sin\theta \, d\theta$$

$$= \frac{4}{\pi}\left[\int_0^{\alpha} KX \sin^2\theta \, d\theta + \int_{\alpha}^{\pi/2} KS \sin\theta \, d\theta\right] \qquad \text{...(13.28)}$$

$$= \frac{4K}{\pi}\left[\frac{X}{2}\alpha - \frac{X}{4}\sin 2\alpha + S\cos\alpha\right]$$

where $\alpha = \sin^{-1} S/X$. Hence, the describing function is given by

$$N(X) = \frac{Y_1}{X} = \frac{2K}{\pi}\left[\sin^{-1}\frac{S}{X} + \frac{S}{X}\sqrt{1-\frac{S^2}{X^2}}\right] \qquad \text{...(13.29)}$$

In this case, $N(X)$ is real number and independent of frequency.

13.6.2 Two-position Relay

The characteristics of a two-position relay are shown in Fig. 13.13.

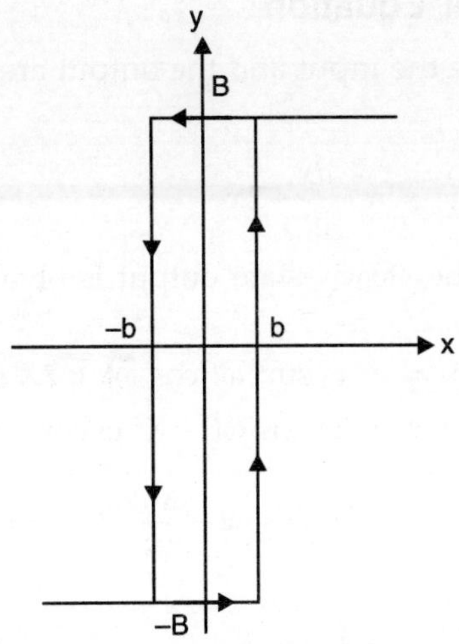

Figure 13.13. *A typical two-position relay*

If the input is $x = X \sin \omega t$, the output is equal to $-B$ for $0 \le \theta \le \sin^{-1} b/X$ and equal to $+B$ for $\sin^{-1} b/X \le \theta \le \pi + \sin^{-1} b/X$, where $\theta = \omega t$. In this case, it is easily seen that the output has the square wave form and it lags behind $x(t)$ by $\sin^{-1} b/X$, as shown in Fig. 13.14.

Due to this phase shift, the fundamental component of the output will have both sine and cosine terms. Consequently, the describing function will be a complex number,

$$N(X) = \frac{4B}{\pi X} e^{-j \sin^{-1} \frac{b}{X}} \qquad \text{...(13.30)}$$

This follows from the fact that if we shift $y(t)$ to the left by the amount $\sin^{-1} b/X$, the resulting square wave will contain only sine terms, with the fundamental component

$$B_1 = \frac{4B}{\pi} \qquad \text{...(13.31)}$$

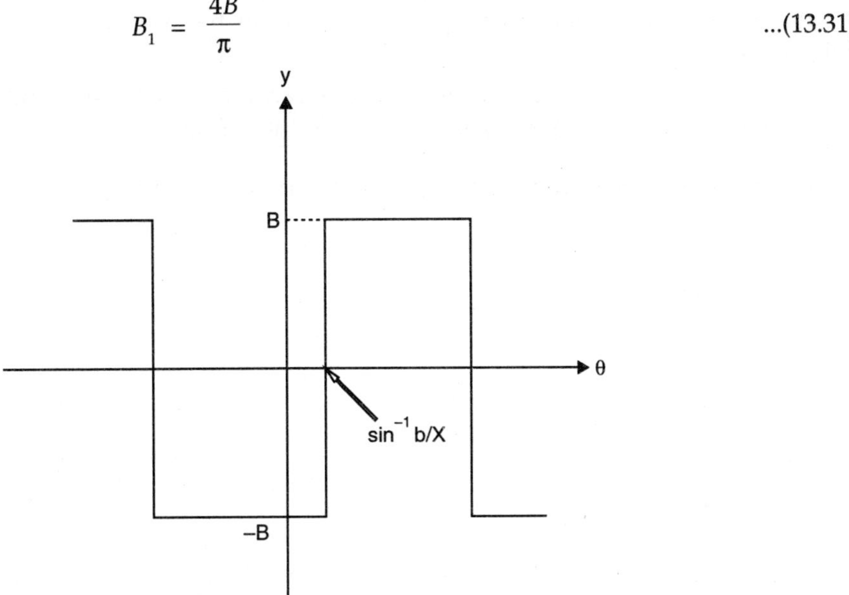

Figure 13.14. *Output of a two-position relay to* x = X *sin* ωt

13.6.3 A Nonlinear Differential Equation

Consider a nonlinear system where the input and the output are related through the differential equation

$$y(t) = x^2 \frac{dx}{dt} + 2x \qquad \text{...(13.32)}$$

For the input $x = X \sin \omega t$, the steady-state output is obtained by direct substitution into Equation (13.32). Thus, we have

$$\begin{aligned} y(t) &= X^3 \omega \sin^2 \omega t \cos \omega t + 2X \sin \omega t \\ &= X^3 \omega \cos \omega t - X^3 \omega \cos^3 \omega t + 2X \sin \omega t \\ &= X^3 \omega \cos \omega t - \frac{X^3 \omega}{4} (\cos 3\,\omega t + 3 \cos \omega t) + 2X \sin \omega t \qquad \text{...(13.33)} \\ &= \frac{X^3 \omega}{4} \cos \omega t + 2X \sin \omega t - \frac{X^3 \omega}{4} \cos 3\omega t \end{aligned}$$

The fundamental component of the output is obtained by combining the sine and cosine terms at that frequency, and is given by

$$Y_1 (t) = \sqrt{\frac{X^6 \omega^2}{16} + 4X^2} \sin \left[\omega t + \tan^{-1} \left(\frac{X^2 \omega}{8} \right) \right]$$

The describing function is now obtained as

$$N (X, \omega) = \sqrt{\left(\frac{X^4 \omega^2}{16} + 4 \right)} e^{j \tan^{-1} \frac{X^2 \omega}{8}} \qquad \text{...(13.34)}$$

In this case, the describing function is a complex number, which is a function of both the amplitude of input and its frequency. This is to be expected in the case of a dynamic nonlinearity.

We shall now give the describing functions of some common static nonlinearities. It will be a good exercise for the student to derive each of them from the first principles.

13.6.4 Describing Functions for Some Common Nonlinearities

1. Saturation

$$N (x) = \frac{2K}{\pi} \left[\sin^{-1} r + r \sqrt{1 - r^2} \right]$$

where $r = \frac{S}{X}$ and $X > S$

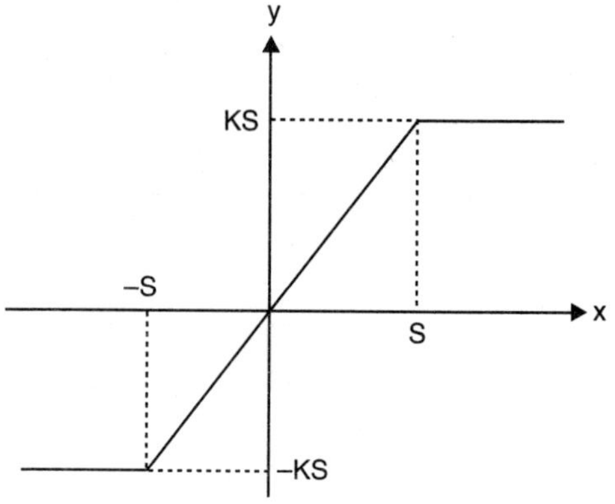

y
KS
–S
S
x
–KS

2. Dead-zone

$$N(X) = \frac{2K}{\pi}\left[\frac{\pi}{2} - \sin^{-1} r + r\sqrt{1-r^2}\right]$$

where $r = \dfrac{B}{X}$ and $B < X$

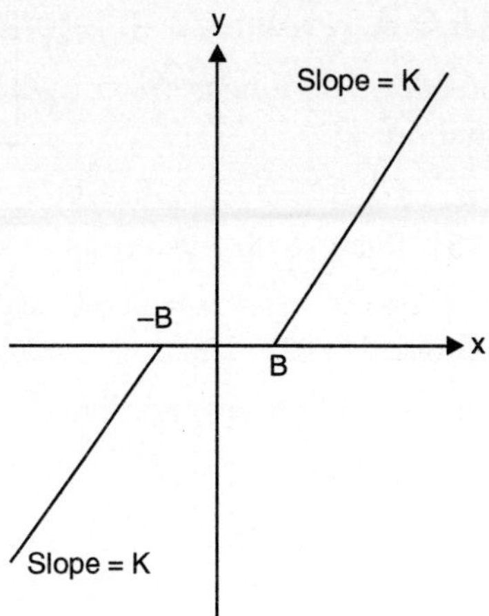

3. Ideal relay

$$N(X) = \frac{4E}{\pi X}$$

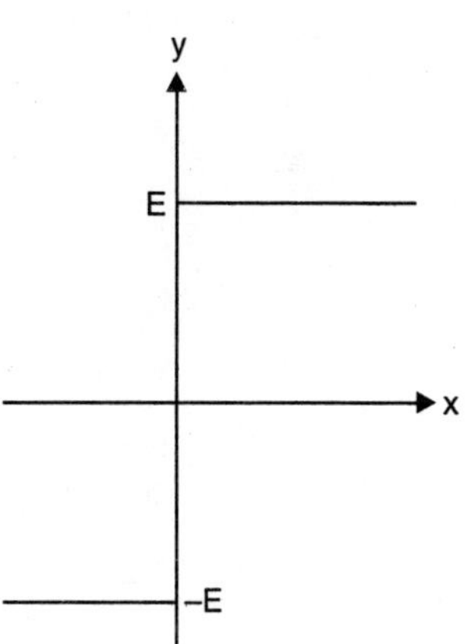

4. Amplifier with dead-zone and saturation

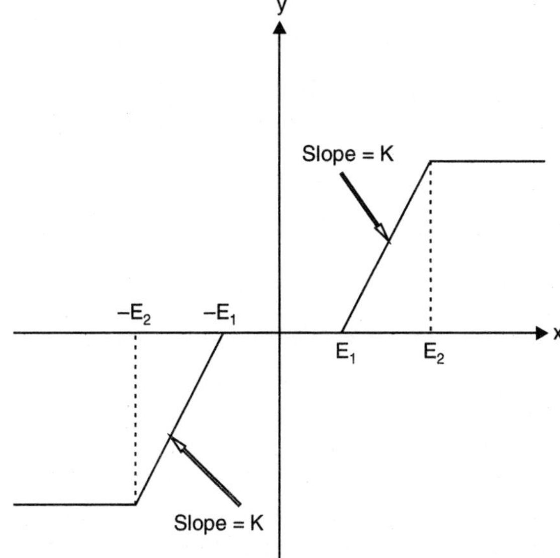

$$N(X) = \frac{2K}{\pi}\left[\sin^{-1}\frac{E_2}{X} - \sin^{-1}\frac{E_1}{X} + \frac{E_2\sqrt{X^2-E_2^2}}{X^2} - \frac{E_1\sqrt{X^2-E_1^2}}{X^2}\right]$$

where $X > E_2$

5. Relay with hysteresis

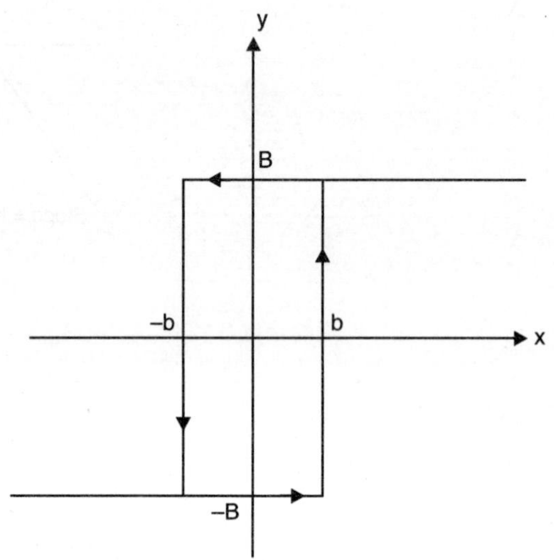

$$N(X) = \frac{4B}{\pi X} e^{-j\sin^{-1}\frac{b}{X}}$$

DRILL PROBLEM 13.2

The output, y (t) of a nonlinear device is related to the input, x (t), through the following differential equation:

$$y = 4\left(\frac{dx}{dt}\right)^2 + 6x + 3x^2\frac{dx}{dt}$$

Determine the describing function for the device.

Ans. $$\sqrt{\left(\frac{9}{16}X^4\omega^2 + 36\right)}\, e^{j\tan^{-1}\frac{X^2\omega}{8}}$$

13.6.5 Stability Analysis with the Describing Function

The main purpose of describing function analysis is to determine the stability of a closed-loop system with one nonlinear element. This is done by plotting on the *G-H* plane the quantity $-1/N(X, \omega)$, where $N(X, \omega)$ is the describing function, along with the polar plot of the transfer function $GH(s)$ of the linear portion. If these two curves intersect for some value of X and ω, as shown in Fig. 13.15, this indicates the possibility of limit cycle oscillations of amplitude X at this frequency due to a loop gain of one with 180° phase-shift.

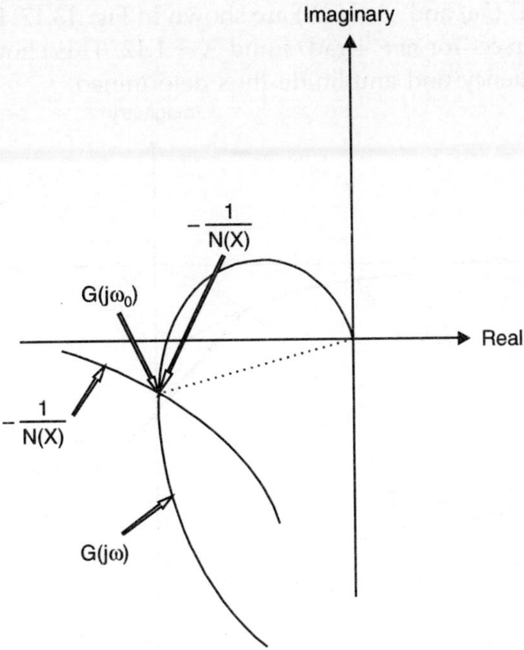

Figure 13.15. *Polar plots of GH (s) and -1/N (X)*

It should be noted, however, that the describing function method is an approximation, since the higher order harmonics have been neglected. Thus, the accuracy is higher in those cases where the linear portion effectively filters out the harmonics. If this is not the case, results may be misleading and should be verified by computer simulation.

EXAMPLE 13.2

Consider the control system shown in Fig. 13.16 where the nonlinear element (*NL*) is a two-position relay of the type shown in Fig. 13.13.

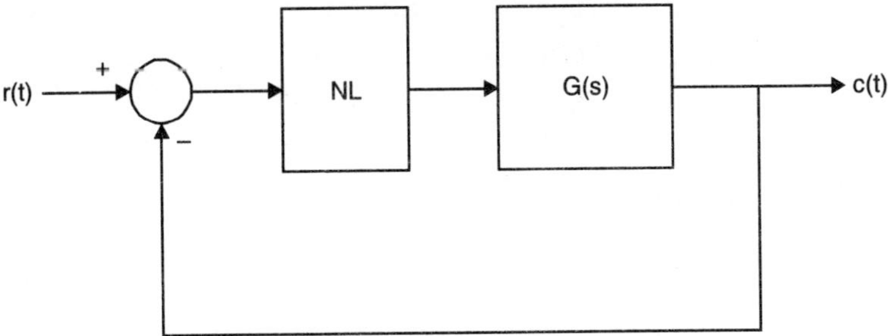

Figure 13.16. *A relay-controlled system*

It will be assumed that the describing function of the relay is given by

$$N(X) = \frac{5}{X} e^{-j\sin^{-1}\frac{0.5}{X}} \qquad \text{...(13.35)}$$

and the transfer function of the linear part of the system is given by

$$G(s) = \frac{4}{s\,(s+1)} \qquad \text{...(13.36)}$$

The polar plots of $G(j\omega)$ and $-1/N(X)$ are shown in Fig. 13.17. From the plot it will be seen that the two curves intersect for $\omega \simeq 4$ rad/s and $X \simeq 1.12$. This shows the possible existence of a limit cycle at the frequency and amplitude thus determined.

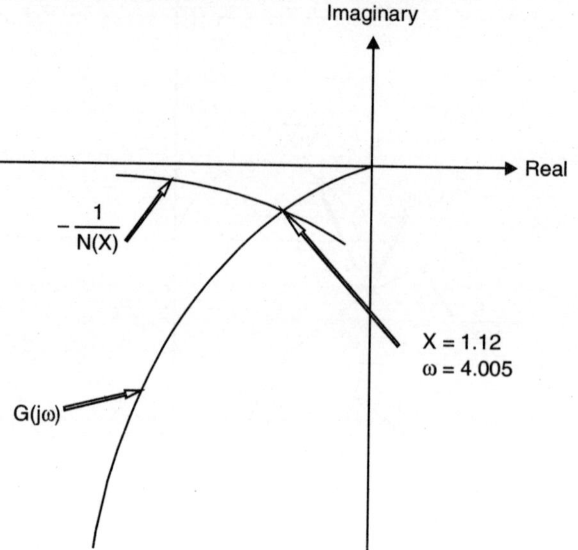

Figure 13.17. *Polar plots of G (jω) and -1/N (X)*

It may be remarked that this method basically allows us to solve graphically the equation

$$G(j\omega)\,N(X) = -1 \quad \text{...(13.37)}$$

It could also have been solved by plotting the gain in decibels against phase shift in degrees for both $G(j\omega)$ and $-1/N(X)$ on a rectangular graph. Such a plot is often easier to draw. For this example, the plots are shown in Fig. 13.18.

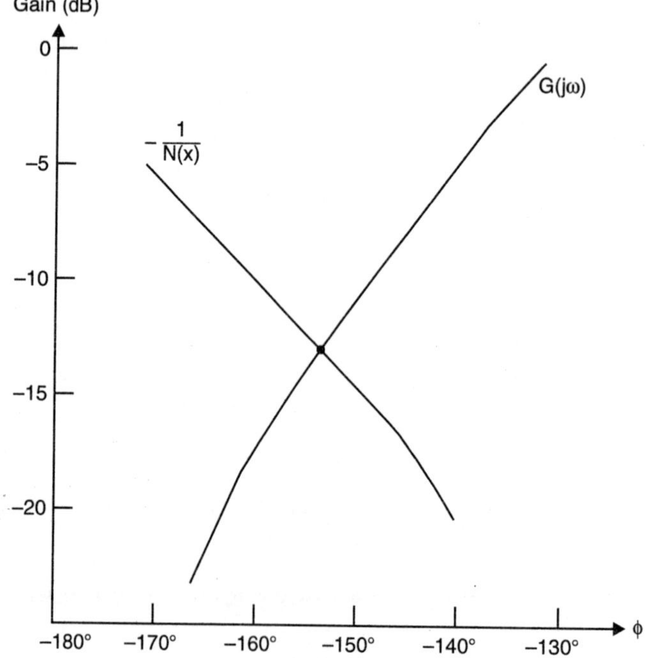

Figure 13.18. *Log-magnitude/phase plots*

DRILL PROBLEM 13.3

A unity-feedback instrument servo is driven by an amplifier that saturates at 70 per cent of the rated voltage of the motor. Assume that the gain of the unsaturated amplifier is 50 and the input to the amplifier is sufficient to saturate it. The transfer function of the linear portion of the system, excluding the amplifier, is given by

$$G(s) = \frac{0.2}{s(s+2)}$$

Utilize the describing function method to determine if a limit cycle exists.

[Hint: Use log-magnitude/phase plots.]

Ans. A limit cycle does not exist.

13.7 THE PHASE-PLANE METHOD

This is basically a graphical method for solving a second-order nonlinear (or linear) differential equation. The coordinates of the plane are x and $\dot{x}$, which are called the phase variables of the system. Given the initial state, this method can be utilized to plot the trajectory in the phase-plane. From a set of trajectories of different initial conditions, we can obtain the 'phase portrait', which provides information about stability and the existence of limit cycles.

There are two well-known methods for obtaining phase-plane plots. These are the *data method* and the *method of isoclines*. We shall discuss the former, which is more general and can be used for obtaining the trajectory for any arbitrary input.

13.7.1 The Delta Method

Consider the general second-order differential equation of the form

$$\frac{d^2x}{dt^2} + f(x, \dot{x}, t) = 0 \qquad ...(13.38)$$

The delta method is based on rearranging the above differential equation in the form

$$\frac{d^2x}{dt^2} + \omega_0^2 (x, \delta) = 0 \qquad ...(13.39)$$

where δ is, in general, a function of x, $\dot{x}$ and t. We normalize the time by defining

$$\tau = \omega_0 t \qquad ...(13.40)$$

which transforms Equation (13.39) into

$$\frac{d^2x_1}{d\tau^2} + x_1 + \delta(x_1, x_2\tau) = 0 \qquad ...(13.41)$$

where $x_1 = x$ and $x_2 = dx/d\tau$. Equation (13.41) may also be written as

$$\frac{d^2x_1}{d\tau^2} = \frac{dx_2}{d\tau} = \frac{dx_2}{dx_1} \cdot \frac{dx_1}{d\tau} = -x_1 - \delta \qquad ...(13.42)$$

or $$\frac{dx_2}{dx_1} = \frac{-x_1 - \delta}{x_2} \qquad ...(13.43)$$

A graphical interpretation is given in Fig. 13.19. For a given initial point P_0, the value of δ is calculated from the known values of $x_1(0)$ and $x_2(0)$. This locates the point Q_0. The slope of the straight-line Q_0P_0 is then obtained as

$$\frac{P_0R_0}{Q_0R_0} = \frac{x_2(0)}{x_1(0) + \delta(0)} \qquad ...(13.44)$$

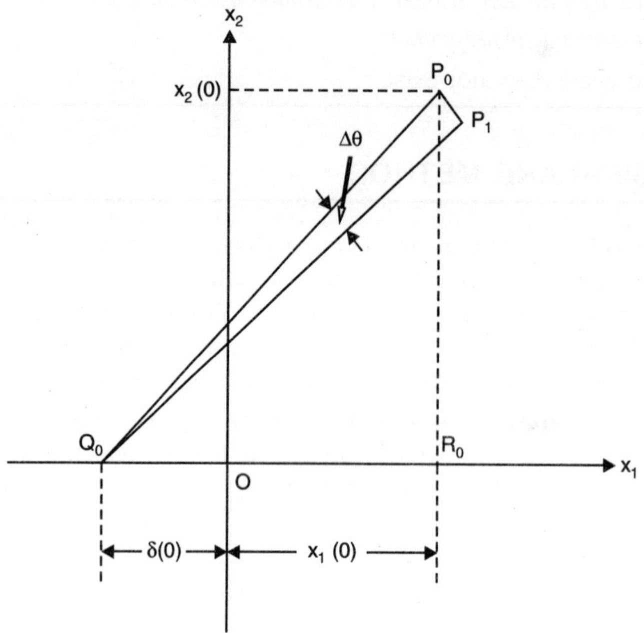

Figure 13.19. *Graphical interpretation of the delta method*

It will be seen that the slope of this line is the negative reciprocal of the slope of the trajectory given by Equation (13.43). In other words, the line Q_0P_0 is perpendicular to the trajectory at the point P_0. It follows that the trajectory near P_0 can be approximated by a small circular arc drawn with centre at Q_0 and radius Q_0P_0. For positive time, the trajectory will be in the clockwise direction. This follows by noting that the slope

$$\frac{dx_2}{dx_1} = \frac{-x_1 - \delta}{x_2} \qquad ...(13.45)$$

is negative for positive x_1 and x_2; that is x_2 decreases as x_1 is increased.

By drawing the circular arc we obtain the next point P_1 on the trajectory, where the angle $P_0Q_0P_1$, denoted as $\Delta\theta$, is small. At the point P_1 we can again calculate the value of δ and, after locating the new centre Q_1, draw another circular, arc.

An interesting feature of the delta method is that the time along the trajectory can be evaluated directly. It is easily shown that the normalized time for the displacement P_0P_1 in Fig. 13.19 will be $\Delta\tau = \Delta\theta$, where $\Delta\theta$ is in radians, and the real time is given by

$$\Delta t = \frac{\Delta\theta}{\omega_0} \text{ seconds} \qquad \text{...(13.46)}$$

From the Taylor series expansion for the increments of x_1 and x_2, the expressions for the errors in approximation are obtained as follows.

$$\varepsilon_{x_1} = \frac{1}{6}\left(\frac{d\delta}{\delta\tau}\right)_0 (\Delta\theta)^3 + O_4(\Delta\theta) \qquad \text{...(13.47)}$$

$$\varepsilon_{x_2} = \frac{1}{2}\left(\frac{d\delta}{d\tau}\right)_0 (\Delta\theta)^2 + \frac{1}{6}\left(\frac{d^2\delta}{d\tau^2}\right)_0 (\Delta\theta)^3 + O_4(\Delta\theta) \qquad \text{...(13.48)}$$

where $\left(\frac{d\delta}{d\tau}\right)_0$ and $\left(\frac{d^2\delta}{d\tau^2}\right)_0$ denote the values of $\frac{d\delta}{d\tau}$ and $\frac{d^2\delta}{d\tau^2}$ at $\tau = 0$, and $O_4(\Delta\theta)$ denotes a quantity if the order of $(\Delta\theta)^4$.

The approximation may be improved if, instead of using the value of δ at τ_0, we use the average value of δ over the interval Δ_τ. This is equivalent to using the modified Euler method. The error at each step is now given by

$$\varepsilon_{x_1} = \frac{1}{12}\left(\frac{d\delta}{\delta\tau}\right)_0 (\Delta\theta)^3 + O_4(\Delta\theta) \qquad \text{...(13.49)}$$

$$\varepsilon_{x_2} = \frac{1}{24}\left(\frac{d^2\delta}{d\tau^2}\right)_0 (\Delta\theta)^3 + O_4(\Delta\theta) \qquad \text{...(13.50)}$$

The procedure is described below, and the graphical construction is shown in Fig. 13.20.

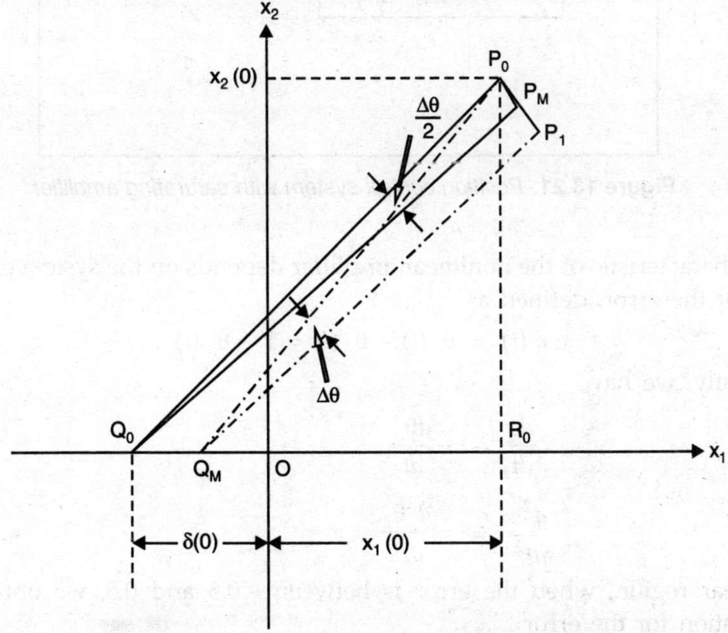

Figure 13.20. *Construction using delta method*

1. Locate the initial point P_0 from x_1 (0) and x_2 (0) at $\tau = \tau_0$.
2. Calculate the initial value of δ at P_0 and locate Q_0 $(-\delta, 0)$ on the x_1 axis.
3. Draw the circular arc P_0P_M with centre at Q_0, the incremental angle being taken as $\Delta\theta/2$.
4. Calculate the value of δ_M by using the intermediate values of x_1 and x_2 at P_M where $\tau = \tau_0 + \Delta\theta/2$. Locate Q_M $(-\delta_M, 0)$ on the x_1-axis.
5. Draw the circular arc P_0P_1 with center at Q_M, the incremental angle being $\Delta\theta$. The arc P_0P_1 represents a portion of the trajectory.
6. This completes one iteration of the procedure. Continue the process by treating P_1 as the new P_0.

A computer program for plotting phase-plane trajectories using the delta method is included on the disk that can be obtained with this book.

EXAMPLE 13.3

Consider the position control system with a saturating amplifier, approximated by straight-line segments, as shown in Fig. 13.21. It is desired to obtain θ_c (t) if θ_r is a unit step for $a = 2$ and $K_m = 1$, assuming zero initial conditions.

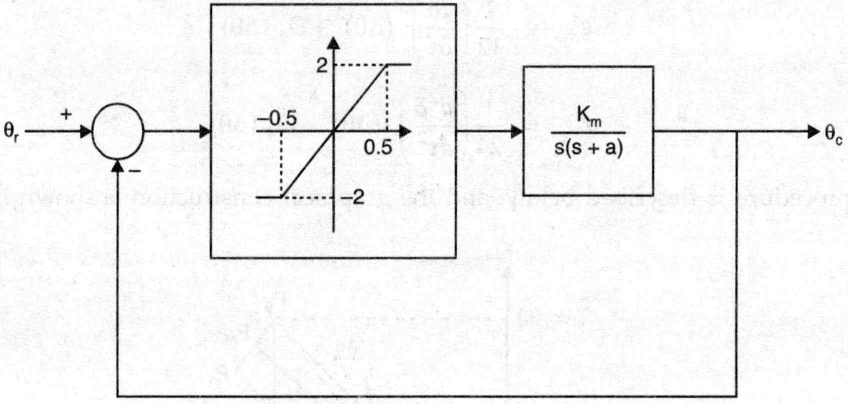

Figure 13.21. *Position control system with saturating amplifier*

SOLUTION

Since the characteristic of the nonlinear amplifier depends on the system error, we shall plot the trajectory for the error, defined as

$$e\ (t) = \theta_r\ (t) - \theta_c\ (t) = 1 - \theta_c\ (t) \qquad \text{...(13.51)}$$

Consequently, we have

$$\frac{de}{dt} = -\frac{d\theta_c}{dt} \qquad \text{...(13.52)}$$

and

$$\frac{d^2e}{dt^2} = -\frac{d^2\theta_c}{dt^2} \qquad \text{...(13.53)}$$

In the linear region, when the error is between -0.5 and 0.5, we obtain the following differential equation for the error

$$\frac{d^2e}{dt^2} + 2\frac{de}{dt} + 4e = 0 \qquad -0.5 \le e \le 0.5 \qquad \text{...(13.54)}$$

In the saturation region, we get

$$\frac{d^2e}{dt^2}+2\frac{de}{dt}+2 = 0 \qquad \text{if } e > 0.5 \qquad ...(13.55)$$

and

$$\frac{d^2e}{dt^2}+2\frac{de}{dt}-2 = 0 \qquad \text{if } e < -0.5 \qquad ...(13.56)$$

The first step is to normalize the time. From Equation (13.54), we note that $\omega_0^2 = 4$. Therefore, we make

$$\tau = 2t \qquad ...(13.57)$$

Then

$$\frac{de}{dt} = \frac{de}{d\tau}\cdot\frac{d\tau}{dt} = 2\frac{de}{d\tau} \qquad ...(13.58)$$

and

$$\frac{d^2e}{dt^2} = \frac{d\tau}{dt}\cdot\frac{d}{d\tau}\left(\frac{de}{dt}\right) = 4\frac{d^2e}{d\tau^2} \qquad ...(13.59)$$

We can now rewrite Equations (13.54), (13.55) and (13.56) for normalized time.

Thus, we get

$$\frac{d^2e}{d\tau^2}+\frac{de}{d\tau}+e = 0 \qquad -0.5 \le e \le 0.5 \qquad ...(13.60)$$

$$\frac{d^2e}{d\tau^2}+\frac{de}{d\tau}+0.5 = 0 \qquad \text{if } e > 0.5 \qquad ...(13.61)$$

and

$$\frac{d^2e}{d\tau^2}+\frac{de}{d\tau}-0.5 = 0 \qquad \text{if } e < -0.5 \qquad ...(13.62)$$

Making $x_1 = e$ and $x_2 = \dot{e}$, we get the following expressions for δ for the three cases:

For $-0.5 \le e \le 0.5$ $\qquad \delta = x_2$ ...(13.63)

for $e > 0.5$, $\qquad \delta = x_2 - x_1 + 0.5$ ...(13.64)

and for $e < 0.5$, $\qquad \delta = x_2 - x_1 - 0.5$ ...(13.65)

Note that since Equations (13.61) and (13.62) did not contain e (or x_1), we had to add and subtract e to get these in the standard form indicated in Equation (13.41).

We can now plot the trajectory in the normalized phase-plane starting from the point $x_1 = 1$, $x_2 = 0$, which is in the saturation region. Note how the boundaries between the linear and nonlinear regions are clearly indicated in the phase-plane. The completed trajectory is shown in Fig. 13.22, where $\Delta\theta$ has been taken as 10°.

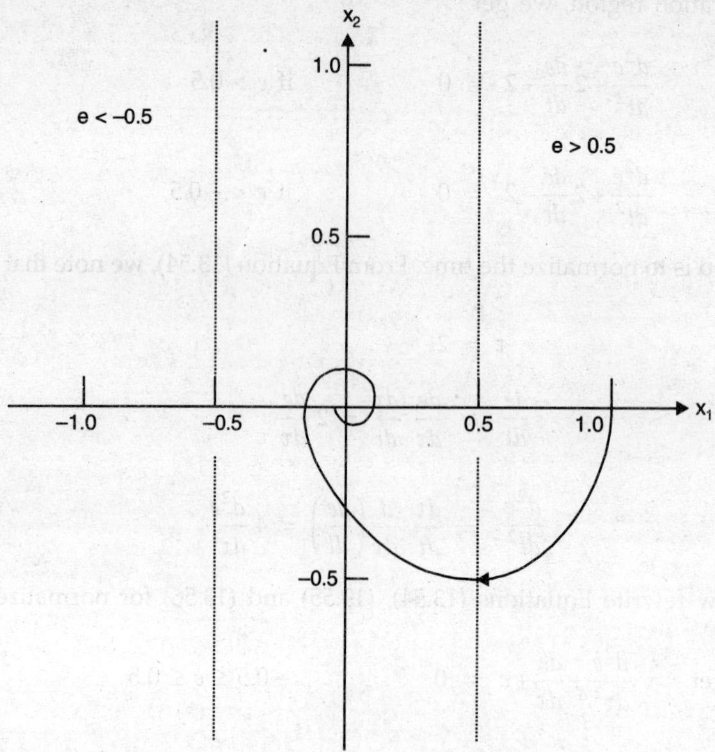

Figure 13.22. *Trajectory for system with saturation*

The choice of $\Delta\theta$ is often quite critical to the accuracy of the solution. This is similar to the problem of selection of the time interval Δt in the numerical solution of a differential equation. The method shown in Fig. 13.20 overcomes this difficulty to a certain extent by improving the accuracy so that the value of $\Delta\theta$ can now be made somewhat larger. In general, one should select $\Delta\theta$ consistent with the graphical accuracy that is possible. With the normalized differential equation, a value of 5° to 10° may be appropriate in most cases for manual plotting, while a value of 0.5° to 1° would be suitable for computer plotting of the trajectory. The computer program on the disk that can be obtained with the book allows the user to select the value of $\Delta\theta$, the minimum allowable value being 0.5°.

A computer plot of family of phase trajectories, known as the *phase portrait,* is very useful for determining the nature of various singular points of a given second-order non-linear differential equation, and has been used extensively for studying such systems. Thus, the method, assisted by a suitable computer program, has considerable potential for practical applications. The main limitation is that the phase-plane method cannot be extended to nonlinear systems described by higher order differential equations.

DRILL PROBLEM 13.4

Draw the phase-plane trajectory for the differential equation

$$\frac{d^2x}{dt^2} + x = 0$$

with the initial conditions $x(0) = 1$ *and* $\frac{dx}{dt}(0) = 0.$

Ans. The trajectory is a circle of unit radius, as shown below.

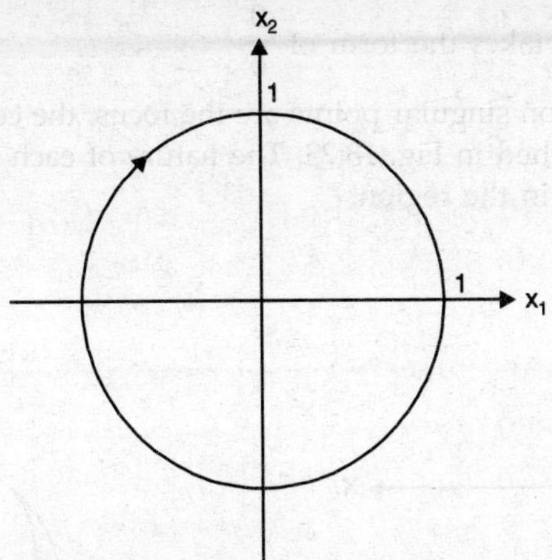

DRILL PROBLEM 13.5

Plot the phase-plane trajectory for the following differential equation $\frac{d^2x}{dt^2} + x\frac{dx}{dt} + x = 0$ *if* $x(0) = 1$ *and* $\frac{dx}{dt}(0) = 0.$

Ans.

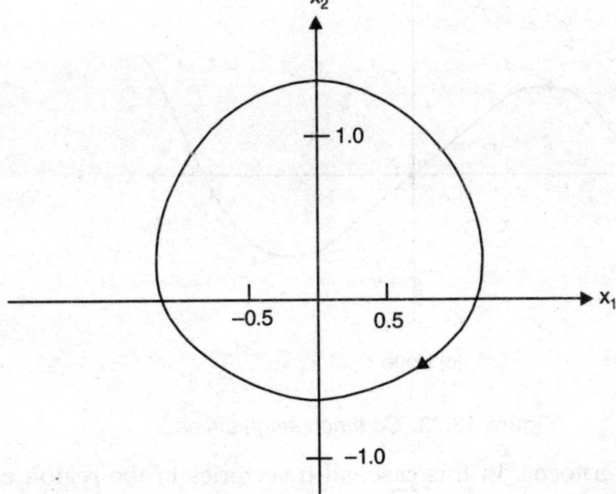

13.7.2 Stability from the Phase Plane

The stability of a system can be determined by examining phase-plane trajectories near each of the points of equilibrium, also called singular points, since the slope of any trajectory at this point is indeterminate, that is $\frac{dx_2}{dx_1}$ takes the form of $\frac{0}{0}$.

Some of the most common singular points are the focus, the center (leading to limit cycles) and the node. These are sketched in Fig. 13.23. The nature of each singularity can be examined by plotting a few trajectories in the region.

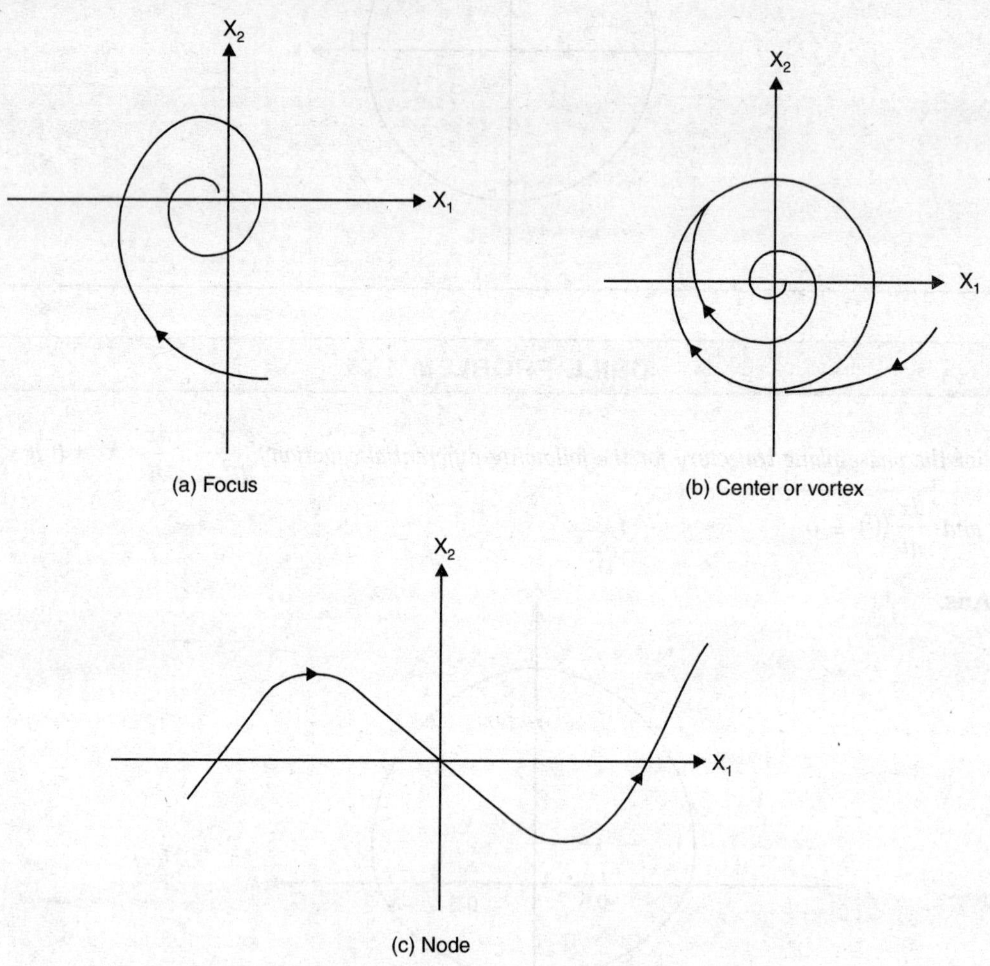

Figure 13.23. *Common singularities*

Figure 13.23(*a*) shows a focus. In this case, all trajectories in the region either converge to the focus or diverge from it; the former represents a stable focus and the latter is an unstable focus. Fig. 13.23(*b*) shows a center or vortex with a stable limit cycle. In this case, all trajectories converge to the closed curve, which represents the limit cycle. Figure 13.23(*c*) shows a node. This is a point of unstable equilibrium, trajectories pass through this point, but do not stop here. A simple example is a bistable multivibrator, in the mode when both transistors are active. This is known to be an unstable mode, as a very small perturbation moves the circuit to one of the two stable modes.

DRILL PROBLEM 13.6

Locate and identify all singularities of the nonlinear differential equation

$$\frac{d^2x}{dt^2} - \left[2 - \left(\frac{dx}{dt}\right)^2\right] x + x^2 = 0$$

Ans. Singularities are at (0, 0) and (2, 0) in the phase-plane.

13.8 STABILITY ANALYSIS USING LYAPUNOV'S DIRECT METHOD

The simple stability criteria developed for linear systems are not applicable to nonlinear systems since the concept of the roots of a characteristic polynomial are no longer valid. As stated in Section 13.3, many different classes of stability have been defined for nonlinear system. In this section, we shall discuss stability in the sense of Lyapunov.

Consider a region ε in the state space enclosing an equilibrium point x_0. Then this is a point of stable equilibrium provided and there is a region δ(ε) contained within ε so that any trajectory starting in the region δ does not leave the region ε.

Note that with this definition it is not necessary for the trajectory to approach the point of equilibrium. It is only required that the trajectory be within the region ε. This permits the existence of oscillations of limited amplitude, like limit cycles.

Lyapunov's direct method provides a means for determining the stability of a system without actually solving for the trajectories in the state space. It is based on the simple concept that the energy stored in a stable system cannot increase with time. Given a set of non-linear state equations, one first defines a scalar function $V(x)$ that this properties similar to energy and then examines its derivative with respect to time.

THEOREM 13.1

A system described by $\dot{x} = f(x)$ is asymptotically stable in the vicinity of the point of equilibrium at the origin of the state space if there exists a scalar function V so that

1. $V(x)$ is continuous and has continuous first partial derivatives at the origin.
2. $V(x) > 0$ for $x \neq 0$ and $V(0) = 0$.
3. $\dot{V}(x) < 0$ for all $x \neq 0$.

Note that these conditions are *sufficient but not necessary* for stability. $V(x)$ is often called a Lyapunov function.

THEOREM 13.2

A system described by $\dot{x} = f(x)$ is unstable in a region Ω about the equilibrium at the origin of the state space if there exists a scalar function V so that

1. $V(x)$ is continuous and has continuous first partial derivatives in Ω.
2. $V(x) \geq 0$ for $x \neq 0$ and $V(0) = 0$.

3. $\dot{V}(x) > 0$ for all $x \neq 0$.

Again, it should be noted that these conditions are *sufficient but not necessary.*

EXAMPLE 13.4

Consider the system described by the equations

$$\dot{x}_1 = x_2 \quad \text{...(13.66)}$$

$$\dot{x}_2 = -x_1 - x_2^3$$

If we make
$$V = x_1^2 + x_2^2 \quad \text{...(13.67)}$$

which satisfies conditions 1 and 2, then we get

$$\dot{V} = 2x_1\dot{x}_1 + 2x_2\dot{x}_2 = 2x_1x_2 + 2x_2\left(-x_1 - x_2^3\right) = -2x_2^4 \quad \text{...(13.68)}$$

It will be seen that $\dot{V} < 0$ for all nonzero values of x_2, and hence the system is asymptotically stable.

EXAMPLE 13.5

Consider the system described by

$$\dot{x}_1 = -x_1(1 - 2x_1x_2)$$

$$\dot{x}_2 = -x_2 \quad \text{...(13.69)}$$

Let
$$V = \frac{1}{2}x_1^2 + x_2^2 \quad \text{...(13.70)}$$

which satisfies conditions 1 and 2, then

$$\dot{V} = x_1\dot{x}_1 + 2x_2\dot{x}_2 = -x_1^2(1 - 2x_1x_2) - 2x_2^2 \quad \text{...(13.71)}$$

Although it is not possible to make a general statement regarding global stability for this case, it is clear that $\dot{V}$ is negative if $1 - 2x_1x_2 > 0$. This defines a region of stability in the state-space, bounded by all points for which $x_1x_2 < 0.5$.

Evidently, the main problem with this approach is the selection of a suitable function $V(x)$ so that its derivative is either positive, definite or negative definite. Unfortunately, there is no general method that will work for every nonlinear system. Gibson [2] has described the variable gradient method for generating Lyapunov functions. It will not be described here but the interested reader may refer to Section 8.1 of the Gibson text.

For a linear system, there is a standard approach for finding a Lyapunov function. Consider an autonomous system described by the equation

$$\dot{x} = Ax \quad \text{...(13.72)}$$

It was seen in Chapter 6 that the necessary and sufficient condition for the system to be stable is that all eigenvalues of A have negative real parts. We shall now consider an alternative approach.

Let the Lyapunov function for this system be given by

$$V = x^T P x \quad \text{...(13.73)}$$

where P is a positive definite symmetric matrix.

The derivative of V with respect to time is then obtained as

$$\dot{V} = \dot{x}^T Px + x^T P\dot{x} = (Ax)^T Px + x^T P(Ax) \quad ...(13.74)$$
$$= x^T (A^TP + PA)\, x$$

Since $\dot{V}$ be negative definite for the system to be stable we must have,

$$A^TP + PA = -Q \quad ...(13.75)$$

where Q is a positive definite matrix.

Instead of first specifying a positive definite matrix P it is convenient to start with a positive definite Q and solve Equation (13.76), which is also called the Lyapunov equation, for P. The system will be stable if a positive definite solution for P can be found.

THEOREM 13.3

Consider a linear system, described by the state equation $\dot{x} = Ax$. A necessary and sufficient condition that the equilibrium state $x = 0$ be asymptotically stable is that for any positive definite symmetric matrix Q, there exists a positive definite symmetric matrix P which satisfies the equation

$$A^TP + PA = -Q$$

Note that the theorem does not depend on the choice of Q; the only requirement is that it be positive definite. A convenient choice is to make Q the identity matrix I.

EXAMPLE 13.6

A linear system is described by the state equation

$$\dot{x} = \begin{bmatrix} 0 & 1 \\ -2 & -3 \end{bmatrix} x$$

Investigate the stability of this system by using Lyapunov's theorem.

SOLUTION

Let

$$P = \begin{bmatrix} p_1 & p_2 \\ p_2 & p_3 \end{bmatrix}.$$

Then the left-hand side of the Lyapunov equation is obtained as

$$A^TP + PA = \begin{bmatrix} 0 & -2 \\ 1 & -3 \end{bmatrix}\begin{bmatrix} p_1 & p_2 \\ p_2 & p_3 \end{bmatrix} + \begin{bmatrix} p_1 & p_2 \\ p_2 & p_3 \end{bmatrix}\begin{bmatrix} 0 & 1 \\ -2 & -3 \end{bmatrix}$$

$$= \begin{bmatrix} -4p_2 & p_1 - 3p_2 - 2p_3 \\ p_1 - 3p_2 - 2p_3 & 2p_2 - 6p_3 \end{bmatrix}$$

Equating the above to the negative of the identity matrix, P is obtained as

$$P = \begin{bmatrix} \frac{5}{4} & \frac{1}{4} \\ \frac{1}{4} & \frac{1}{4} \end{bmatrix}$$

which is seen to be positive definite. Hence, this system is asymptotically stable.

DRILL PROBLEM 13.7

Consider a linear system described by the second-order differential equation

$$\frac{d^2x}{dt^2} + \alpha\frac{dx}{dt} + x = 0$$

Detemine (a) a suitable Lyapunov function for this system, and hence (b) the range of values of K and α for the system to be stable.

Ans. (*a*) $V = \left(K + \frac{1}{4}\alpha^2\right)x_1^2 + \left(\frac{1}{2}\alpha x_1 + x_2\right)^2$, where $x_1 = x$ and $x_2 = \frac{dx}{dt}$

(*b*) Stable for $K > 0$ and $\alpha > 0$.

DRILL PROBLEM 13.8

Determine a suitable Lyapunov function for the differential equation for a nonlinear position control system

$$\frac{d^2x}{dt^2} + \alpha\frac{dx}{dt} + K\left(\frac{dx}{dt}\right)^2 + x = 0$$

Ans. $V = x_1^2 + x_2^2$, where $x_1 = x$ and $x_2 = \frac{dx}{dt}$

SUMMARY

- We have discussed briefly three approaches to the study of nonlinear systems. The method of linearization is often used to obtain linear incremental models, which can be tested for local stability about points of equilibrium. One must be careful in applying this method, which is based on neglecting the higher order terms in the Taylor series expansion about that point. The describing function method is attractive since it attempts to extend the familiar frequency response concepts to nonlinear systems, but it is only approximate because of the assumption that the effect of the harmonics can be neglected. The phase-plane approach gives us the approximate solution of a nonlinear differential equation in a graphical form and allows complete stability analysis by examining the points of singularity. It is possible to implement this method on a personal computer. However, it is limited to second-order differential equation.
- Lyapunov's theorems provide an excellent approach to determining the stability of a system using energy-like functions, and are intuitively very appealing. Practical application is limited due to the lack of a general rule for obtaining Lyapunov function for nonlinear systems.

Problems

1. The characteristic curve for a nonlinear spring is shown in Fig. P13.1. Derive an expression for its describing function.

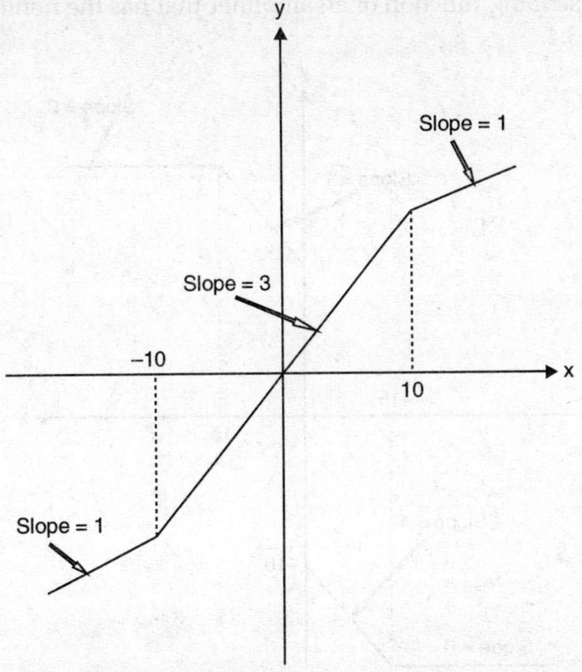

Figure P13.1. *Characteristic curve for a nonlinear spring*

2. The block diagram of a position control system is shown in Fig. P13.2. The transfer function of the servomotor is given by

$$G(s) = \frac{Ke^{-0.1s}}{s(s+10)}$$

It is driven by an amplifier with a gain of 20 in the linear range, which saturates when its input exceeds 1.

(*a*) Investigate the stability of the system when $K = 10$.

(*b*) What is the largest value of K for no limit cycle to exist?

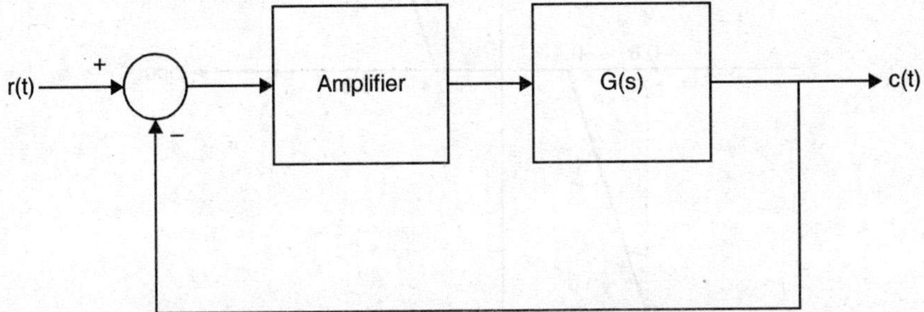

Figure P13.2. *Position control system with saturating amplifier*

3. Determine the describing function of a nonlinear device of its input $x(t)$ and has output $y(t)$ is related through the following differential equation:

$$y\ (t) = \left(\frac{dx}{dt}\right)^3 + x^2 \frac{dx}{dt}$$

4. Determine the describing function of an amplifier that has the nonlinear gain characteristic shown in Fig. P13.4.

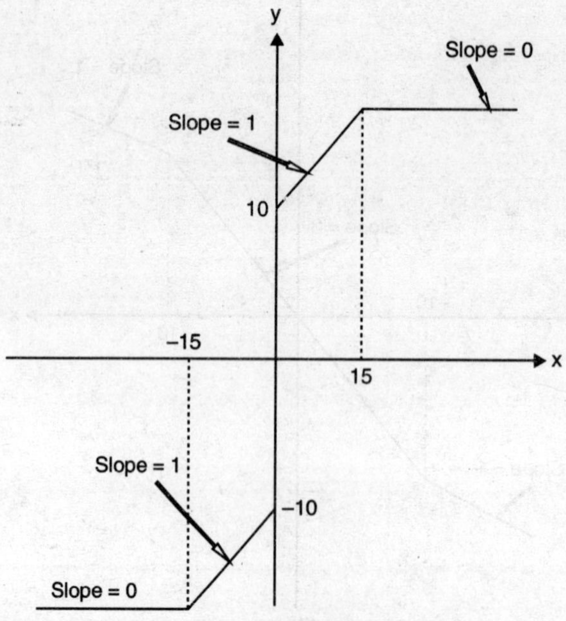

Figure P13.4. *A nonlinear amplifier*

5. The amplifier in the position control system shown in Fig. P13.2 has the characteristic shown in Fig. P13.5. The transfer function is the same as in Problem 2.

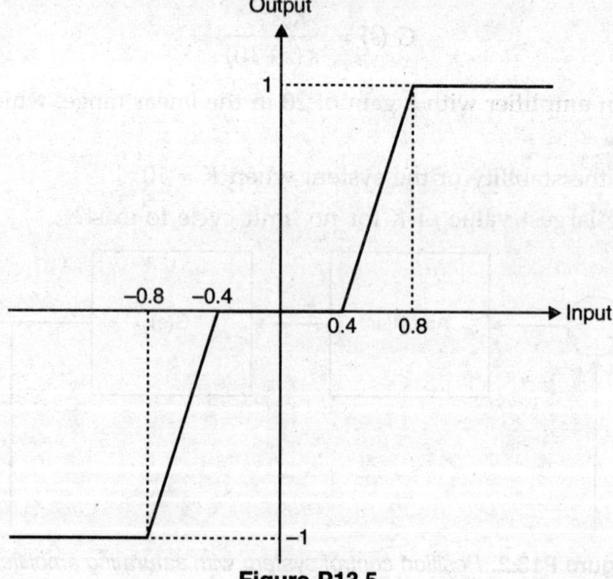

Figure P13.5

(*a*) Investigate the stability of the system.

(*b*) What is the maximum value of *K* for no limit cycle to exist?

6. Draw the phase-plane trajectory of the response of the position control system considered in Problem 2 if the input is a step of size 2 and the transfer function has no delay.
7. Draw the phase-plane trajectory of the response of the position control system considered in Problem 5 if the input is a step 1.5, and the transfer function has no delay.
8. A unity feedback position-control system has an effective motor-load inertia $J = 1$ and an effective friction coefficient $f = 10$. The torque developed by the motor is 100 times the square of the error. Plot the phase-plane trajectory in response to a step input of 1 rad.
9. The differential equation of a nonlinear device is given below. Use Lyapunov's method to determine whether the system is stable.

$$\frac{d^2x}{dt^2} + 2x^2 \frac{dx}{dt} + x = 0$$

10. Verify that (*a*) $V_1 = 0.5x_1^4 + x_2^2$ and (*b*) $V_2 = x_1^4 + 2x_2^2 + 2x_1x_2 + x_1^2$ are suitable Lyapunov functions for the system described by the equations

$$\dot{x}_1 = x_2$$
$$\dot{x}_2 = -x_2 - x_1^3$$

11. The input-output characteristic of an amplifier are shown in Fig. P13.11. Determine the describing function.

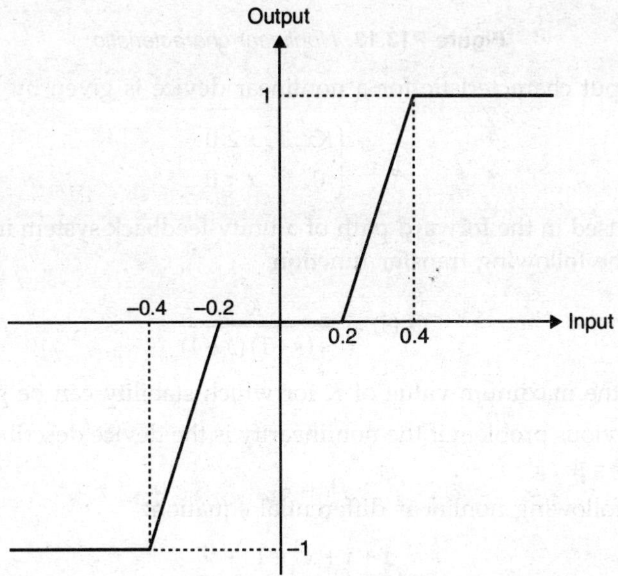

Figure P13.11. *Input-output characteristic of an amplifier*

12. Derive the describing function for a nonlinear device with the input-output characteristic given below (where x is the input and y is the output)

$$y = \begin{cases} Kx, & x \geq 0 \\ 0, & x < 0 \end{cases}$$

13. Derive the describing function for the nonlinearity shown in Fig. P13.13.

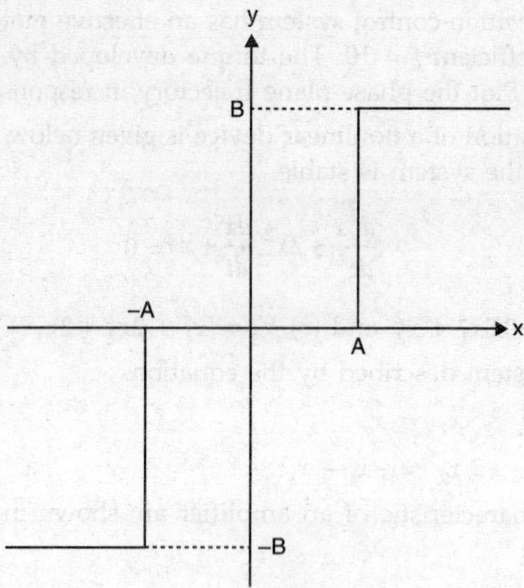

Figure P13.13. *Nonlinear characteristic*

14. The input-output characteristic for a nonlinear device is given by

$$y = \begin{cases} Kx, & x \geq 0 \\ 0, & x < 0 \end{cases}$$

The device is used in the forward path of a unity-feedback system in cascade with a linear system with the following transfer function:

$$G(s) = \frac{K}{s(s+1)(s+4)}$$

What will be the maximum value of K for which stability can be guaranteed?

15. Repeat the previous problem if the nonlinearity is the device described in Problem 13, with $A = 0.2$ and $B = 4$.

16. Consider the following nonlinear differential equation:

$$\ddot{x} + \dot{x} + x^2 - 1 = 0$$

Draw the phase-plane trajectory if the initial conditions are $x(0) = 2$ and $\dot{x}(0) = 0$.

17. Repeat Problem 16 for the following nonlinear differential equation, with the same initial conditions:

$$\ddot{x} + x^2(\dot{x} - 1) + x = 0.$$

18. Determine a suitable Lyapunov function for investigating the stability of the system described in Problem 16.

19. Determine a suitable Lyapunov function for investigating the stability of the system described in Problem 17.

20. Derive the describing function for the nonlinearity shown in Fig. P13.20.

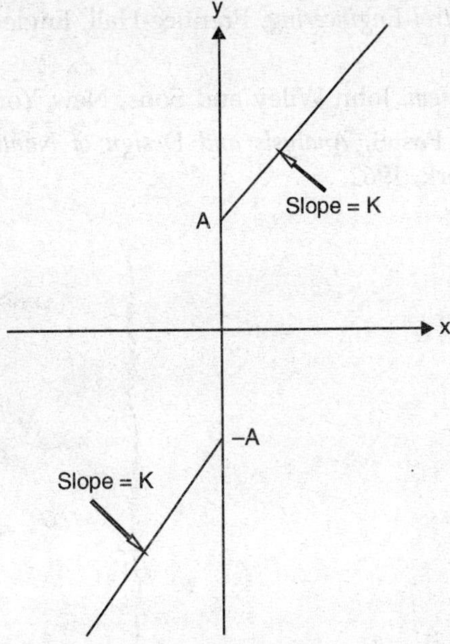

Figure P13.20. *Nonlinear characteristic in Problem 20*

21. The state equations for a nonlinear system are given below. Use Lyapunov's method to determine the range of values of k over which the system will be stable at the origin.

$$\dot{x}_1 = -kx_1\left(x_1^2 + x_2\right)^2 + x_2$$

$$\dot{x}_2 = kx_1\left(x_1^2 + x_2\right)^2 - x_1$$

22. The state equations for a linear system are given by $\dot{x} = Ax$, where

$$A = \begin{bmatrix} -5 & 2 \\ 1 & -3 \end{bmatrix}$$

Investigate the stability of the system by finding a suitable Lyapunov function.

23. Repeat Problem 22 for the following A-matrix:

$$A = \begin{bmatrix} -5 & 1 & -2 \\ 1 & 0 & -1 \\ 3 & 2 & 4 \end{bmatrix}$$

References

1. Atherton, D.P., *Nonlinear Control Enginnering,* Van Nostrand Reinhold, London, 1982.
2. Gibson, J.E., *Nonlinear Automatic Control,* McGraw-Hill, New York, 1963.
3. Minorsky, N., *Theory of Nonlinear Control Systems,* McGraw-Hill, New York, 1969.
4. Ogata, K., *Modern Control Engineering,* Prentice-Hall, Englewood Cliffs, New Jersey, 1990, second edition.
5. Sinha, N.K., *Linear System,* John Wiley and Sons, New York, 1991.
6. Thaler G.J., and M.P. Pastel, *Analysis and Design of Nonlinear Feedback Control Systems,* McGraw-Hill, New York, 1962.

CHAPTER 14 Epilogue

In the previous chapters of this book an attempt was made to present the basic concepts of control theory as developed over the years. Since this work is intended as a text-book for a course to be covered in the equivalent of two quarters, it was not possibel to describe all aspects of the subject. Many important areas have been left out. Some of these will now be mentioned briefly. It is hoped that this will arouse the interest of the reader for further studies in this fascinating area.

Most practical control systems are subject to random disturbances. For example, an antenna in a radar-controlled system may be subject to gusts of wind, or a navigation control system may be affected by waves. Noise may also be introduced by sensors or generated internally in amplifiers. To get the best results in such cases we must try to filter the noise as much as possible. This is the subject of stochastic control and requires a thorough understanding of the theory of random processes. An interesting development in this area has been the theory of optimal filtering and prediction, proposed independently by Wiener in the United States and Kolmogorov in the U.S.S.R. in the 1940's. It was followed in the 1960s by the well-known Kalman filter theory, which allows us to obtain the optimal estimates of the states of a system from noise-contaminated measurements of its inputs and outputs.

In a large number of control systems the parameters of the dynamic equations may vary considerably. For example, the differential equations for the motion of an airplane near the sea level are quite different from those of the same airplane flying at the height of 10 km above the sea level. As another example, the dynamic equations of a nuclear power plant undego considerable change with the amount of power developed. For the best performance, the controller for such a system must adapt itself to changes in the parameters. This has led to the development of the theory of adaptive control. As a result of the microelectronics revolution, it is now quite practical to use microprocessors for implementing adaptive control in these cases.

In all of our deliberations it was assumed that the model of the plant to be controlled is known precisely. In practice, this is seldom the case. Often one has to determine the model of the system from its response to test inputs. This is called the problem of system indentification, and has been a subject of considerable research. Complications are introduced by the presence of noise in the measurements, with the result that there is always some uncertainty in the values of the parameters of the model. Furthermore, most physical systems have some inherent nonlinearities, with the result that exact analysis is often very involved. One of the topics of considerable amount of current research is the theory of the design of robust control systems, the performance of which will not be affected by small variations in parameters.

In Chapter 11 we had a glimpse of the theory of digital control. This subject has attracted considerable attention due to the microelectronics revolution and the consequent availability of inexpensive microprocessors. Several books have been appeared on the subject during the past ten years, and many engineering schools have graduate courses on digital control. In fact, some schools now offer a course on this subject at the undergraduate level.

In Chapter 12 we presented a rather brief discussion .of the analysis and design of control systems using the state-space representation. This is a very powerful approach, particularly in the case if systems with many inputs and outputs. Recently it has been found that a combination of state-space and transfer function methods leads to a greater insight into the behavior of such systems. One example of this approach was seen in Chapter 10, where the poles of the transfer function of the closed-loop system were placed at specified locations by the design of suitable compensator instead of the more commonly used approach based on state-variable feedback together with an asymptotic state observer. It has also been found that the design of compensators for multivariable systems can be carried out with greater ease if they are represented in the form of polynomial matrices.

Some other areas of great interest to control engineers are optimal control theory and the general theory of nonlinear systems. Both of these subjects require a better background in some areas of mathematics than expected in a typical undergraduate program in engineering. On the other hand, one cannot ignore the fact that most real systems have some nonlinearities.

There are a number of excellent books that deal with these advanced topics in sufficient detail, For the enthusiastic and interested reader, some of these reference books are listed below.

Suggestions for Further Reading

1. Åström, K.J., *Introduction to Stochastic Control Theory*, Academic Press, New York, 1970.
2. Åström K.J., and B. Wittenmark, *Computer Controlled System: Theory and Practice*, Prentice-Hall, Englewood Cliffs, N.J., 1984.
3. Atherton, D.P., *Nonlinear Control*, Van Nostrand Reinhold, London, 1982.
4. Barnett, S., *Polynomials and Linear Control System*, Marcel Dekker, New York, 1983.
5. Franklin, G.F., J.D. Powell and M.L. Workman, *Digital Control of Dynamic System*, Addison-Wesley Publishing Company, Reading, Mass., 1990, second edition.
6. Isermann, R., *Digital Control Systems*, Springer, New York, 1983.
7. Kailath, T., *Linear Control System*, Prentice-Hall, Englewood Cliffs, N.J., 1980.
8. Kwakernak H., and R. Sivan, *Linear Optimal Control Systems*, Wiley-Interscience, New York, 1972.

9. Middleton R.H., and G.C. Goodwin, *Digital Control and Estimation*, Prentice-Hall, Englewood Cliffs, N.J., 1990.

10. Saridis, G.N., *Self-Organizing Control of Stochastic System*, Marcell Dekker, New York, 1977.

11. Sinha N.K., and B. Kuszta, *Modeling and Identification of Dynamic Systems*, Van Nostrand Reinhold, New York, 1983.

12. Steingel, R.F., *Stochastic Optimal Control*, John Wiley and Sons, New York, 1986.

13. Vidyasagar, M., *Control System Synthesis:* A *Factorization Approach*, MIT Press, Cambridge, Mass., 1985.

14. Wolovich, W.A., *Linear Multivariable Systems*, Springer-Verlag, New York, 1974.

❑❑❑

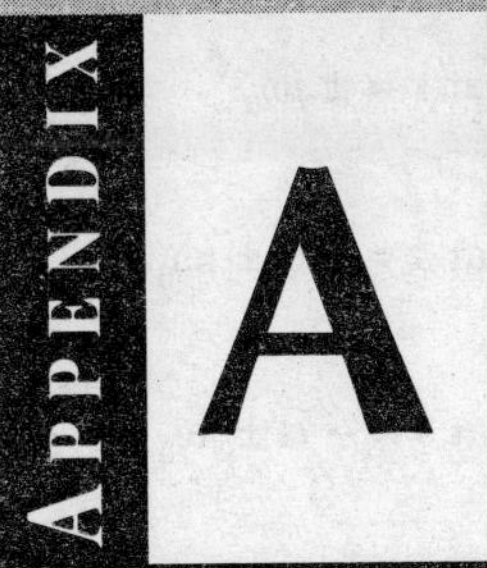

Appendix A Review of Laplace Transforms

A.1 DEFINITION

The Laplace transform of a time function $f(t)$ is defined as

$$F(s) = \mathscr{L}[f(t)] \triangleq \int_{0^-}^{\infty} f(t)\, e^{-st}\, dt \qquad \text{...(A.1)}$$

where $s = \sigma + j\omega \triangleq$ complex frequency.

Note that in this definition, the lower limit in the integral has been taken as 0^-, that is, an infinitesimal time before 0. This has been done to ensure that if the time function contains an impulse at $t = 0$, this is taken into account in the Laplace transform.

This integral will exist if

$$\lim_{t\to\infty} e^{-\sigma t} f(t) = 0$$

for some finite σ.

A.2 SHORT TABLE OF LAPLACE TRANSFORM PAIRS

TABLE A.1

$f(t)$	$F(s)$	*Comments*
Unit impulse $\delta(t)$	1	
Unit step u_t	$\frac{1}{s}$	Pole at the origin
Exponential $e^{-\alpha t}$	$\frac{1}{s+\alpha}$	Pole at $s = -\alpha$

$\cos \omega_0 t$	$\dfrac{s}{s^2+\omega_0^2}$	Poles at $s = \pm j\omega_0$
$\sin \omega_0 t$	$\dfrac{\omega_0}{s^2+\omega_0^2}$	Poles at $s = \pm j\omega_0$
$e^{-\alpha t} \cos \omega_0 t$	$\dfrac{s+\alpha}{(s+\alpha)^2+w_0^2}$	Poles at $s = -\alpha \pm j\omega_0$
$e^{-\alpha t} \sin \omega_0 t$	$\dfrac{\omega_0}{(s+\alpha)^2+w_0^2}$	Poles at $s = -\alpha \pm j\omega_0$
t	$\dfrac{1}{s^2}$	Two poles at the origin
$t^n e^{-\alpha t}$	$\dfrac{n!}{(s+\alpha)^{n+1}}$	$n + 1$ poles at $s = -\alpha$

A.3 THEOREMS

1. $\mathcal{L}[e^{-\alpha t} f(t)] = F(s + \alpha)$ Translation in the s domain ...(A.2)
2. $\mathcal{L}[f(t - T)\, u_{t-T}] = e^{-sT} F(s)$ Translation in the time domain ...(A.3)
3. $\mathcal{L}[tf(t)] = -\dfrac{dF}{ds}$ Multiplication by t ...(A.4)
4. $\mathcal{L}\left[\dfrac{df}{dt}\right] = sF(s) - f(0)$ Differentiation in the time domain ...(A.5)
5. $\lim_{t\to 0} f(t) = \lim_{s\to\infty} [sF(s)]$ Initial value theorem ...(A.6)
6. $\lim_{tr\to\infty} f(t) = \lim_{s\to 0} [sF(s)]$ Final value theorem ...(A.7)
   (valid only all the poles of $F(s)$ have negative real parts, with the exception of one pole at the origin)

A.4 INVERSE LAPLACE TRANSFORMATION

The inverse Laplace transform of $F(s)$ is defined as

$$f(t) = \frac{1}{2\pi j} \int_{c-j\infty}^{c+j\infty} F(s)\, e^{jst}\, ds \qquad \text{...(A.8)}$$

In practice, if $F(s)$ is a rational function of s (that is, the ratio of two polynomials in s), it is easier to use partial fraction expansion. If $F(s)$ is strictly proper, that is, the degree of the numerator is less than that of the denominator, we can obtain the following expansion for the case of distinct poles

$$F(s) = \frac{K(s-z_1)(s-z_2)\ldots(s-z_n)}{(s-p_1)(s-p_2)\ldots(s-p_n)} \quad n > m \quad \ldots(A.9)$$

$$= \frac{A_1}{s-p_1} + \frac{A_2}{s-p_2} + \ldots + \frac{A_k}{s-p_k} + \ldots + \frac{A_n}{s-p_n}$$

where the residue, A_k, $k = 1, 2, \ldots, n$, can be evaluated as

$$A_k = [(s - p_k)\, F(s)]_{s = p_k} \quad \ldots(A.10)$$

Computation of the residues from Equation (A.10) can also be given the following graphical interpretation

$$A_k = K \cdot \frac{\text{Product of directed distances from each zero to } p_k}{\text{Product of directed distances from all other poles to } p_k} \quad \ldots(A.11)$$

If a pole, p_k has multiplicity $r > 1$, the partial fraction contains additional terms of the form

$$F(s) = \frac{A_{k1}}{s-p_k} + \frac{A_{k2}}{(s-p_k)^2} + \ldots + \frac{A_{kr}}{(s-p_k)^r} \quad \ldots(A.12)$$

where

$$A_{k,r} = \left[(s-p_k)^r\, F(s)\right]_{s=p_k}$$

$$A_{k,r-1} = \left[\frac{d}{ds}(s-p_k)^r\, F(s)\right]_{s=p_k}$$

$$A_{k,r-i} = \frac{1}{(r-i)!}\left[\frac{d^{r-i}}{ds^{r-i}}\left((s-p_k)^r\, F(s)\right)\right]_{s=p_k}$$

for $\quad i = 2$ to $r - 1$

The residues can also be easily evaluated using the program Residue. EXE on the disk available with the book, and described in Appendix E. The inverse Laplace transform can then be obtained for each term in the partial fraction expansion using the following Table A.2:

TABLE A.2

$F(s)$	$f(t)$
$\frac{A}{s+\alpha}$	$Ae^{-\alpha t}$
$\frac{C+jD}{s+\alpha+j\beta} + \frac{C-jD}{s+\alpha-j\beta}$	$e^{-\alpha t}\,(2C \cos \beta t + 2D \sin \beta t)$
$\frac{A}{(s+\alpha)^{n+1}}$	$\frac{At^n\, e^{-\alpha t}}{n!}$
$\frac{C+jD}{(s+\alpha+j\beta)^{n+1}} + \frac{C-jD}{(s+\alpha-j\beta)^{n-1}}$	$\frac{2\,t^n\, e^{-\alpha t}}{n!}(C \cos \beta t + D \sin \beta t)$

The following example will illustrate the procedure.

EXAMPLE A.1

Consider the rational function $F(s)$ given in Equation (A.13).

$$F(s) = \frac{10\left(s^2+2s+5\right)}{(s+1)\left(s^2+6s+25\right)} = \frac{10\,(s+1+j2)\,(s+1-j2)}{(s+1)\,(s+3+j4)\,(s+3-j4)}$$

$$= \frac{A_1}{s+1} + \frac{A_2}{s+3-j4} + \frac{A_3}{s+3+j4} \qquad \text{...(A.13)}$$

The residues are evaluated as shown below:

$$A_1 = \frac{10\,(j\,2)\,(-j\,2)}{(2-j\,4)\,(2+j\,4)} = 2$$

$$A_2 = \frac{10\,(-2+j2)\,(-2+j6)}{j8\,(-2+j4)} = 4+j3, \quad A_3 = A_2^* = 4-j3$$

Graphical evaluation of residues is illustrated in Fig. A.1.

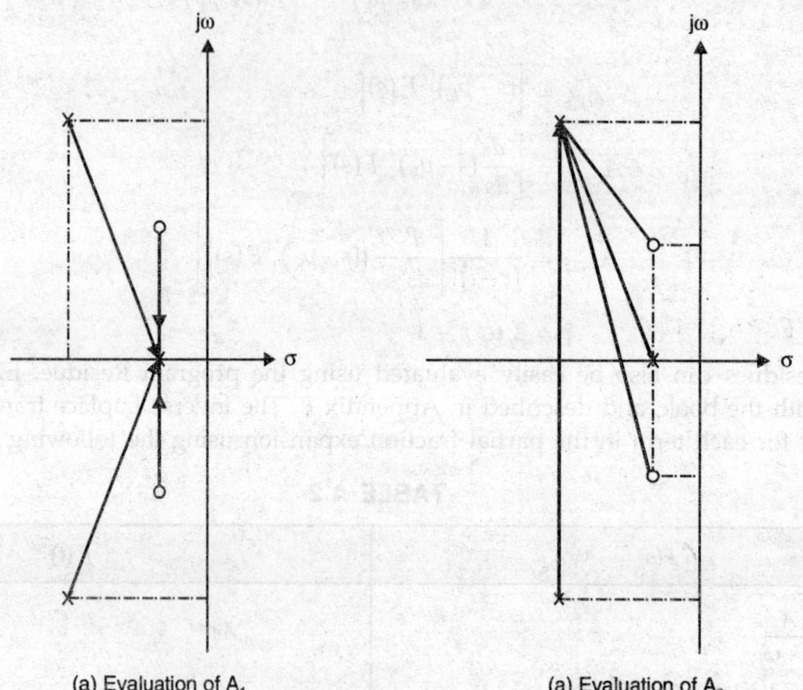

Figure A.1. *Graphical evaluation of residues*

Hence, the inverse transform is given by

$$f(t) = 2\,e^{-t} + e^{-3t}\,(8\cos 4t - 6\sin 4t) \qquad \text{...(A.14)}$$

Note how the entries from Table A.2 have been used after evaluation of the residues to obtain the inverse Laplace transform directly.

EXAMPLE A.2

Consider

$$F(s) = \frac{10\,(s+2)}{(s+1)^2\,(s+3)} \qquad \text{...(A.15)}$$

The partial fraction expansion can be written as

$$F(s) = \frac{A_{1,1}}{s+1} + \frac{A_{1,2}}{(s+1)^2} + \frac{A_2}{s+3}$$

Then

$$A_{1,2} = \left[(s+1)^2\,F(s)\right]_{s=-1} = \left[\frac{10\,(s+2)}{s+3}\right]_{s=-1} = 5 \qquad \text{...(A.16)}$$

$$A_{1,1} = \frac{d}{ds}\left[(s+1)^2\,F(s)\right]_{s=-1} = \frac{d}{ds}\left[\frac{10\,(s+2)}{s+3}\right]_{s=-1}$$

$$= \left[\frac{10}{(s+3)^2}\right]_{s=-1} = 2.5 \qquad \text{...(A.17)}$$

$$A_2 = \left[(s+3)\,F(s)\right]_{s=-3} = \left[\frac{10\,(s+2)}{(s+1)^2}\right]_{s=-3} = 2.5 \qquad \text{...(A.18)}$$

The resulting inverse transform is now obtained as

$$f(t) = 5\,te^{-t} + 2.5\,e^{-t} - 2.5\,e^{-3t} \qquad \text{...(A.19)}$$

A.5 THE TRANSFER FUNCTION

The transfer function of a linear system is defined as the ratio of the Laplace transforms of the output and the input assuming zero initial conditions.

EXAMPLE A.3

For the *RC* network shown in Fig. A. 2., the transfer function is obtained as the ratio of the Laplace transform of $v_1(t)$, and $v_2(t)$, and is given by

$$G(s) = \frac{V_2(s)}{V_1(s)} = \frac{\frac{1}{sC}}{R + \frac{1}{sC}} = \frac{1}{1+sCR} \qquad \text{...(A.20)}$$

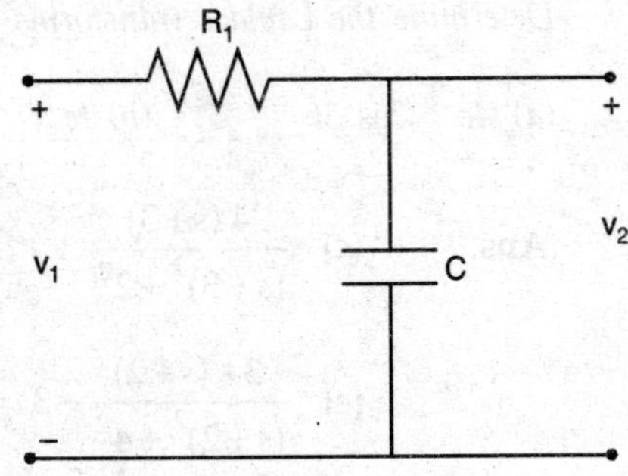

Figure A.2. *An RC network*

A.6 FREQUENCY RESPONSE FROM TRANSFER FUNCTION

If we apply a sinusoidal input to a stable linear system, the steady-state output is also a sinusoid of the same frequency. Replacing s by $j\omega$ in the transfer function gives $G(j\omega) = M(\omega)\, e^{j(w)}$, which is a complex number for a given ω. Thus if the input is a sine wave of frequency ω_1, the steady-state output will be a sine wave of frequency ω_1, with magnitude $M(\omega_1)$ times that of the input and leading it by the angle $\phi(\omega_1)$. The function $G(j\omega)$ is called the frequency response of the system. This is discussed in more detail in Chapter 8, where a graphical interpretation is also shown, similar to Equation (A.11). Here we shall discuss how to obtain the steady-state response of a linear system to a periodic input expressed as a sum of sinusoids.

EXAMPLE A.4

Consider
$$G(s) = \frac{Y(s)}{U(s)} = \frac{20(s+4)}{s+2} \qquad \text{...(A.21)}$$

where the input is
$$u(t) = 2 + 4\cos(3t - 0.3) \qquad \text{...(A.22)}$$

and we are required to determine the steady-state output $y_{ss}(t)$.

Since the input has two components, of angular frequency 0 to 3, we evaluate

$$G(j0) = \left[\frac{20(s+4)}{s+2}\right]_{s=j0} = 40 \qquad \text{...(A.23)}$$

and

$$G(j3) = \left[\frac{20(s+4)}{s+2}\right]_{s=j3} = \frac{340}{13} - j\frac{120}{13} = 27.735\, e^{j0.339} \qquad \text{...(A.24)}$$

Consequently, the steady-state output is given by

$$\begin{aligned} y_{ss}(t) &= (40 \times 2) + (4 \times 27.735)\cos(3t - 0.3 - 0.339) \qquad \text{...(A.25)} \\ &= 80 + 111.94\cos(3t - 0.693) \end{aligned}$$

DRILL PROBLEM A.1

Determine the Laplace transforms of the following:

(a) $4e^{-3t}\cos 3t$ (b) te^{-2t} (c) $\frac{d}{dt}\left(3e^{-2t}\cos 4t\right)$

Ans. (a) $\frac{4(s+3)}{(s+3)^2+2^2}$ (b) $\frac{1}{(s+2)^2}$

(c) $\frac{3s(s+2)}{(s+2)^2+4^2} - 3 = \frac{-6s-60}{s^2+4s+20}$

DRILL PROBLEM A.2

Determine the inverse Laplace transform of the following:

(a) $\frac{10\,(s+1)}{s\,(s+2)\,(s+5)}$ (b) $\frac{600\,(s+2)\left(s^2+2s+5\right)}{(s+3)\left(s^2+4\right)\left(s^2+6s+25\right)}$

(c) $\frac{36\,(s+1)}{s\,(s+2)^2\,(s+3)}$ (d) $\frac{12\,(s+3)}{s^3\,(s+2)^2}$

Ans. (a) $1+\frac{5}{3}e^{-2t}-\frac{8}{3}e^{-5t}$

(b) $-\frac{30}{13}e^{-3t}+\frac{44}{13}\cos 2t+\frac{28}{13}\sin 2t-\frac{14}{13}e^{-3t}\cos 4t+\frac{148}{13}e^{-3t}\sin 4i$

(c) $3 - 27\,e^{-2t} + 24\,e^{-3t} + 18\,te^{-t}$

(d) $4.5\,t^2 - 6\,t + 3.75 - 1.5\,te^{-2t} - 3.75\,e^{-2t}$

DRILL PROBLEM A.3

The transfer function of a linear system is given by

$$G(s) = \frac{Y(s)}{U(s)} = \frac{200\,(s+2)}{(s+4)^2\left(s^2+2s+10\right)}$$

Determine the steady-state output if $u(t) = 5 + 3\cos(3t + \pi/4)$

Ans. $12.5 + 14.226\cos(3t - 0.925)$

A.7 PARSEVAL'S FORMULA

If $F(s)$ is the Laplace transform of $f(t)$ then

$$\int_0^{\infty} f^2(t)\,dt = \frac{1}{2\pi j}\int_{-j\infty}^{j\infty} F(s)\,F(-s)\,ds \qquad \text{...(A.26)}$$

This integral is very useful for calculating the integral square value of the error in a closed-loop system (see Section 5.8 in Chapter 5). The integral will be finite only if $F(s)$ has all its poles in the finite portion of the left half of ths s-plane, and in that case we may use the residue theorem to evaluate the integral on the right-hand side. Hence, we may write

$$\int_0^{\infty} f^2(t)\,dt = \text{Sum of the residues of } F(s)\cdot F(-s) \text{ at all of its poles in the left half of the } s\text{-plane} \qquad \text{...(A.27)}$$

EXAMPLE A.5

Consider the function

$$f(t) = 4\, e^{-3t} \qquad \text{...(A.28)}$$

Then

$$F(s) = \frac{4}{s+2} \qquad \text{...(A.29)}$$

and

$$F(s)\, F(-s) = \frac{4}{s+2} \cdot \frac{4}{-s+2} = \frac{-16}{(s+2)(s-2)} \qquad \text{...(A.30)}$$

From Equation (A.27),

$$\int_0^\infty f^2(t)\, dt = \left[\frac{-16}{s-2}\right]_{s=-2} = 4 \qquad \text{...(A.31)}$$

This can be easily verified by actual integration of 16 e^{-4t}.

DRILL PROBLEM A.4

Use Parseval's formula to evaluate the integral given by Equation (A.26) if

(a) $f(t) = 4\, e^{-2t} \cos 3t$

(b) $f(t) = 5e^{-3t} + 3e^{-t} \sin 2t$

Ans. *(a)* 34/13, *(b)* 31/240

❑❑❑

APPENDIX

B Review of the Theory of Matrices

B.1 MATRICES VIEWED AS TRANSFORMATIONS

It is expected that the reader has come across the basic theory of matrices and is familiar with the rules for multiplication of matrices. It is common practice to define a matrix as "a collection of elements arranged in rectangular (or square) array." In general, a matrix will have n rows and m columns, where m and n are positive integers. A matrix with only one column (that is, $m = 1$) is called a column vector (or simply a vector). A matrix with one row is called a row vector.

It is often helpful to regard a matrix as a transformation. For example, consider the equation

$$y = Ax \qquad ...(B.1)$$

where A is an $n \times m$ matrix, x is a column vector with n rows, and y is a column vector with m rows. We can say that the matrix A has changed the vector x into the vector y. In particular, if A is an $n \times n$ (square) matrix, then both x and y are n-dimensional vectors.

A square matrix is said to be nonsingular of its inverse exist. This is possible if and one of the following conditions is satisfied:

1. The determinant of the matrix is not zero.
2. All the columns of the matrix are linearly independent.
3. All the rows of the matrix are linearly independent.

To calculate the inverse of a matrix, we first obtain its adjoint and then divide each element of the adjoint by the determinant of the matrix. The adjoint of a matrix is obtained replacing each element of its the transpose by the corresponding cofactor.

One should use a computer for inverting matrices of order higher that three. Such computer programs are part of most computer libraries.

B.2 EIGENVALUES AND EIGENVECTORS

For the case when A is a square matrix it is of interest to find vectors that are not changed in direction by the transformation indicated by Equation (B.1). Such vectors are called eigenvectors of A and represent the normal coordinates of the transformation A. For these, we have

$$y = Ax = \lambda x \qquad \text{...(B.2)}$$

where the scalar λ is called an eigenvalue of A.

We can rearrange Equation (B.2) as

$$(\lambda I - A)\, x = 0 \qquad \text{...(B.3)}$$

where I is the identity matrix of dimension $n \times n$.

A trivial solution of Equation (B.3) is the null vector, in which each component of x is zero. A nontrivial solution for x is obtained if and only if the determinant given by

$$\det(\lambda I - A) = 0 \qquad \text{...(B.4)}$$

The determinant in Equation (B.4) is a polynomial of degree n in λ, and is called the characteristic polynomial of A. Hence, it will have n roots. This implies that every $n \times n$ matrix will have n eigenvalues, although some of these may turn out to be equal, corresponding to repeated roots of Equation (B.4). Evaluation of the eigenvalues and eigenvectors will be shown by the following example.

EXAMPLE B.1

Consider the 2 × 2 matrix

$$A = \begin{bmatrix} 0 & 1 \\ -6 & -5 \end{bmatrix} \qquad \text{...(B.5)}$$

Here,

$$\lambda I - A = \lambda \begin{bmatrix} 1 & 0 \\ 0 & 1 \end{bmatrix} - \begin{bmatrix} 0 & 1 \\ -6 & -5 \end{bmatrix} = \begin{bmatrix} \lambda & -1 \\ 6 & \lambda + 5 \end{bmatrix} \qquad \text{...(B.6)}$$

and

$$\det(\lambda I - A) = \lambda^2 + 5\lambda + 6 = (\lambda + 2)(\lambda + 3) \qquad \text{...(B.7)}$$

with roots at $\lambda_1 = -2$ and $\lambda_2 = -3$.

These are the eigenvalues for this matrix. We shall now determine the eigenvectors v_1 and v_2 corresponding to the two eigenvalues. We have

$$(\lambda_1 I - A)\, v_1 = \begin{bmatrix} -2 & -1 \\ 6 & 3 \end{bmatrix} \begin{bmatrix} v_{11} \\ v_{22} \end{bmatrix} = 0 \qquad \text{...(B.8)}$$

We note that the two equations for v_{11} and v_{12} are not independent. From either of these equations we get the relationship

$$2v_{11} + v_{12} = 0 \qquad \text{...(B.9)}$$

If we arbitrarily select $v_{11} = 1$, we get $v_{12} = -2$, so that

$$\begin{bmatrix} v_1 \\ v_2 \end{bmatrix} = \begin{bmatrix} 1 \\ -2 \end{bmatrix} \qquad \text{...(B.10)}$$

is an eigenvector of A corresponding to λ_1.

The nonuniqueness of the values v_{11} and v_{12} is due to the fact that replacing x by kx in Equation (B.2), where k is a constant scalar, simply scales the vector x without changing its direction. Hence, any scalar multiple of v_1 is also an eigenvector of A corresponding to λ_1.

Similarly, an eigenvector corresponding to λ_2 is found to be

$$v_2 = \begin{bmatrix} 1 \\ -3 \end{bmatrix} \quad \text{...(B.11)}$$

A simple way to find eigenvectors of a matrix is first to obtain the adjoint of $(\lambda I - A)$. Then any column of the adjoint of $(\lambda_i I - A)$ is an eigenvector corresponding to λ_i.

EXAMPLE B.2

For the problem considered in Example B.1, the adjoint of $(\lambda I - A)$ is given by

$$\text{adj}\,(\lambda I - A) = \begin{bmatrix} \lambda + 5 & 1 \\ -6 & \lambda \end{bmatrix} \quad \text{...(B.12)}$$

Consequently an eigenvector corresponding to $\lambda_1 = -2$ is obtained as

$$\text{either} \begin{bmatrix} 3 \\ -6 \end{bmatrix} \text{ or } \begin{bmatrix} 1 \\ -2 \end{bmatrix} \quad \text{...(B.13)}$$

EXAMPLE B.3

Consider the third-order matrix

$$A = \begin{bmatrix} -2 & -5 & -5 \\ 1 & -1 & 0 \\ 0 & 1 & 0 \end{bmatrix} \quad \text{...(B.14)}$$

Here, we have

$$\det\,(\lambda I - A) = \begin{vmatrix} \lambda + 2 & 5 & 5 \\ -1 & \lambda + 1 & 0 \\ 0 & -1 & \lambda \end{vmatrix} = (\lambda + 2)\,(\lambda + 1)\,\lambda + (5\lambda + 5)$$

$$= (\lambda + 1)\,(\lambda^2 + 2\lambda + 5) \quad \text{...(B.15)}$$

Hence, the eigenvalues are $\lambda_1 = -1$, $\lambda_2 = -1 + j2$, and $\lambda_3 = -1 - j2$.

The adjoint of $(\lambda I - A)$ is obtained as

$$\text{adj}\,(\lambda I - A) = \begin{bmatrix} \lambda^2 + \lambda & -(5\lambda + 5) & -(5\lambda + 5) \\ \lambda & \lambda^2 + 2\lambda & -5 \\ 1 & \lambda + 2 & \lambda^2 + 3\lambda + 7 \end{bmatrix} \quad \text{...(B.16)}$$

Eigenvectors corresponding to the three eigenvalues are easily calculated from any column of this matrix by substituting λ_i for λ in Equation (B.16). Hence, we get

$$v_1 = \begin{bmatrix} 0 \\ -1 \\ 1 \end{bmatrix} \quad v_2 = \begin{bmatrix} -3 - j4 \\ -1 + j2 \\ 1 \end{bmatrix} \quad v_3 = \begin{bmatrix} -3 + j4 \\ -1 - j2 \\ 1 \end{bmatrix} \qquad \text{...(B.17)}$$

From this example, it is evident that the calculation of the eigenvalues and eigen-vectors can be a tedious task for matrices of high order. The algorithm described in Chapter 3 (Leverrier's algorithm) can be utilized for obtaining the characteristic polynomial as well as the adjoint of $(\lambda I - A)$. In practice, finding the roots of the characteristic polynomial is not a good general method for computing the eigenvalues of a matrix. The *Q-R* algorithm is usually more suitable (for details see Wilkinson [1]). Standard program packages are available [3]:

Some properties of the eigenvalues of a matrix are given below:

1. A matrix of order n will have n eigenvalues, which may not all be distinct (*i.e.*, some eigenvalues may be repeated).
2. The eigenvalues of a matrix and its transpose are identical.
3. The sum of the eigenvalues of a matrix is equal to its trace. (The trace of a matrix is defined as the sum of the elements on its main diagonal).
4. If all the elements of a matrix are real, then each of its eigenvalues will be either real or one of a complex conjugate pair.
5. A symmetric matrix with real elements can have only real eigenvalues.
6. The product of the eigenvalues of a matrix is equal to its determinant.
7. The elements on the diagonal of a triangular (and a diagonal) matrix are its eigenvalues.
8. For any nonsingular matrix P, the matrix $B = P^{-1} AP$ has the same eigenvalues as A.

B.3 DIAGONALIZATION OF A MATRIX

Let a square matrix A of order n have the eigenvalues $\lambda_1, \lambda_2, \ldots, \lambda_n$, and let the corresponding eigenvectors be $v_1, v_2, \ldots, v_n$. Then, we have

$$\begin{aligned} A\, v_1 &= \lambda_1\, v_1 \\ A\, v_2 &= \lambda_2\, v_2 \\ &\vdots \\ A\, v_n &= \lambda_n\, v_n \end{aligned} \qquad \text{...(B.18)}$$

These equations can be combined together to obtain

$$A\,[v_1 \quad v_2 \quad \cdots \quad v_n] = [v_1 \quad v_2 \quad \cdots \quad v_n] \leftarrow \qquad \text{...(B.19)}$$

where Λ is the diagonal matrix given by

$$\Lambda = \begin{bmatrix} \lambda_1 & 0 & 0 & \cdots & 0 \\ o & \lambda_2 & 0 & \cdots & 0 \\ \vdots & \vdots & \vdots & \cdots & \cdots \\ 0 & 0 & 0 & \cdots & \lambda_n \end{bmatrix} \qquad \text{...(B.20)}$$

Let $$M = [v_1 \;\; v_2 \;\; \dots \;\; v_n] \qquad \text{...(B.21)}$$

be defined as the modal matrix for A. Note that the columns of M are the n eigenvectors of A, and hence, M is an $n \times n$ matrix. We may now write Equation (B.19) as

Thus, $$AM = M\Lambda \qquad \text{...(B.22)}$$

$$\Lambda = M^{-1} AM \qquad \text{...(B.23)}$$

is the diagonalized form of A.

It may be noted that this is based on the assumption that M is nonsingular, *i.e.*, the various eigenvectors are linearly independent. In general, this will be true only if all the eigenvalues of A are distinct, or if A is a symmetric matrix. Matrices with repeated eigenvalues can be diagonalized if and only if for each eigenvalue λ_i of multiplicity m_i, the rank of the matrix $(\lambda_i I - A)$ is equal to $n - m_i$. This will always be the case if A is a symmetric matrix.

EXAMPLE B.4

For the matrix in Example B.1 the eigenvalues were calculated as $\lambda_1 = -2$ and $\lambda_2 = -3$. The corresponding eigenvectors are found as (see Example B.2)

$$v_1 = \begin{bmatrix} 1 \\ -2 \end{bmatrix} \quad \text{and} \quad v_2 = \begin{bmatrix} 1 \\ -3 \end{bmatrix} \qquad \text{...(B.24)}$$

Hence, $$M = \begin{bmatrix} 1 & 1 \\ -2 & -3 \end{bmatrix} \qquad \text{...(B.25)}$$

and $$M^{-1} = \begin{bmatrix} 3 & 1 \\ -2 & -1 \end{bmatrix} \qquad \text{...(B.26)}$$

It is easily verified that

$$M^{-1} AM = \begin{bmatrix} -2 & 0 \\ 0 & -3 \end{bmatrix} \qquad \text{...(B.27)}$$

B.4 THE JORDAN FORM

If an eigenvalue λ_i of an $n \times n$ matrix A has multiplicity $m_i > 1$, then A cannot be diagonalized unless the rank of $(\lambda_i I - A)$ is equal to $(n - m_i)$. If this condition is not satisfied, we can transform A to the Jordan form, which is very close to the diagonal form, by using generalized eigenvectors. Some typical Jordan matrices are shown below.

$$J_1 = \begin{bmatrix} \lambda_1 & 1 & 0 & \vdots & 0 & 0 & \vdots & 0 \\ 0 & \lambda_2 & 1 & \vdots & 0 & 0 & \vdots & 0 \\ 0 & 0 & \lambda_1 & \vdots & 0 & 0 & \vdots & 0 \\ \cdots & \cdots & \cdots & \vdots & \cdots & \cdots & \vdots & \cdots \\ 0 & 0 & 0 & \vdots & \lambda_2 & 1 & \vdots & 0 \\ 0 & 0 & 0 & \vdots & 0 & \lambda_2 & \vdots & 0 \\ \cdots & \cdots & \cdots & \vdots & \cdots & \cdots & \vdots & \cdots \\ 0 & 0 & 0 & \vdots & 0 & 0 & \vdots & \lambda_3 \end{bmatrix} \quad \text{...(B.28)}$$

$$J_2 = \begin{bmatrix} \lambda_1 & 1 & 0 & \vdots & 0 & 0 \\ 0 & \lambda_1 & 1 & \vdots & 0 & 0 \\ 0 & 0 & \lambda_1 & \vdots & 0 & 0 \\ \cdots & \cdots & \cdots & \vdots & \cdots & \cdots \\ 0 & 0 & 0 & \vdots & \lambda_1 & 1 \\ 0 & 0 & 0 & \vdots & 0 & \lambda_1 \end{bmatrix} \quad \text{...(B.29)}$$

It may be noted that J_1 has two Jordan blocks, one for λ_1 and the other for λ_2. On the other hand, J_2 has two Jordan blocks for the same eigenvalue λ_1, which has multiplicity five. This situation will arise only if the rank of $(\lambda_i I - A)$ is $n - 2$ so that we get two independent eigenvectors. In general, the number of Jordan blocks for a given eigenvalue will be equal to r if the rank of $(\lambda_i I - A)$ is $n - r$.

A vector v is said to be a generalized eigenvector of rank k of A associated with λ_i if and only if

and
$$\left.\begin{aligned} (\lambda_i I - A)^k v &= 0 \\ (\lambda_i I - A)^{k-1} v &\neq 0 \end{aligned}\right\} \quad \text{...(B.30)}$$

It may be noted that for $k = 1$, we get v as an eigenvector, consistent with the definition given in Equation (B.3). This justifies the use of the term 'generalized eigenvectors.'

In general, if the multiplicity of λ_i is m_i and the rank of $(\lambda_i I - A)$ is n-1, then we should be able to get one eigenvector and m_i-1 generalized eigenvectors that will be linearly independent. On the other hand, if the rank of $(\lambda_i - A)$ is n-r, then we can get r eigenvectors, leading to r Jordan blocks. For each Jordan block we obtain sets of generalized eigenvectors following Equation (B.30), and the total number of linearly independent generalized eigenvectors will be m_i-r.

The procedure will be illustrated by an example.

EXAMPLE B.5

Consider the matrix

$$A = \begin{bmatrix} 0 & 1 & 0 \\ 0 & 0 & 1 \\ -2 & -5 & -4 \end{bmatrix} \quad \text{...(B.31)}$$

In this case, the characteristic polynomial is given by

$$\det(\lambda I - A) = \begin{bmatrix} \lambda & -1 & 0 \\ 0 & \lambda & -1 \\ 2 & 5 & \lambda+4 \end{bmatrix} = \lambda^3 + 4\lambda^2 + 5\lambda + 2$$

$$= (\lambda + 1)^2 (\lambda + 2) \quad \text{...(B.32)}$$

Hence the eigenvalues are $\lambda_1 = -1$, with multiplicity two, and $\lambda_3 = -2$.

The adjoint of $(\lambda I - A)$ is given by

$$\text{adj}(\lambda I - A) = \begin{bmatrix} \lambda^2 + 4\lambda + 5 & \lambda + 4 & 1 \\ -2 & \lambda^2 + 4\lambda & \lambda \\ -2\lambda & -5\lambda - 2 & \lambda^2 \end{bmatrix} \quad \text{...(B.33)}$$

Consequently, the eigenvectors corresponding to λ_1 and λ_3 are obtained as

$$v_1 = \begin{bmatrix} 1 \\ -1 \\ 1 \end{bmatrix} \text{ and } v_2 = \begin{bmatrix} 1 \\ -2 \\ 4 \end{bmatrix} \quad \text{...(B.34)}$$

The vector v_2, must be a generalized eigenvector of order 2, corresponding to λ_1. First we note that

$$(\lambda_1 I - A)^2 = \begin{bmatrix} -1 & -1 & 0 \\ 0 & -1 & -1 \\ 2 & 5 & 3 \end{bmatrix}^2 = \begin{bmatrix} 1 & 2 & 1 \\ -2 & -4 & -2 \\ 4 & 8 & 4 \end{bmatrix} \quad \text{...(B.35)}$$

It is seen by inspection that the rank of $(\lambda_1 I - A)^2$ is one, since any column (or row) of this matrix is a multiple of the first column (or row). Accordingly, we can find two linearly independent nontrivial solutions of

$$(\lambda_1 I - A)^2 v = 0 \quad \text{..(B.36)}$$

One obvious solution is v_1, given in Equation (B.34), that will satisfy Equation (B.36), since

$$(\lambda_1 I - A)\, v_1 = 0 \quad \text{...(B.37)}$$

A second solution is obtained from

$$(\lambda_1 I - A)\, v_2 = v_1 \quad \text{...(B.38)}$$

which satisfies Equation (B.30) for $k = 2$. This following solution for v_2 is obtained.

$$v_2 = \begin{bmatrix} 0 \\ -1 \\ 2 \end{bmatrix} \qquad \text{...(B.39)}$$

Note that in obtaining the solution for v_2 in this manner, we did not have to calculate the square of the matrix $(\lambda_1 - A)$. This is desirable from the computational point of view, as it reduces the amount of round-off error involved in squaring the matrix.

Furthermore, this solution for v_2 is not unique, since any linear combination of v_1 and v_2 will also be a generalized eigenvector of A corresponding to the eigenvalue λ_2. This is in addition to the fact that v_1 is not unique (it can be multiplied by a scalar).

We can now construct the modal matrix as

$$M = [v_1 \quad v_2 \quad v_3] = \begin{bmatrix} 1 & 0 & 1 \\ -1 & -1 & -2 \\ 1 & 2 & 4 \end{bmatrix} \qquad \text{...(B.40)}$$

It is easily verified that

$$J = M^{-1} AM$$

$$= \begin{bmatrix} 0 & -2 & -1 \\ 2 & -3 & -1 \\ 1 & 2 & 1 \end{bmatrix} \begin{bmatrix} 0 & 1 & 0 \\ 0 & 0 & 1 \\ -2 & 5 & -4 \end{bmatrix} \begin{bmatrix} 1 & 0 & 1 \\ -1 & -1 & -2 \\ 1 & 2 & 4 \end{bmatrix}$$

$$= \begin{bmatrix} -1 & 1 & \vdots & 0 \\ 0 & -1 & \vdots & 0 \\ \cdots & \cdots & \vdots & \cdots \\ 0 & 0 & \vdots & -2 \end{bmatrix} \qquad \text{...(B.41)}$$

DRILL PROBLEM B.1

Determine the diagonal or Jordan form for the following matrices

(a) $\begin{bmatrix} 3 & 0 & 1 \\ 0 & 4 & 0 \\ 1 & 0 & 3 \end{bmatrix}$ *(b)* $\begin{bmatrix} 5 & 6 & 0 \\ 0 & 1 & 0 \\ -4 & 4 & 1 \end{bmatrix}$

Ans. (*a*) $\begin{bmatrix} 2 & 0 & 0 \\ 0 & 4 & 0 \\ 0 & 0 & 4 \end{bmatrix}$ (*b*) $\begin{bmatrix} 5 & 0 & 0 \\ 0 & 1 & 1 \\ 0 & 0 & 1 \end{bmatrix}$

B.5 CAYLEY-HAMILTON THEOREM

This theorem states that if

$$\Delta(\lambda) \triangleq \det(\lambda I - A) = \lambda^n + a_1 \lambda^{n-1} + \ldots + a_{n-1} \lambda + a_n \quad \text{...(B.42)}$$

then the matrix polynomial

$$\Delta(A) \triangleq A^n + a_1 A^{n-1} + \ldots + a_{n-1} A + a_n I = 0 \quad \text{...(B.43)}$$

This is a very useful result and convenient for calculation A^k where $k > n$, since we can express A^k as a linear combination of $\{A^n, A^{n-1}, \ldots, A, I\}$. This is done be dividing the polynomial λ^k by $\Delta(\lambda)$ and utilizing the remainder.

EXAMPLE B.6

Consider the matrix A given in Equation (B.31). We shall use Cayley-Hamilton theorem to calculate A^{10}.

First note that from Example B.5

$$\Delta(\lambda) = \lambda^3 + 4\lambda^2 + 5\lambda + 2 \quad \text{...(B.44)}$$

Using long division we obtain

$$\lambda^{10} = (\lambda^7 - 4\lambda^6 + 11\lambda^5 - 26\lambda^4 + 57\lambda^3 - 120\lambda^2 + 24\lambda - 502)\,\Delta(\lambda) + 1013\lambda^2 + 2016\lambda + 1004 \quad \text{...(B.45)}$$

Consequently, we have

$$A^{10} = 1013A^2 + 2016A + 1004I \quad \text{...(B.46)}$$

which is evaluated more easily. The final answer is

$$A^{10} = \begin{bmatrix} 1004 & -2016 & 1013 \\ -2026 & -4061 & -2036 \\ 4072 & 8154 & 4083 \end{bmatrix} \quad \text{...(B.47)}$$

We can also use Cayley-Hamilton theorem to calculate the inverse of a matrix. This is done by simple multiplying both sides of Equation (B.43) by A^{-1} and rearranging so that we can express A^{-1} as a linear combination of $\{A^n, A^{n-1}, \ldots, A, I\}$.

Cayley-Hamilton theorem can also be utilized to obtain a unique polynomial of degree n-1 or less, in A for the matrix e^{At}. This method, known as Sylvester's interpolation formula is a useful approach, in addition to those described in Chapter 3 (Section 3.7). The interested reader is referred to the book by Ogata [2] for further details.

B.6 QUADRATIC FORMS

For an $n \times n$ real symmetric matrix A, and an n-dimensional vector x, the quantity $x^T Ax$ is called a *quadratic form.* Note that this is a scalar quantity.

A real symmetric matrix A is said to be *positive definite* if

$$x^T Ax > 0 \quad \text{for all } x \neq 0 \quad \text{...(B.48)}$$

A real symmetric matrix A is said to be *positive semi-definite* if

$$x^T Ax \geq 0 \quad \text{for all } x \neq 0 \quad \text{...(B.49)}$$

A real symmetric matrix is positive definite if all its eigenvalues are positive.

A necessary and sufficient condition for a real symmetric matrix to be positive definite is that its determinant be positive and all sucessive principal minors have positive determinants.

References

1. Lastman, G.J., and N.K. Sinha, *Microcomputer-based Numerical Methods for Science and Engineering,* Holt, Rinehart and Winston, New York, 1989.
2. Ogata, K., *State Space Analysis of Control Systems,* Prentice-Hall, Englewood-Cliffs, N.J., 1967.
3. Wilkinson, J.H., *The Algebraic Eigenvalue Problem,* Clarendon Press, Oxford, U.K., 1965.

APPENDIX C Review of the Theory of z-Transforms

The role of the z-transform in the analysis and design of discrete-time systems is similar to that of the Laplace transform in continuous-time systems. The main idea here is to replace a continuous-time signal by a sequence of equally spaced impulses, where the strength (or area) of each impulse is equal to the magnitude of the signal at the sampling instant. In this appendix we shall review the basic theory of the z-transform.

C.1 DEFINITION

Consider a sequence of unit impulses T seconds apart, shown in Fig. C.1 and defined by

$$\delta_T(t) = \sum_{n=0}^{\infty} \delta(t - nT) \quad \text{...(C.1)}$$

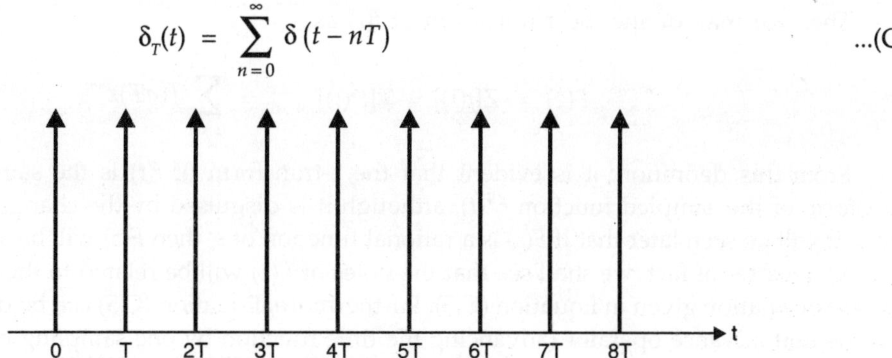

Figure C.1. *Sequence of input impulses T seconds apart*

The Laplace transform of this sequence of unit impulses is given by

$$I(s) = \mathcal{L}[\delta_T(t)] = \sum_{n=0}^{\infty} e^{-sTn} \quad \text{...(C.2)}$$

Now consider a function $f(t)$ modulating these impulses as shown in Fig. C.2. This gives rise to the impulse train

$$f^*(t) = \sum_{n=0}^{\infty} f(nT)\,\delta(t - nT) \qquad \text{...(C.3)}$$

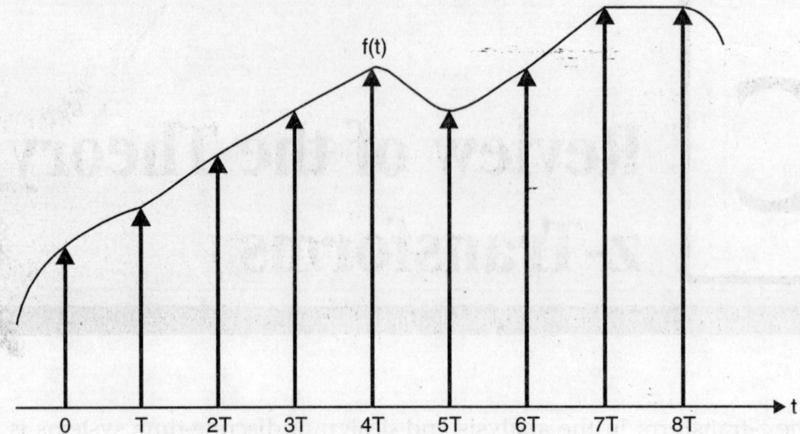

Figure C.2. *Impulse modulated by f(t)*

Taking the Laplace transform, we get

$$F^*(s) = \mathcal{L}[f^*(t)] = \sum_{n=0}^{\infty} f(nT)\, e^{-sTn} \qquad \text{...(C.4)}$$

Note that both $I(s)$ and $F^*(s)$ involve e^{-sT}, making them irrational functions of s. A simplification is obtained if we use the substitution

$$z = e^{sT} \qquad \text{...(C.5)}$$

Then we may define the z-transform of $f(t)$ as

$$F(z) = Z[f(t)] = \mathcal{L}[f^*(t)]_{z\,=\,e^{sT}} = \sum_{n=0}^{\infty} f(nT) z^{-n} \qquad \text{...(C.6)}$$

From this definition, it is evident that the z-transform of $f(t)$ is the same as the Laplace transform of the sampled function $f^*(t)$, although it is disguised by the change in variable from s to z. It will be seen later that if $F(s)$ is a rational function of s, then $F(z)$ will be a rational function of z. As a matter of fact, we shall see that the poles of $F(z)$ will be related to those of $F(s)$ through the transformation given in Equation (C.5). Furthermore, Equation (C.5) can be utilized to interpet z as the unit advance operator (advancing the time function by one sampling interval). We shall now obtain the z-transforms of some standard functions.

C.2 SOME SIMPLE *z*-TRANSFORM PAIRS

These will be derived using the definitions in Equation (C.6).

(*a*) *Unit Step*: This is defined as

$$f(t) = 1 \quad \text{for} \quad t \ge 0 \qquad \text{...(C.7)}$$

Its z-transform is obtained as

$$F(z) = \sum_{n=0}^{\infty} f(nT)z^{-n} = 1 + z^{-1} + z^{-2} + \ldots$$

$$= \frac{1}{1-z^{-1}} = \frac{z}{z-1} \quad \ldots(C.8)$$

Note that the location of the pole of the z-transform is related to that of the pole of the Laplace transform of the step function through Equation (C.5).

(b) Exponential function: Consider the exponential function

$$f(t) = e^{at} \quad \ldots(C.9)$$

Its z-transform is obtained as

$$F(z) = \sum_{n=0}^{\infty} e^{naT} z^{-n} = 1 + e^{aT} z^{-1} + e^{2aT} z^{-2} + \ldots \quad \ldots(C.10)$$

$$= \frac{1}{1-e^{aT}z^{-1}} = \frac{z}{z-e^{aT}}$$

It will be seen again that the location of the pole of $F(s)$ relates to that of the pole of the Laplace transform of $f(t)$ in the s-plane, through the mapping given by Equation (C.5).

(c) Sine and Cosine functions: If in Equation (C.9) we replace a by $j\omega$, we get

$$Z[e^{j\omega t}] = Z[\cos \omega t + j \sin \omega t] = \frac{z}{z - e^{j\omega T}}$$

$$= \frac{z}{(z - \cos \omega T) - j \sin \omega T}$$

$$= \frac{z(z - \cos \omega T) + jz \sin \omega T}{(z - \cos \omega T)^2 + (\sin \omega T)^2} \quad \ldots(C.11)$$

Separating the real and imaginary parts of Equation (C.11) leads to

$$Z[\cos \omega t] = \frac{z(z - \cos \omega T)}{z^2 - 2z \cos \omega T + 1} \quad \ldots(C.12)$$

and

$$Z[\sin \omega t] = \frac{z \sin \omega T}{z^2 - 2z \cos \omega T + 1} \quad \ldots(C.13)$$

(d) Ramp function: Consider the unit ramp function

$$f(t) = t \quad \ldots(C.14)$$

Its z-transform is obtained as

$$F(z) = \sum_{n=0}^{\infty} nTz^{-n} = Tz^{-1} + 2Tz^{-2} + 3Tz^{-3} + \ldots$$

$$= Tz^{-1} (1 + 2z^{-1} + 3z^{-2} + \ldots) \quad \ldots(C.15)$$

$$= \frac{Tz^{-1}}{\left(1-z^{-1}\right)^2} = \frac{Tz}{\left(z-1\right)^2}$$

We shall now study the mapping between the *s*-plane and the *z*-plane.

C.3 MAPPING BETWEEN THE *s*-PLANE AND THE *z*-PLANE

Equation (C.5) represents a transformation or mapping between the *s*-plane and the *z*-plane, as shown in Fig. C.3.

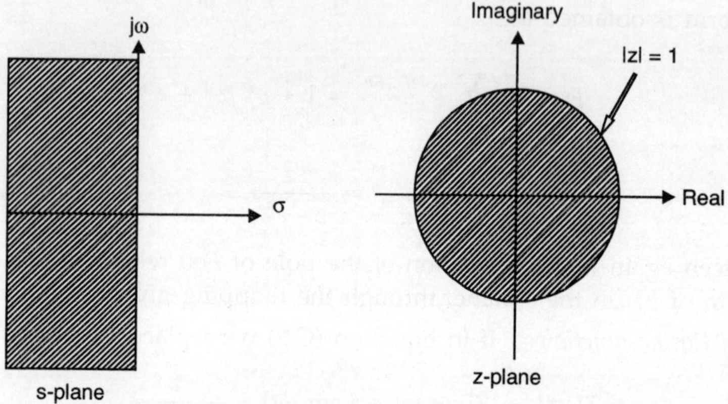

Figure C.3. *Mapping from the* s-plane *to the z-plane*

It may be observed that the *jω*-axis of the *s*-plane maps into the unit circle of the *z*-plane and the origin of the *s*-plane maps into the point 1 + *j*0 in the *z*-plane. We also see that the poles of *F*(*s*) and *F*(*z*) will be related through Equation (C.5) for all the elementary functions considered in Section C.2. We can extend this to obtain the *z*-transform of a function *f*(*t*) if its Laplace transform *F*(*s*) is known. First we note from Equation (C.10) that

$$Z\left[\frac{A}{s+a}\right] = \frac{Az}{z-e^{-aT}} \quad \text{...(C.16)}$$

Consequently, if we have $F(s) = \sum_{k=1}^{n} \frac{A_k}{s+a_k}$ ...(C.17)

then it follows that $F(z) = \sum_{k=1}^{n} \frac{A_k z}{z-e^{-a_k T}}$ ...(C.18)

Equation (C.18) provides us with a simple procedure for obtaining the *z*-transform of a function if its Laplace transform is known. The only case where difficulty may arise is that of repeated poles. This is overcome by noting that

$$\frac{K}{(s+a)^{r+1}} = (-1)^r \frac{K}{r!} \frac{d^r}{da^r}\left(\frac{1}{s+a}\right) \quad \text{...(C.19)}$$

Hence, if $$F(s) = \frac{K}{(s+a)^{r+1}} \qquad ...(C.20)$$

then $$F(z) = (-1)^r \frac{K}{r!} \frac{d^r}{da^r}\left(\frac{z}{z-e^{-aT}}\right) \qquad ...(C.21)$$

DRILL PROBLEM C.1

Determine the z-transforms of the following for T = 0.2.

(a) $\frac{10}{s(s+2)}$ (b) $\frac{4s}{(s+1)(s^2+4)}$ (c) $\frac{25(s+2)}{s(s+1)(s^2+6s+25)}$ (d) $\frac{1}{(s+3)^2}$

Ans. (a) $\frac{1.648z}{(z-1)(z-0.6703)}$ (b) $\frac{0.7049z^2-0.7068z}{(z-0.8187)(z^2-1.8421z+1)}$

(c) $\frac{0.3453z^3-0.0613z^2-0.1255z}{(z-1)(z-0.8187)(z^2-0.76472z+0.2019)}$

(d) $\frac{0.1098z}{(z-0.5488)^2}$

C.4 SOME USEFUL THEOREMS

We shall now present some useful theorems in z-transform theory. It is interesting to compare them with similar theorems from the theory of Laplace transforms.

(a) Functions with exponential damping: If $F(z)$ is the z-transform of $f(t)$ then

$$Z[e^{-at} f(t)] = \sum_{n=0}^{\infty} e^{-anT} f(nT) z^{-n} = \sum_{n=0}^{\infty} f(nT)\left(ze^{-aT}\right)^{-n} \qquad ...(C.22)$$

$$= F\,[ze^{aT}]$$

This relates easily to the corresponding theorem for Laplace transforms in view of Equation (C.5).

(b) Multiplication of a function by t: If $F(z)$ is the z-transform of $f(t)$ then

$$Z[tf(t)] = -Tz\frac{dF(z)}{dz} \qquad ...(C.23)$$

This is easily proved, as shown below:

$$\frac{dF(z)}{dz} = \frac{d}{dz}\sum_{n=0}^{\infty} f(nT)z^{-n} = -\sum_{n=0}^{\infty} nf(nT)z^{-n-1} \qquad ...(C.24)$$

$$= -\frac{1}{T}\sum_{n=0}^{\infty} z^{-1}\, nT\, f(nT)z^{-n} = -\frac{1}{Tz} Z[tf(nT)]$$

(c) *Initial value theorem:* If $F(z)$ is the z-transform of $f(t)$ then

$$f(0) = \lim_{z\to\infty} F(z) \qquad ...(C.25)$$

This follows directly from the definition given in Equation (C.6).

(d) *Final value theorem:* If $F(z)$ is the z-transform of $f(t)$ then

$$\lim_{z\to\infty} f(nT) = \lim_{z\to 1}\left[(z-1)F(z)\right] \qquad ...(C.26)$$

This theorem makes it possible to determine the final value of $f(nT)$ without actually determining the inverse z-transform of $F(z)$.

As in the case of Laplace transform theory, this theorem must be applied with caution to only those cases where the final value exists. If $F(z)$ has any pole outside the unit circle, the final value will be infinite, and the theorem cannot be applied. Similarly, if there are poles on the unit circle, with the exception of a simple pole at $z = 1$, the final value is indeterminate, and the theorem is again inapplicable. If all the poles are strictly inside the unit circle (*i.e.,* there is no pole at $z = 1$), then Equation (C.26) gives zero final value. With a simple pole at $z = 1$, Equation (C.26) gives a constant as the final value.

DRILL PROBLEM C.2

Determine the z-transform of the following if T = 0.5:

(a) $10 \cos (3t + 60°)$, *(b)* $t\, e^{-3t}$, *(c)* $10\, e^{-2t} \cos (3t + 60°)$

Ans. *(a)* $\dfrac{5z^2 - 8.9922z}{z^2 - 0.1415z + 1}$, *(b)* $\dfrac{0.1116z}{(z-0.2231)^2}$, *(c)* $\dfrac{5z^2 - 3.3081z}{z^2 - 0.052z + 0.1353}$

C.5 EVALUATION OF THE INVERSE *z*-TRANSFORM

Since $F(z)$ is actually $F^*(s)$, the inverse transform gives us the function $f^*(t)$, which contains $f(nT)$ at the sampling instants only. To emphasize this fact, we shall always denote the inverse z-transform of $F(z)$ as $f(nT)$. The two most straightforward methods for evaluating the inverse z-transform will now be discussed.

C.5.1 The Method of Partial Fraction Expansion

This is based on the use of Equation (C.2). It may be noted that it is more desirable to expand $F(z)/z$ into partial fractions, in order that we may get z in the numerator after the expansion. The following recognitions are useful:

$F(z)$	$f(nT)$
$\dfrac{Kz}{z-a}$	$K\, a^n$

$\dfrac{(C+jD)z}{z-re^{j\phi}}+\dfrac{(C-jD)z}{z-re^{-j\phi}}$	$2\,r^n\,(C\cos n\phi - D\sin n\phi)$
$\dfrac{Kz}{(z-a)^r}\quad r = 2, 3, \ldots$	$\dfrac{Kn(n-1)\cdots(n-r+2)}{(r-1)!\,a^{r-1}}\,a^n$

Partial fraction of $F(z)/z$ can be carried out in the same way as done in the case of inverse Laplace transformation. The program Residue. EXE on the disk that can be obtained to accompany this book, can be used for this purpose. The only difference is that in order to use these recognitions, complex poles have to be expressed in the polar form.

EXAMPLE C.1

Consider

Then $$F(z) = \frac{3z^2-4z}{(z-2)(z^2-2z-2)} \qquad \text{...(C.27)}$$

$$\frac{F(z)}{z} = \frac{3z-4}{(z-2)(z^2-2z+2)} \qquad \text{...(C.28)}$$

$$= \frac{1}{z-2}+\frac{-0.5-j1}{z-\sqrt{2}e^{j\pi/4}}+\frac{-0.5+j1}{z-\sqrt{2}e^{-j\pi/4}}$$

Consequently, $$f(nT) = 2^n - 2^{\frac{n}{2}}\left(\cos\frac{n\pi}{4} - 2\sin\frac{n\pi}{4}\right) \qquad \text{...(C.29)}$$

C.5.2 The Method of Long Division

Since $F(z)$ is a rational function of z, it can be expanded into a power series of the form

$$F(z) = \sum_{n=0}^{\infty} a_n\, z^{-n} \qquad \text{...(C.30)}$$

by long division of the numerator by the denominator, thus yielding $f(nT) = a_n$. This is seen to be very straightforward, and can be carried out conveniently on either a computer or a programmable pocket calculator.

EXAMPLE C.2

We shall consider again the previous example.

$$F(z) = \frac{3z^2-4z}{(z-2)(z^2-2z+2)} = \frac{3z^2-4z}{z^3-4z^2+6z-4} \qquad \text{...(C.31)}$$

$$= 3z^{-1} + 8z^{-2} + 14z^{-3} + 20z^{-4} + 28z^{-5} + \ldots$$

It may be verified that the first five terms of Equation (C.31) match exactly with the values obtained from Equation (C.29) for the corresponding values of n.

Although this method of inverse transformation is quite straightforward, it is not in a closed form. Hence, it is not convenient for obtaining values of $f(nT)$ for very large n. For example, if we want to determine the value of $f(100T)$, with this method we would have to obtain the first 100 terms of the infinite power series indicated by Equation (C.29).

DRILL PROBLEM C.3

Determine the inverse z-transform of each of the following:

(a) $\dfrac{2z}{(z-1)(z-0.5)}$ (b) $\dfrac{4z^3 - 21z^2 + 29z}{(z-2)(z-3)^2}$

(c) $\dfrac{6z^2 - 5z}{(z-1)(z^2 - z + 1)}$ (d) $\dfrac{8z^2 - 3z}{(z-2)(z^2 - 0.4z + 0.16)}$

Answers: (a) $4 - 2\,(0.5)^n$ (b) $3\,(2)^n + 3^n + \dfrac{2n}{3}(3)^n$ (c) $1 - \cos\dfrac{n\pi}{3} + 6.3029\sin\dfrac{n\pi}{3}$

(d) $3.8691\,(2)^n - (0.4)^n\,(3.869\cos 1.0472n - 2.9899\sin 1.0472n)$

C.6 *z*-TRANSFORMS AND DIFFERENCE EQUATIONS

The relationship between the z-transform and difference equations is similar to that between Laplace transforms and differential equations. Just as Laplace transformation converts linear differential equations with constant coefficients into algebraic equations in terms of the complex variable *s*, we shall see that z-transformation will change linear difference equations with constant coefficients into algebraic equations in terms of *z*. This is possible because of the shifting property, that is multiplication by *z* equivalent to a forward shift, so that

$$Z[f(t+T)] = zF(z) - z\,f(0) \qquad \text{...(C.32)}$$

which is similar to the relationship

$$\mathcal{L}\left[\frac{df}{dt}\right] = s\,F(s) - f(0) \qquad \text{...(C.33)}$$

Consequently, the complete solution of a difference equation is obtained by finding the inverse z-transform. This is demonstrated through the following example:

EXAMPLE C.3

Consider the second-order difference equation

$$y(nT + 2T) - 0.8\,y(nT + T) + 0.25\,y(nT) = u(nT) \qquad \text{...(C.34)}$$

with initial conditions $y(0) = 4$, $y(T) = 2$. It is desired to determine the output sequence for the input

$$u(nT) = 10(0.5)^n \qquad \text{...(C.35)}$$

Taking the z-transform of both sides of Equation (C.34), for $u(uT)$ as in Equation (C.35), we obtain

$$(z^2 - 0.8z + 0.25)\,Y(z) = U(z) + z^2y(0) + zy(T) - 0.8zy(0)$$

$$= \frac{10z}{z-0.5} + 4z^2 - 1.2z \qquad \text{...(C.36)}$$

$$Y(z) = \frac{4z^3 - 3.2z^2 + 10.6z}{(z-0.5)\left(z^2 - 0.8z + 0.25\right)}$$

Expanding $Y(z)/z$ into partial fractions, we get

$$\frac{Y(z)}{z} = \frac{100}{z-0.5} + \frac{-48+j16}{z-0.5\,e^{j0.6435}} + \frac{-48-j16}{z-0.5\,e^{-j0.6435}} \qquad \text{...(C.37)}$$

Taking the inverse transform, the output sequence is obtained as

$$y(nT) = 100(0.5)^n - (0.5)^n\,(96\cos 0.6435n + 32\sin 0.6435n) \qquad \text{...(C.38)}$$

DRILL PROBLEM C.4

Determine the initial values of the time functions corresponding to each of the following z-transforms:

$$(a)\ \frac{4z^3 - 5z^2 + 8z}{(z-1)(z-0.5)^2} \qquad (b)\ \frac{z(z-0.1)(z-0.5)}{\left(z^2+0.5z+1\right)\left(z^2-0.2z+0.1\right)}$$

Ans. (*a*) 4 (*b*) 0.

DRILL PROBLEM C.5

Determine the final values of the time functions corresponding to the z-transforms given in Drill Problem C.4.

Ans. (*a*) 28, (*b*) Indeterminate due to pair of complex poles on the unit circle.

DRILL PROBLEM C.6

Solve the following difference equation if $y(0) = 2$ and $y(T) = 5$:

$$y(nT + 2T) + 0.6\,y(nT + T) + 0.08\,y(nT) = 4$$

Ans. $\frac{50}{21} + \frac{259}{21}(-0.2)^n - \frac{267}{21}(-0.4)^n$

C.7 INPUT-OUTPUT RELATIONSHIP

Consider the block diagram shown in Fig. C.4, where the input $u(t)$ is applied to a linear system with transfer function $G(s)$ after sampling, that is, the input to $G(s)$ is a sequence of impulses.

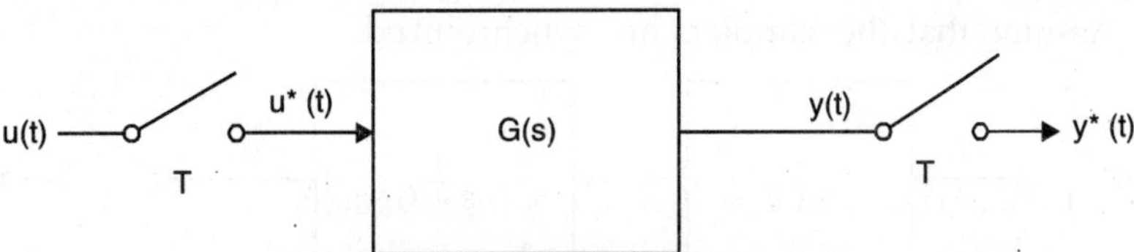

Figure C.4. *A sampled-data system*

The output, $y(t)$ is a continuous-time function. A fictitious sampler has been introduced to obtain the sampled output, $y^*(t)$. The two samplers have the same sampling period and assumed to be synchronized.

Since, by definition,

$$u^*(t) = \sum_{n=0}^{\infty} u(nT)\,\delta\,(t - nT) \qquad \text{...(C.39)}$$

the continuous output $y(t)$ is the sum of the responses of the system to the sequence of impulses. This follows from the principle of superposition. If $g(t)$ represents the impulse response of the system, *i.e.*, the inverse Laplace transform of $G(s)$, then

$$y(t) = u(0)\,g(t)\,u_t + u(T)\,g(t - T)\,u_{t-T} + u(2T)\,g(t - 2T)\,u_{t-2T} + \ldots \qquad \text{...(C.40)}$$

$$= \sum_{n=0}^{\infty} u(nT)\,g(t - nT)\,u_{t-nT}$$

where u_{t-nT} is the unit step occurring at $t = nT$.

The output at the kth sampling instant is given by

$$y(kT) = \sum_{i=0}^{k} u(iT)\,g(kT - iT) \qquad \text{...(C.41)}$$

Note that this is a finite sum, since $g(t)$ must be zero for $t < 0$. This follows from the fact that the system cannot respond to the impulse before it is applied. This is called the property of 'causality', that is, the cause must precede the effect. Define

$$G\,(z) = \sum_{n=0}^{\infty} g(nT)\,z^{-n} \qquad \text{...(C.42)}$$

as the z-transform of the impulse response $g(t)$. Then

$$G\,(z)\,U\,(z) = [g(0) + g(T)\,z^{-1} + g(2T)\,1z^{-2} + \ldots)\,[u(0) + u(T)\,z^{-1} + u(2T)\,z^{-2} + \ldots]$$

$$= g(0)\,u(0) + [g(0)\,u(T) + g(T)\,u(0)]\,z^{-1} + \ldots \qquad \text{...(C.43)}$$

$$= \sum_{n=0}^{\infty}\left[\sum_{i=0}^{n} g(iT)\,u(kT - iT)\right] z^{n} = \sum_{n=0}^{\infty} y(nT)\,z^{-n}$$

Thus, we get the input-output relationship

$$Y(z) = G(z)\,U(z) \qquad \text{...(C.44)}$$

and $G(z)$ is called the z (or pulse) transfer function of $G(s)$.

EXAMPLE C.4

Determine $y(nT)$ for the system shown in the block diagram of Fig. C.5 if $u(t)$ is a unit step and $T = 0.1$ second. Assume that the samplers are synchronized.

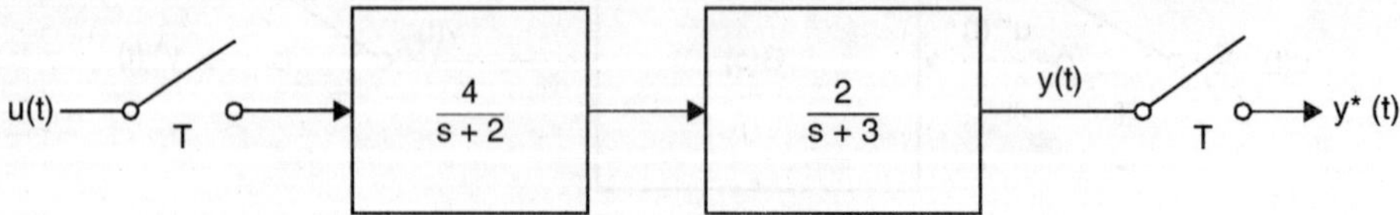

Figure C.5. *Block diagram*

Solution: A partial fraction expansion of the transfer function yields

$$G(s) = \frac{8}{(s+2)(s+3)} = \frac{8}{s+2} - \frac{8}{s+3} \qquad \text{...(C.45)}$$

Hence, we obtain the z-transfer function

$$G(z) = \frac{8z}{z-e^{-2T}} - \frac{8z}{z-e^{-3T}} = \frac{8z\left(e^{-2T}-e^{-3T}\right)}{\left(z-e^{-2T}\right)\left(z-e^{-3T}\right)} \qquad \text{...(C.46)}$$

For the given input, we have

$$U(z) = \frac{z}{z-1} \qquad \text{...(C.47)}$$

Consequently, the z-transform of the output is given by

$$Y(z) = G(z)\,U(z) = \frac{8z^2\left(e^{-0.2}-e^{-0.3}\right)}{(z-1)\left(z-e^{-0.2}\right)\left(z-e^{-0.3}\right)} \qquad \text{...(C.48)}$$

$$= \frac{13.2667z}{z-1} - \frac{36.133z}{z-e^{-0.2}} + \frac{22.866z}{z-e^{-0.3}}$$

and $$y(nT) = 13.2667 - 36.133\,e^{-0.2n} + 22.866\,e^{-0.3n} \qquad \text{...(C.49)}$$

EXAMPLE C.5

Determine $y(nT)$ for the block diagram shown in Fig. C.6 if the input is a unit step and $T = 0.1$. Assume that all the samplers are synchronized.

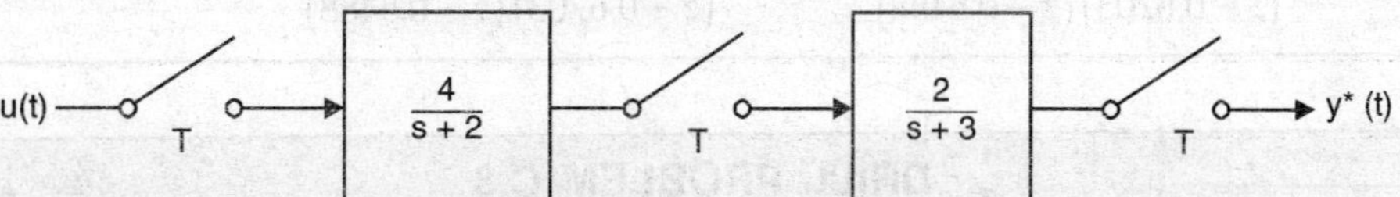

Figure C.6. *Block diagram*

Solution: In this case, since we have a sampler between the two analog blocks, we must find the z-transfer functions of these blocks separately and multiply them. Using the results from the previous problem, we have

$$G(z) = \frac{4z}{z-e^{-0.2}} \cdot \frac{2z}{z-e^{-0.3}} = \frac{8z^2}{\left(z-e^{-0.2}\right)\left(z-e^{-0.3}\right)} \qquad \text{...(C.50)}$$

The z-transform of the output is obtained as

$$Y(z) = G(z)\,U(z) = \frac{8z^3}{(z-1)\left(z-e^{-0.2}\right)\left(z-e^{-0.3}\right)} \qquad \text{...(C.51)}$$

$$= \frac{170.279z}{z-1} - \frac{379.7z}{z-e^{-0.2}} + \frac{217.421z}{z-e^{-0.3}}$$

so that $$y(nT) = 170.279 - 379\,e^{-0.2n} + 217.421\,e^{-0.3n} \qquad \text{...(C.52)}$$

These examples show that when a number of continuous-times systems are connected in cascade, without any sampler between them, the z-transfer function of the combination is the z-transform of the overall continuous-time transfer function. On the other hand, if there are samplers between the continuous-time blocks, the z-transfer function of the combination is the product of the z-transfer functions of the various blocks separated by the samplers.

DRILL PROBLEM C.7

Determine the transfer function of each of the sampled data systems shown in Fig. C.7, assuming that the samplers are synchronized and T = 0.2 second.

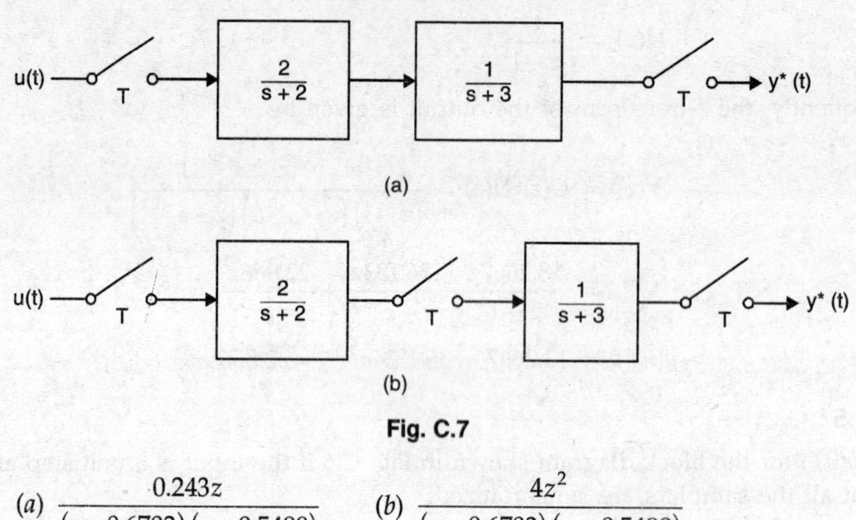

Fig. C.7

Ans. (*a*) $\dfrac{0.243z}{(z-0.6703)(z-0.5488)}$ (*b*) $\dfrac{4z^2}{(z-0.6703)(z-0.5488)}$

DRILL PROBLEM C.8

For each of the two systems in Drill Problem C.7, determine the response if the input is given by $u(t) = 10\, e^{-t}$.

Ans. (*a*) $y(nT) = 49.6682\ e^{-0.2n} - 90.3331\ e^{-0.4n} + 40.6649\ e^{-0.6n}$

(*b*) $y(nT) = 16.7344\ e^{-0.2n} - 24.9168\ e^{-0.4n} + 91.1835\ e^{-0.6n}$

DRILL PROBLEM C.9

Repeat the previous problem if the input is given by $u(t) = 10 \cos 2t$.

Ans. (*a*) $y(nT) = 25.591\ e^{-0.6n} - 23.2933^{-0.4} - 2.298 \cos 0.2n + 9.465 \sin 0.2n$

(*b*) $y(nT) = 127.172\ e^{-0.6n} - 172.887^{-0.4n} + 85.715 \cos 0.2n$

C.8 A SHORT TABLE OF LAPLACE AND z-TRANSFORMS

$f(t)$	$F(s)$	$F(z)$
Unit impulse	1	1
Unit step	$\frac{1}{s}$	$\frac{z}{z-1}$
$e^{-\alpha t}$	$\frac{1}{s+\alpha}$	$\frac{z}{z-e^{-\alpha T}}$
t	$\frac{1}{s^2}$	$\frac{Tz}{(z-1)^2}$
$\cos \beta t$	$\frac{s}{s^2+\beta^2}$	$\frac{z(z-\cos \beta T)}{z^2-2z\cos \beta T+1}$
$\sin \beta t$	$\frac{\beta}{s^2+\beta^2}$	$\frac{z \sin \beta T}{z^2-2z\cos \beta T+1}$
$e^{-\alpha t} \cos \beta t$	$\frac{s+\alpha}{(s+\alpha)^2+\beta^2}$	$\frac{z(z-e^{-\alpha T}\cos \beta T)}{z^2-2z\,e^{-\alpha T}\cos \beta T+e^{-2\alpha T}}$
$e^{-\alpha t} \sin \beta t$	$\frac{\beta}{(s+\alpha)^2+\beta^2}$	$\frac{ze^{-\alpha T}\sin \beta T}{z^2-2ze^{-\alpha T}\cos \beta T+e^{-2\alpha T}}$
$tf(t)$	$-\frac{dF(s)}{ds}$	$-zT\cdot\frac{dF(z)}{dz}$
$e^{-\alpha t} f(t)$	$F(s+\alpha)$	$F(ze^{\alpha T})$

References

1. Benjamin C. Kuo, *Digital Control Systems,* Holt, Rinehart and Winston, Inc., New York, 1990.
2. Gene F., Franklin, J. David, Powell, and Michael L. Workman, *Digital Control of Dynamic Systems,* Addison-Wesley Publishing Company, Reading, Mass., Second Edition, 1990.
3. Sinha, Naresh K., *Linear Systems,* John Wiley and Sons, Inc., New York, 1991.

❑❑❑

State Equations for Electrical Networks

The reader is probably familiar with the methods for writing loop or node equations for electrical networks. In this appendix, we shall describe a procedure for writing a set of state equations for a linear electrical network that contains resistors, inductors, capacitors, and voltage as well as current sources. The procedure is based on certain concepts of network topology which will first be presented.

D.1 SOME CONCEPTS FROM NETWORK TOPOLOGY

The topological properties of a network depend on the number of nodes and how they are interconnected. These are independent of the nature of circuit elements in the network. Therefore, it is desirable to represent a network by its *graph*, which connects the various *nodes* by line segments corresponding to the *branches* in the network. For example, Fig. D.1 shows an electrical network and its graph.

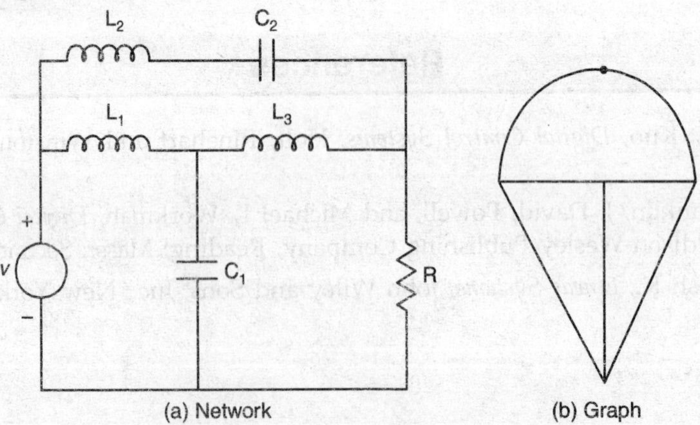

Figure D.1. *Network with associated graph*

A *tree* of the graph is defined as a set of branches that connects all nodes without forming any closed path. In general many different trees can be obtained for any graph. Some possible trees for the graph in Fig. D.1(b) are shown in Fig. D.2.

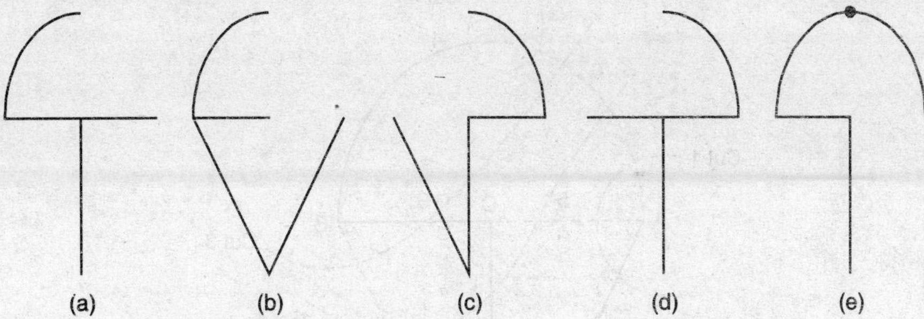

Figure D.2. *Some possible trees for the graph in Fig. D.1(b)*

The branches of a graph that do not form part of a particular tree are said to be the *links* for that tree. Each link forms a closed path with some branches of the tree. Figure D.3 shows the links as dashed lines for each of the trees depicted in Fig. D.2.

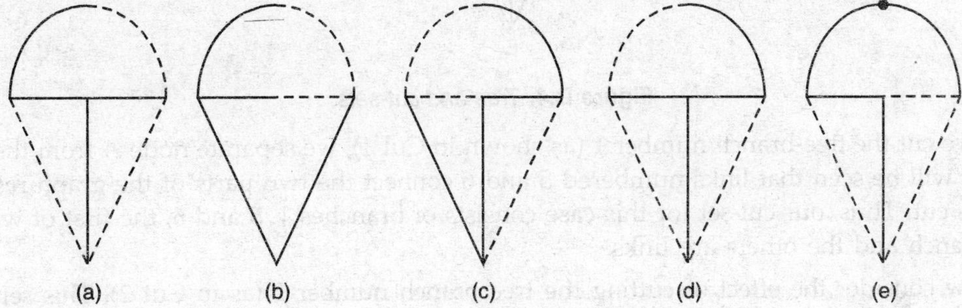

Figure D.3. *Trees and links for the graph in Fig. D.1(b)*

It follows that for any network with b branches and n nodes, any tree will consist of $n-1$ branches and the number of links, l, is given by

$$l = b - (n - 1) \qquad \text{...(D.1)}$$

The reader may recall that the number of independent node-pairs (that is, the number of independent node voltages) in a network in $n-1$, whereas the number of independent loop currents is l.

We shall now introduce the concepts of *tie-sets* and *cut-sets*. We have already seen that every link will form a closed-path together with some branches of a given tree. These tree branches and the link are said to form a tie-set. Evidently, the number of tie-sets for any tree will be equal to l, the number of links.

A cut-set is defined as the minimum number of branches of a graph that must be cut in order to divide it into two separate (unconnected) parts. It is easier to visualize these in terms of trees. Since a tree connects all nodes of the graph without forming a closed path, we separate the graph into two parts by cutting any branch of the tree. Thus, this tree-branch and those links that connect the two parts of the graph, form a cut-set. This is called a *fundamental cut-set* for the tree, and as there are $n-1$ tree branches, there will be n-1 fundamental cut-sets for any given tree. For example, we shall consider one of the trees in Fig. D.3, and find the fundamental cut-sets. For convenience, the branches and nodes are numbered as in Fig. D.4.

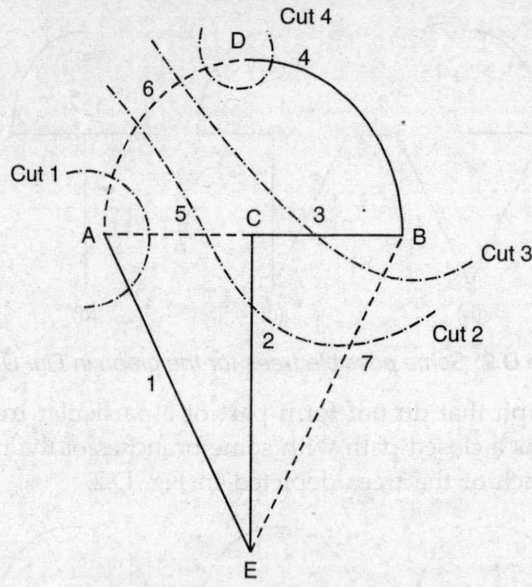

Figure D.4. *Tree and cut-sets*

If we cut the tree-branch number 1 (as shown in Cut 1), we separate node A from the other nodes. It will be seen that links numbered 5 and 6 connect the two parts of the graph resulting from this cut. Thus, our cut-set for this case consists of branches 1, 5 and 6, the first of which is a tree branch and the others are links.

Now consider the effect of cutting the tree branch number 2 (as in Cut 2). This separates nodes A and B from the rest of the graph. The branches separating the two parts are 2, 5, 6 and 7. This cut-set consists of one tree-branch and three links. Similarly, cutting the branch 3 produces a cut-set consisting of branches 3, 6 and 7. Finally, cutting branch 4 leads to a cut-set consisting of branches 4 and 6.

It is easily seen that Kirchhoff's voltage law (KVL) must be satisfied for every tie-set and Kirchhoff's current law (KCL) must be satisfied for every cut-set. Consequently, from any tree, we get l independent KVL equations and $n - 1$ independent KCL equations. We shall now see how to get a set of first-order differential equations for a network. The number of such equations (the dimension of the state vector) is normally equal to the number of storage elements (inductors and capacitors) in the network, except for 'degenerate' cases, when a tie-set consists of only capacitors and voltage sources, or a cut-set contains only inductors and current sources. For the former, the voltages across all capacitors in the tie-set cannot be independent, and one of them has to be expressed in terms of the others. Similarly, for the latter, the current in each inductor cannot be independent. We shall only describe the step-by-step procedure for the normal case and illustrate it by examples. For the other case (including networks containing controlled sources) the reader can see the references listed at the end of this appendix.

D.2 STEPS IN WRITING STATE EQUATIONS

1. Select a *proper* tree containing all voltage sources and capacitors, but no inductors or current sources. The proper tree may include resistors.

2. Write Kirchhoff's voltage law equations for the various tie-sets formed by the links containing inductors. Each of these equations will contain only one derivative term.
3. Write Kirchhoff's current law equations for the various cut-sets containing a capacitor. Since all the capacitors are on tree branches, there will be only one capacitor in any cut-set. Consequently, there will be only one derivative term in the Kirchhoff's current law equation for each cut-set, corresponding to the voltage across that capacitor.
4. Use the equations for the tie-sets for resistive links or cut-sets with resistive tree-branches to eliminate the variables that are not state variables (that is currents through resistors or voltages across resistors).
5. The state equations can now be rearranged as a single vector differential equation.

Example D.1

Consider the network shown in Fig. D.1(*a*). A proper tree is shown in Fig. D.5, which contains the voltage source, all capacitors and no inductors. The assumed directions for the currents and polarities of voltages are indicated. There are five state variables, the current through each of the inductors and the voltage across each of the capacitors. The voltage v of the source is the given input. The voltage v_R across the resistor is not a state variable, and will have to be eliminated.

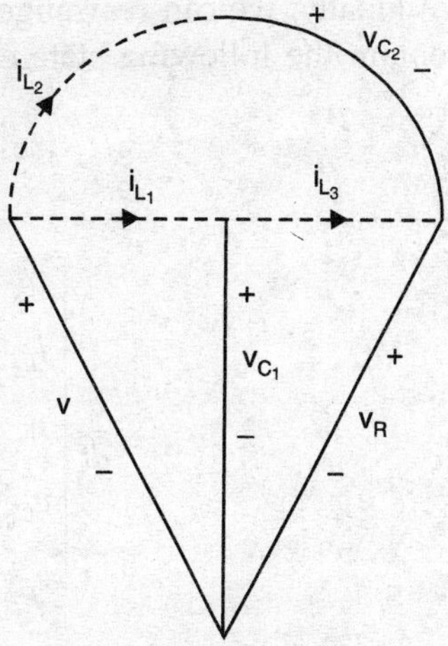

Figure D.5. *Proper tree for network*

We shall first write the tie-set equations. From the tie-set formed by the inductor L_1, we get the following *KVL* equation

$$L_1 \frac{di_{L_1}}{dt} = v - v_{C_1} \qquad \text{...(D.2)}$$

For the tie-set formed by L_2, we obtain

$$L_2 \frac{di_{L_2}}{dt} = v - v_{C_2} - v_R \qquad \text{...(D.3)}$$

Similarly for the tie-set formed by L_3, we have

$$L_3 \frac{di_{L_3}}{dt} = v_{C_1} - v_R \qquad \text{...(D.4)}$$

These are all first-order differential equations, containing only one derivative term. However, the last two equations contain v_R, which is not a state variable, and will have to be eliminated later.

We shall now write the *KCL* equations for the cut-sets containing the two capacitors. These are

$$C_1 \frac{dv_{C_1}}{dt} = i_{L_1} - i_{L_3} \qquad \text{...(D.5)}$$

and

$$C_2 \frac{dv_{C_2}}{dt} = i_{L_2} \qquad \text{...(D.6)}$$

To eliminate v_R from Equations (D.3) and (D.4), we write the *KCL* equation for the cut-set for the tree-branch containing the resistor. This gives us

$$\frac{v_R}{R} = i_{L_2} + i_{L_3} \qquad \text{...(D.7)}$$

or, $$v_R = R\,(i_{L_2} + i_{L_3}) \qquad \text{...(D.8)}$$

Substituting Equation (D.8) into (D.3) and (D.4) eliminates v_R from these equations. This leads to the following equations:

and $$L_2 \frac{di_{L_2}}{dt} = v - v_{C_2} - R\,(i_{L_2} + i_{L_3}) \qquad \text{...(D.9)}$$

$$L_3 \frac{di_{L_3}}{dt} = v_{C_1} - R\,(i_{L_2} + i_{L_3}) \qquad \text{...(D.10)}$$

Finally, we can rearrange Equations (D.2), (D.9), (D.10), (D.5) and (D.6) in the matrix form to obtain the following state equations:

$$\frac{d}{dt}\begin{bmatrix} i_{L_1} \\ i_{L_2} \\ i_{L_3} \\ v_{C_1} \\ v_{C_2} \end{bmatrix} = \begin{bmatrix} 0 & 0 & 0 & -\frac{1}{L_1} & 0 \\ 0 & \frac{R}{L_2} & \frac{R}{L_2} & 0 & -\frac{1}{L_2} \\ 0 & \frac{R}{-L_3} & \frac{R}{L_3} & -\frac{1}{L_3} & 0 \\ \frac{1}{C_1} & 0 & -\frac{1}{C_1} & 0 & 0 \\ 0 & \frac{1}{C_2} & 0 & 0 & 0 \end{bmatrix} \begin{bmatrix} i_{L_1} \\ i_{L_2} \\ i_{L_3} \\ v_{C_1} \\ v_{C_2} \end{bmatrix} + \begin{bmatrix} \frac{1}{L_1} \\ \frac{1}{L_2} \\ 0 \\ 0 \\ 0 \end{bmatrix} v \qquad \text{...(D.11)}$$

EXAMPLE D.2

Consider the network shown in Fig. D.6. It is desired to write a set of state equations for this network.

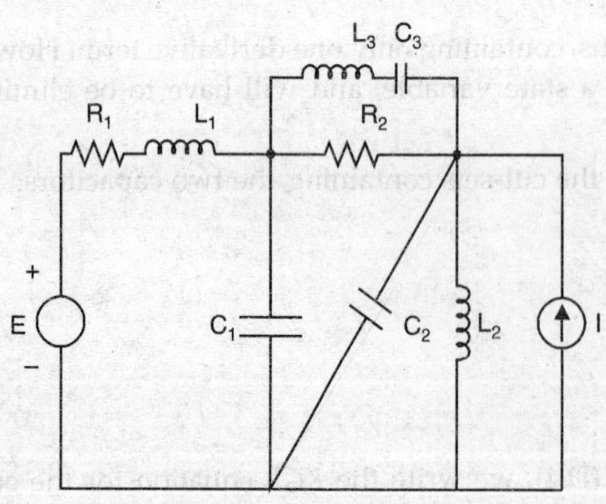

Figure D.6. *Network*

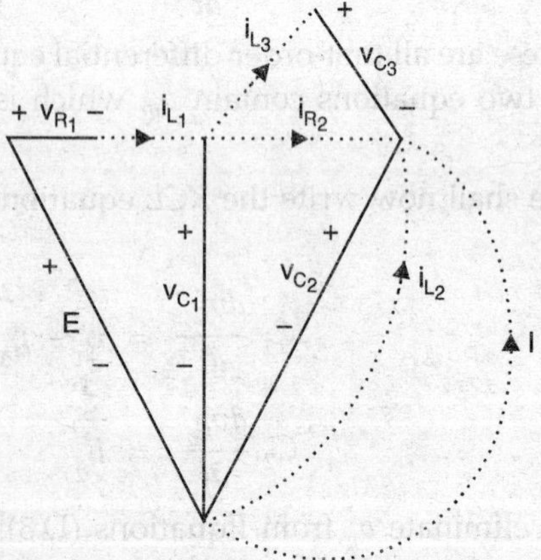

Figure D.7. *Proper tree*

A 'proper' tree is shown in Fig. D.7 with the assumed directions of link currents and the polarities of tree-branch voltages. The tree contains all capacitors and the voltage source, and excludes the current source and all inductors. It will be seen that the tree has three capacitors and a resistor in addition to the voltage source. The five links consist of three inductors, a resistor and a current source. We shall start by writing the *KVL* equations for the tie-sets formed by the inductors. These are given below:

$$L_1 \frac{di_{L_1}}{dt} = E - v_{R_1} - v_{C_1} \quad \text{...(D.12)}$$

$$L_2 \frac{di_{L_2}}{dt} = -v_{C_2} \quad \text{...(D.13)}$$

and

$$L_3 \frac{di_{L_3}}{dt} = v_{C_1} - v_{C_2} - v_{C_3} \quad \text{...(D.14)}$$

The *KCL* equations for the cut-sets containing each of the three capacitors are:

$$C_1 \frac{dv_{C_1}}{dt} = i_{L_1} - i_{L_3} - i_{R_2} \quad \text{...(D.15)}$$

$$C_2 \frac{dv_{C_2}}{dt} = I + i_{L_2} + i_{R_2} + i_{L_3} \quad \text{...(D.16)}$$

$$C_3 \frac{dv_{C_3}}{dt} = i_{L_3} \quad \text{...(D.17)}$$

Note that Equation (D.12) contains the voltage across R_1 which is not a state variable. Similarly, Equations. (D.15) and (D.16) contain the current through R_2. These are readily eliminated by writing the *KCL* equation for the cut-set containing R_1 and the *KVL* equation for the tie-set containing R_2, respectively. These are given below.

$$i_{L_1} = \frac{v_{R_1}}{R_1} \quad \text{...(D.18)}$$

and

$$R_2\, i_{R_2} = v_{C_1} - v_{C_2} \quad \text{...(D.19)}$$

Substituting these in Equations (D.12), (D.15) and (D.16), we get the six state equations. Arranging them in the matrix form, we finally obtain

$$\frac{d}{dt}\begin{bmatrix} i_{L_1} \\ i_{L_2} \\ i_{L_3} \\ v_{C_1} \\ v_{C_2} \\ v_{C_3} \end{bmatrix} = \begin{bmatrix} -\frac{R}{L_1} & 0 & 0 & -\frac{1}{L_1} & 0 & 0 \\ 0 & 0 & 0 & 0 & -\frac{1}{L_2} & 0 \\ 0 & 0 & 0 & \frac{1}{L_3} & -\frac{1}{L_3} & -\frac{1}{L_3} \\ \frac{1}{C_1} & 0 & -\frac{1}{C_1} & -\frac{1}{C_1R_2} & \frac{1}{C_1R_2} & 0 \\ 0 & \frac{1}{C_2} & \frac{1}{C_2} & \frac{1}{C_2R_2} & -\frac{1}{C_2R_2} & 0 \\ 0 & 0 & -\frac{1}{C_3} & 0 & 0 & 0 \end{bmatrix} \begin{bmatrix} i_{L_1} \\ i_{L_2} \\ i_{L_3} \\ v_{C_1} \\ v_{C_2} \\ v_{C_3} \end{bmatrix} + \begin{bmatrix} \frac{1}{L_1} & 0 \\ 0 & 0 \\ 0 & 0 \\ 0 & 0 \\ 0 & \frac{1}{C_2} \\ 0 & 0 \end{bmatrix} \begin{bmatrix} E \\ I \end{bmatrix} \quad \text{...(D.20)}$$

In all of these examples, we were able to select a 'proper' tree. This will not be possible if there are tie-sets containing only inductors and voltage sources or cut-sets containing only capacitors and current sources. In such cases, the number of state variable is smaller than the number of storages elements in the network. Special techniques are required to eliminate these variables so that we may obtain a set of first-order differential equations. Also, we have not considered networks that have mutual inductances or those that contain controlled sources. The details for writing the state equations in such cases are available in reference [2].

References

1. Charles A. Desoer and Ernest S. Kuh, *Basic Circuit Theory*, McGraw-Hill Book Company, New York, 1969.
2. Leon O. Chua and Pen-Min Lin, *Computer-Aided Analysis of Electronic Circuits*, Prentice-Hall, Inc,. Englewood Cliffs, New Jersey, 1975.
3. Norman Balabanian and Theodore A. Bickart, *Linear Network Theory*, Matrix Publishers, Inc., Beaverton, Oregon, 1981.

Appendix E Computational Aspects

Realistic problems in control systems require a great deal of computation. To remove some of the drudgery involved in this process, computer programs to accompany this book can be obtained on a floppy disk at a nominal cost from the publisher. Alternatively, a student can freely copy the programs from the instructor of the course who will be supplied a disk by the publisher. These programs can be run on any IBM-compatible computer. Two versions are available, (*i*) those which can be run only on machines containing numeric coprocessors, and (*ii*) those which do not require the numeric coprocessor. The former run faster.

The programs are listed in the table on the following page. To run any of these programs simple type the name of the program. Then just answer the questions asked by the program. For example, if you want to run the program for determining stability using the Routh criterion, type stability and press the return key. This will cause the program to be run. Each program will prompt the user to enter the information required for each case.

	Name of Program	*Purpose*
1.	Simeq.EXE	To solve a set of linear simultaneous equations.
2.	Residue.EXE	For partial fraction expansion of a rational function in connection with inverse Laplace or z-transformation.
3.	Transfer.EXE	To determine the transfer function of a linear system from its state equations.
4.	Response.EXE	To obtain the response of a linear continuous-time system to a unit impulse or step input.
5.	Roots.EXE	To calculate the roots of a polynomial.
6.	Transition.EXE	To calculate the state transition equations of a discrete-time system from the state equations of the corresponding continuous-time system.
7.	Stability.EXE	Determination of stability using the Routh criterion. It can also shift the axis to determine relative stability and transform from the z-plane to the w-plane for sampled-data systems.

(Contd.)

8.	Rootloc.EXE	To calculate points on the root locus.
9.	Locus.EXE	To do complete root locus plots.
10.	Freqresp.EXE	To calculate frequency response from transfer function.
11.	FRP.EXE	To calculate and plot the frequency response (including Bode, polar and log gain-phase plots).
12.	Compensator.EXE	To calculate the transfer function of a lead, lag or lag-lead compensator that will provide a specified gain and phase-shift at a given frequency.
13.	Nichols.EXE	Compensation with Nichols chart.
14.	Pole-placement.EXE	Design of compensators that will place the poles of a controllable and observable system at specified locations.
15.	Statefeed.EXE	To calculate the state feedback vector that will place the poles of a controllable system at specified locations.
16.	Observer.EXE	To design an asymptotic state observer.
17.	Phaseplane.EXE	To obtain the phase-plane trajectory for a second-order nonlinear system using the delta method.

Please note that although the programs have been tested carefully by the author, no warranty, express or implied, is made to their accuracy, nor shall their distribution with the book constitute such warranty, and no responsibility is assumed by the author.

In particular, if you use this program to solve a problem whose incorrect solution may lead to injury to a person or loss of property, it is at your own risk.

Answers to Selected Problems

CHAPTER 2

1. (*a*) $\dfrac{s^2+3s+2}{s^2+3.2s+2}$, (*b*) $\dfrac{1}{s^3+5s^2+6s+1}$

2. $\dfrac{1}{(s+1)^3} = \dfrac{1}{s^3+3s^2+3s+1}$

3. (*a*) $2 + 2.83 \cos(2t + 0.542)$, (*b*) $2 + 0.155 \cos(2t - 2.41)$

4. $G(s) = \dfrac{0.7259}{s(s+0.578)}$

5. $G(s) = \dfrac{10/3}{s^2+2s+4/3}$, $x(t) = 5 - 5e^{-t} \cos \dfrac{t}{\sqrt{3}} - 5\sqrt{3}e^{-t} \sin \dfrac{t}{\sqrt{3}}$

6. $\dfrac{K_m}{s(sL_f+R_f)(Js+D)}$

7. $\dfrac{1}{(s+5)(0.12s^2+2.803s+0.08)}$

8. $\dfrac{K_1}{Js^2+Ds+K_2s}$

9. $K_2 = \omega_0^2 M_2$

10. $\dfrac{10}{s^3+5s^2+3s+10}$

11. (*a*) $\dfrac{G_1G_2G_3+G_4(1+G_2H_2+G_1G_2H_1)}{1+G_2H_2+G_1G_2H_1+G_2G_3H_3}$

(*b*) $\dfrac{G_4+G_1G_2G_3}{1+G_1G_2H_3+G_2G_3H_2+G_1G_2G_3H_1+G_4H_1-G_4H_2G_2H_3}$

12. $\dfrac{M}{Ms^2+2Ds+2K}$

13. $G_e = G_1 G_2$, $H_e = \dfrac{H}{G_1}$

14. $G_e = G_1 G_2$, $H_e = HG_2$

CHAPTER 3

1. Let $y_1 = x_1$, $y_2 = \dot{x}_1$, $y_3 = x_2$ and $y_4 = \dot{x}_2$. Then

$$\dot{y}_1 = y_2$$

$$\dot{y}_2 = \frac{1}{M_1}f - \frac{K_1}{M_1}y_1 - \frac{D_1}{M_1}y_2 - \frac{K_2}{M_1}(y_1 - y_2)$$

$$\dot{y}_3 = y_4$$

$$\dot{y}_4 = -\frac{K_2}{M_2}(y_3 - y_1)$$

2.

$$\begin{bmatrix}\dot{x}_1\\ \dot{x}_2\\ \dot{x}_3\end{bmatrix} = \begin{bmatrix}-5 & -3 & -1\\ 1 & 0 & 0\\ 0 & 1 & 0\end{bmatrix}\begin{bmatrix}x_1\\ x_2\\ x_3\end{bmatrix} + \begin{bmatrix}1\\ 0\\ 0\end{bmatrix}u$$

$$y = [0 \quad 0 \quad 10]\begin{bmatrix}x_1\\ x_2\\ x_3\end{bmatrix}$$

3.

$$\begin{bmatrix}\dot{x}_1\\ \dot{x}_2\\ \dot{x}_3\end{bmatrix} = \begin{bmatrix}0 & 1 & 0\\ 0 & 0 & 1\\ -1 & -3 & -5\end{bmatrix}\begin{bmatrix}x_1\\ x_2\\ x_3\end{bmatrix} + \begin{bmatrix}0\\ 0\\ 1\end{bmatrix}u$$

$$y = [10 \quad 0 \quad 0]\begin{bmatrix}x_1\\ x_2\\ x_3\end{bmatrix}$$

4. (*a*) $\dfrac{s^2 + 5s + 4}{s^3 + 6s^2 + 10s + 2}$ (*b*) $F = \begin{bmatrix}0.4318 & 0.0481 & 0.2498\\ 0.1535 & 0.2301 & 0.0481\\ 0.2979 & 0.2016 & 0.6815\end{bmatrix}$ $G = \begin{bmatrix}0.4078\\ 0.0687\\ 0.4970\end{bmatrix}$.

8. (*a*)

$$\dot{x} = \begin{bmatrix}-7 & -18 & -20 & -8\\ 1 & 0 & 0 & 0\\ 0 & 1 & 0 & 0\\ 0 & 0 & 1 & 0\end{bmatrix}x + \begin{bmatrix}1\\ 0\\ 0\\ 0\end{bmatrix}u \text{ for controller form,}$$

$$y = [8 \quad 47 \quad 90 \quad 54]\; x$$

(*b*) $\dot{x} = \begin{bmatrix} -7 & 1 & 0 & 0 \\ -18 & 0 & 1 & 0 \\ -20 & 0 & 0 & 1 \\ -8 & 0 & 0 & 0 \end{bmatrix} x + \begin{bmatrix} 8 \\ 47 \\ 90 \\ 54 \end{bmatrix} u$ for observer form,

$y = [1 \quad 0 \quad 0 \quad 0]\, x$

10. $\dfrac{d}{dt}\begin{bmatrix} i_f \\ i_a \\ \omega \end{bmatrix} = \begin{bmatrix} -5 & 0 & 0 \\ \dfrac{200}{3} & -\dfrac{70}{3} & -\dfrac{10}{3} \\ 0 & 0.025 & -0.025 \end{bmatrix}\begin{bmatrix} i_f \\ i_a \\ \omega \end{bmatrix} + \begin{bmatrix} 5 \\ 0 \\ 0 \end{bmatrix} v(t)$

11. (*a*) $\dot{x} = \begin{bmatrix} -6 & -16 & -16 \\ 1 & 0 & 0 \\ 0 & 0 & 1 \end{bmatrix} x + \begin{bmatrix} 1 \\ 0 \\ 0 \end{bmatrix} u$ (*b*) $\dot{x} = \begin{bmatrix} -6 & 1 & 0 \\ -16 & 0 & 1 \\ -16 & 0 & 0 \end{bmatrix} x + \begin{bmatrix} 3 \\ 5 \\ 13 \end{bmatrix} u$

$y = [3 \quad 5 \quad 13]\, x$ $y = [1 \quad 0 \quad 0]\, x$

(*c*) $\dot{x} = \begin{bmatrix} -2 & 0 & 0 \\ 0 & -2 & -2 \\ 0 & 2 & -2 \end{bmatrix} x + \begin{bmatrix} 1 \\ 1 \\ 0 \end{bmatrix} u$

$y = [3.75 \quad -0.75 \quad -3.75]\, x$

12. (*a*) $\dot{x} = \begin{bmatrix} -6 & -16 & -16 \\ 1 & 0 & 0 \\ 0 & 0 & 1 \end{bmatrix} x + \begin{bmatrix} 1 \\ 0 \\ 0 \end{bmatrix} u$ (*b*) $\dot{x} = \begin{bmatrix} -6 & 1 & 0 \\ -16 & 0 & 1 \\ -16 & 0 & 0 \end{bmatrix} x + \begin{bmatrix} 3 \\ 2 \\ 9 \end{bmatrix} u$

$y = [3 \quad 2 \quad 9]\, x + u$ $y = [1 \quad 0 \quad 0]\, x + u$

(*c*) $\dot{x} = \begin{bmatrix} -2 & 0 & 0 \\ 0 & -2 & -2 \\ 0 & 2 & -2 \end{bmatrix} x + \begin{bmatrix} 1 \\ 1 \\ 0 \end{bmatrix} u$

$y = [3.75 \quad -0.75 \quad -3.75]\, x$

13. (*a*) $\begin{bmatrix} 2e^{-t} - e^{-2t} & 2e^{-t} - 2e^{-2t} \\ e^{-2t} - e^{-t} & 2e^{-2t} - e^{-t} \end{bmatrix}$ (*b*) $\begin{bmatrix} e^{t} & 0.5\left(e^{t} - e^{-t}\right) \\ 0 & e^{-t} \end{bmatrix}$

(*c*) $\begin{bmatrix} e^{-t} & \dfrac{1}{3}\left(e^{-t} - e^{-4t}\right) & \dfrac{1}{27}\left(2e^{t} - 3e^{-4t} + e^{-10t}\right) \\ 0 & e^{-4t} & 0 \\ 0 & 0 & e^{-10t} \end{bmatrix}$

14. (*a*) $\begin{bmatrix} e^{2t} & te^{2t} \\ 0 & e^{2t} \end{bmatrix}$ (*b*) $\begin{bmatrix} e^{2t} & te^{2t} & \frac{1}{2}t^2e^{2t} \\ 0 & e^{2t} & te^{2t} \\ 0 & 0 & e^{2t} \end{bmatrix}$ (*c*) $\begin{bmatrix} e^{2t} & te^{2t} & 0 \\ 0 & e^{2t} & 0 \\ 0 & 0 & e^{2t} \end{bmatrix}$

15. $\dfrac{X_1(s)}{U(s)} = \dfrac{s^3+3}{s^4+4s^2-9}$

16. $F = \begin{bmatrix} 1.0598 & 0.1987 & -0.0079 & 0.0395 \\ 0.5961 & 0.9809 & -0.1184 & 0.3895 \\ -0.0079 & -0.0395 & 0.9414 & 0.1908 \\ -0.1184 & -0.3895 & -0.5723 & 0.8624 \end{bmatrix} \quad G = \begin{bmatrix} 0.0199 \\ 0.1987 \\ -0.0026 \\ -0.0039 \end{bmatrix}$

17. $\begin{bmatrix} \dfrac{9s+34}{s^3+15s^2+54s+40} & \dfrac{-7s-14}{s^3+15s^2+54s+40} \\ \dfrac{5s^2+41s+86}{s^3+15s^2+54s+40} & \dfrac{-3s^2-15s-26}{s^3+15s^2+54s+40} \end{bmatrix}$

19. (*a*) $\dot{x} = \begin{bmatrix} -10.2 & 1 & 7.5 \\ 1 & 0 & 0 \\ 0 & 0 & 1 \end{bmatrix} x + \begin{bmatrix} 1 \\ 0 \\ 0 \end{bmatrix} u, \quad y = [0 \;\; 60 \;\; 4]\, x$

(*b*) $\dot{x} = \begin{bmatrix} -10.2 & 1 & 0 \\ -9.5 & 0 & 1 \\ -7.5 & 0 & 0 \end{bmatrix} x + \begin{bmatrix} 1 \\ 64 \\ 4 \end{bmatrix} u, \quad y = [1 \;\; 0 \;\; 0]\, x$

20. (*a*) $\dot{x} = \begin{bmatrix} -5.2 & 1 & 0 \\ -1 & 0 & 1 \\ 0 & 0 & 0 \end{bmatrix} x + \begin{bmatrix} 0 \\ 0 \\ 5 \end{bmatrix} u, \quad y = [1 \;\; 0 \;\; 0]\, x$

(*b*) $F = \begin{bmatrix} 0.59096 & 0.07785 & 0.00423 \\ -0.07785 & 0.99577 & 0.00999 \\ 0 & 0 & 1 \end{bmatrix}, \quad G = \begin{bmatrix} 0.000735 \\ 0.024981 \\ 0.5 \end{bmatrix}$

21. (*a*)

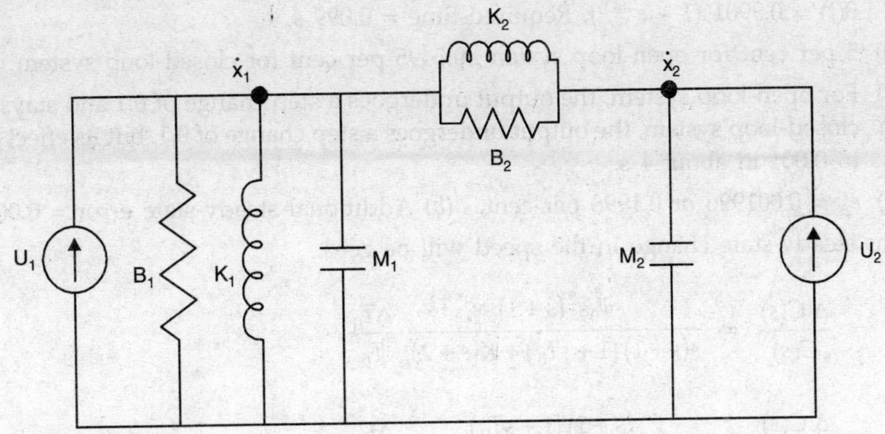

(*b*) $$\frac{d}{dt}\begin{bmatrix} x_1 \\ \dot{x}_1 \\ x_2 \\ \dot{x}_2 \end{bmatrix} = \begin{bmatrix} 0 & 1 & 0 & 0 \\ -\dfrac{K_1+K_2}{M_1} & -\dfrac{B_1+B_2}{M_1} & \dfrac{K_2}{M_1} & \dfrac{B_2}{M_1} \\ 0 & 0 & 0 & 1 \\ \dfrac{K_2}{M_2} & \dfrac{B_2}{M_2} & -\dfrac{K_2}{M_2} & -\dfrac{B_2}{M_2} \end{bmatrix}\begin{bmatrix} x_1 \\ \dot{x}_1 \\ x_2 \\ \dot{x}_2 \end{bmatrix} + \begin{bmatrix} 0 & 0 \\ \dfrac{1}{M_1} & 0 \\ 0 & 0 \\ 0 & \dfrac{1}{M_2} \end{bmatrix}\begin{bmatrix} u_1 \\ u_2 \end{bmatrix}$$

CHAPTER 4

1. (*i*) $$S_{K_e}^{T(s)} = \frac{60s^2+23s+1}{60s^2+23s+12K_e+1}$$

(*ii*) $$\frac{C(s)}{D(s)} = G_d(s) = \frac{K_e(1+3s)}{60s^2+23s+12K_e+1}$$

$$C(s) = T(s)\,R(s) + G_d(s)\,D(s)$$

(*iii*) c_{ss} = 59.95 km/h

2. (*a*) Δc_{ss} = 0.0025 km/h. (*b*) Δc_{ss} = 2.995 km/h

3. (*a*) $S_{K_a}^{G(s)} = 1, \quad S_{K_a}^{T(s)} = \dfrac{s+0.6}{s+20.6}$

(*b*) e_{ss} = 0.029 rad/s (*c*) $\dfrac{1}{20.6}(1-e^{-20.6t})$ rad/s

4. (*a*) $S_K^{T(s)} = \dfrac{s(s+20)}{s^2+20s+400}$, (*b*) $S_K^{T(s)} = \dfrac{0.1s^3+3s^2+20s}{0.1s^3+3s^2+20s+400}$

5. (*a*) For open-loop system, $\theta(t) = 1 - e^{-0.1t}$, Required-time = 10 s. For closed-loop system, $\theta(t) = 0.9901\ (1 - e^{-10.1t})$, Required-time = 0.099 s.
   (*b*) 5 per cent for open-loop system, 0.0495 per cent for closed-loop system.
   (*c*) For open-loop system, the output undergoes a step change of 0.1 and stays there. For closed-loop system, the output undergoes a step change of 0.1, but its effect is reduced to 0.001 in about 4 s.
6. (*a*) e_{ss} = 0.001996 or 0.1996 per cent, (*b*) Additional steady-state error = 0.003992
7. Ths steady-state change in the speed will be zero.
8. $$\frac{\Delta C(s)}{C(s)} = \frac{T_0 s^2(s+1)}{s(s+1)(1+sT_0)+K(s+2)} \cdot \frac{\Delta T_0}{T_0}$$
9. $$\frac{\Delta C(s)}{C(s)} = \frac{s(s+1)(1+sT_0)}{s(s+1)(1+sT_0)+K(s+2)} \cdot \frac{\Delta K}{K}$$
10. (*a*) $e_{ss} = -1$, (*b*) Δe_{ss} = 0.05 (or 5 per cent), (*c*) Δe_{ss} = 0.05
11. $S_K^{T(s)} = \dfrac{s(s+2)}{s^2+3.4s+10}$ $\quad S_p^{T(s)} = \dfrac{-2s}{s^2+3.4s+10}$ $\quad S_\alpha^{T(s)} = \dfrac{-1.4s}{s^2+3.4s+10}$
12. (*a*) – 0.01 (*b*) 0.01 (*c*) 0.007

CHAPTER 5

1. (*a*) The second specification (on settling time) cannot be satisfied.
   (*b*) K = 2.8615 will satisfy the specification on maximum overshoot. Settling time is 4 s; time to reach the first peak is 2.3026 s.
2. (*a*) (*i*) 0, (*ii*) 0.2, (*iii*) ∞ (*b*) (*i*) 0, (*ii*) 0, (*iii*) $\dfrac{4}{9}$
3. (*a*) (*i*) 0.6, (*ii*) ∞ (*b*) (*i*) 0, (*ii*) $\dfrac{16}{9}$
4. α = 0.1739, e_{ss} = 0.3739
5. (*a*) Settling time = 8 s, maximum overshoot = 63.2 per cent, $e_{ss} = \dfrac{1}{12}$
   (*b*) Settling time = 2.667 s, maximum overshoot = 22.11 per cent, e_{ss} = 0.25
   (*c*) α = 3.1569, settling time = 1.9245 s, maximum overshoot = 9.48 per cent, and e_{ss} =0.3464
6. K_1 = 10.4, K_2 = 1, settling time = 0.8 s, maximum overshoot = 16.3 per cent
7. (*a*) K = 4, z = 2; (*b*) 0.125; (*c*) 0.125
8. K = 100, α = 0.08
11. $K = \dfrac{106}{17}$, $p = \dfrac{4}{17}$, $z = \dfrac{25}{106}$, e_{ss} = 1.02

12. (*a*) $1 - 0.08397\, e^{-0.3406t} - e^{-0.004699t}\,(0.91603 \cos 0.3831t + 0.08589 \sin 0.3831t)$
(*b*) 45.365; (*c*) 0.5

13. (*a*) $1 - 0.2095\, e^{-0.3122t} - e^{-0.1439t}\,(0.7905 \cos 0.3734t + 0.4798 \sin 0.3734t)$
(*b*) 2.7375; (*c*) 2.5

14. (*a*) $1 - e^{-10t}\,(\cos 17.3205t + 0.57725 \sin 17.3205t)$;
(*b*) 0.05; (*c*) 0.05; (*d*) 1000

15. (*a*) $1 - 1.11828\, e^{-6.50455t} + 0.11823\, e^{-61.49545t}$
(*b*) 0.08235; (*c*) 0.17; (*d*) a solution does not exist for positive K.

16. The responses are very close. For $t > 3$ s, the difference is less than 1.3 per cent.

CHAPTER 6

1. (*a*) Stable, (*b*) unstable, (*c*) unstable, (*d*) unstable.
2. (*a*) $0 < K < 66$, (*b*) $0 < K < 20$, (*c*) unstable for all K.
3. (*a*) $4.125 < K < 30$, (*b*) not possible, (*c*) not possible
4. (*a*) $0 < K < 12$, (*b*) not possible
5. (*a*) $0 < \alpha < 1$, (*b*) $0.1772 < \alpha < 0.605$
6. (*a*) $K > 0$, (*b*) $K > 3.75$
7. Unstable
8. (*a*) None, (*b*) none
9. $K > -4.4635$
10. $-3.8927 < K < 3.636$
11. $K > 3.4884$
12. (*a*) $K > 0$, (*b*) not possible for any K
13. $y_{ss} = 3$
14. $y_{ss} = 3t - 17$
15. Not stable for the given range of parameters
16. Stable for $0 < K < 148.75$
17. Stable for $K > 0$.

CHAPTER 7

1. (*a*) Yes, $K = 2.125$, (*b*) no, (*c*) yes, $K = 90$, (*d*) no.

3. $\alpha = 0.1162$, $S_{\alpha}^{q_1} = 0.5774\, e^{j\frac{\pi}{6}}$

4. (*b*) Two solutions are possible, (*i*) $Z = 14.454$, (*ii*) $Z = 0.22304$
(*c*) (*i*) $1 - 0.4036\, e^{-283.314t} - e^{-3.115t}\,(0.5964 \cos 7.1378t - 0.2747 \sin 7.1378t)$, 22.1 per cent
(*ii*) $1 - 0.9894\, e^{101.04t} - e^{-0.5943t}\,(0.0106 \cos 1.3617t - 0.0205 \sin 1.3617t)$, 0.78 per cent

5. (*a*) 0.6

(*b*) (*i*) 0.7438, (*ii*) 0.5115

(*c*) (*i*) $1 - 1.2769\, e^{-1.4476t} + e^{-2.276t}\,(0.2769 \cos 3.9424t - 0.30898 \sin 3.4924t)$

(*ii*) $c(t) = 1 - 1.2156\, e^{-2.609t} + e^{-1.6954t}\,(0.2156 \cos 2.9366t - 0.9556 \sin 2.9366t)$

(*d*) (*i*) $S_{\alpha}^{q_1} = 3.198\, e^{j1.887}$, (*ii*) $S_{\alpha}^{q_1} = 2.8807\, e^{j2.396}$

6. $\alpha = 0.4537$, $e_{ss} = 0.3102$

7. $\alpha = 1.2371$, $S_{\alpha}^{q_1} = 1.2193\, e^{-j0.0691}$

8. $K = 9.6192$, $c(t) = 0.5869 + 0.4006\, e^{-37.0184t} - e^{-4.16t}\,(0.863 \cos 4.16t - 0.7698 \sin 4.16t)$

9. There are two solutions.

(*i*) $p = 9.3259$, $c(t) = 1 - 0.096\, e^{-0.8497t} - e^{-5.2381t}\,(0.904 \cos 9.073t + 0.531 \sin 9.073t)$

(*ii*) $p = 78.9228$, $c(t) = 1 + 0.0155\, e^{-77.7365t} + 1.0265\, e^{-1.5932t} \cos (2.7594t - 3.2885)$

10. The damping ratio is maximum for $K = 3.6$, with $\zeta = 0.0558$

$c(t) = 1 + 0.173\, e^{-1.638t} - 0.0564\, e^{-4.282t} - 1.1166\, e^{-0.04t} \cos 0.715t - 0.0039\, e^{-0.04t} \sin 0.715t$

11. The best damping (considering all poles) is obtained for $\alpha = 0.318$.

$c(t) = 1 + 0.2758 e^{-0.392t} - 0.1177 e^{-4.823t} - e^{0.392t}\,(1.1581 \cos 1.365t + 0.6695 \sin 1.365t)$

12. There are two solutions.

(*i*) $k_v = 0.3777$, $e_{ss} = 1.1777$, maximum overshoot = 20 per cent

(*ii*) $k_v = 30.7303$, $e_{ss} = 31.53$, no overshoot

13. $k_a = 7.20937$

$c(t) = 1 - 0.000021 e^{-77.681t} - e^{0.1974t}\,(0.999979 \cos 0.3107t + 0.582694 \sin 0.3107t)$ maximum overshoot $\simeq$ 18 per cent

14. $k_v = 11.05$, $e_{ss} = 3.5125$

15. $K = 17.1$ causes the complex poles to have damping ratio of 0.707. The dominant pole, then, is real, at $s = -1.166$. The complex poles are at $-5.417 \pm j5.417$.

16. Sensitivity of the real dominant pole to K is -1.0225. Sensitivity of the upper complex pole to K is $0.5119 + j1.3196$.

17. (*b*) The poles are at -0.4518 and $-2.7741 \pm j2,7924$

(*c*) $S_{p}^{q_1} = 1.1215 - j0.3783$

18. (*a*) The poles are at -4 and $-0.5 \pm j3.1225$

(*b*) $S_{p_1}^{q_1} = -0.4091 - j0.1383$

19. $S_{z}^{q_1} = -0.1818 - j0.2038$

20. $S_{z}^{q_1} = 0.15 - j0.41$

CHAPTER 8

2. $G(s) = \dfrac{100}{s\left(s^2 + 5s + 100\right)}$

5. $\dfrac{31.62s\,(s+300)^2}{(s+20)\,(s+50)\,(s+5692.1)}$

6. $\dfrac{5\,(s+2)\,(s+10)}{s}$

8. 13 dB

15. $\dfrac{10}{s(s+2)}$

16. $\dfrac{10}{s^2 + 4s + 10}$

CHAPTER 9

1. (*a*) 2.14 dB, 6.85°; (*b*) 3.93 dB, 16.88°
2. (*a*) 28.18; (*b*) 17.95
3. (*a*) Stable; (*b*) unstable
4. (*b*) Gain must be increased by 8.98 dB. Then, M_m = 3.24 dB, ω_m = 9.5 rad/s
   (*c*) Gain must be increased by 8.87 dB. Then, M_m = 3 dB, ω_m = 9.48 rad/s
5. 11.63
6. (*a*) 39°, (*b*) 1.85, (*c*) 1.55, (*d*) 8 rad/s
7. (*a*) Stable with gain margin = 7 dB, (*b*) 2.088
8. K = 13.8 for M_m to be 3 dB. Phase margin = 30.2°
9. (*a*) Gain margin is ∞ for all positive K, (*b*) 28.28, (*c*) 34
10. (*a*) Stable with gain margin 12 dB. (*b*) K_p = 8.8 for phase margin to be 45°.
11. (*a*) 0.889, (*b*) 2.7 dB
12. 1.35, 41.6°
13. 0.122, 42°
14. 1.2, 52.5°
15. 142, 43°
16. 133, 2.5 dB
17. Stable

CHAPTER 10

1. (*a*) $K = 28.45, \quad \omega_n = 1.5$ rad/s
   (*b*) $K = 3327, \quad \omega_n = 17.8$ rad/s

2. (*a*) Lag compensator, $G_c(s) = \dfrac{s + 0.06}{s + 0.01}$

   (*b*) Lag compensator, $G_c(s) = \dfrac{s + 0.5}{s + 0.1}$

3. (*a*) $G_c(s) = \dfrac{6.6819\,(s + 3.9669)}{s + 26.506}$ with $K = 250$

   (*b*) With $K = 250$, $G_c(s) = \dfrac{0.4329 + 0.1674}{s + 0.1674}$

4. $G_c(s) = \dfrac{7.6573\,(s + 0.58)}{s + 4.4416}$

5. (*a*) $K = \dfrac{224}{3}, \quad \alpha = 24$

   (*b*) $1 - 0.1026\, e^{-28t} + 1.2943\, e^{-4t} \cos\left(4\sqrt{3}t - 3.9462\right)$

6. (*a*) With compensation, gain margin = 15.25 dB, phase margin = 52°, $M_m = 1.1$ dB, $\omega_m = 5.5$ rad/s.

   (*b*) Without compensation, gain margin = 7 dB, phase margin = 12°, $M_m = 12$ dB, $\omega_m = 2.4$ rad/s.

7. $K = 250$

   (*a*) $\dfrac{5.9648\,(s + 4.8926)}{s + 29.1832}$ (*b*) $\dfrac{0.3749\,(s + 0.6593)}{s + 0.2472}$

8. One solution is $G_c(s) = \dfrac{(s + 3)(s + 5)}{(s + 40)^2}$ with $T(s) = \dfrac{24{,}000}{(s + 60)\left(s^2 + 20s + 400\right)}$.

   Sensitivity of the upper complex pole to K is $S_K^{q_1} = 13.093\, e^{j1.2373}$

9. (*a*) 4.5 to 8 rad/s, (*b*) Below 2.3 rad/s, (*c*) 2.4 to 3.5 rad/s

10. $G_u(s) = \dfrac{0.26625s + 36.5318}{s^2 + 9s + 20}, \quad G_c(s) = \dfrac{-33.803s^2 + 3.34s + 11.76}{s^2 + 9s + 20}, \quad K = \dfrac{40}{3}$

11. $K = 128{,}000, \quad G_c(s) = \dfrac{(s + 4)(s + 8)}{s^2 + 120s + 6400}$ (Using pole-zero cancellation)

12. $K = 36.19, \quad G_c(s) = \dfrac{0.05632s + 0.012656}{s + 0.012656}$ $(\omega_c = 1$ rad/s$)$

13. $K = 200, \quad G_c(s) = \dfrac{0.05271\,(s+0.3541)}{s+01866} \quad (\omega_c = 1 \text{ rad/s})$

$c(t) = 1 + 0.95\, e^{-0.5632t} - e^{-0.7277t}\,(1.95 \cos 0.8294t + 0.9918 \sin 0.8294t)$

14. $K = 200, \quad G_c(s) = \dfrac{3\,(s+2.5)}{s+7.5}$ makes $M_m = 2.21$ dB, $\omega_m = 5.8$ rad/s

$c(t) = 1 + 0.2605\, e^{-2.5589t} - e^{-3.4706t}\,(1.02604 \cos 10.2654t + 0.3407 \sin 10.2654t)$

15. $G_p(s) = \dfrac{20}{s\,(s+2)}, \quad G_c(s) = \dfrac{20\,(s+2)}{s+20}, \quad T(s) = \dfrac{400}{s^2+20s+400}$

$c(t) = 1 - e^{\,10t}(\cos 17.321t + 0.57735 \sin 17.321t)$

16. $K_p = 9.5, \quad G_c(s) = \dfrac{5.5977s + 14.2803}{s+14.2803} \quad (\omega_c = 5 \text{ rad/s})$

17. $K_A = 10.823, \quad G_c(s) = \dfrac{4.5397(s+12.595)}{s+57.177} \quad (\omega_c = 20 \text{ rad/s})$

18. $K = 100, \quad G_c(s) = \dfrac{7200\,(s+25)\left(s^2+2s+26\right)}{s\,(s+4)\,(s+12)\left(s^2+244s+28{,}800\right)}$ (cascade compensator)

19. $T(s) = \dfrac{7.8125\,(s+4)\,(s+12)}{(s+7.5)\,1\left(s^2+10s+50\right)} \quad K = 3.90625, \quad G_u(s) = \dfrac{-3.33875s - 44.76675}{(s+10)^2}$

$G_c(s) = \dfrac{-0.78864s^2 + 11.8391s + 4.26237}{(s+10)^2}$

20. $K = 2.748244, \quad c(t) = 1 - 0.1187e^{-0.2561t} + 0.0733e^{-0.574t} - 0.9547\, e^{-1.1844t} \cos 1.1841t$
$+ 1.3773\, e^{-1.1844t} \sin 1.1841t$

$e_{ss} = 0.158$

21. $K = 2.4496644, \ \alpha = 0.6794521, \ e_{ss} = 0.2$

$c(t) = 1 - 0.13581e^{-0.2496644t} - e^{-t}(0.86419 \cos t - 1.55156 \sin t)$

22. $K = 53, \quad \delta = \dfrac{130}{53} = 2.4528, \quad G_c(s) = \dfrac{0.245283\,(s+10)}{s+2.4528}, \quad G_u(s) = \dfrac{0.142399}{s+2.4528}$

23. $K = 256$, $\alpha = 0.0546875$, transfer function of the lag compensator is

$G_c(s) = \dfrac{s+0.0625}{s+0.01}$

CHAPTER 11

1. (a) 10.3445 (b) $1 - (0.9536)^n \{\cos(0.4455n) - 0.10289 \sin(0.4455n)\}$

2. $$\frac{10.322\,(z - 0.818731)}{z + 0.935528}$$

3. (a) $$\frac{0.3075\,(z - 0.6988)}{z - 0.9074}$$

   (b) $c(nT) = 0.9085 - 0.0664\,(0.7188)^n - 0.8421\,(-0.4148)^n$

4. (a) $K < 2.449$, (b) $G_c(z) = \dfrac{3.41782\,(z - 0.739915)}{z - 0.111072}$

5. $$G_c(z) = \frac{(z - 0.9048374)(z - 0.9801987)}{(z-1)(0.0485z + 0.046166)}$$

7. $$G_u(z) = \frac{-0.995706z - 0.009719}{(z-0.1)^2} \qquad G_y(z) = \frac{0.262142z^2 - 0.322918z + 0.141151}{(z-0.1)^2}$$

8. $$G_c(z) = \frac{1067.76\,(z - 0.818731)}{K\,(z + 0.935525)}$$

9. $K = 100$, $G_c(z) = \dfrac{3.85\,(z - 0.740775)}{(z + 0.001926)}$

   $c(nT) = 1 - (0.597724)^n\,[1.27096 \cos(0.884738n) - 0.745658 \sin(0.884738n)] + 0.269159\,(0.697283)^n$

10. Adjust K_A to make the open-loop *dc* gain equal to 20.

    $$G_c(z) = \frac{0.213903\,(z - 0.316758)}{z - 0.853868}$$

    $$c(nT) = 0.9524 - 0.019885\,(0.775205)^n - (0.57901)^n\,0.972268 \cos(0.856964n) - (0.57901)^n\,0.50331 \sin(0.856964n)$$

11. $$G_u(z) = \frac{0.691778}{z - 0.15} \qquad G_c(z) = \frac{42.0197z - 36.834657}{z - 0.15}$$

12. $$G_u(z) = \frac{0.685196}{z - 0.15} \qquad G_c(z) = \frac{2.306353z - 1.460607}{z - 0.15}$$

    $$c(nT) = 0.96987 - (0.28284)^n\,0.96987 \cos\left(\frac{n\pi}{4}\right) - (0.28284)^n\,3.19772 \sin\left(\frac{n\pi}{4}\right)$$

13. $K = 0.0004664$ for phase margin to be 45°

14. $K = 0.0125$, $G_c(z) = \dfrac{0.00211965\,(z - 0.994557372)}{z - 0.999988506}$

$$c(nT) = 1 - 0.0693\ (0.8901)^n + 0.5691\ (0.9923)^n - (0.98353)^n\ 1.52868 \cos (3.12093n) - (0.98353)^n\ 1.36802 \sin (3.12093n)$$

15. $G_u(z) = \dfrac{0.4133}{z - 0.1}, \quad G_c(z) = \dfrac{0.1188z + 0.2217}{z - 0.1}$

16. (*a*) $K = 1$ for gain margin equal to 6 dB

(*b*) $G_c(z) = \dfrac{0.50561\,(z - 0.68988)\,(z - 0.95708)}{(z + 0.25572)\,(z - 0.99109)}$

CHAPTER 12

1. Not completely controllable, observable, stable.

2. $D(z) = \dfrac{1.58199 - 0.58199z^{-1}}{1 + 0.41801z^{-1}}$

3. $k^T = [-4.64\ \ -0.96\ \ 5.36]$

4. $G_c(s) = \dfrac{9s^2 + 100s + 298}{12.5\left(3s^2 + 16s + 24\right)}$

5. $G(s) = \dfrac{s^2 - 3s - 4}{s^3 - 2s^2 + 4s - 4} \quad Kk^T = \begin{bmatrix} \dfrac{12}{11} & \dfrac{54}{11} & -8 \end{bmatrix}$

6. $Kk^T = [0.169991\ \ 1.100326\ \ 5.546071]$

7. $L = [-7996.0392\ \ 47.3508\ \ 13.598]^T$

8. $L = [149.8421\ \ 57.3158\ \ -34.2632]^T$

9. $G_u(s) = \dfrac{-0.36s^? - 1127.645s - 288.349}{s^3 + 30s^2 + 325s + 1250}, \quad G_c(s) = \dfrac{-115.004s^2 - 118.869s + 1193.608}{s^3 + 30s^2 + 325s + 1250}$

10. $G_u(s) = \dfrac{-18.6695s - 50.8063}{s^2 + 20s + 125}, \quad G_y(s) = \dfrac{-18.3095s^2 - 41.7663s + 115.6989}{s^2 + 20s + 125}$

11. $u(0) = 11.4592, \quad u(T) = -25.3942, \quad u(2T) = 13.9235$

$$D(z) = \frac{11.4592 - 25.394z^{-1} + 13.9235z^{-2}}{1 + 0.6654z^{-1} + 0.850459z^{-2}}$$

12. $Kk^T = [5.87415\ \ -9.35706\ \ 17.8538]$

$$c(nT) = 1.022359(0.25)^n - 0.000182 - (0.28284)^n\ 1.022178 \cos n\pi/4 - (0.288284)^n\ 0.088588 \sin n\pi/4$$

13. $L = [-6235.6\ \ 17.245\ \ 2.8121]^T$

14. $G_u(z) = \dfrac{13310.01z - 6839.23}{z^2 - 0.22z + 0.12}, \quad G_y(z) = \dfrac{-92182.38z^2 + 204306.5z - 111984.225}{z^2 - 0.22z + 0.12}$

15. $Kk^T = \begin{bmatrix} -\frac{4}{9} & -\frac{10}{9} & -\frac{184}{9} & -\frac{55}{9} \end{bmatrix}$

16. $L = [14 \quad 83 \quad -308 \quad -974]^T$

17. $Kk^T = \begin{bmatrix} \frac{13}{3} & 5 & \frac{5}{2} & -\frac{5}{6} \end{bmatrix}$

18. $Kk^T = [0.0001878 \quad 19.26578 \quad -845.8537 \quad 820.3408]$

19.
$$G_u(z) = \frac{-75.222z^2 - 309.179z - 77.5577}{z^3 - 0.3z^2 + 0.04z - 0.002}$$

$$G_c(z) = \frac{-236370.3z^3 + 693812.15z^2 - 691608.626z + 231140}{z^3 - 0.3z^2 + 0.04z - 0.002}$$

20. (*a*) Yes, (*b*) $Kk^T = [-0.1823 \quad 1.048 \quad 5.8764 \quad 0.4158]$

21. $Kk^T = [-450 \quad 46.8 \quad 843.75]$

22. $Kk^T = [-551.25 \quad 64.8 \quad 843.75]$

23. The state x_3 is uncontrollable. All states are observable.

24. (*a*)
$$x(k+1) = \begin{bmatrix} 0.8 & 0.21 & -0.01 \\ 1 & 0 & 0 \\ 0 & 1 & 0 \end{bmatrix} x(k) + \begin{bmatrix} 1 \\ 0 \\ 0 \end{bmatrix} u(k)$$
$$y(k) = [1 \quad \alpha \quad 0]$$

(*b*) Unobservable if $\alpha = -1, -0.41421$ or 0.241421. [The transfer function has pole-zero cancellation for these values of α.]

(*c*)
$$x(k+1) = \begin{bmatrix} 0.8 & 1 & 0 \\ 0.21 & 0 & 1 \\ -0.01 & 0 & 0 \end{bmatrix} x(k) + \begin{bmatrix} 1 \\ \alpha \\ 0 \end{bmatrix} u(k)$$
$$y(k) = [1 \quad 0 \quad 0]$$

(*d*) Uncontrollable if α is $-1, -0.041421$ or 0.241421. [The transfer has pole-zero cancellation for these values of α].

26. $Kk^T = [-9.38035 \quad -1.7]$

27. (*a*) $k_1 = \sqrt{3}, \quad k_2 = 1$ (*b*) $\sqrt{3}$.

CHAPTER 13

1. $1 + \frac{4}{\pi}\sin^{-1}\frac{10}{X} + \frac{40}{\pi X^2}\sqrt{X^2 - 100}$

2. (*a*) Unstable, (*b*) $K < 5.682$ for stability.

3. $\left(\frac{3}{4}X^2\omega^3 + \frac{1}{4}X^2\omega\right)e^{j\frac{\pi}{2}}$

4. $\frac{2}{\pi}\left[\sin^{-1}\frac{15}{X} + \frac{15}{X^2}\sqrt{X^2 - 225} + \frac{20}{X}\right]$

5. (*a*) Stable (*b*) *K* must be less than 14.53 for no limit cycle to exist.

9. Stable

11. $\frac{40}{\pi}\left[\sin^{-1}\frac{1}{X} - \sin^{-1}\frac{1}{2X} + \frac{\sqrt{X^2 - 1}}{X^2} - \frac{\sqrt{X^2 - 0.25}}{2X^2}\right]$

12. $0.5K$

13. $\frac{B}{\pi}\frac{\sqrt{X^2 - A^2}}{X^2}$

14. *K* must be less than 10 for no limit cycle to exist.

15. *K* must be less than 5π (*i.e.*, less than 15.7) for no limit cycle to exist.

20. $\frac{4A}{\pi X} + K$

21. A suitable Lyapunov function for this case is $V = x_1^4 + 2x_2^2$, which shows that the system is stable at the origin.

22. The system is stable, since the solution of the Lyapunov equation for $Q = I$ gives the positive definite matrix *P* given below:

$$P = \begin{bmatrix} 0.110577 & 0.052885 \\ 0.052885 & 0.201923 \end{bmatrix}$$

23. The system is unstable, since solution of the Lyapunov equation for $Q = I$ gives the following negative definite matrix *P*:

$$P = \begin{bmatrix} -0.113070 & -0.374164 & -0.230395 \\ -0.374164 & -0.245593 & -0.062918 \\ -0.230395 & -0.062918 & -0.255927 \end{bmatrix}$$

❑❑❑

Index